HANDBOOK OF
SPECIFIC LOSSES
IN FLOW SYSTEMS

PLENUM PRESS DATA DIVISION

NEW. YORK

1966

HANDBOOK OF SPECIFIC LOSSES IN FLOW SYSTEMS

Robert P. Benedict and Nicola A. Carlucci

Westinghouse Electric Corporation, Steam Divisions, Philadelphia, Pennsylvania

ISBN-13:978-1-4684-6065-0

e-ISBN-13:978-1-4684-6063-6

DOI: 10.1007/978-1-4684-6063-6

Softcover reprint of the hardcover 1st edition 1966

Library of Congress Catalog Card Number 65-25129

Preface

EVEN WHEN one is willing to estimate the various loss coefficients in a given system, it is not always an easy matter to determine the flow rate and/or the total pressure drop across the system. While there are gas dynamics books that contain Fanno tables which involve flow with losses, such tables are never specific; that is, the conventional tabulations are never given in terms of specific loss coefficients or specific total pressure ratios.

The tables contained in this book are unique in this respect. The user can establish from these tables not only the various state point functions, but the total pressure losses as well. (The total pressure ratio is shown to be the only true indication of loss in a flow system.)

Both compressible and constant-density solutions are presented. Tables for fluids of various ratios of specific heats are included. Use of these tables is not restricted to constant-area systems, nor does their use require iterative procedures. For compressible flows, tables of solutions for both the subsonic and supersonic regimes are given. The loss coefficients obtained from these tables are unique in that they are shown to be additive in series systems. This permits the investigator to evaluate a flow system either as a series of components or in its entirety.

These tables will be of interest primarily to engineers working with actual flow systems. In the field of engineering education, these tables will be useful reference material for courses in fluid mechanics, thermodynamics, aerodynamics, and chemical engineering.

Without these tables, solutions to flow with losses require a tedious trial-and-error (iterative) procedure, which in practice can be accomplished in a practical sense only by the use of a high-speed digital computer.

The book is divided into seven parts. In Chapter 1 we present a generalized solution for compressible flow with losses. The conventional Darcy approximations to such problems are also reviewed here. Generalized compressible flow loss curves and tables are developed in Chapter 2 for both the subsonic and supersonic flow regimes. Numerical tables are presented in the Appendix. In Chapter 3, numerical examples are given to illustrate the use of these tables and curves for compressible fluids. The development of a generalized constant-density solution to flow with losses is presented in Chapter 4, along with several numerical examples involving liquids. Chapters 5 and 6 concern application of the compressible flow loss tables for determining loss coefficients and their combining characteristics in series systems, and for dealing with the specific problem of flow in systems involving abrupt changes in flow area.

The authors would like to acknowledge the cooperation and generosity of the Westinghouse Electric Corporation's Steam Divisions in making available both computer facilities and time for the completion of this work and granting permission to publish these tables.

The Authors

Lester, Pennsylvania
March, 1966

NOTE: The text in this book is based primarily on the authors' papers:
Flow with Losses
(*Trans. ASME, J. Engrg. for Power,* Jan. 1965, p. 37)
On the Determination and Combination of Loss
Coefficients for Compressible Fluid Flows
(*Trans. ASME, J. Engrg. for Power*, Jan. 1966, p. 67)
Flow Losses in Abrupt Enlargements and Contractions
(*Trans. ASME, J. Engrg. for Power*, Jan. 1966, p. 73)

Contents

Notation ix

Chapter 1

Generalized Compressible Flow with Losses 1
    Introduction 1
    General Remarks 1
    Darcy Static-Pressure Approximation 1
    Darcy Total-Pressure Approximation 3
    Generalized Solution for Compressible Adiabatic Flow with Losses 3
    References 5

Chapter 2

Generalized Compressible Flow Loss Curves and Tables . . . 7
    General Remarks 7
    Generalized Fanno Flow Curves 7
    Generalized Fanno Flow Tables 7
    Comparisons of Exact Analytic Solutions with Darcy Approxima-
        tions 7

Chapter 3

Specific Numerical Examples of Flow with Losses 11

Chapter 4

Generalized Solution for Constant-Density Flow with Losses . . 17
    General Remarks 17
    Constant-Density Loss Solution 17
    Numerical Examples 19

Chapter 5

On Combining Loss Coefficients 21
    General Remarks 21
    Constant-Density Loss Coefficients 21
    Compressible-Flow Loss Coefficients 21

Chapter 6

Losses across Abrupt Enlargements and Contractions . . . 23
    General Remarks 23
    Abrupt Enlargements 23
    Abrupt Contractions 25

Appendix

Generalized Compressible Flow Loss Tables 29
    Table I $\gamma = 1.10$ Subsonic 29
    Table II $\gamma = 1.20$ Subsonic 43
    Table III $\gamma = 1.30$ Subsonic 57
    Table IV $\gamma = 1.40$ Subsonic 72
    Table V $\gamma = 1.67$ Subsonic 88

    Table VI $\gamma = 1.10$ Supersonic 105
    Table VII $\gamma = 1.20$ Supersonic 124
    Table VIII $\gamma = 1.30$ Supersonic 143
    Table IX $\gamma = 1.40$ Supersonic 161
    Table X $\gamma = 1.67$ Supersonic 178

Notation

a = Acceleration
A = Area
c = Specific heat capacity
C = Constant, coefficient
d = Exact differential
D = Diameter
E = Error function
f = Darcy friction factor
F = Frictional head, force
g = Acceleration gravity
h = Specific enthalpy, fluid head
k = Incompressible loss coefficient
K = Compressible loss coefficient
L = Length
M = Mass
Ma = Mach number
p = Absolute pressure
Q = External heat transfer/lb, volumetric flow rate
R = Pressure ratio
R_g = Gas constant
Re = Reynolds number
s = Specific entropy
t = Time
T = Absolute temperature
u = Specific internal energy
v = Specific volume
V = Directed velocity
w = Specific weight
W = Weight
$W/\Delta t$ = Flow rate

$\overline{W}$ = External work/lb
Z = Potential head
SPR = Static pressure ratio
TPR = Total pressure ratio
α = Flow number
β = Elevation factor, diameter ratio
γ = Isentropic exponent
Γ_{isen} = Generalized compressible flow function
Γ_{inc} = Generalized constant-density flow function
δ = Inexact differential
Δ = Finite difference
ε = Roughness
ρ = Fluid density

Subscripts

1 = Inlet
2 = Exit
x, y, z = Arbitrary stations
c = Contraction
d = Discharge
n = General station
p = Constant pressure
s = Entropy, static
t = Total
v = Constant volume
V = Velocity
isen = Isentropic
inc = Incompressible
ref = Reference

Note: $\overline{K}$ and $\overline{k}$ represent equivalent values of loss coefficients. An asterisk * signifies conditions at the critical state.

Chapter 1

Generalized Compressible Flow with Losses

INTRODUCTION

A common problem in engineering concerns the determination of pressure drop in a system involving losses. Confusion begins with the question, "Which pressure drop is significant, total or static?" We will show that the total-pressure drop is the only one of significance in determining losses in a general flow involving a compressible fluid in a variable-area passage. Confusion reigns as one asks, "Which formulation of Δp, or Δp_t (as the case may be), is to be used?" We will show that the conventional Darcy approximations are not reliable for compressible fluid flows. Confusion is done away with when the exact analytical solution is obtained.

Our solution of this problem is based primarily on the works of Frössel [1],* Egli [2], and Benedict and Steltz [3]. The first two of these references concern constant-area flow only, and give no indication of the total-pressure drop across the system, their results being in terms of the hybrid ratio of exit static pressure to inlet total pressure (p_2/p_{t1}). The last reference, while removing the restriction of constant area, and while generalizing the compressible flow function, indicates no means for obtaining the required ratio of inlet total pressure to exit total pressure (p_{t1}/p_{t2}). We will show that, whenever the losses can be estimated in the form of loss coefficients, K, and/or friction factors, f, the total-pressure drop can be established for any adiabatic system within an uncertainty which is dependent only on the validity of the loss coefficients chosen.

The several Darcy approximations are first discussed. The exact general solutions are then derived in detail. These solutions are tabulated and presented graphically. Sample problems are then given to illustrate the graphical and tabular solutions as presented here.

*Numbers in brackets designate references at the end of this chapter.

GENERAL REMARKS

The general energy equation (see, for example, [4])

$$\delta Q + \delta \overline{W} = du + p\,dv + v\,dp + \frac{V\,dV}{g} + dz \qquad (1)$$

combined with the first law (see, for example, [4]) of thermodynamics

$$\delta Q + \delta F = du + p\,dv \qquad (2)$$

yields, for the case of a flow in the absence of external work, the general expression for head loss

$$\delta F = -v\,dp - \frac{V\,dV}{g} - dz \qquad (3)$$

For a compressible fluid, the effects of a change in elevation are usually negligible, and equation (3) reduces to

$$\delta F = -v\,dp - \frac{V\,dV}{g} \qquad (4)$$

DARCY STATIC-PRESSURE APPROXIMATION

Most texts on the subjects of hydraulics, fluid mechanics, and thermodynamics give the Darcy–Weisbach equation for frictional head loss in the form

$$\delta F = f\frac{dx}{D}\frac{V^2}{2g} \qquad (5)$$

Equation (5) has also been given in the form of a static-pressure drop as

$$\Delta p_s = f\frac{L}{D}\frac{V^2}{2gv} \qquad (6)$$

We maintain that equation (5) is beyond reproach, since it serves to define the friction factor, f, and we continue

1

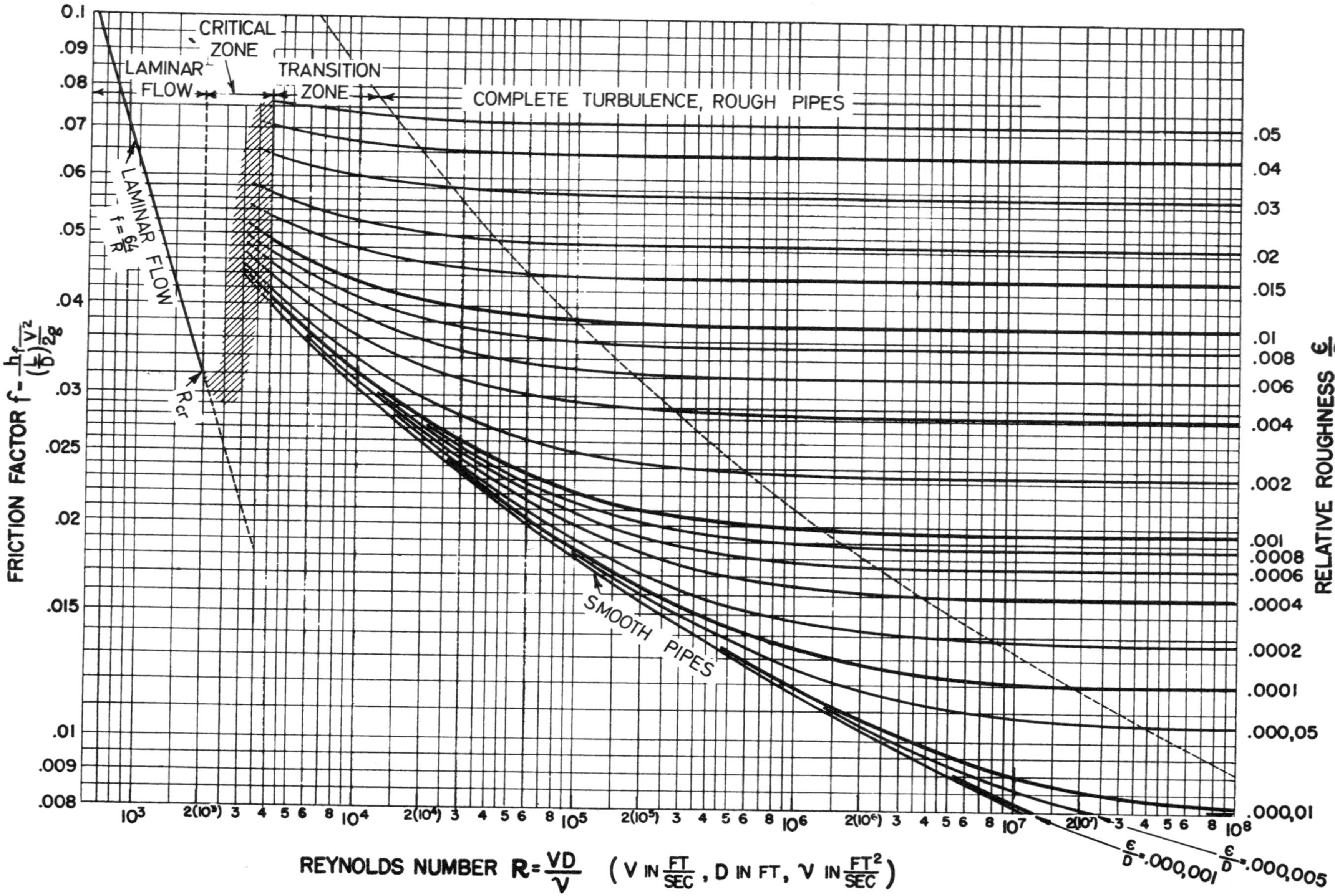

Figure 1. Friction Factors (reproduced from Trans. ASME, Nov. 1944, p. 671, by L. F. Moody).

to observe this definition in this paper. Hence, Fig. 1, which is reproduced from the well-known Moody paper [5], and which indicates f as a function of Reynolds number (Re) and relative roughness, ε/D, is presented here for completeness. However, we strongly caution against the indiscriminate use of equation (6). For, comparing equations (3) and (6), we note that, even in the constant-density case, equation (6) overlooks changes in elevation and any fluid accelerations such as would accompany area changes. In the case of compressible flows we note, on comparing equations (4) and (6), that fluid accelerations arising from changes in specific volume and/or changes in area are entirely overlooked by equation (6). In addition, whenever specific volume varies, the user is in a quandary as to which velocity and which specific volume to use in equation (6) (usually one settles for V_2 and v_2, i.e., one obtains a loss based on exit conditions). Thus equation (6) precisely defines the frictional static-pressure drop only when conditions of constant density, constant area, and constant elevation prevail. But, in any general compressible or constant-density flow, equation (6) must be recognized as a static-pressure approximation only.

DARCY TOTAL-PRESSURE APPROXIMATION

Some references [6, 7] suggest that the frictional pressure drop is best given in the form

$$\Delta p_t = f \frac{L}{D} \frac{V^2}{2gv} \tag{7}$$

where f is defined as in equation (5). Equation (7) represents an attempt to account for accelerations of the fluid. However, we strongly caution against the indiscriminate use of equation (7); for, comparing equations (3) and (7), we note that in the constant-density case, equation (7) overlooks changes in elevation. In the compressible case we note, on comparing equations (4) and (7), that fluid accelerations still are not entirely accounted for because of variations in specific volume. Thus equation (7) precisely defines the frictional total-pressure drop only when conditions of constant density and constant elevation prevail. [Note that when the area is also constant, equation (7) reduces to equation (6).] But, in any general compressible or constant-density flow, equation (7) must be recognized as a total pressure approximation only.

GENERALIZED SOLUTION FOR COMPRESSIBLE ADIABATIC FLOW WITH LOSSES

Turning our attention from the approximations of equations (6) and (7), we next seek the general solution for flow with losses. Briefly, this complete solution should include the acceleration term, it should recognize variations in density with flow, it should open up the possibility of treating variable-area problems, and, since it should apply in the supersonic as well as in the subsonic regime, the normal shock process (always a possibility in supersonic flow with losses) should also be considered.

Such a generalized approach was given in an earlier work [3], where it was demonstrated that the expression

$$\left(\frac{p_{t1}}{p_{t2}}\right)\left(\frac{A_1}{A_2}\right)\Gamma_{isen1} = \Gamma_{isen2} \tag{8}$$

applied to all one-dimensional, compressible, steady, adiabatic flows of a perfect gas. In this same reference [3], it was also shown that the effect of an area variation can always be represented by an isentropic change of state from a given initial state (corresponding to the initial area) to a new initial state (corresponding to the final area). One then proceeds from the area-modified initial state to the final state via a constant-area process. Thus the compressible, adiabatic-flow process with losses can be handled in two steps:

1. Solve for Γ'_{isen1} according to the relation

$$\left(\frac{A_1}{A_2}\right)\Gamma_{isen1} = \Gamma'_{isen1} \tag{9}$$

where Γ'_{isen1} represents the generalized isentropic flow function at the revised initial state and accounts for the area change.
2. Solve for Γ_{isen2} according to

$$\left(\frac{p_{t1}}{p_{t2}}\right)\Gamma'_{isen1} = \Gamma_{isen2} \tag{10}$$

where Γ_{isen2} represents the generalized flow function at the final area, and accounts for the loss in total pressure.

There is only one hitch in the solution; while (A_1/A_2) can be specified, and hence Γ'_{isen1} can be obtained easily by equation (9), (p_{t1}/p_{t2}) is the very unknown we seek to determine, and therefore equation (10) cannot be solved immediately. Even iteratively, equation (10) defies solution without further information, for note that for every guess of (p_{t1}/p_{t2}) there is a corresponding value of Γ_{isen2}; and while it is true that there are intelligent bounds on the (p_{t1}/p_{t2}) guess, these bounds limit but do not particularize Γ_2.

To pinpoint (p_{t1}/p_{t2}), we are in need of more information, which we introduce in the form of pertinent loss coefficients of the system. Lest this be thought a weakness of this solution, let us point out strongly that without estimates of loss coefficients the problem is untractable since the system could vary from a smooth nozzle with a minimum of losses to a barely porous plug with alarmingly high losses. Thus we lack only the constant-area solution in terms of a loss coefficient, K, and Γ_{isen} to account for total pressure losses. Once obtained, we will combine this constant-area solution with the Γ-solution of equations (9) and (10) to obtain the complete solution. In the following section, we develop the constant-area solution [1, 2] to equation (4).

An expression for the velocity at a point (where the effects of friction are necessarily meaningless) can always be obtained by integrating equation (4) between static and total state. There results

$$V^2 = \frac{2g\gamma}{\gamma - 1}[p_t v_t - pv] \tag{11}$$

Introducing continuity

$$\frac{W}{\Delta t} = \frac{AV}{v} \tag{12}$$

into equation (11) to eliminate v, we obtain

$$V^2 + \left(\frac{2g\gamma pA}{W/\Delta t(\gamma - 1)}\right)V - \left(\frac{2g\gamma p_t v_t}{\gamma - 1}\right) = 0 \tag{13}$$

This quadratic equation in velocity has the real solution

$$V = \frac{2(gp_t v_t)^{\frac{1}{2}}}{[(R/\alpha)^2 + 2(\gamma - 1)/\gamma]^{\frac{1}{2}} + R/\alpha} \tag{14}$$

where

$R = p/p_{t1}$, a variable depending only on the static pressure (for given inlet total pressure)

$\alpha = \dfrac{W/\Delta t}{Ap_{t1}}\left(\dfrac{p_t v_t}{g}\right)^{\frac{1}{2}}$, a constant (for a given area and given inlet total pressure)

$p_t v_t = R_g T_t$, a constant (for a given total temperature, i.e., for the adiabatic case)

An expression for the head loss (δF) of equation (4) has already been given in equation (5). There is no loss in generality in maintaining this definition of f; on the contrary, there is much to be gained in making use of the bulk of information already published on the friction factor and the loss coefficient.

When equations (14) and (5) are combined according to equation (4), with specific volume again eliminated by use of equation (12), there results

$$dK = -\frac{1}{\alpha}\left\{\left[\left(\frac{R}{\alpha}\right)^2 + \frac{2(\gamma - 1)}{\gamma}\right]^{\frac{1}{2}} + \frac{R}{\alpha}\right\}dR - 2\frac{dV}{V} = f\frac{dx}{D} \tag{15}$$

Equation (15) can be integrated between arbitrary limits, noting that when fL/D plays the role of the loss coefficient, we have assumed that the friction factor does not

vary with length. If f is a function of L, an average value of the friction factor may be used. The last term in equation (15) is expressed in terms of equation (14) as

$$\int_1^2 \frac{dV}{V} = \ln 2(gp_t v_t)^{\frac{1}{2}}\Big|_1^2 - \ln\left\{\left[\left(\frac{R}{\alpha}\right)^2 + \frac{2(\gamma - 1)}{\gamma}\right]^{\frac{1}{2}} + \frac{R}{\alpha}\right\}\Big|_1^2 \tag{16}$$

where the first term of equation (16) drops out for the adiabatic case under consideration here.

Combining equations (15) and (16), we obtain

$$K = \frac{1}{2}\left(\frac{R}{\alpha}\right)\left[\left(\frac{R}{\alpha}\right)^2 + \frac{2(\gamma - 1)}{\gamma}\right]^{\frac{1}{2}}\Big|_2^1 + \frac{1}{2}\left(\frac{R}{\alpha}\right)^2\Big|_2^1$$
$$-\left(\frac{\gamma + 1}{\gamma}\right)\ln\left\{\left[\left(\frac{R}{\alpha}\right)^2 + \frac{2(\gamma - 1)}{\gamma}\right]^{\frac{1}{2}} + \frac{R}{\alpha}\right\}\Big|_2^1 \tag{17}$$

This equation relates the loss coefficient, K, the flow number, α, and the hybrid pressure ratio, $p_2/p_{t1} = R_2$, in terms of the inlet pressure ratio, $p_1/p_{t1} = R_1$, and the isentropic exponent, γ, of the fluid. The flow number, in turn, is defined in terms of the inlet-pressure ratio by the familiar [8]

$$\alpha = \left[\left(\frac{2\gamma}{\gamma - 1}\right)R_1^{2/\gamma}[1 - R_1^{(\gamma - 1)/\gamma}]\right]^{\frac{1}{2}} \tag{18}$$

Equations (17) and (18) are solved simultaneously to yield α as a function of R_2, the parameter being constant K. Such solutions exactly represent the adiabatic flow with losses in constant-area passages. It is an easy matter [3, 8] to express α in terms of the more general Γ-function since

$$\Gamma_{\text{isen}} = \frac{\alpha}{\alpha^*} \tag{19}$$

where α^* represents the isentropic critical flow number defined as

$$\alpha^* = \left[\gamma\left(\frac{2}{\gamma + 1}\right)^{(\gamma + 1)/(\gamma - 1)}\right]^{\frac{1}{2}} \tag{20}$$

A useful linear relation which defines the locus of maximum flows possible, and which therefore separates the subsonic and supersonic regimes, is obtained from the foregoing equations as

$$R^* = \alpha^* \Gamma_{\text{isen}}\left[\frac{2}{\gamma(\gamma + 1)}\right]^{\frac{1}{2}} \tag{21}$$

where R^* represents the p/p_{t1} value at the maximum flow point.

Now the effects of area change and the total-pressure drop are determined by combining the K,α-solution of equations (17) and (19) with the Γ-solution of equations (9) and (10), as will be demonstrated in detail in the sample problems of Chapter 3.

REFERENCES

1. W. Frössel, *Flow in Smooth Straight Pipes at Velocities Above and Below Sound Velocity*, NACA TM 844, January, 1938.
2. A. Egli, The Leakage of Gases through Narrow Channels, *J. Appl. Mech., Trans. ASME*, Vol. 59, 1937, p. A–63.
3. R. P. Benedict and W. G. Steltz, A Generalized Approach to One-Dimensional Gas Dynamics, *J. Eng. for Power, Trans. ASME*, Vol. 84, Series A, 1962, p. 49. See also, Thermodynamics of Compressible Fluids, *Electro-Technology*, Feb. 1963, p. 85; and *Handbook of Generalized Gas Dynamics*, Plenum Press, New York, 1966.
4. R. P. Benedict, Essentials of Thermodynamics. *Electro-Technology*, July 1962, p. 107.
5. L. F. Moody, Friction Factors for Pipe Flow, *Trans. ASME*, Vol. 66, 1944, p. 671.
6. *Aero-Space Applied Thermodynamics Manual*, SAE, Committee A-9, Section 1, January, 1962, p. A-8.
7. *Airplane Heating and Ventilating Equipment*, SAE Aero Information Report 23, October, 1951, p. 5.
8. R. P. Benedict, Comparisons Between Compressible and Incompressible Treatments of Compressible Fluids, *J. Basic Eng., Trans. ASME*, Vol. 86, 1964, p. 527.

Chapter 2

Generalized Compressible Flow Loss Curves and Tables

GENERAL REMARKS

The solutions to equations (9), (10), (17), and (19) can be presented in graphical form, facilitating interpolation, or in tabular form, where a greater number of significant figures can be preserved. In this chapter, both forms of solution are discussed, and in addition, comparisons are drawn between the exact analytic solutions and the Darcy approximations.

GENERALIZED FANNO FLOW CURVES

In the generalized Fanno flow curves of Figs. 2, 3, and 4, we present representative solutions to equations (9), (10), (17), and (19). The outer envelope represents the isentropic-flow solutions, the heavy solid lines within the envelope represent certain constant-loss-coefficient solutions, and the dotted parameters represent constant-total-pressure-ratio solutions; all pertain to the constant-area case. Methods for handling variable-area problems will be discussed in Chapter 3.

One simply enters the proper γ-curve with the information available and obtains (without iteration) the required solutions.

GENERALIZED FANNO FLOW TABLES

The generalized Fanno flow tables present various point functions (p/p_t, M, T/T_t, Γ_{isen}, K^*) in terms of p_2/p_{t1} and p_{t1}/p_{t2} for selected values of the loss coefficient, K.

One simply enters the proper γ-table with the information available and obtains (without iteration) the required solutions.

Computer-generated numerical tables of specific losses in compressible flow systems appear in the Appendix in the following order.

Table I	$\gamma = 1.1$	
Table II	$\gamma = 1.2$	
Table III	$\gamma = 1.3$	subsonic
Table IV	$\gamma = 1.4$	
Table V	$\gamma = 1.67$	
Table VI	$\gamma = 1.1$	
Table VII	$\gamma = 1.2$	
Table VIII	$\gamma = 1.3$	supersonic
Table IX	$\gamma = 1.4$	
Table X	$\gamma = 1.67$	

A useful quantity (α^*) which allows one to obtain the table entry Γ from the flow number α is tabulated for the various γ-tables:

γ	α^*
1.1	0.62836
1.2	0.64853
1.3	0.66726
1.4	0.68473
1.67	0.72666

COMPARISONS OF EXACT ANALYTIC SOLUTIONS WITH DARCY APPROXIMATIONS

In Fig. 5, we present comparisons of the static-pressure approximation of equation (6) (which is based on the Darcy formulation only) with the actual static-pressure drop for the constant-area, one-dimensional, subsonic flow of a perfect, compressible fluid in terms of identical inlet conditions and given loss coefficients. These comparisons are drawn in terms of an error function defined as

$$E_s = \frac{(p_2/p_1)_{\text{actual}} - (p_2/p_1)_{\text{Darcy}}}{(p_2/p_1)_{\text{actual}}} \qquad (22)$$

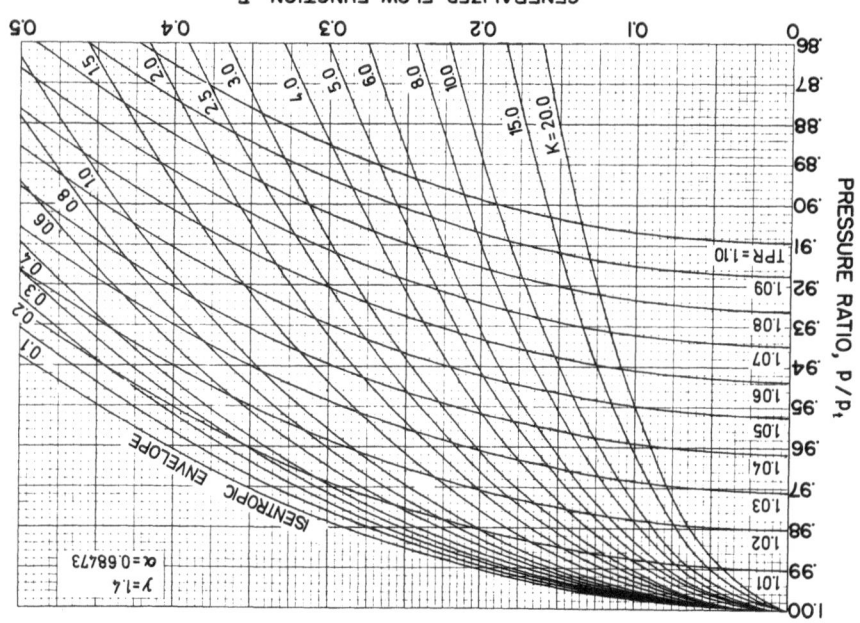

Figure 3. Amplified Portion of $\gamma = 1.4$ Fanno Curve. Parametric Solutions of Compressible-Flow Loss Coefficients and Total-Pressure Ratios as Functions of p/p_t and Γ.

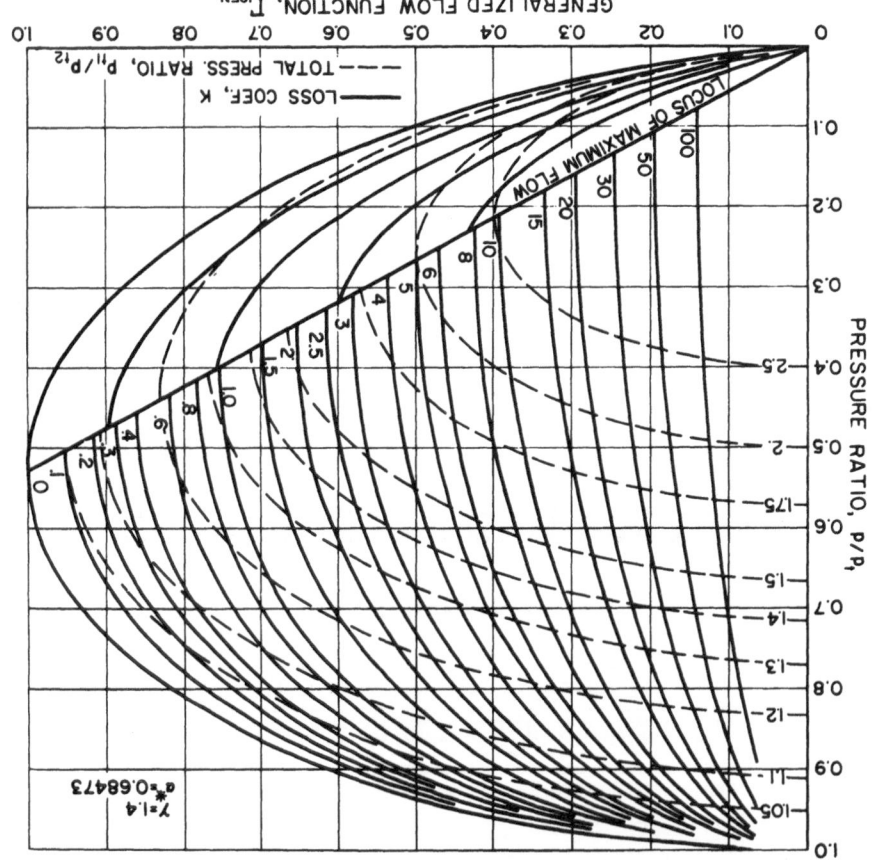

Figure 2. Generalized Fanno Flow Curve ($\gamma = 1.4$).

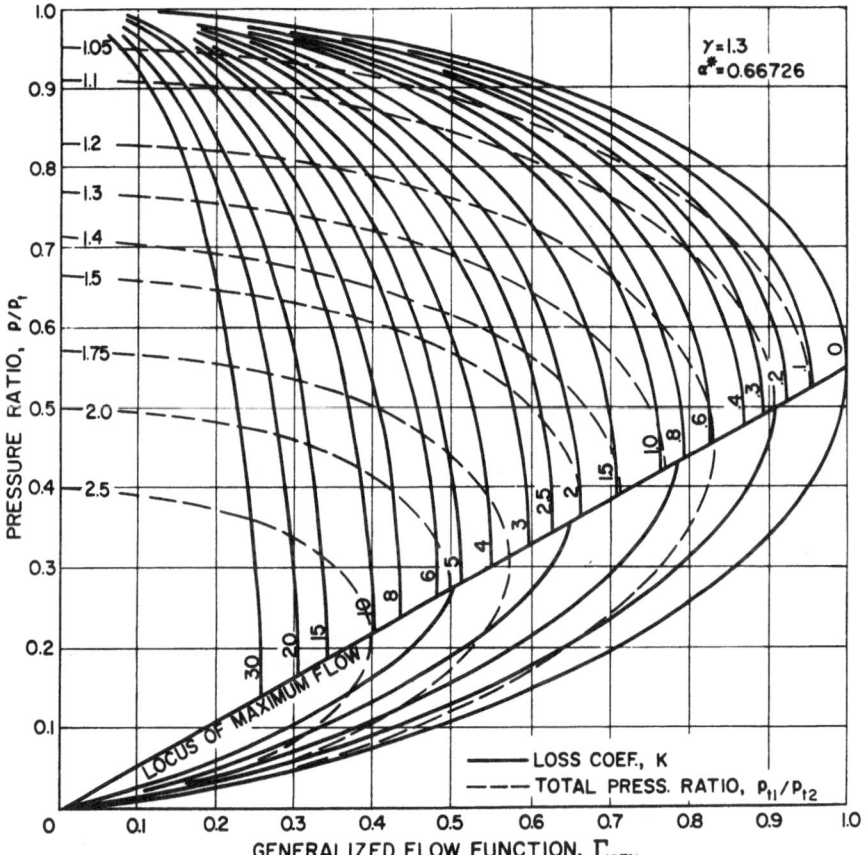

Figure 4. Generalized Fanno Flow Curve ($\gamma = 1.3$).

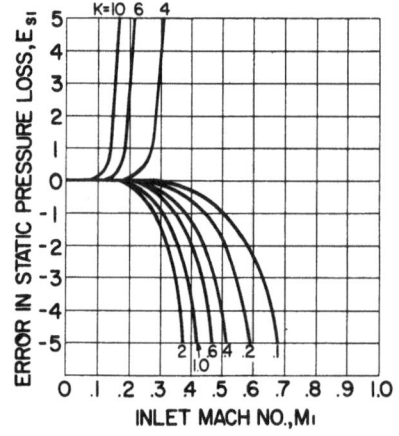

Figure 5. Error in Static-Pressure Approximation Based on Inlet Conditions.

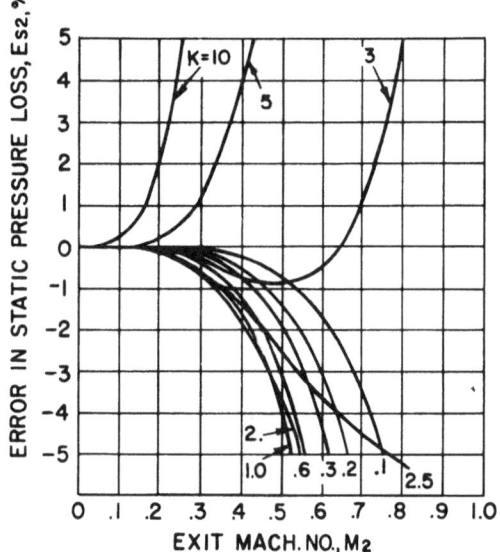

Figure 6. Error in Static-Pressure Approximation Based on Exit Conditions.

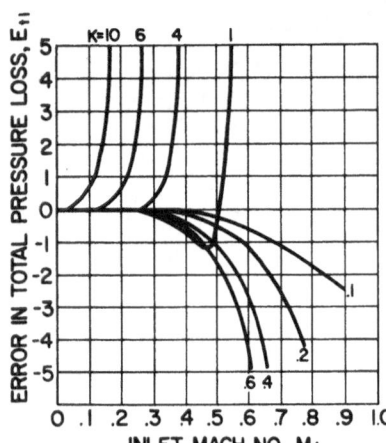

Figure 7. Error in Total-Pressure Approximation Based on Inlet Conditions.

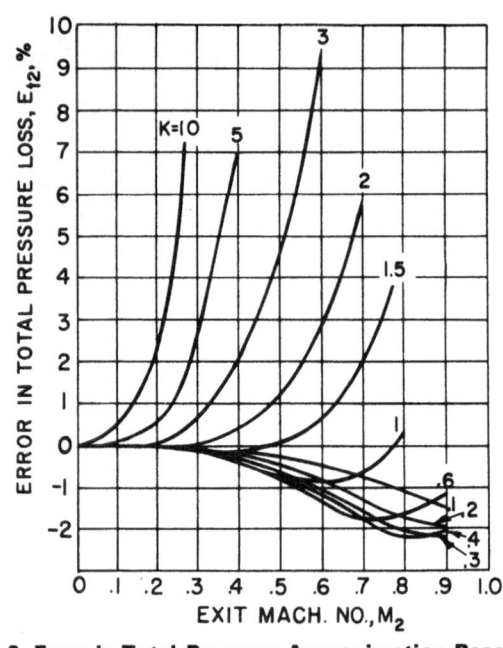

Figure 8. Error in Total-Pressure Approximation Based on Exit Conditions.

where the quantity $(p_2/p_1)_{\text{Darcy}}$ is obtained by an iterative procedure based on equation (6).

In Fig. 6, we present similar comparisons for identical exit conditions and given loss coefficients. The first comparison, i.e., the one based on inlet conditions, is the more significant in most actual flow problems; however, Figs. 5 and 6 both show the complete failure of the static-pressure approximation of equation (6).

In Fig. 7, we present comparisons of the total-pressure approximation of equation (7) (which is based on the Darcy formulation only) with the actual total-pressure drop for the constant-area, one-dimensional, subsonic flow of a perfect, compressible fluid in terms of identical inlet conditions and given loss coefficients.

These comparisons are drawn in terms of an error function defined as

$$E_t = \frac{(p_{t2}/p_{t1})_{\text{actual}} - (p_{t2}/p_{t1})_{\text{Darcy}}}{(p_{t2}/p_{t1})_{\text{actual}}} \qquad (23)$$

where the quantity $(p_{t2}/p_{t1})_{\text{Darcy}}$ is obtained by an iteration based on equation (7).

In Fig. 8, we present similar comparisons for identical exit conditions and given loss coefficients. Once again, the comparison based on inlet conditions is the more meaningful in most actual flow problems; however, Figs. 7 and 8 both indicate the complete failure of the total-pressure approximation of equation (7).

Chapter 3

Specific Numerical Examples of Flow with Losses

Given Loss Coefficient and Flow Rate

Example 1 (Fig. 9)

Consider the adiabatic flow of air ($\gamma = 1.4$, $R = 53.3$ ft/deg R) from an inlet total pressure of 50 psia and an inlet total temperature of 540° R; find the static- and total-pressure drops through a constant-area duct (25 in.^{2}) if the flow rate is 11.567 lb/sec and the loss coefficient is 2.

Solution

At

$$\Gamma_{\text{isen}} = \frac{\alpha}{\alpha^*} = \frac{\dfrac{W/\Delta t}{Ap_{t1}}\left(\dfrac{RT_t}{g}\right)^{\frac{1}{2}}}{0.68473} = 0.40412$$

obtain from Table IV

$p_1/p_{t1} = 0.96$	$p_1 = 48.00$ psia
$p_2/p_{t1} = 0.86936$	$p_2 = 43.47$ psia
$p_{t1}/p_{t2} = 1.09461$	$p_{t2} = 45.68$ psia

Thus

$$p_1 - p_2 = 4.53 \text{ psi}$$
$$p_{t1} - p_{t2} = 4.32 \text{ psi}$$

NOTE: Static-pressure drop always exceeds total-pressure drop in constant-area compressible flow because of fluid acceleration.

From Fig. 2 (graph): At

$$\Gamma_{\text{isen}} = \frac{\alpha}{\alpha^*} = 0.40412$$

obtain

$$p_1/p_{t1} = 0.96$$
$$p_2/p_{t1} = 0.87$$
$$p_{t1}/p_{t2} = 1.09$$

These values are seen to give substantially the same results as the tabular method.

Given Loss Coefficient and Ratio of Exit Static Pressure to Inlet Total Pressure

Example 2 (Fig. 10)

Consider the adiabatic flow of air from an inlet total pressure of 25.488 psia and an inlet total temperature of 540° R. Find the flow rate and the static- and total-pressure drops through a convergent passage which discharges to the atmosphere if the area ratio is $A_1/A_2 = 100$, the exit area is 1 in.^{2}, and the loss coefficient, based on the exit area, is 0.2.

Solution

At

$$p_2/p_{t1} = \frac{14.7}{25.488} = 0.57672$$

obtain from Table IV

$\Gamma'_{\text{isen}} = 0.90497$	$W/\Delta t = 0.528$ lb/sec
$p_{t1}/p_{t2} = 1.08231$	$p_{t2} = 23.55$ psia

$\Gamma'_{\text{isen 1}}$ is so labeled because we are dealing with a variable-area problem. Thus, by equation (9),

$$\Gamma_{\text{isen 1}} = \Gamma'_{\text{isen 1}}(A_2/A_1) = 0.00905$$

Again referring to Fig. 2 and Table IV this indicates that $p_1/p_{t1} \approx 1$. Hence

$$p_1 - p_2 = 25.49 - 14.70 = 10.79 \text{ psi}$$
$$p_{t1} - p_{t2} = 1.94 \text{ psi}$$

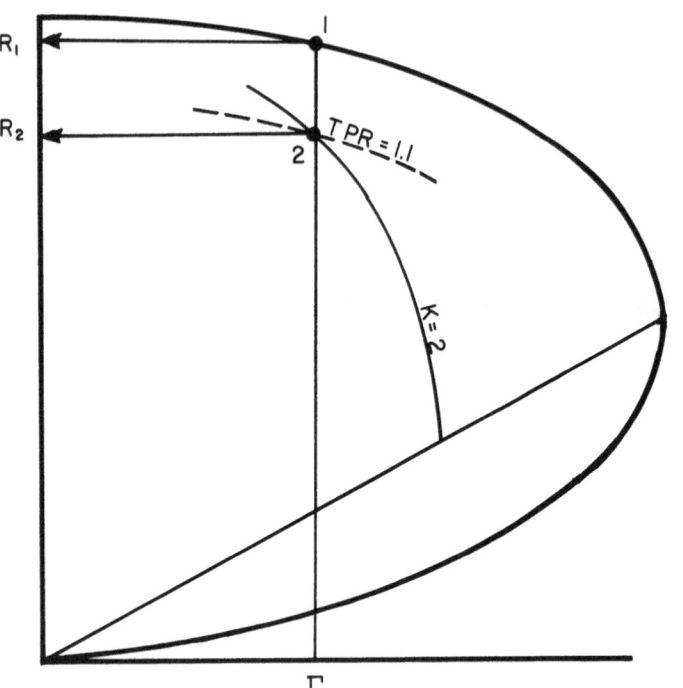

Figure 9. Generalized Fanno Curve for Example 1.

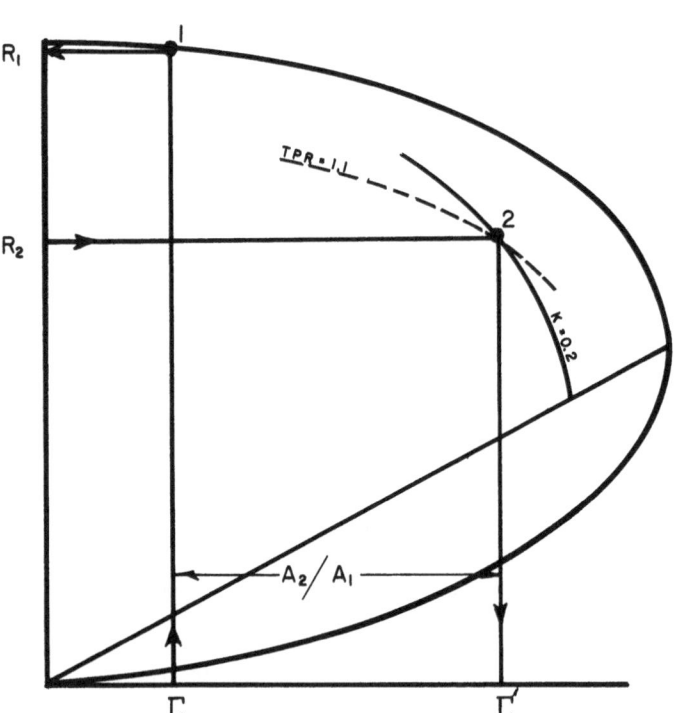

Figure 10. Generalized Fanno Curve for Example 2.

Given Inlet Conditions and Loss Coefficients in the Variable-Area Case

Example 3 (Fig. 11)

Consider the adiabatic flow of air from an inlet static pressure of 46.5 psia, an inlet total pressure of 50 psia, and an inlet total temperature of 540° R. Find the flow rate and the static- and total-pressure drops through a sudden enlargement where $A_1 = A_2 = 1$ in.², $A_3 = A_4 = 2$ in.², the loss coefficient to the enlargement is 4, the loss coefficient across the step change is 1 (based on the exit area), and the loss coefficient after the enlargement is 0.6.

Solution

From 1 to 2, at $p_1/p_{t1} = 0.93$ obtain from Table IV

$$p_{t1}/p_{t2} = 1.68333 \qquad p_{t2} = 29.704 \text{ psia}$$
$$p_2/p_{t1} = 0.44507 \qquad p_2 = 22.254 \text{ psia}$$
$$\Gamma_{\text{isen}1} = 0.52555$$

Thus

$$p_1 - p_2 = 24.246 \text{ psi}$$
$$p_{t1} - p_{t2} = 20.296 \text{ psi}$$
$$W/\Delta t = 0.602 \text{ lb/sec}$$

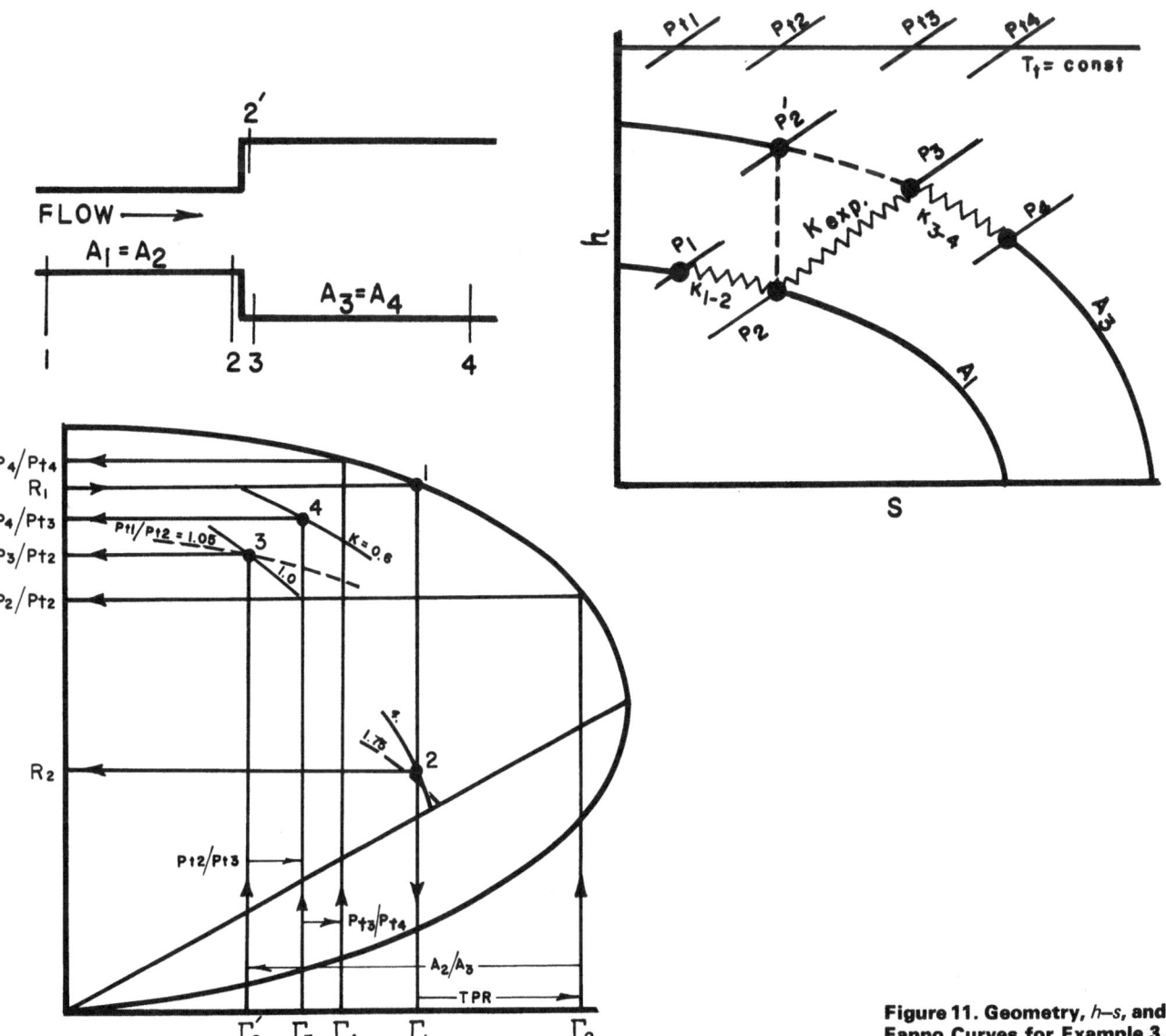

Figure 11. Geometry, *h–s*, and Fanno Curves for Example 3.

and

$$p_2/p_{t2} = 0.7492 \qquad \Gamma_2 = 0.8849$$

From 2 to 3, at $\Gamma'_2 = \Gamma_2(A_2/A_3) = 0.44245$ obtain from Table IV

$$p_{t2}/p_{t3} = 1.05438 \qquad p_{t3} = 28.172 \text{ psia}$$
$$p_3/p_{t2} = 0.89701 \qquad p_3 = 26.645 \text{ psia}$$

Thus

$$p_2 - p_3 = -4.391 \text{ psi}$$
$$p_{t2} - p_{t3} = 1.532 \text{ psi}$$

From 3 to 4, at $\Gamma_3 = \Gamma'_2 \, (p_{t2}/p_{t3}) = 0.46651$ obtain from Table IV

$$p_{t3}/p_{t4} = 1.03565 \qquad p_{t4} = 27.202 \text{ psia}$$
$$p_4/p_{t3} = 0.90920 \qquad p_4 = 25.614 \text{ psia}$$

Alternatively, the equivalent-loss coefficient of this system could be formed, based on the exit area, if p_4/p_{t1} were measured to be 0.51228 (as in the preceding), and if $\Gamma_1 = 0.52555$ (as in the preceding). Under these conditions, $\Gamma'_1 = 0.26278$, referred to the exit area, and Fig. 2 indicates a $\overline{K} = 21$ and a $p_{t1}/p_{t4} = 1.84$ in agreement with the step-by-step method.

Given Exit Conditions and Loss Coefficient

Example 4

Consider the adiabatic flow of air through a 10-in. duct of length 200 ft with a friction factor of 0.0159. The exit conditions are static pressure, 94 psia; total pressure 100 psia; total temperature 540° R. Find the inlet conditions.

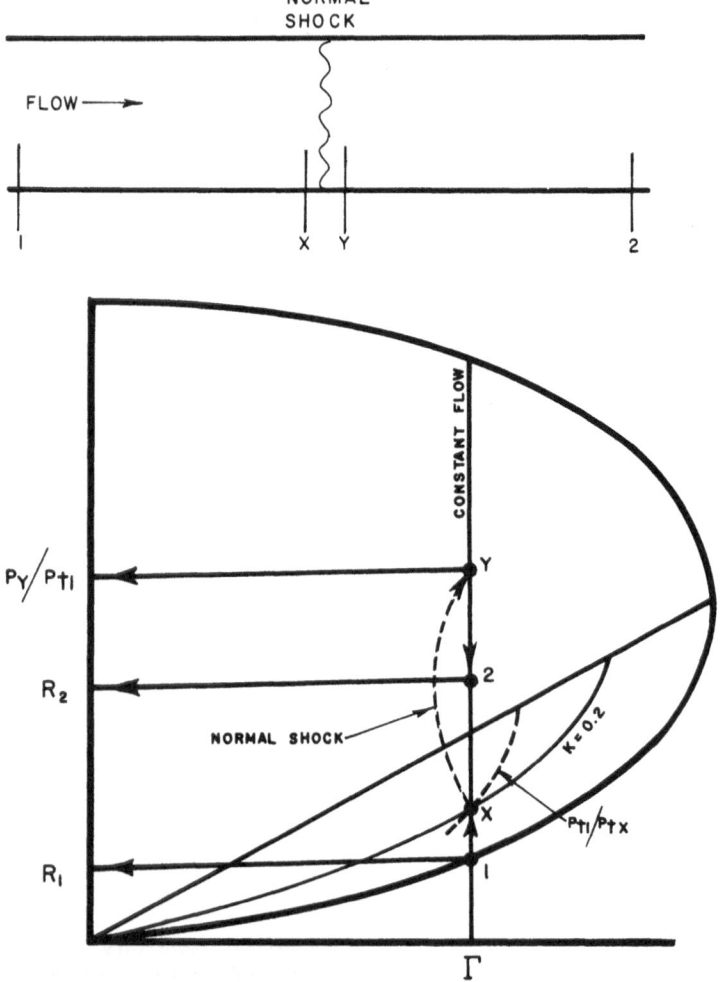

Figure 12. Geometry and Fanno Curve for Example 5.

Solution from Table IV

At

$$p_2/p_{t2} = 0.94 \qquad K_2^* = 5.365$$

but

$$K_{1-2} = fL/D = 3.813$$

thus

$$K_1^* = K_{1-2} + K_2^* = 9.178 \qquad p_1/p_{t1} = 0.96$$

Given Inlet Conditions and Loss Coefficients for Normal Shock Process

Example 5 (Fig. 12)

Consider the supersonic flow of air in a constant-area, adiabatic passage from an inlet pressure ratio of 0.04. Exit conditions are such that a normal shock takes place when $fL/D = 0.2$ from the inlet station. Find the static- and total-pressure drops across the system from inlet to a point where $fL/D = 0.4$ after the shock.

Solution

At

$$p_1/p_{t1} = 0.04 \qquad K = 0.2$$

obtain from Table IX

$$M_1 = 2.746 \qquad \Gamma_1 = 0.30065$$
$$p_x/p_{t1} = 0.06880 \qquad p_{t1}/p_{tx} = 2.10631$$

At

$$p_x/p_{tx} = \frac{p_x}{p_{t1}} \frac{p_{t1}}{p_{tx}} = 0.145 \qquad M_x = 1.92$$

obtain from Table IX

$$p_y/p_x = 4.12805 \qquad p_{tx}/p_{ty} = 1.31796$$

Thus

$$p_y/p_{t1} = (p_y/p_x)(p_x/p_{t1}) = 0.2840$$

and from

$$p_y/p_{ty} = (p_y/p_x)(p_x/p_{tx})(p_{tx}/p_{ty}) = 0.79$$

obtain from Table IV

$$p_2/p_{ty} = 0.61199 \qquad p_{ty}/p_{t2} = 1.12796$$

and

$$p_2/p_{t1} = (p_2/p_{ty})(p_{ty}/p_{tx})(p_{tx}/p_{t1}) = 0.22$$
$$p_1 - p_2 = (0.04 - 0.22)p_{t1} = -0.18p_{t1}$$

NOTE: The necessary increase in static pressure which occurs whenever decelerating from supersonic to subsonic flow:

$$p_{t1} - p_{t2} = [1 - (p_{tx}/p_{t1})(p_{ty}/p_{tx})(p_{t2}/p_{ty})]p_{t1} = 0.681p_{t1}$$

NOTE: The necessary decrease in total pressure which occurs in any process with losses.

Cases Involving Other Than Known Inlet Conditions and Loss Coefficients

The same solution presented here has also been used, without any change in the approach, for the following inputs: p_{t1}/p_{t2} and $W/\Delta t$; p_2/p_{t1} and p_{t1}/p_{t2}; p_2/p_{t1} and p_1/p_{t1}; p_2/p_{t1} and $W/\Delta t$. In these cases, the loss coefficient (K) is the main parameter to be determined.

Chapter 4

Generalized Solution for Constant-Density Flow with Losses

GENERAL REMARKS

It is true that constant-density flow with losses can be handled completely by the Darcy equation (5) and the general energy equation (3). However, for the sake of completeness, and to allow comparisons between compressible and constant-density solutions, and because of the symmetry of the solutions, we present in this chapter a brief review of a generalized solution for this type of flow.

CONSTANT-DENSITY LOSS SOLUTION

In an earlier work* it was demonstrated that the expression

$$\left(\frac{p_{t1}}{p_{t2}}\right)\left(\frac{A_1}{A_2}\right)^2 \Gamma_{\text{inc }1} = \Gamma_{\text{inc }2} \tag{24}$$

applied to all one-dimensional, constant-density, steady-flow processes. In this same reference it was also shown that the effect of area and elevation variations can always be represented by an isentropic change of state from a given initial state (corresponding to the initial area and initial elevation) to a new initial state (corresponding to the final area and the final elevation). One then proceeds from the area-elevation-modified initial state to the final state via a constant-area, constant-elevation process. Thus the constant-density flow process with losses can be handled in two steps: apply isentropic multipliers and then account for losses.

As before, the isentropic multipliers can always be specified; that is, from geometry one obtains $(A_1/A_2)^2$,

---

*W. G. Steltz and R. P. Benedict, Some Generalizations in One-Dimensional Constant Density Fluid Dynamics, *J. Eng. for Power, Trans. ASME*, Vol. 84, Series A, 1962, p. 44. See also, Thermodynamics of Constant-Density Fluids, *Electro-Technology*, April 1963, p. 70.

and from an integration of equation (3) between initial and final states, one obtains for the isentropic case

$$\left(\frac{p_{t1}}{p_{t2}}\right)_{\substack{\text{isen} \\ \Delta z}} = \frac{1}{1 + (z_1 - z_2)/p_{t1}v} = \frac{1}{1 + \beta} \tag{25}$$

However, the total-pressure ratio which represents the losses is the very unknown we seek, and more information is required before a solution is possible. Again, we turn to the well-known loss coefficient and/or friction factor for this additional information. It is with this constant-area solution, in terms of the constant-density flow function (Γ_{inc}), the pressure ratio, p/p_{t1}, and the loss coefficient, k, that we now concern ourselves.

An integration of equation (3) for the constant-area case yields

$$k\frac{V^2}{2g} = v(p_2 - p_1) + (z_2 - z_1) \tag{26}$$

Upon dividing through by the factor $p_{t1}v$, equation (26) can also be expressed as

$$k = \frac{2}{\alpha^2}(R_1 - R_2 + \beta) = \frac{2}{\alpha^2}(1 - R_2 + \beta) - 1 \tag{27}$$

and thus we obtain the useful relation

$$R_1 = \frac{R_2 + k - \beta}{1 + k} \tag{28}$$

where, as before,

$R = p/p_{t1}$
β = elevation factor defined in equation (25)
$\alpha = V/(p_{t1}vg)^{\frac{1}{2}}$, hybrid flow number

The flow number (α) can also be given in terms of the inlet-pressure ratio by the familiar

$$\alpha = [2(1 - R_1)]^{\frac{1}{2}} \tag{29}$$

Equation (29) is solved simultaneously with either

equation (27) or (28) to yield α as a function of R_2, the parameter being constant k. Such solutions exactly represent constant-density flows with losses in constant-area passages. The flow number of equation (29) is related to the more general constant-density flow function as follows:

$$\Gamma_{\text{inc}} = \left(\frac{\alpha}{\alpha^*}\right)^2 \qquad (30)$$

where α^* represents the isentropic maximum flow number (based on $R_1 = 0$) defined as

$$\alpha^* = [2]^{\frac{1}{2}} \qquad (31)$$

Another useful relation which follows directly from the foregoing equation is

$$R_2 = -\Gamma_{\text{inc}} + p_{t2}/p_{t1} = -(k+1)\Gamma_{\text{inc}} + 1 \qquad (32)$$

which is seen to describe linear relationships between Γ_{inc} and R_2, where p_{t2}/p_{t1} represents both the R_2 and the Γ_{inc} intercepts for constant-total-pressure-loss parameters, and where $-(1 + k)$ represents the slope of constant-loss-coefficient parameters.

Use of these constant-density relations will be demonstrated by sample problems. Representative solutions for all these equations are presented graphically in the generalized constant-density-flow curve in Fig. 13. Tabular solutions are, of course, not required for the simplistic constant-density-flow problems.

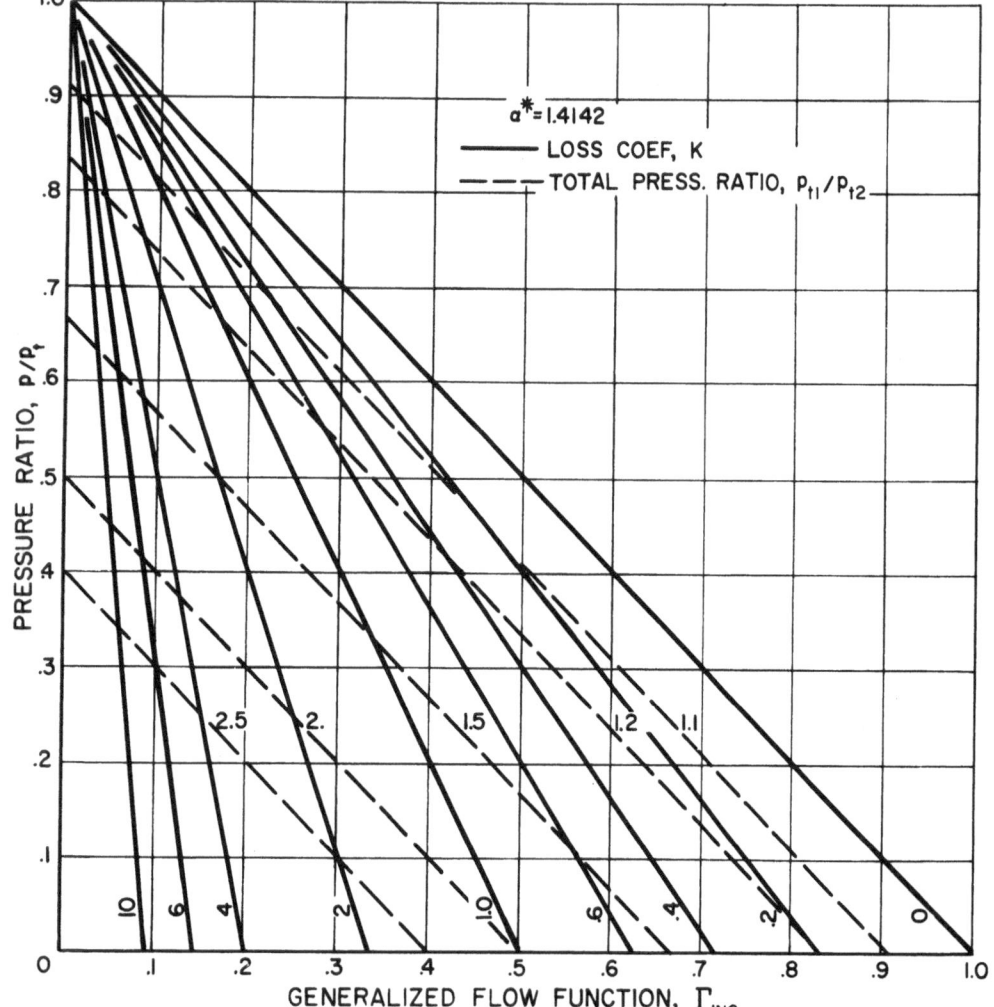

Figure 13. Generalized Constant-Density Flow Curve.

18

NUMERICAL EXAMPLES

Example 1

Consider the flow of water ($w = 62.4$ pcf) from an inlet total pressure of 50 psia. Find the static- and total-pressure drops through a constant-area (5 in.^{2}), horizontal pipe if the flow rate is 100 lb/sec and the loss coefficient is 1.

Solution

At

$$\alpha = \frac{W/\Delta t}{Ap_{t1}}\left(\frac{p_{t1}v}{g}\right)^{\ddagger} = 0.75717$$

obtain by equation (29)

$$p_1/p_{t1} = 0.71333 \qquad p_1 = 35.67 \text{ psia}$$

by equation (30)

$$\Gamma_{inc} = \left(\frac{\alpha}{\alpha^*}\right)^2 = 0.28667$$

and by equation (32)

$$p_2/p_{t1} = 0.42666 \qquad p_2 = 21.33 \text{ psia}$$
$$p_{t2}/p_{t1} = 0.71333 \qquad p_{t2} = 35.66 \text{ psia}$$

Thus

$$p_1 - p_2 = 14.34 \text{ psi}$$
$$p_{t1} - p_{t2} = 14.34 \text{ psi}$$

NOTE: Static-pressure drop always equals total-pressure drop in constant-area, constant-elevation, constant-density flow because fluid velocity (and hence the dynamic pressure) is constant. Use of Fig. 13 yields essentially the same results as the numerical method.

Example 2

Consider the leakage of oil ($w = 50$ pcf) from a tank where the total pressure is 29.4 psia. Find the flow rate and the static- and total-pressure drops through an orificelike passage which discharges to the atmosphere if the exit area is 1 in.^{2} and the loss coefficient is 0.2.

Solution

At

$$p_2/p_{t1} = \frac{14.7}{29.4} = 0.5$$

obtain by equation (32)

$$p_{t2}/p_{t1} = 0.917 \qquad p_{t2} = 27.0 \text{ psia}$$

$$\Gamma'_{inc} = 0.417 \qquad \alpha_2 = 0.913 \qquad W/\Delta t = 16.54 \text{ lb/sec.}$$

The foregoing is so labeled because we are dealing with a variable-area problem; i.e.,

$$A_2/A_1 \approx 0$$

Thus

$$\Gamma_{inc\,1} \approx 0$$

and

$$p_1/p_{t1} \approx 1$$

Hence

$$p_1 - p_2 = 29.4 - 14.7 = 14.7 \text{ psi}$$
$$p_{t1} - p_{t2} = 2.4 \text{ psi}$$

Example 3

Consider the flow of water under the same conditions and geometry as in Example 3 of Chapter 3 (i.e., a sudden enlargement problem). Find the flow rate, the overall-loss coefficient (based on exit area), and the static- and total-pressure drops of the system.

Solution

From

$$\Gamma_{inc\,1} = 1 - p_1/p_{t1} = 0.07 \quad W/\Delta t = 9.88 \text{ lb/sec}$$

The total head loss in the system is given by

$$F = k_{1-2}\frac{V_2^2}{2g} + k_{enlargement}\frac{V_3^2}{2g} + k_{3-4}\frac{V_4^2}{2g}$$

$$F = \left[k_{1-2}\left(\frac{A_4}{A_2}\right)^2 + k_{enlargement} + k_{3-4}\right]\frac{V_4^2}{2g} = \bar{k}\frac{V_4^2}{2g}$$

Thus, $\bar{k} = 17.6$, and the problem now reduces to one of constant area (the exit area must be used).

From

$$\Gamma_{inc\,4} = (A_1/A_4)^2\Gamma_{inc\,1} = 0.0175$$

obtain by equation (32)

$$p_4/p_{t1} = 0.6745 \qquad p_4 = 33.72 \text{ psia}$$
$$p_{t4}/p_{t1} = 0.6920 \qquad p_{t4} = 34.60 \text{ psia}$$

Hence

$$p_1 - p_4 = 12.78 \text{ psi}$$
$$p_{t1} - p_{t4} = 15.40 \text{ psi}$$

Chapter 5

On Combining Loss Coefficients

GENERAL REMARKS

Rarely does a flow system under evaluation consist of a single loss element. One therefore is confronted with the task of evaluating the pressure loss across each of the loss elements in sequence if the overall flow loss is desired. This approach is not only tedious but must be repeated if the flow conditions are altered. It becomes apparent that the possibility of combining loss coefficients to obtain one overall coefficient for the entire flow system is desirable. Such a coefficient, once established, will allow one to evaluate the flow loss of the entire system quite simply at any flow condition.

In this chapter the combining features of both the constant-density and compressible-flow loss coefficients are examined.

CONSTANT-DENSITY LOSS COEFFICIENTS

Considering first the more familiar constant-density loss coefficient (k), we have the general expression

$$k_{z,y} = \frac{2gv}{V_y^2}(p_z - p_y) \tag{33}$$

with the loss coefficient based on exit velocity, as is conventional. The overall loss coefficient for an arbitrary system is then

$$k_{0,n} = \frac{2gv}{V_n^2}(p_0 - p_n) = C_n(p_0 - p_n) \tag{34}$$

while the loss coefficients of the various elements of the flow system are

$$k_{0,1} = \frac{2gv}{V_1^2}(p_0 - p_1) = C_n\left(\frac{A_1}{A_n}\right)^2(p_0 - p_1)$$

$$k_{1,2} = \frac{2gv}{V_2^2}(p_1 - p_2) = C_n\left(\frac{A_2}{A_n}\right)^2(p_1 - p_2) \tag{35}$$

$$k_{n-1,n} = \frac{2gv}{V_n^2}(p_{n-1} - p_n) = C_n\left(\frac{A_n}{A_n}\right)^2(p_{n-1} - p_n)$$

A comparison of equations (34) and (35) indicates that

$$k_{0,n} = \left[k_{0,1}\left(\frac{A_n}{A_1}\right)^2 + k_{1,2}\left(\frac{A_n}{A_2}\right)^2 + \ldots + k_{n-1,n}\right] \tag{36}$$

Thus we conclude in the constant-density case that loss coefficients are additive when adjusted to a common reference area. Incidentally, it follows that head losses as defined by equation (4) are also additive in the constant-density case, although it should be noted that equation (36) is entirely independent of this combining characteristic of head losses.

COMPRESSIBLE-FLOW LOSS COEFFICIENTS

Considering now the compressible-flow loss coefficient (K), we find upon examination of equation (17) that area is inexorably involved in the definition of K (through the flow number, α), and hence resists factoring. Thus we can expect no simple combining relation for compressible-flow loss coefficients such as was given by equation (36) for the constant-density case. However, the problem becomes tractable if all loss coefficients are *initially* adjusted to a common reference area.

This entails an evaluation of K according to equation (17) based on the pertinent exit area, and then a reevaluation of K based on the common reference area while maintaining the original total-pressure drop and, of course, the same flow rate.

Under these conditions, equation (17) can be expressed in general terms as

$$K_{z,y}]_{ref} = f(p_z) - f(p_y) - \left(\frac{\gamma + 1}{\gamma}\right)\ln\frac{F(p_z)}{F(p_y)} \tag{37}$$

while the overall loss coefficient for an arbitrary system is

$$K_{0,n}]_{ref} = f(p_0) - f(p_n) - \left(\frac{\gamma + 1}{\gamma}\right)\ln\frac{F(p_0)}{F(p_n)} \tag{38}$$

and the loss coefficients of the various parts of the system are

$$K_{0,1}]_{ref} = f(p_0) - f(p_1) - \left(\frac{\gamma + 1}{\gamma}\right) \ln \frac{F(p_0)}{F(p_1)}$$

$$K_{1,2}]_{ref} = f(p_1) - f(p_2) - \left(\frac{\gamma + 1}{\gamma}\right) \ln \frac{F(p_1)}{F(p_2)} \qquad (39)$$

$$K_{n-1,n}]_{ref} = f(p_{n-1}) - f(p_n) - \left(\frac{\gamma + 1}{\gamma}\right) \ln \frac{F(p_{n-1})}{F(p_n)}$$

A comparison of equations (38) and (39) indicates that

$$K_{0,n}]_{ref} = K_{0,1}]_{ref} + K_{1,2}]_{ref} + \ldots + K_{n-1,n}]_{ref} \qquad (40)$$

Thus we conclude in the compressible case also that loss coefficients are additive when adjusted to a common reference area. However, note that while equation (40) is the compressible counterpart of equation (36), these relations are far from being identical. Incidentally, it follows that head losses as defined by equation (4) are not additive in the compressible-flow case.

Recent experimental studies* have verified this combining feature of the compressible-flow loss coefficient.

---

*R. P. Benedict, "On the Determination and Combination of Loss Coefficients for Compressible Fluid Flows," *J. Engrg. for Power, Trans. ASME* **88A**, 67 (1966).

Chapter 6

Losses across Abrupt Enlargements and Contractions

GENERAL REMARKS

One's ability to evaluate the pressure loss associated with a given flow system is dependent upon the available knowledge of the loss coefficients of the specific elements comprising the flow system. Therefore the success of such an evaluation is dictated by the validity of the established values of these coefficients. Although numerous references can be found specifying values for these loss elements, most are vague when specifying their limitations. Also, since a large part of these data was obtained for constant-density fluids, the investigator is usually confronted with the problem of how applicable they are when compressible fluids are involved. Two of the more common loss elements encountered in flow systems are the abrupt enlargement and contraction elements. Recent studies,* both analytical and experimental, performed on these elements for both compressible and constant-density fluids are presented in this chapter.

ABRUPT ENLARGEMENTS

Figure 14 depicts a two-dimensional flow model of an abrupt enlargement element. For this element the loss is considered to occur between position 1, the point of flow separation, and position 2, the point of maximum static pressure recovery. Starting with equation (4) and integrating between arbitrary limits yields the general head-loss equation for this element:

$$h_{\substack{loss \\ 1,2}} = \int_1^2 \delta F = \int_2^1 v\, dp + \frac{V_1^2 - V_2^2}{2g} \tag{41}$$

The momentum balance

$$\Sigma F_z = Ma = \left(\frac{M}{\Delta t}\right) \Delta V \tag{42}$$

*R. P. Benedict, N. A. Carlucci, and S. D. Swetz, *J. Engrg. for Power, Trans. ASME* **88A**, 73 (1966).

can be written in differential form for the fluid element between the system limits 1 and 2 in Fig. 14 as

$$[p - (p + dp)]A_2 = (\rho_2 A_2 V_2)\, dV \tag{43}$$

by making use of the continuity equation

$$\frac{M}{\Delta t} = \rho A V = \text{constant} \tag{44}$$

and the assumptions that p at 1 acts over the area A_2 and that the shearing stresses along the boundary from 1 to 2 are negligible. Equation (43) can be integrated to give

$$\int_2^1 v_2\, dp = \frac{V_2(V_2 - V_1)}{g} \tag{45}$$

When the fluid is taken to have a constant density, i.e., $v_2 = v_1 = v$, equations (41) and (45) can be combined directly to yield the well-known Carnot-Borda equation

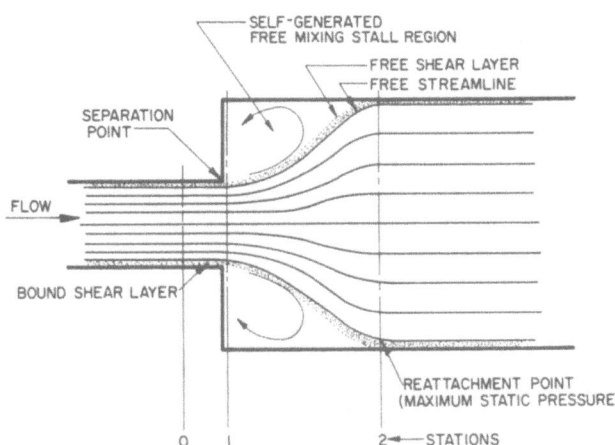

Figure 14. Flow Model of an Abrupt Enlargement.

$$h_{\text{loss}\atop 1,2} = \frac{(V_1 - V_2)^2}{2g} \qquad (46)$$

This equation is often used to describe the head loss involved when a constant-density flowing fluid encounters an abrupt increase in the area of its system boundaries.

When equation (46) is written in terms of a general loss coefficient k_x defined by the equation

$$h_{\text{loss}} = k_x \frac{V_x^2}{2g} \qquad (47)$$

a specific loss coefficient for an abrupt enlargement, based on the velocity in the smaller pipe (Fig. 14), can be given as

$$(k_{1,2})_1 = \left(1 - \frac{A_0}{A_2}\right)^2 = (1 - 2\beta^2 + \beta^4) \qquad (48)$$

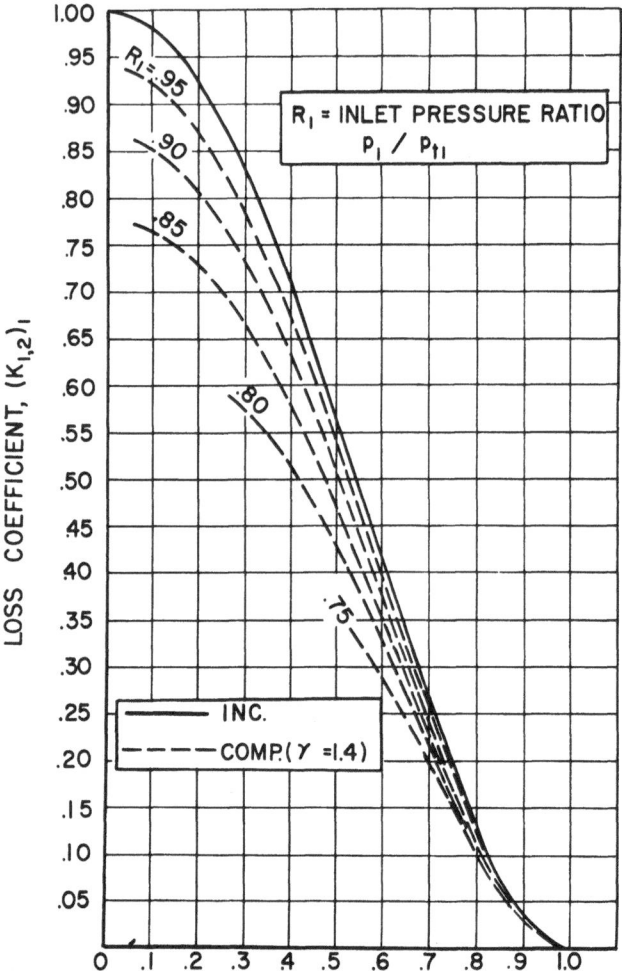

Figure 15. Compressible and Constant-Density Loss Coefficients for Abrupt Enlargements.

where $\beta = D_0/D_2$, the diameter ratio, and where $(k_{1,2})_1$ is read as the loss coefficient, k, from 1 to 2 based on the velocity at 1. Another significant parameter in considering flow with losses, although one seldom discussed, is the total-pressure ratio (TPR = p_{t1}/p_{t2}), which gives a clear thermodynamic indication of the loss in terms of entropy. Thus, according to work previously cited,* we have for constant-density, constant-elevation, adiabatic flow, with losses

$$S_2 - S_1 = c \ln \left[1 + \frac{p_{t1}}{cT_1 w}\left(1 - \frac{p_{t2}}{p_{t1}}\right)\right] \qquad (49)$$

A general expression for the constant-density TPR also can be given in terms of loss coefficients as

$$\left(\frac{1}{\text{TPR}}\right)_{\text{inc}} = \frac{p_{t2}}{p_{t1}} = 1 - (1 - R_1)k_1 = \frac{1}{1 + (1 - R_2)k_2} \qquad (50)$$

where R_1 is the inlet pressure ratio (p_1/p_{t1}), k_1 is the loss coefficient expressed in terms of the inlet velocity, and R_2 and k_2 refer to similar quantities at the exit state. Equation (50), which is obtained by combining energy (41), continuity (44), and the relevant definition of total pressure $(p_t = p + wV^2/2g)$, can be expressed explicitly for the abrupt enlargement of equation (48) as

$$\left(\frac{1}{\text{TPR}}\right)_{\text{inc}\atop\text{step-up}} = 1 - (1 - R_1)(1 - 2\beta^2 + \beta^4) \qquad (51)$$

A final parameter of interest in considering flow through an abrupt enlargement is the static-pressure ratio (SPR = p_2/p_1), which is a direct indication of the static-pressure recovery across the step-up. The constant-density, static-pressure ratio also can be given explicitly for the abrupt enlargement, making use of the same relations used to obtain (51), as

$$(\text{SPR})_{\text{inc}} = \frac{p_2}{p_1} = 1 + \left(\frac{1 - R_1}{R_1}\right)(2\beta^2 - 2\beta^4) \qquad (52)$$

where it too is seen to be a function of β and R_1 only.

For compressible fluids, the $\int v\, dp$ of equation (41) cannot be evaluated for a general flow process. Hence, an experimental study is necessary if specific values of the compressible-flow loss coefficient are desired. Experimentally established loss coefficients for the abrupt enlargement element are presented in Fig. 15. It will be observed that the compressible-flow loss coefficient is not solely dependent upon geometry but it is also influenced by the inlet pressure ratio (p/p_t). The reason for the termination of the compressible curves is that at these values of diameter ratio and flow the resulting loss yields critical conditions at the exit. A more useful parameter than the more commonly used

---

*See footnote on page 17.

loss coefficient to describe the loss of an abrupt enlargement is the TPR previously defined. It has been experimentally demonstrated that this parameter is essentially independent of the fluid and can be applied to yield the pressure loss when either compressible or constant-density fluids are involved. Figure 16 presents this parameter *versus* diameter ratio for various values of the inlet pressure ratio.

ABRUPT CONTRACTIONS

A two-dimensional flow model of an abrupt contraction element is depicted in Fig. 17. A study of this figure will disclose that the loss of this element basically consists of an inlet loss, between positions 3 and 4, and an abrupt enlargement loss, between positions 4 and 5. Although the analysis of this system appears quite similar to that of the abrupt enlargement, it is not possible to obtain an explicit analytic expression for the loss coefficient of an abrupt contraction. However, analytic expressions based on the use of one or two experimentally determined coefficients can be given. One expression for the overall loss coefficient for an abrupt contraction, which is simply the sum of the inlet and enlargement losses, has been given in terms of the velocity at position 5 as

$$(k_{3,5})_5 = \frac{1}{C_v^2 C_c^2} - \frac{2}{C_c} + 1 \qquad (53)$$

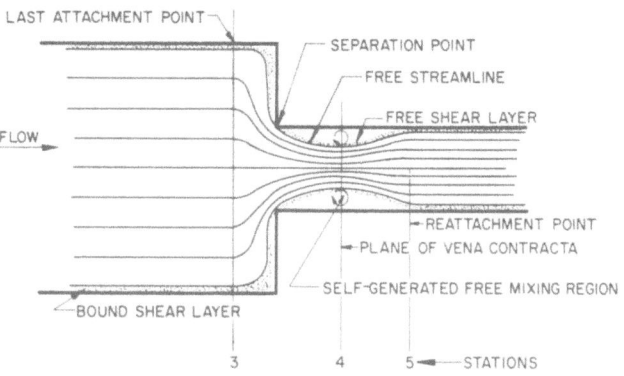

Figure 17. Flow Model of an Abrupt Contraction.

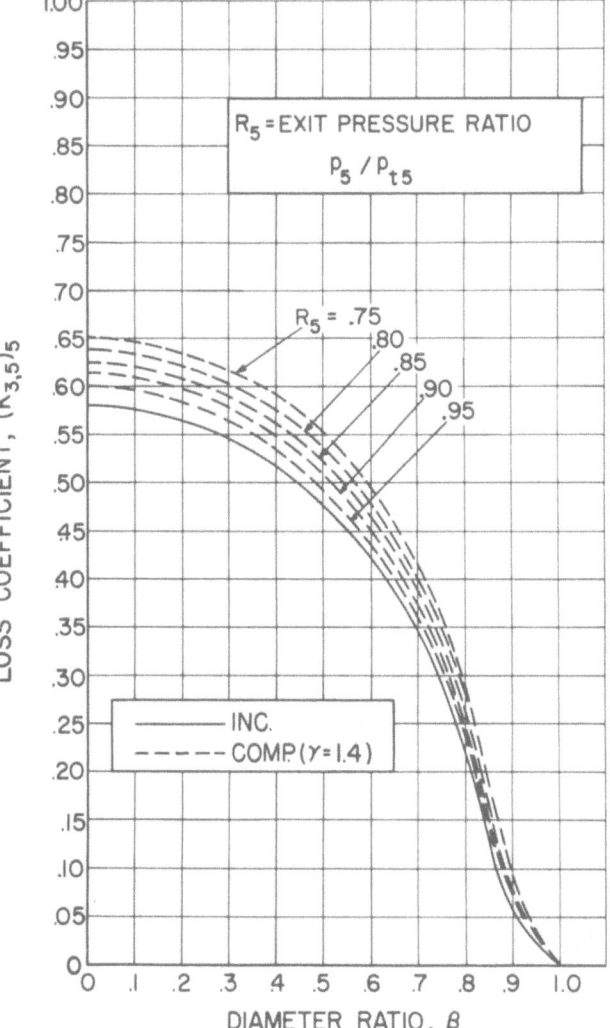

Figure 18. Compressible and Constant-Density Loss Coefficients for Abrupt Contractions.

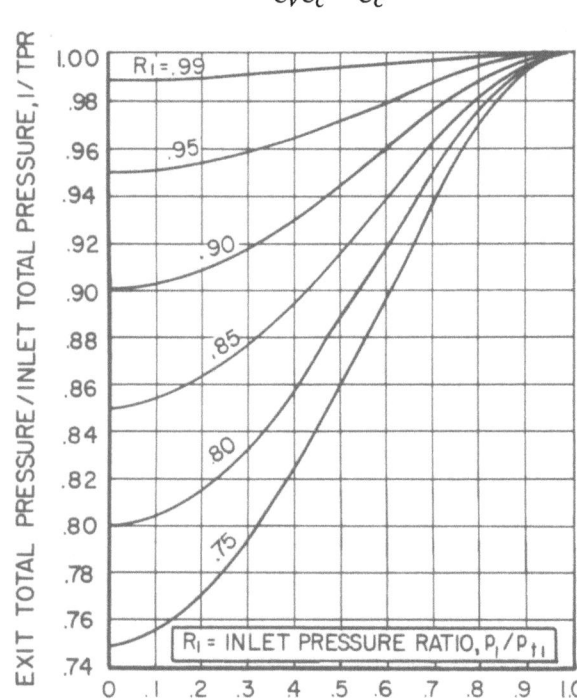

Figure 16. General Loss Curves for Abrupt Enlargements.

25

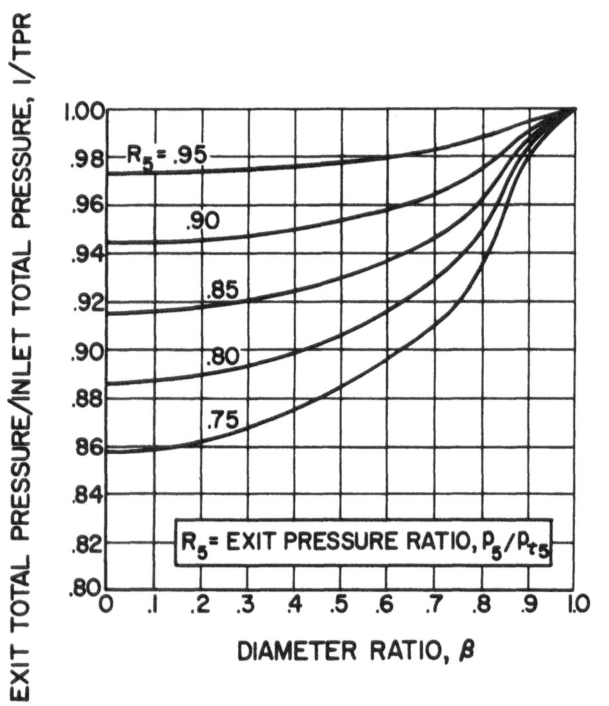

Figure 19. General Loss Curves for Abrupt Contractions.

In this expression the velocity coefficient (C_V) and the contraction coefficient are defined as follows:

$$C_V \equiv \frac{V_4}{V_4'} \qquad (54)$$

$$C_C \equiv \frac{A_4}{A_5} \qquad (55)$$

Note that the above determination of the abrupt contraction loss coefficient for a constant-density fluid hinges on a knowledge of the two coefficients, C_V from 3 to 4, and C_C between 4 and 5. The validity of equation (53) is also seriously restricted by the plenum flow assumption.

A second expression for the overall loss coefficient for an abrupt contraction can be given in terms of a discharge coefficient defined as

$$C_D \equiv \frac{Q_{\text{actual}}}{Q_{\text{ideal}}} \qquad (56)$$

The ideal volumetric flow rate, Q_{ideal}, is simply $A_5 V_5'$, where the ideal velocity, V_5', is defined by the continuity relation

$$\frac{A_5}{A_3} = \beta^2 = \frac{V_3}{V_5} = \frac{V_3'}{V_5'} \qquad (57)$$

and the velocity coefficient of equation (54), applied now to stations 3 and 5. Noting that at 3 and 5 the pipe flows full, we can set $C_D = C_{V5}$ and obtain the loss coefficient for an abrupt contraction as

$$(k_{3,5})_5 = \left(\frac{1}{C_D^2} - 1\right)(1 - \beta^4) \qquad (58)$$

by way of the energy relation of equation (4) written in terms of stations 3 and 5 as

$$\frac{p_3 - p_5}{w} = \frac{V_5^2 - V_3^2}{2g} + (k_{3,5})_5 \frac{V_5^2}{2g} = \frac{(V_5')^2 - (V_3')^2}{2g} \qquad (59)$$

Note that the determination of the abrupt contraction loss coefficient by equation (58) depends on a knowledge of one coefficient only, namely, the experimentally determinable C_D, and because it is also free of the plenum flow restriction of equation (53), it is the preferred constant-density contraction counterpart of enlargement equation (48).

The constant-density total-pressure ratio is as significant in considerations of the abrupt contraction as it is for the abrupt enlargement, and for the same reasons. For the step-down, TPR is obtained by applying equation (50) to this situation to yield

$$\left(\frac{1}{\text{TPR}}\right)_{\substack{\text{inc} \\ \text{step-down}}} = \frac{p_{t5}}{p_{t3}} = \frac{1}{1 + (1 - R_5)(k_{3,5})_5} \qquad (60)$$

where R_5 is the exit pressure ratio (p_5/p_{t5}). Equation (60) is the contraction counterpart of enlargement equation (51).

Experimentally established values of the constant-density loss coefficient for the abrupt contraction are presented in Fig. 18. As was the case for the abrupt enlargement, the compressible-flow loss coefficient for the abrupt contraction must also be experimentally determined. These values of the compressible-flow loss coefficient are also presented in Fig. 18. Similar to the case of the abrupt enlargement, these coefficients are dependent upon the inlet-pressure ratio in addition to the diameter ratio. Of interest also is the fact that the constant-density loss coefficient has been found experimentally to exceed the commonly considered limiting value of 0.5.

Those desiring to obtain an estimate of the total pressure loss of the abrupt contraction directly without first determining the loss coefficient should refer to Fig. 19. The TPR parameter obtained from this figure can be applied regardless of fluid.

Appendix

TABLE I. Specific Heat Ratio = 1.10

R₂ and TPR versus Loss Coefficient K

p/p_t	MACH	T/T_t	GAMMA	K_{crit}	.1	.2	.4	.6	1.0	2.0	4.0	6.0	10.0
.999	.0426	.9999	.07112	492.88948	.99890	.99880	.99860	.99840	.99800	.99700	.99499	.99297	.98893
					1.00010	1.00020	1.00040	1.00060	1.00100	1.00201	1.00403	1.00606	1.01016
.998	.0603	.9998	.10051	243.51870	.99780	.99760	.99720	.99680	.99599	.99398	.98994	.98589	.97773
					1.00020	1.00040	1.00080	1.00120	1.00201	1.00403	1.00811	1.01223	1.02064
.997	.0739	.9997	.12302	160.56198	.99670	.99640	.99579	.99519	.99398	.99095	.98487	.97875	.96639
					1.00030	1.00060	1.00120	1.00181	1.00302	1.00606	1.01224	1.01853	1.03148
.996	.0854	.9996	.14195	119.17007	.99560	.99519	.99439	.99358	.99197	.98792	.97977	.97155	.95489
					1.00040	1.00080	1.00161	1.00241	1.00403	1.00812	1.01643	1.02496	1.04269
.995	.0955	.9995	.15860	94.38176	.99450	.99399	.99298	.99197	.98995	.98487	.97463	.96428	.94324
					1.00050	1.00100	1.00201	1.00302	1.00505	1.01018	1.02068	1.03152	1.05428
.994	.1046	.9995	.17362	77.88776	.99339	.99279	.99157	.99036	.98792	.98181	.96946	.95695	.93142
					1.00060	1.00121	1.00242	1.00363	1.00607	1.01226	1.02499	1.03822	1.06630
.993	.1130	.9994	.18740	66.13065	.99229	.99158	.99017	.98874	.98590	.97874	.96426	.94956	.91943
					1.00070	1.00141	1.00282	1.00424	1.00710	1.01436	1.02936	1.04506	1.07876
.992	.1209	.9993	.20020	57.33076	.99119	.99038	.98875	.98713	.98386	.97566	.95903	.94210	.90726
					1.00080	1.00161	1.00323	1.00486	1.00813	1.01647	1.03380	1.05205	1.09169
.991	.1282	.9992	.21220	50.49827	.99009	.98917	.98734	.98551	.98183	.97257	.95376	.93457	.89489
					1.00091	1.00181	1.00364	1.00547	1.00917	1.01860	1.03829	1.05919	1.10514
.990	.1352	.9991	.22353	45.04363	.98898	.98797	.98593	.98389	.97979	.96946	.94846	.92696	.88232
					1.00101	1.00202	1.00405	1.00609	1.01021	1.02075	1.04286	1.06650	1.11913
.989	.1418	.9990	.23427	40.58994	.98788	.98676	.98451	.98226	.97774	.96635	.94312	.91928	.86954
					1.00111	1.00222	1.00446	1.00671	1.01126	1.02291	1.04749	1.07397	1.13370
.988	.1482	.9989	.24452	36.88504	.98678	.98555	.98310	.98063	.97569	.96322	.93774	.91152	.85653
					1.00121	1.00242	1.00487	1.00733	1.01231	1.02509	1.05220	1.08162	1.14891
.987	.1543	.9988	.25433	33.75670	.98567	.98434	.98168	.97900	.97364	.96007	.93233	.90368	.84328
					1.00131	1.00263	1.00528	1.00795	1.01336	1.02728	1.05697	1.08945	1.16479
.986	.1602	.9987	.26375	31.08084	.98457	.98313	.98026	.97737	.97158	.95692	.92687	.89576	.82977
					1.00141	1.00283	1.00569	1.00858	1.01442	1.02949	1.06182	1.09747	1.18142
.985	.1658	.9986	.27282	28.76588	.98346	.98192	.97884	.97574	.96951	.95375	.92138	.88774	.81598
					1.00152	1.00304	1.00611	1.00920	1.01548	1.03172	1.06675	1.10570	1.19885
.984	.1713	.9985	.28157	26.74457	.98236	.98071	.97741	.97410	.96744	.95057	.91584	.87963	.80190
					1.00162	1.00324	1.00652	1.00983	1.01655	1.03397	1.07176	1.11413	1.21716
.983	.1766	.9984	.29004	24.96454	.98125	.97950	.97599	.97246	.96537	.94738	.91026	.87143	.78749
					1.00172	1.00345	1.00694	1.01046	1.01762	1.03624	1.07684	1.12279	1.23642
.982	.1818	.9984	.29824	23.38563	.98015	.97829	.97456	.97082	.96329	.94417	.90464	.86313	.77275
					1.00182	1.00366	1.00735	1.01109	1.01870	1.03852	1.08202	1.13167	1.25673
.981	.1868	.9983	.30620	21.97570	.97904	.97707	.97313	.96917	.96120	.94095	.89897	.85472	.75763
					1.00193	1.00386	1.00777	1.01173	1.01979	1.04083	1.08728	1.14080	1.27820
.980	.1917	.9982	.31393	20.70907	.97793	.97586	.97170	.96752	.95911	.93772	.89325	.84621	.74211
					1.00203	1.00407	1.00819	1.01236	1.02087	1.04315	1.09263	1.15018	1.30096
.979	.1965	.9981	.32146	19.56570	.97683	.97465	.97027	.96587	.95702	.93447	.88748	.83758	.72614
					1.00213	1.00428	1.00861	1.01300	1.02197	1.04549	1.09807	1.15983	1.32514
.978	.2012	.9980	.32880	18.52818	.97572	.97343	.96884	.96422	.95491	.93120	.88167	.82883	.70969
					1.00223	1.00448	1.00903	1.01364	1.02306	1.04786	1.10361	1.16977	1.35093
.977	.2058	.9979	.33595	17.58291	.97461	.97221	.96740	.96256	.95281	.92792	.87580	.81995	.69270
					1.00234	1.00469	1.00945	1.01429	1.02417	1.05024	1.10925	1.18000	1.37853
.976	.2103	.9978	.34294	16.71816	.97350	.97100	.96597	.96091	.95070	.92463	.86988	.81095	.67511
					1.00244	1.00490	1.00988	1.01493	1.02528	1.05265	1.11499	1.19055	1.40817
.975	.2147	.9977	.34976	15.92418	.97239	.96978	.96453	.95924	.94858	.92132	.86390	.80180	.65684
					1.00254	1.00511	1.01030	1.01558	1.02639	1.05507	1.12084	1.20144	1.44016
.974	.2190	.9976	.35644	15.19275	.97128	.96856	.96309	.95758	.94646	.91799	.85786	.79251	.63781
					1.00265	1.00532	1.01073	1.01623	1.02751	1.05752	1.12680	1.21268	1.47487
.973	.2232	.9975	.36298	14.51689	.97018	.96734	.96164	.95591	.94433	.91465	.85177	.78306	.61791
					1.00275	1.00553	1.01115	1.01688	1.02863	1.05999	1.13288	1.22431	1.51274
.972	.2274	.9974	.36938	13.89067	.96907	.96612	.96020	.95424	.94219	.91129	.84561	.77344	.59698
					1.00286	1.00574	1.01158	1.01753	1.02976	1.06248	1.13907	1.23633	1.55435
.971	.2315	.9973	.37565	13.30883	.96796	.96490	.95876	.95257	.94005	.90791	.83939	.76366	.57485
					1.00296	1.00595	1.01201	1.01818	1.03090	1.06499	1.14539	1.24879	1.60047

TABLE I. Specific Heat Ratio = 1.10

R₁ and TPR versus Loss Coefficient K

p/p,	MACH	T/T,	GAMMA	K_crit	.1	.2	.4	.6	1.0	2.0	4.0	6.0	10.0
.970	.2355	.9972	.38180	12.76681	.96684	.96368	.95731	.95089	.93791	.90452	.83311	.75368	.55126
					1.00306	1.00616	1.01244	1.01884	1.03204	1.06753	1.15184	1.26171	1.65211
.969	.2395	.9971	.38784	12.26088	.96573	.96245	.95586	.94921	.93575	.90111	.82675	.74351	.52586
					1.00317	1.00637	1.01287	1.01950	1.03318	1.07009	1.15842	1.27512	1.71066
.968	.2434	.9970	.39377	11.78755	.96462	.96123	.95441	.94753	.93359	.89768	.82032	.73313	.49813
					1.00327	1.00658	1.01330	1.02016	1.03434	1.07267	1.16515	1.28907	1.77813
.967	.2472	.9970	.39959	11.34390	.96351	.96001	.95295	.94584	.93143	.89423	.81382	.72252	.46723
					1.00338	1.00679	1.01373	1.02083	1.03549	1.07528	1.17201	1.30358	1.85762
.966	.2510	.9969	.40531	10.92715	.96240	.95878	.95150	.94415	.92926	.89076	.80724	.71167	.43166
					1.00348	1.00700	1.01417	1.02149	1.03666	1.07792	1.17904	1.31872	1.95433
.965	.2547	.9968	.41094	10.53503	.96129	.95755	.95004	.94246	.92708	.88728	.80058	.70055	.38812
					1.00359	1.00722	1.01460	1.02216	1.03783	1.08058	1.18621	1.33453	2.07823
.964	.2584	.9967	.41648	10.16548	.96017	.95633	.94858	.94076	.92490	.88377	.79383	.68915	.32500
					1.00369	1.00743	1.01504	1.02283	1.03900	1.08326	1.19356	1.35107	2.25512
.963	.2620	.9966	.42192	9.81670	.95906	.95510	.94712	.93907	.92271	.88024	.78700	.67744	
					1.00380	1.00764	1.01548	1.02350	1.04018	1.08597	1.20108	1.36841	
.962	.2656	.9965	.42728	9.48691	.95794	.95387	.94566	.93736	.92051	.87670	.78007	.66539	
					1.00390	1.00786	1.01592	1.02418	1.04137	1.08871	1.20878	1.38662	
.961	.2692	.9964	.43256	9.17476	.95683	.95264	.94419	.93566	.91831	.87313	.77305	.65296	
					1.00401	1.00807	1.01636	1.02486	1.04256	1.09148	1.21667	1.40580	
.960	.2727	.9963	.43776	8.87879	.95572	.95141	.94273	.93395	.91609	.86954	.76592	.64012	
					1.00412	1.00829	1.01680	1.02554	1.04376	1.09428	1.22476	1.42606	
.959	.2762	.9962	.44288	8.59791	.95460	.95018	.94126	.93224	.91388	.86593	.75869	.62682	
					1.00422	1.00850	1.01724	1.02622	1.04497	1.09710	1.23307	1.44750	
.958	.2796	.9961	.44793	8.33094	.95348	.94894	.93979	.93052	.91165	.86229	.75134	.61301	
					1.00433	1.00872	1.01768	1.02690	1.04618	1.09996	1.24159	1.47029	
.957	.2830	.9960	.45291	8.07697	.95237	.94771	.93831	.92880	.90942	.85863	.74387	.59860	
					1.00444	1.00893	1.01813	1.02759	1.04740	1.10285	1.25035	1.49458	
.956	.2863	.9959	.45782	7.83503	.95125	.94648	.93684	.92708	.90718	.85495	.73628	.58353	
					1.00454	1.00915	1.01857	1.02828	1.04862	1.10576	1.25936	1.52061	
.955	.2896	.9958	.46266	7.60439	.95013	.94524	.93536	.92535	.90493	.85124	.72855	.56767	
					1.00465	1.00937	1.01902	1.02897	1.04986	1.10871	1.26863	1.54863	
.954	.2929	.9957	.46744	7.38422	.94902	.94400	.93388	.92362	.90267	.84751	.72068	.55088	
					1.00475	1.00958	1.01947	1.02967	1.05110	1.11169	1.27818	1.57898	
.953	.2962	.9956	.47215	7.17388	.94790	.94277	.93240	.92188	.90041	.84375	.71266	.53297	
					1.00486	1.00980	1.01992	1.03037	1.05234	1.11471	1.28802	1.61209	
.952	.2994	.9955	.47680	6.97277	.94678	.94153	.93091	.92015	.89814	.83997	.70448	.51368	
					1.00497	1.01002	1.02037	1.03106	1.05359	1.11776	1.29817	1.64852	
.951	.3026	.9954	.48140	6.78027	.94566	.94029	.92942	.91840	.89586	.83615	.69613	.49260	
					1.00508	1.01024	1.02082	1.03177	1.05485	1.12084	1.30866	1.68908	
.950	.3057	.9953	.48594	6.59588	.94454	.93905	.92794	.91666	.89358	.83231	.68759	.46910	
					1.00518	1.01046	1.02127	1.03247	1.05612	1.12396	1.31950	1.73492	
.949	.3089	.9953	.49042	6.41912	.94342	.93780	.92644	.91491	.89128	.82844	.67886	.44206	
					1.00529	1.01067	1.02173	1.03318	1.05739	1.12712	1.33073	1.78786	
.948	.3120	.9952	.49485	6.24951	.94230	.93656	.92495	.91316	.88898	.82454	.66991	.40903	
					1.00540	1.01089	1.02218	1.03389	1.05867	1.13032	1.34237	1.85111	
.947	.3151	.9951	.49922	6.08669	.94118	.93532	.92345	.91140	.88666	.82061	.66074	.36199	
					1.00551	1.01111	1.02264	1.03460	1.05996	1.13355	1.35445	1.93199	
.946	.3181	.9950	.50354	5.93025	.94006	.93407	.92196	.90964	.88434	.81665	.65131		
					1.00562	1.01133	1.02310	1.03532	1.06126	1.13683	1.36701		
.945	.3211	.9949	.50782	5.77982	.93894	.93283	.92045	.90787	.88201	.81266	.64160		
					1.00572	1.01156	1.02356	1.03604	1.06256	1.14014	1.38009		
.944	.3241	.9948	.51204	5.63509	.93782	.93158	.91895	.90610	.87967	.80863	.63160		
					1.00583	1.01178	1.02402	1.03676	1.06387	1.14350	1.39374		
.943	.3271	.9947	.51622	5.49575	.93669	.93033	.91745	.90432	.87733	.80457	.62125		
					1.00594	1.01200	1.02448	1.03748	1.06519	1.14690	1.40801		
.942	.3300	.9946	.52035	5.36153	.93557	.92908	.91594	.90254	.87497	.80047	.61054		
					1.00605	1.01222	1.02495	1.03821	1.06652	1.15035	1.42295		
.941	.3330	.9945	.52444	5.23214	.93445	.92783	.91443	.90076	.87260	.79633	.59941		
					1.00616	1.01244	1.02541	1.03894	1.06785	1.15384	1.43866		

TABLE I. Specific Heat Ratio = 1.10

R₂ and TPR versus Loss Coefficient K

p/p_t	MACH	T/T_t	GAMMA	K_crit	.1	.2	.4	.6	.8	1.0	2.0	3.0	4.0
.940	.3359	.9944	.52848	5.10736	.93332	.92658	.91291	.89897	.88475	.87022	.79216	.70152	.58781
					1.00627	1.01267	1.02588	1.03967	1.05409	1.06920	1.15738	1.27653	1.45519
.939	.3388	.9943	.53247	4.98694	.93220	.92533	.91140	.89718	.88267	.86784	.78795	.69460	.57567
					1.00638	1.01289	1.02634	1.04040	1.05512	1.07055	1.16097	1.28428	1.47268
.938	.3416	.9942	.53643	4.87066	.93107	.92408	.90988	.89538	.88058	.86544	.78369	.68754	.56290
					1.00649	1.01311	1.02681	1.04114	1.05615	1.07191	1.16461	1.29225	1.49123
.937	.3445	.9941	.54034	4.75834	.92994	.92282	.90835	.89358	.87848	.86303	.77940	.68033	.54939
					1.00660	1.01334	1.02728	1.04188	1.05719	1.07328	1.16830	1.30044	1.51099
.936	.3473	.9940	.54422	4.64978	.92882	.92156	.90683	.89178	.87638	.86062	.77506	.67297	.53499
					1.00671	1.01356	1.02776	1.04263	1.05824	1.07466	1.17204	1.30888	1.53217
.935	.3501	.9939	.54805	4.54480	.92769	.92031	.90530	.88997	.87427	.85819	.77067	.66543	.51950
					1.00682	1.01379	1.02823	1.04337	1.05929	1.07604	1.17584	1.31757	1.55502
.934	.3529	.9938	.55185	4.44324	.92656	.91905	.90377	.88815	.87215	.85575	.76624	.65771	.50260
					1.00693	1.01401	1.02870	1.04412	1.06034	1.07744	1.17970	1.32654	1.57985
.933	.3557	.9937	.55560	4.34493	.92544	.91779	.90224	.88633	.87002	.85330	.76176	.64979	.48384
					1.00704	1.01424	1.02918	1.04488	1.06140	1.07884	1.18362	1.33581	1.60715
.932	.3584	.9936	.55932	4.24975	.92431	.91653	.90070	.88450	.86789	.85084	.75723	.64165	.46239
					1.00715	1.01447	1.02966	1.04563	1.06247	1.08025	1.18759	1.34539	1.63761
.931	.3611	.9935	.56301	4.15754	.92318	.91527	.89917	.88267	.86575	.84836	.75264	.63327	.43659
					1.00726	1.01470	1.03014	1.04639	1.06354	1.08168	1.19163	1.35532	1.67240
.930	.3638	.9934	.56666	4.06816	.92205	.91400	.89762	.88083	.86360	.84588	.74800	.62463	.40169
					1.00737	1.01492	1.03062	1.04716	1.06462	1.08311	1.19574	1.36562	1.71383
.929	.3665	.9933	.57027	3.98152	.92092	.91274	.89608	.87899	.86144	.84338	.74330	.61571	
					1.00748	1.01515	1.03110	1.04792	1.06570	1.08455	1.19991	1.37632	
.928	.3692	.9932	.57385	3.89748	.91979	.91147	.89453	.87715	.85927	.84087	.73854	.60646	
					1.00759	1.01538	1.03158	1.04869	1.06679	1.08600	1.20416	1.38746	
.927	.3719	.9931	.57739	3.81593	.91865	.91021	.89298	.87529	.85710	.83835	.73372	.59686	
					1.00770	1.01561	1.03207	1.04946	1.06789	1.08747	1.20848	1.39908	
.926	.3745	.9930	.58090	3.73677	.91752	.90894	.89143	.87344	.85492	.83581	.72883	.58685	
					1.00782	1.01584	1.03256	1.05024	1.06899	1.08894	1.21287	1.41123	
.925	.3772	.9929	.58438	3.65992	.91639	.90767	.88987	.87157	.85272	.83326	.72388	.57637	
					1.00793	1.01607	1.03305	1.05102	1.07010	1.09042	1.21734	1.42398	
.924	.3798	.9928	.58783	3.58526	.91526	.90640	.88831	.86970	.85052	.83070	.71885	.56537	
					1.00804	1.01630	1.03353	1.05180	1.07121	1.09192	1.22190	1.43738	
.923	.3824	.9927	.59125	3.51271	.91412	.90512	.88675	.86783	.84831	.82813	.71374	.55372	
					1.00815	1.01653	1.03403	1.05259	1.07234	1.09342	1.22654	1.45153	
.922	.3850	.9926	.59463	3.44220	.91299	.90385	.88518	.86595	.84609	.82554	.70856	.54133	
					1.00826	1.01676	1.03452	1.05338	1.07346	1.09494	1.23127	1.46653	
.921	.3875	.9925	.59799	3.37364	.91185	.90257	.88361	.86406	.84386	.82293	.70329	.52801	
					1.00838	1.01699	1.03501	1.05417	1.07460	1.09646	1.23610	1.48253	
.920	.3901	.9924	.60131	3.30695	.91072	.90130	.88204	.86217	.84162	.82032	.69794	.51352	
					1.00849	1.01723	1.03551	1.05497	1.07574	1.09800	1.24102	1.49969	
.919	.3926	.9924	.60461	3.24206	.90958	.90002	.88046	.86027	.83937	.81768	.69248	.49748	
					1.00860	1.01746	1.03601	1.05577	1.07688	1.09955	1.24605	1.51826	
.918	.3952	.9923	.60787	3.17892	.90844	.89874	.87888	.85837	.83711	.81503	.68693	.47924	
					1.00872	1.01769	1.03651	1.05657	1.07804	1.10111	1.25118	1.53859	
.917	.3977	.9922	.61111	3.11744	.90730	.89746	.87730	.85645	.83484	.81237	.68128	.45756	
					1.00883	1.01793	1.03701	1.05738	1.07920	1.10268	1.25643	1.56126	
.916	.4002	.9921	.61432	3.05758	.90616	.89618	.87571	.85454	.83256	.80969	.67551	.42919	
					1.00894	1.01816	1.03751	1.05819	1.08037	1.10427	1.26180	1.58730	
.915	.4027	.9920	.61751	2.99927	.90503	.89489	.87412	.85261	.83027	.80699	.66962		
					1.00906	1.01840	1.03802	1.05900	1.08155	1.10587	1.26729		
.914	.4052	.9919	.62066	2.94245	.90389	.89361	.87253	.85068	.82797	.80427	.66360		
					1.00917	1.01863	1.03852	1.05982	1.08273	1.10748	1.27292		
.913	.4076	.9918	.62379	2.88709	.90274	.89232	.87093	.84874	.82566	.80154	.65744		
					1.00928	1.01887	1.03903	1.06065	1.08392	1.10910	1.27868		
.912	.4101	.9917	.62690	2.83311	.90160	.89103	.86933	.84680	.82333	.79879	.65114		
					1.00940	1.01911	1.03954	1.06147	1.08512	1.11074	1.28460		
.911	.4125	.9916	.62997	2.78048	.90046	.88974	.86772	.84485	.82100	.79602	.64467		
					1.00951	1.01934	1.04005	1.06230	1.08632	1.11239	1.29067		

TABLE I. Specific Heat Ratio = 1.10

p/p_t	MACH	T/T_t	GAMMA	K_{crit}	.1	.2	.4	.6	.8	.9	1.0	1.5	2.0
.910	.4150	.9915	.63303	2.72915	.89932	.88845	.86611	.84289	.81865	.80610	.79324	.72307	.63803
					1.00963	1.01958	1.04057	1.06314	1.08754	1.10051	1.11406	1.19240	1.29692
.909	.4174	.9914	.63605	2.67906	.89818	.88716	.86450	.84092	.81629	.80353	.79043	.71880	.63120
					1.00974	1.01982	1.04108	1.06398	1.08876	1.10195	1.11574	1.19572	1.30334
.908	.4198	.9913	.63906	2.63020	.89703	.88586	.86288	.83895	.81391	.80093	.78760	.71447	.62417
					1.00986	1.02006	1.04160	1.06482	1.08999	1.10340	1.11743	1.19910	1.30996
.907	.4222	.9912	.64203	2.58251	.89589	.88457	.86126	.83697	.81153	.79832	.78476	.71008	.61690
					1.00997	1.02030	1.04212	1.06567	1.09123	1.10486	1.11914	1.20253	1.31679
.906	.4246	.9911	.64499	2.53594	.89474	.88327	.85964	.83498	.80913	.79570	.78189	.70562	.60938
					1.01009	1.02054	1.04264	1.06652	1.09247	1.10634	1.12086	1.20602	1.32384
.905	.4270	.9910	.64792	2.49048	.89359	.88197	.85801	.83298	.80672	.79306	.77900	.70109	.60158
					1.01020	1.02078	1.04316	1.06738	1.09373	1.10782	1.12260	1.20957	1.33113
.904	.4294	.9909	.65083	2.44608	.89245	.88067	.85637	.83097	.80429	.79040	.77609	.69649	.59346
					1.01032	1.02102	1.04369	1.06824	1.09499	1.10932	1.12436	1.21317	1.33869
.903	.4317	.9908	.65371	2.40270	.89130	.87937	.85474	.82896	.80185	.78772	.77315	.69181	.58498
					1.01044	1.02126	1.04421	1.06910	1.09626	1.11083	1.12613	1.21685	1.34655
.902	.4341	.9907	.65657	2.36031	.89015	.87806	.85309	.82694	.79940	.78502	.77019	.68705	.57608
					1.01055	1.02151	1.04474	1.06997	1.09755	1.11235	1.12792	1.22059	1.35473
.901	.4364	.9906	.65941	2.31890	.88900	.87676	.85145	.82491	.79693	.78231	.76721	.68221	.56670
					1.01067	1.02175	1.04527	1.07085	1.09884	1.11388	1.12972	1.22440	1.36326
.900	.4387	.9905	.66223	2.27842	.88785	.87545	.84980	.82287	.79444	.77957	.76420	.67728	.55674
					1.01079	1.02199	1.04580	1.07172	1.10014	1.11543	1.13154	1.22828	1.37220
.899	.4411	.9904	.66502	2.23884	.88670	.87414	.84814	.82082	.79194	.77682	.76117	.67225	.54609
					1.01090	1.02224	1.04634	1.07261	1.10145	1.11699	1.13338	1.23224	1.38159
.898	.4434	.9903	.66780	2.20014	.88555	.87283	.84648	.81877	.78943	.77404	.75811	.66711	.53458
					1.01102	1.02248	1.04687	1.07350	1.10277	1.11856	1.13524	1.23628	1.39151
.897	.4457	.9902	.67055	2.16229	.88440	.87151	.84481	.81670	.78689	.77125	.75502	.66187	.52196
					1.01114	1.02273	1.04741	1.07439	1.10410	1.12015	1.13712	1.24040	1.40205
.896	.4480	.9901	.67328	2.12526	.88324	.87020	.84314	.81462	.78435	.76843	.75190	.65651	.50782
					1.01126	1.02297	1.04795	1.07529	1.10544	1.12175	1.13902	1.24462	1.41334
.895	.4502	.9900	.67599	2.08904	.88209	.86888	.84147	.81254	.78178	.76558	.74876	.65102	.49148
					1.01137	1.02322	1.04850	1.07619	1.10679	1.12337	1.14093	1.24893	1.42556
.894	.4525	.9899	.67868	2.05359	.88093	.86756	.83979	.81044	.77920	.76272	.74558	.64540	.47138
					1.01149	1.02347	1.04904	1.07710	1.10815	1.12500	1.14287	1.25334	1.43903
.893	.4548	.9898	.68135	2.01890	.87978	.86624	.83810	.80834	.77659	.75983	.74237	.63962	.44249
					1.01161	1.02371	1.04959	1.07801	1.10952	1.12664	1.14483	1.25785	1.45439
.892	.4570	.9897	.68400	1.98494	.87862	.86492	.83641	.80622	.77397	.75692	.73913	.63369	
					1.01173	1.02396	1.05014	1.07893	1.11090	1.12831	1.14681	1.26248	
.891	.4593	.9896	.68663	1.95169	.87746	.86359	.83472	.80410	.77133	.75398	.73585	.62758	
					1.01185	1.02421	1.05069	1.07986	1.11230	1.12998	1.14881	1.26723	
.890	.4615	.9895	.68925	1.91914	.87631	.86226	.83302	.80196	.76867	.75101	.73254	.62128	
					1.01197	1.02446	1.05124	1.08079	1.11370	1.13168	1.15084	1.27211	
.889	.4638	.9894	.69184	1.88726	.87515	.86093	.83131	.79981	.76599	.74802	.72919	.61477	
					1.01209	1.02471	1.05180	1.08172	1.11512	1.13339	1.15288	1.27712	
.888	.4660	.9893	.69441	1.85604	.87399	.85960	.82960	.79765	.76329	.74500	.72580	.60802	
					1.01220	1.02496	1.05236	1.08266	1.11655	1.13512	1.15496	1.28229	
.887	.4682	.9892	.69696	1.82544	.87282	.85827	.82788	.79548	.76057	.74195	.72238	.60100	
					1.01232	1.02521	1.05292	1.08361	1.11799	1.13686	1.15705	1.28761	
.886	.4704	.9891	.69950	1.79547	.87166	.85693	.82616	.79330	.75782	.73886	.71891	.59368	
					1.01244	1.02547	1.05348	1.08456	1.11945	1.13862	1.15918	1.29312	
.885	.4726	.9890	.70201	1.76610	.87050	.85560	.82443	.79111	.75506	.73575	.71540	.58602	
					1.01256	1.02572	1.05404	1.08552	1.12092	1.14041	1.16133	1.29881	
.884	.4748	.9889	.70451	1.73732	.86934	.85426	.82270	.78890	.75227	.73261	.71184	.57797	
					1.01268	1.02597	1.05461	1.08649	1.12240	1.14221	1.16350	1.30472	
.883	.4770	.9888	.70699	1.70910	.86817	.85292	.82096	.78668	.74945	.72943	.70823	.56945	
					1.01280	1.02623	1.05518	1.08746	1.12389	1.14403	1.16571	1.31086	
.882	.4792	.9887	.70946	1.68145	.86700	.85157	.81921	.78445	.74661	.72621	.70458	.56038	
					1.01293	1.02648	1.05575	1.08843	1.12540	1.14587	1.16794	1.31727	
.881	.4813	.9885	.71190	1.65433	.86584	.85023	.81746	.78221	.74374	.72296	.70087	.55062	
					1.01305	1.02674	1.05633	1.08942	1.12692	1.14773	1.17021	1.32397	

TABLE I. Specific Heat Ratio = 1.10

R₁ and TPR versus Loss Coefficient K

p/Pₜ	MACH	T/Tₜ	GAMMA	K_{crit}	.1	.2	.4	.6	.7	.8	.9	1.0	1.5
.880	.4835	.9884	.71433	1.62774	.86467	.84888	.81570	.77995	.76088	.74085	.71967	.69711	.54001
					1.01317	1.02699	1.05691	1.09041	1.10879	1.12846	1.14962	1.17251	1.33102
.879	.4857	.9883	.71674	1.60166	.86350	.84753	.81393	.77767	.75830	.73792	.71634	.69329	.52826
					1.01329	1.02725	1.05749	1.09140	1.11004	1.13001	1.15152	1.17484	1.33848
.878	.4878	.9882	.71913	1.57608	.86233	.84617	.81216	.77539	.75571	.73497	.71297	.68941	.51493
					1.01341	1.02751	1.05807	1.09241	1.11131	1.13158	1.15345	1.17720	1.34643
.877	.4900	.9881	.72151	1.55099	.86116	.84482	.81038	.77308	.75309	.73199	.70955	.68547	.49913
					1.01353	1.02777	1.05865	1.09342	1.11258	1.13316	1.15540	1.17960	1.35502
.876	.4921	.9880	.72387	1.52637	.85999	.84346	.80859	.77077	.75045	.72898	.70609	.68146	.47871
					1.01365	1.02802	1.05924	1.09444	1.11387	1.13477	1.15738	1.18204	1.36446
.875	.4942	.9879	.72621	1.50221	.85882	.84210	.80680	.76844	.74779	.72593	.70259	.67737	.44059
					1.01378	1.02828	1.05983	1.09546	1.11516	1.13638	1.15939	1.18452	1.37541
.874	.4964	.9878	.72853	1.47851	.85764	.84073	.80500	.76609	.74511	.72285	.69903	.67321	
					1.01390	1.02854	1.06043	1.09649	1.11647	1.13802	1.16141	1.18703	
.873	.4985	.9877	.73084	1.45525	.85647	.83937	.80319	.76372	.74240	.71974	.69542	.66898	
					1.01402	1.02880	1.06102	1.09753	1.11779	1.13967	1.16347	1.18959	
.872	.5006	.9876	.73314	1.43241	.85529	.83800	.80138	.76134	.73967	.71659	.69175	.66465	
					1.01415	1.02907	1.06162	1.09858	1.11912	1.14134	1.16556	1.19220	
.871	.5027	.9875	.73542	1.41000	.85411	.83663	.79956	.75894	.73691	.71340	.68803	.66024	
					1.01427	1.02933	1.06222	1.09963	1.12046	1.14303	1.16767	1.19484	
.870	.5048	.9874	.73768	1.38799	.85294	.83526	.79773	.75653	.73412	.71017	.68425	.65572	
					1.01439	1.02959	1.06283	1.10070	1.12182	1.14474	1.16982	1.19754	
.869	.5069	.9873	.73992	1.36639	.85176	.83388	.79589	.75409	.73131	.70690	.68040	.65110	
					1.01452	1.02986	1.06343	1.10177	1.12319	1.14647	1.17199	1.20029	
.868	.5090	.9872	.74216	1.34517	.85058	.83250	.79404	.75164	.72847	.70358	.67648	.64637	
					1.01464	1.03012	1.06405	1.10285	1.12457	1.14822	1.17420	1.20309	
.867	.5111	.9871	.74437	1.32433	.84939	.83112	.79219	.74916	.72560	.70022	.67249	.64151	
					1.01476	1.03039	1.06466	1.10394	1.12597	1.14999	1.17645	1.20595	
.866	.5131	.9870	.74657	1.30387	.84821	.82974	.79032	.74667	.72270	.69681	.66842	.63652	
					1.01489	1.03065	1.06528	1.10503	1.12738	1.15179	1.17873	1.20887	
.865	.5152	.9869	.74876	1.28377	.84703	.82835	.78845	.74416	.71976	.69336	.66427	.63139	
					1.01501	1.03092	1.06590	1.10614	1.12880	1.15361	1.18105	1.21186	
.864	.5173	.9868	.75093	1.26402	.84584	.82696	.78657	.74162	.71680	.68985	.66004	.62609	
					1.01514	1.03119	1.06652	1.10725	1.13024	1.15545	1.18340	1.21491	
.863	.5193	.9867	.75308	1.24462	.84466	.82557	.78468	.73906	.71380	.68628	.65571	.62062	
					1.01526	1.03146	1.06714	1.10838	1.13170	1.15731	1.18580	1.21803	
.862	.5214	.9866	.75522	1.22555	.84347	.82417	.78279	.73648	.71076	.68265	.65127	.61495	
					1.01539	1.03172	1.06777	1.10951	1.13317	1.15921	1.18825	1.22124	
.861	.5234	.9865	.75735	1.20682	.84228	.82277	.78088	.73388	.70768	.67897	.64673	.60907	
					1.01552	1.03199	1.06841	1.11065	1.13465	1.16113	1.19074	1.22452	
.860	.5255	.9864	.75946	1.18841	.84109	.82137	.77896	.73125	.70457	.67522	.64208	.60293	
					1.01564	1.03227	1.06904	1.11181	1.13616	1.16308	1.19328	1.22790	
.859	.5275	.9863	.76156	1.17032	.83990	.81996	.77703	.72859	.70141	.67139	.63729	.59652	
					1.01577	1.03254	1.06968	1.11297	1.13768	1.16505	1.19587	1.23138	
.858	.5295	.9862	.76364	1.15254	.83871	.81855	.77510	.72591	.69821	.66750	.63236	.58978	
					1.01589	1.03281	1.07033	1.11415	1.13921	1.16706	1.19851	1.23497	
.857	.5316	.9861	.76571	1.13506	.83751	.81714	.77315	.72320	.69497	.66352	.62729	.58266	
					1.01602	1.03309	1.07097	1.11533	1.14077	1.16910	1.20121	1.23867	
.856	.5336	.9860	.76777	1.11787	.83632	.81573	.77119	.72046	.69168	.65946	.62204	.57509	
					1.01615	1.03336	1.07162	1.11653	1.14235	1.17117	1.20398	1.24251	
.855	.5356	.9859	.76981	1.10097	.83512	.81431	.76922	.71770	.68833	.65531	.61660	.56697	
					1.01628	1.03364	1.07228	1.11774	1.14395	1.17328	1.20681	1.24650	
.854	.5376	.9858	.77184	1.08436	.83392	.81289	.76724	.71490	.68494	.65107	.61096	.55817	
					1.01640	1.03391	1.07294	1.11896	1.14556	1.17543	1.20971	1.25066	
.853	.5396	.9856	.77385	1.06803	.83272	.81146	.76525	.71207	.68149	.64672	.60508	.54847	
					1.01653	1.03419	1.07360	1.12019	1.14720	1.17761	1.21269	1.25501	
.852	.5416	.9855	.77585	1.05196	.83152	.81003	.76324	.70921	.67798	.64225	.59892	.53754	
					1.01666	1.03447	1.07426	1.12143	1.14886	1.17984	1.21576	1.25960	
.851	.5436	.9854	.77784	1.03616	.83032	.80860	.76123	.70631	.67441	.63766	.59246	.52475	
					1.01679	1.03475	1.07493	1.12269	1.15055	1.18211	1.21892	1.26449	

TABLE I. Specific Heat Ratio = 1.10

R₁ and TPR versus Loss Coefficient K

p/p_t	MACH	T/T_t	GAMMA	K_{crit}	.1	.2	.4	.5	.6	.7	.8	.9	1.0
.850	.5456	.9853	.77981	1.02062	·.82911	.80716	.75920	.73253	.70337	.67077	.63294	.58564	.50871
					1.01692	1.03503	1.07561	1.09862	1.12397	1.15226	1.18443	1.22217	1.26977
.849	.5476	.9852	.78178	1.00534	.82791	.80572	.75716	.73009	.70040	.66706	.62807	.57839	.48384
					1.01705	1.03531	1.07628	1.09956	1.12525	1.15399	1.18679	1.22554	1.27564
.848	.5496	.9851	.78372	0.99030	.82670	.80428	.75510	.72762	.69739	.66328	.62304	.57060	
					1.01717	1.03559	1.07696	1.10052	1.12655	1.15575	1.18921	1.22904	
.847	.5516	.9850	.78566	0.97551	.82549	.80283	.75304	.72513	.69433	.65942	.61783	.56217	
					1.01730	1.03587	1.07765	1.10148	1.12787	1.15754	1.19168	1.23268	
.846	.5535	.9849	.78758	0.96095	.82429	.80138	.75096	.72261	.69124	.65547	.61241	.55288	
					1.01743	1.03615	1.07834	1.10245	1.12920	1.15935	1.19422	1.23648	
.845	.5555	.9848	.78949	0.94663	.82307	.79993	.74886	.72007	.68809	.65143	.60676	.54242	
					1.01756	1.03644	1.07904	1.10343	1.13054	1.16120	1.19682	1.24049	
.844	.5575	.9847	.79139	0.93254	.82186	.79847	.74675	.71750	.68490	.64729	.60086	.53018	
					1.01770	1.03672	1.07974	1.10441	1.13190	1.16308	1.19949	1.24474	
.843	.5594	.9846	.79327	0.91868	.82065	.79700	.74463	.71490	.68166	.64304	.59465	.51487	
					1.01783	1.03701	1.08044	1.10541	1.13328	1.16500	1.20224	1.24932	
.842	.5614	.9845	.79514	0.90503	.81943	.79554	.74249	.71228	.67836	.63868	.58809	.49132	
					1.01796	1.03730	1.08115	1.10642	1.13468	1.16694	1.20507	1.25439	
.841	.5633	.9844	.79700	0.89160	.81821	.79406	.74033	.70963	.67500	.63419	.58111		
					1.01809	1.03758	1.08186	1.10743	1.13610	1.16893	1.20800		
.840	.5653	.9843	.79885	0.87838	.81699	.79259	.73816	.70694	.67159	.62956	.57362		
					1.01822	1.03787	1.08258	1.10845	1.13753	1.17096	1.21103		
.839	.5672	.9842	.80069	0.86537	.81577	.79111	.73597	.70423	.66811	.62477	.56549		
					1.01835	1.03816	1.08331	1.10949	1.13898	1.17303	1.21418		
.838	.5692	.9841	.80251	0.85256	.81455	.78962	.73376	.70148	.66456	.61981	.55652		
					1.01848	1.03845	1.08403	1.11053	1.14046	1.17515	1.21748		
.837	.5711	.9840	.80432	0.83995	.81333	.78813	.73153	.69869	.66094	.61466	.54638		
					1.01862	1.03875	1.08477	1.11159	1.14196	1.17732	1.22094		
.836	.5730	.9838	.80612	0.82754	.81210	.78664	.72929	.69587	.65724	.60929	.53448		
					1.01875	1.03904	1.08551	1.11266	1.14348	1.17954	1.22461		
.835	.5749	.9837	.80791	0.81532	.81087	.78514	.72702	.69301	.65345	.60367	.51941		
					1.01888	1.03933	1.08625	1.11373	1.14502	1.18182	1.22856		
.834	.5769	.9836	.80968	0.80329	.80964	.78363	.72474	.69011	.64958	.59777	.49513		
					1.01902	1.03963	1.08701	1.11481	1.14659	1.18416	1.23294		
.833	.5788	.9835	.81145	0.79144	.80841	.78212	.72243	·68717	.64560	.59154			
					1.01915	1.03993	1.08776	1.11593	1.14819	1.18657			
.832	.5807	.9834	.81320	0.77977	.80718	.78061	.72011	.68419	.64152	.58490			
					1.01928	1.04022	1.08853	1.11704	1.14981	1.18906			
.831	.5826	.9833	.81494	0.76828	.80594	.77909	.71776	.68115	.63733	.57778			
					1.01942	1.04052	1.08930	1.11817	1.15146	1.19163			
.830	.5845	.9832	.81667	0.75696	.80470	.77756	.71538	.67807	.63300	.57005			
					1.01955	1.04082	1.09007	1.11931	1.15315	1.19430			
.829	.5864	.9831	.81839	0.74582	.80346	.77603	.71299	.67493	.62854	.56152			
					1.01969	1.04112	1.09085	1.12046	1.15487	1.19709			
.828	.5883	.9830	.82009	0.73484	.80222	.77450	.71057	.67174	.62392	.55187			
					1.01982	1.04143	1.09164	1.12164	1.15663	1.20002			
.827	.5902	.9829	.82179	0.72402	.80098	.77296	.70812	.66849	.61913	.54056			
					1.01996	1.04173	1.09244	1.12282	1.15842	1.20311			
.826	.5921	.9828	.82347	0.71337	.79973	.77141	.70564	.66518	.61414	.52624			
					1.02010	1.04203	1.09324	1.12402	1.16025	1.20642			
.825	.5940	.9827	.82514	0.70288	.79849	.76986	.70314	.66180	.60893	.50318			
					1.02023	1.04234	1.09405	1.12524	1.16213	1.21008			
.824	.5959	.9826	.82681	0.69254	.79724	.76830	.70061	.65834	.60347				
					1.02037	1.04265	1.09487	1.12648	1.16406				
.823	.5978	.9824	.82846	0.68236	.79598	.76673	.69804	.65481	.59771				
					1.02051	1.04296	1.09569	1.12773	1.16604				
.822	.5997	.9823	.83010	0.67232	.79473	.76516	.69545	.65120	.59160				
					1.02064	1.04327	1.09653	1.12900	1.16808				
.821	.6015	.9822	.83173	0.66244	.79347	.76358	.69282	.64750	.58507				
					1.02078	1.04358	1.09737	1.13030	1.17019				

34

TABLE I. Specific Heat Ratio = 1.10

R₂ and TPR versus Loss Coefficient K

p/p,	MACH	T/T,	GAMMA	K_crit	.04	.06	.08	.09	.1	.2	.3	.4	.5
.820	.6034	.9821	.83335	0.65270	.80914	.80359	.79795	.79510	.79222	.76199	.72848	.69015	.64370
					1.00815	1.01233	1.01659	1.01874	1.02092	1.04389	1.06940	1.09822	1.13162
.819	.6053	.9820	.83496	0.64310	.80804	.80244	.79674	.79386	.79096	.76040	.72644	.68745	.63979
					1.00820	1.01241	1.01669	1.01886	1.02106	1.04420	1.06994	1.09907	1.13295
.818	.6071	.9819	.83655	0.63364	.80694	.80129	.79554	.79263	.78969	.75880	.72439	.68471	.63577
					1.00825	1.01249	1.01680	1.01899	1.02120	1.04452	1.07049	1.09994	1.13432
.817	.6090	.9818	.83814	0.62432	.80584	.80013	.79433	.79139	.78843	.75720	.72232	.68192	.63162
					1.00830	1.01257	1.01691	1.01911	1.02134	1.04483	1.07104	1.10082	1.13570
.816	.6109	.9817	.83972	0.61513	.80474	.79898	.79312	.79015	.78716	.75559	.72023	.6791C	.62733
					1.00835	1.01265	1.01702	1.01924	1.02148	1.04515	1.07159	1.10171	1.13712
.815	.6127	.9816	.84128	0.60608	.80364	.79782	.79191	.78891	.78589	.75397	.71812	.67622	.62288
					1.00841	1.01273	1.01713	1.01936	1.02162	1.04547	1.07215	1.10260	1.13856
.814	.6146	.9815	.84284	0.59716	.80254	.79667	.79069	.78767	.78461	.75234	.71600	.67330	.61826
					1.00846	1.01281	1.01724	1.01949	1.02176	1.04579	1.07271	1.10351	1.14004
.813	.6164	.9814	.84439	0.58836	.80144	.79551	.78948	.78642	.78334	.75070	.71385	.67033	.61344
					1.00851	1.01289	1.01735	1.01961	1.02190	1.04611	1.07328	1.10443	1.14155
.812	.6183	.9812	.84592	0.57970	.80034	.79435	.78826	.78518	.78206	.74906	.71169	.66730	.60838
					1.00856	1.01297	1.01746	1.01974	1.02204	1.04644	1.07385	1.10536	1.14309
.811	.6201	.9811	.84745	0.57115	.79923	.79319	.78704	.78393	.78078	.74740	.70950	.66422	.60306
					1.00862	1.01305	1.01757	1.01986	1.02218	1.04676	1.07442	1.10630	1.14468
.810	.6220	.9810	.84896	0.56273	.79813	.79203	.78582	.78267	.77949	.74574	.70729	.66106	.59743
					1.00867	1.01313	1.01768	1.01999	1.02232	1.04709	1.07500	1.10725	1.14630
.809	.6238	.9809	.85047	0.55443	.79702	.79087	.78460	.78142	.77821	.74407	.70506	.65785	.59143
					1.00872	1.01321	1.01779	1.02012	1.02247	1.04742	1.07559	1.10822	1.14798
.808	.6256	.9808	.85196	0.54624	.79591	.78970	.78337	.78016	.77692	.74240	.70281	.65456	.58496
					1.00877	1.01329	1.01790	1.02024	1.02261	1.04775	1.07618	1.10920	1.14971
.807	.6275	.9807	.85345	0.53817	.79481	.78854	.78215	.77890	.77563	.74071	.70053	.65119	.57791
					1.00883	1.01338	1.01802	1.02037	1.02275	1.04808	1.07677	1.11020	1.15151
.806	.6293	.9806	.85492	0.53022	.79370	.78737	.78092	.77764	.77433	.73901	.69822	.64775	.57006
					1.00888	1.01346	1.01813	1.02050	1.02290	1.04841	1.07737	1.11121	1.15338
.805	.6311	.9805	.85639	0.52237	.79259	.78620	.77969	.77638	.77303	.73731	.69589	.64421	.56110
					1.00893	1.01354	1.01824	1.02063	1.02304	1.04875	1.07797	1.11223	1.15534
.804	.6329	.9804	.85785	0.51464	.79148	.78503	.77845	.77511	.77173	.73559	.69353	.64057	.55039
					1.00899	1.01362	1.01835	1.02076	1.02318	1.04908	1.07858	1.11328	1.15740
.803	.6348	.9803	.85929	0.50702	.79037	.78386	.77721	.77384	.77043	.73386	.69114	.63682	.53616
					1.00904	1.01370	1.01846	1.02088	1.02333	1.04942	1.07919	1.11434	1.15962
.802	.6366	.9801	.86073	0.49950	.78925	.78268	.77598	.77257	.76912	.73213	.68872	.63295	
					1.00909	1.01379	1.01858	1.02101	1.02347	1.04976	1.07981	1.11542	
.801	.6384	.9800	.86216	0.49209	.78814	.78151	.77474	.77129	.76781	.73038	.68627	.62895	
					1.00915	1.01387	1.01869	1.02114	1.02362	1.05011	1.08044	1.11652	
.800	.6402	.9799	.86357	0.48478	.78702	.78033	.77349	.77001	.76649	.72862	.68378	.62481	
					1.00920	1.01395	1.01880	1.02127	1.02377	1.05045	1.08107	1.11764	
.799	.6420	.9798	.86498	0.47757	.78591	.77915	.77225	.76873	.76518	.72685	.68126	.62050	
					1.00926	1.01403	1.01892	1.02140	1.02391	1.05079	1.08171	1.11878	
.798	.6438	.9797	.86638	0.47046	.78479	.77797	.77100	.76745	.76385	.72507	.67870	.61599	
					1.00931	1.01412	1.01903	1.02153	1.02406	1.05114	1.08235	1.11995	
.797	.6456	.9796	.86777	0.46345	.78367	.77679	.76975	.76616	.76253	.72327	.67609	.61128	
					1.00937	1.01420	1.01915	1.02166	1.02421	1.05149	1.08301	1.12115	
.796	.6474	.9795	.86915	0.45654	.78255	.77561	.76849	.76487	.76120	.72146	.67345	.60631	
					1.00942	1.01429	1.01926	1.02179	1.02436	1.05184	1.08367	1.12237	
.795	.6492	.9794	.87052	0.44972	.78143	.77442	.76724	.76358	.75987	.71964	.67076	.60104	
					1.00947	1.01437	1.01938	1.02193	1.02450	1.05220	1.08433	1.12363	
.794	.6510	.9792	.87188	0.44300	.78031	.77323	.76598	.76228	.75854	.71781	.66803	.59542	
					1.00953	1.01445	1.01949	1.02206	1.02465	1.05255	1.08501	1.12492	
.793	.6528	.9791	.87324	0.43637	.77919	.77204	.76472	.76098	.75720	.71596	.66524	.58934	
					1.00958	1.01454	1.01961	1.02219	1.02480	1.05291	1.08569	1.12626	
.792	.6546	.9790	.87458	0.42983	.77807	.77085	.76345	.75968	.75585	.71410	.66240	.58269	
					1.00964	1.01462	1.01972	1.02232	1.02495	1.05327	1.08638	1.12765	
.791	.6564	.9789	.87591	0.42338	.77694	.76966	.76219	.75837	.75451	.71222	.65951	.57528	
					1.00969	1.01471	1.01984	1.02246	1.02510	1.05364	1.08707	1.12909	

TABLE I. Specific Heat Ratio = 1.10

R₂ and TPR versus Loss Coefficient K

p/p_t	MACH	T/T_t	GAMMA	K_{rit}	.01	.02	.04	.06	.08	.09	.1	.2	.3
.790	.6582	.9788	.87724	0.41701	.78652	.78299	.77582	.76846	.76091	.75706	.75315	.71033	.65655
					1.00240	1.00482	1.00975	1.01479	1.01996	1.02259	1.02525	1.05400	1.08778
.789	.6600	.9787	.87856	0.41074	.78549	.78193	.77469	.76727	.75964	.75575	.75180	.70842	.65352
					1.00241	1.00485	1.00980	1.01488	1.02007	1.02272	1.02541	1.05437	1.08850
.788	.6618	.9786	.87986	0.40455	.78445	.78087	.77356	.76607	.75837	.75443	.75044	.70649	.65043
					1.00242	1.00488	1.00986	1.01496	1.02019	1.02286	1.02556	1.05474	1.08922
.787	.6635	.9785	.88116	0.39844	.78342	.77981	.77243	.76486	.75709	.75311	.74907	.70455	.64725
					1.00244	1.00490	1.00991	1.01505	1.02031	1.02299	1.02571	1.05511	1.08996
.786	.6653	.9783	.88245	0.39242	.78239	.77874	.77130	.76366	.75580	.75179	.74771	.70259	.64400
					1.00245	1.00493	1.00997	1.01513	1.02043	1.02313	1.02586	1.05549	1.09071
.785	.6671	.9782	.88373	0.38648	.78136	.77768	.77017	.76246	.75452	.75046	.74633	.70061	.64064
					1.00246	1.00495	1.01002	1.01522	1.02055	1.02326	1.02602	1.05586	1.09147
.784	.6689	.9781	.88500	0.38062	.78033	.77661	.76904	.76125	.75323	.74912	.74496	.69861	.63719
					1.00248	1.00498	1.01008	1.01530	1.02067	1.02340	1.02617	1.05624	1.09224
.783	.6706	.9780	.88627	0.37483	.77930	.77555	.76790	.76004	.75194	.74779	.74357	.69659	.63363
					1.00249	1.00501	1.01013	1.01539	1.02078	1.02354	1.02633	1.05663	1.09303
.782	.6724	.9779	.88752	0.36913	.77827	.77448	.76677	.75882	.75064	.74645	.74219	.69455	.62994
					1.00250	1.00503	1.01019	1.01548	1.02090	1.02367	1.02648	1.05701	1.09382
.781	.6742	.9778	.88877	0.36350	.77723	.77342	.76563	.75761	.74934	.74510	.74079	.69248	.62611
					1.00252	1.00507	1.01025	1.01556	1.02102	1.02381	1.02664	1.05741	1.09464
.780	.6759	.9777	.89000	0.35795	.77620	.77235	.76449	.75639	.74804	.74375	.73939	.69039	.62213
					1.00253	1.00509	1.01030	1.01565	1.02114	1.02395	1.02679	1.05780	1.09547
.779	.6777	.9776	.89123	0.35248	.77517	.77128	.76335	.75517	.74673	.74240	.73799	.68828	.61796
					1.00254	1.00511	1.01036	1.01574	1.02127	1.02409	1.02695	1.05819	1.09632
.778	.6795	.9774	.89245	0.34707	.77413	.77022	.76221	.75395	.74542	.74104	.73658	.68615	.61359
					1.00256	1.00514	1.01041	1.01583	1.02139	1.02423	1.02711	1.05859	1.09718
.777	.6812	.9773	.89367	0.34174	.77310	.76915	.76106	.75273	.74411	.73968	.73517	.68399	.60898
					1.00257	1.00517	1.01047	1.01591	1.02151	1.02437	1.02727	1.05899	1.09807
.776	.6830	.9772	.89487	0.33649	.77207	.76808	.75992	.75150	.74279	.73831	.73375	.68179	.60407
					1.00258	1.00520	1.01053	1.01600	1.02163	1.02451	1.02742	1.05940	1.09898
.775	.6847	.9771	.89606	0.33130	.77103	.76701	.75877	.75027	.74146	.73694	.73232	.67957	.59881
					1.00260	1.00522	1.01058	1.01609	1.02175	1.02465	1.02758	1.05981	1.09992
.774	.6865	.9770	.89725	0.32618	.77000	.76594	.75762	.74904	.74014	.73556	.73089	.67732	.59311
					1.00261	1.00525	1.01064	1.01618	1.02187	1.02479	1.02774	1.06022	1.10089
.773	.6882	.9769	.89843	0.32113	.76896	.76487	.75648	.74780	.73881	.73418	.72946	.67504	.58684
					1.00262	1.00528	1.01070	1.01627	1.02200	1.02493	1.02790	1.06064	1.10189
.772	.6900	.9767	.89960	0.31615	.76793	.76379	.75532	.74656	.73747	.73279	.72801	.67272	.57977
					1.00264	1.00531	1.01075	1.01636	1.02212	1.02507	1.02807	1.06106	1.10292
.771	.6917	.9766	.90076	0.31123	.76689	.76272	.75417	.74532	.73613	.73140	.72656	.67037	.57151
					1.00265	1.00533	1.01081	1.01644	1.02225	1.02521	1.02823	1.06148	1.10401
.770	.6935	.9765	.90191	0.30638	.76586	.76165	.75302	.74408	.73479	.73000	.72510	.66797	.56109
					1.00266	1.00536	1.01087	1.01653	1.02237	1.02536	1.02839	1.06191	1.10517
.769	.6952	.9764	.90306	0.30159	.76482	.76057	.75186	.74283	.73344	.72859	.72364	.66554	.54484
					1.00268	1.00539	1.01093	1.01662	1.02249	1.02550	1.02855	1.06235	1.10642
.768	.6970	.9763	.90420	0.29687	.76378	.75950	.75070	.74158	.73209	.72718	.72217	.66306	
					1.00269	1.00542	1.01098	1.01671	1.02262	1.02564	1.02872	1.06279	
.767	.6987	.9762	.90533	0.29221	.76275	.75842	.74954	.74033	.73073	.72577	.72069	.66054	
					1.00270	1.00544	1.01104	1.01680	1.02275	1.02579	1.02888	1.06323	
.766	.7004	.9761	.90645	0.28761	.76171	.75734	.74838	.73907	.72936	.72434	.71921	.65796	
					1.00272	1.00547	1.01110	1.01689	1.02287	1.02593	1.02905	1.06368	
.765	.7022	.9759	.90756	0.28308	.76067	.75627	.74722	.73781	.72799	.72291	.71771	.65533	
					1.00273	1.00550	1.01116	1.01698	1.02300	1.02608	1.02921	1.06413	
.764	.7039	.9758	.90866	0.27860	.75963	.75519	.74605	.73654	.72662	.72148	.71621	.65265	
					1.00275	1.00553	1.01121	1.01708	1.02313	1.02623	1.02938	1.06459	
.763	.7056	.9757	.90976	0.27418	.75859	.75411	.74488	.73528	.72524	.72004	.71470	.64990	
					1.00276	1.00556	1.01127	1.01717	1.02325	1.02637	1.02955	1.06506	
.762	.7074	.9756	.91085	0.26983	.75755	.75303	.74371	.73401	.72386	.71859	.71318	.64708	
					1.00277	1.00558	1.01133	1.01726	1.02338	1.02652	1.02972	1.06553	
.761	.7091	.9755	.91193	0.26552	.75651	.75195	.74254	.73273	.72246	.71713	.71166	.64419	
					1.00279	1.00561	1.01139	1.01735	1.02351	1.02667	1.02989	1.06601	

TABLE I. Specific Heat Ratio = 1.10

R₁ and TPR versus Loss Coefficient K

p/p_t	MACH	T/T_t	GAMMA	K_{crit}	.01	.02	.04	.06	.07	.08	.09	.1	.2
.760	.7108	.9754	.91300	0.26128	.75547	.75086	.74137	.73146	.72632	.72106	.71567	.71012	.64122
					1.00280	1.00564	1.01145	1.01744	1.02052	1.02364	1.02682	1.03006	1.06649
.759	.7125	.9752	.91407	0.25709	.75443	.74978	.74019	.73017	.72498	.71966	.71420	.70857	.63816
					1.00281	1.00567	1.01151	1.01754	1.02063	1.02377	1.02697	1.03023	1.06699
.758	.7143	.9751	.91513	0.25296	.75339	.74870	.73901	.72889	.72364	.71825	.71272	.70702	.63500
					1.00283	1.00570	1.01157	1.01763	1.02074	1.02390	1.02712	1.03040	1.06749
.757	.7160	.9750	.91618	0.24889	.75235	.74761	.73783	.72760	.72229	.71684	.71123	.70545	.63173
					1.00284	1.00572	1.01162	1.01772	1.02085	1.02403	1.02727	1.03057	1.06799
.756	.7177	.9749	.91722	0.24486	.75131	.74653	.73665	.72630	.72093	.71541	.70973	.70388	.62834
					1.00286	1.00575	1.01168	1.01781	1.02096	1.02416	1.02742	1.03075	1.06851
.755	.7194	.9748	.91825	0.24089	.75027	.74544	.73546	.72501	.71957	.71398	.70823	.70229	.62481
					1.00287	1.00578	1.01174	1.01791	1.02107	1.02429	1.02758	1.03092	1.06904
.754	.7211	.9747	.91928	0.23698	.74923	.74435	.73428	.72370	.71820	.71255	.70672	.70070	.62112
					1.00288	1.00581	1.01180	1.01800	1.02119	1.02443	1.02773	1.03110	1.06958
.753	.7228	.9745	.92030	0.23311	.74818	.74326	.73309	.72240	.71683	.71110	.70520	.69909	.61724
					1.00290	1.00584	1.01186	1.01810	1.02130	1.02456	1.02788	1.03127	1.07013
.752	.7246	.9744	.92131	0.22930	.74714	.74217	.73189	.72109	.71545	.70965	.70366	.69747	.61315
					1.00291	1.00587	1.01192	1.01819	1.02141	1.02469	1.02804	1.03145	1.07069
.751	.7263	.9743	.92232	0.22554	.74609	.74108	.73070	.71977	.71407	.70819	.70212	.69584	.60880
					1.00293	1.00589	1.01198	1.01829	1.02153	1.02483	1.02819	1.03163	1.07127
.750	.7280	.9742	.92331	0.22182	.74505	.73999	.72950	.71845	.71268	.70672	.70057	.69419	.60414
					1.00294	1.00592	1.01204	1.01838	1.02164	1.02496	1.02835	1.03181	1.07186
.749	.7297	.9741	.92430	0.21816	.74400	.73889	.72830	.71712	.71128	.70525	.69901	.69253	.59907
					1.00295	1.00595	1.01210	1.01848	1.02176	1.02510	1.02851	1.03199	1.07247
.748	.7314	.9739	.92528	0.21454	.74296	.73780	.72710	.71579	.70987	.70376	.69744	.69086	.59348
					1.00297	1.00598	1.01216	1.01857	1.02187	1.02523	1.02867	1.03217	1.07310
.747	.7331	.9738	.92626	0.21098	.74191	.73670	.72589	.71445	.70846	.70227	.69585	.68917	.58716
					1.00298	1.00601	1.01222	1.01867	1.02199	1.02537	1.02882	1.03235	1.07375
.746	.7348	.9737	.92722	0.20746	.74087	.73561	.72468	.71311	.70705	.70077	.69425	.68747	.57968
					1.00300	1.00604	1.01229	1.01877	1.02210	1.02551	1.02898	1.03254	1.07444
.745	.7365	.9736	.92818	0.20398	.73982	.73451	.72347	.71176	.70562	.69926	.69265	.68575	.57012
					1.00301	1.00607	1.01235	1.01887	1.02222	1.02565	1.02914	1.03272	1.07516
.744	.7382	.9735	.92913	0.20056	.73877	.73341	.72225	.71041	.70419	.69774	.69102	.68402	.55347
					1.00302	1.00610	1.01241	1.01896	1.02234	1.02578	1.02931	1.03291	1.07595
.743	.7399	.9734	.93008	0.19718	.73772	.73231	.72104	.70905	.70275	.69620	.68939	.68227	
					1.00304	1.00613	1.01247	1.01906	1.02246	1.02592	1.02947	1.03310	
.742	.7416	.9732	.93102	0.19384	.73667	.73121	.71981	.70768	.70130	.69466	.68774	.68050	
					1.00305	1.00615	1.01253	1.01916	1.02258	1.02606	1.02963	1.03329	
.741	.7433	.9731	.93194	0.19055	.73562	.73010	.71858	.70631	.69984	.69311	.68608	.67871	
					1.00307	1.00618	1.01260	1.01926	1.02269	1.02621	1.02980	1.03348	
.740	.7450	.9730	.93287	0.18730	.73457	.72900	.71735	.70493	.69837	.69154	.68440	.67690	
					1.00308	1.00621	1.01266	1.01936	1.02281	1.02635	1.02996	1.03367	
.739	.7467	.9729	.93378	0.18409	.73352	.72789	.71612	.70355	.69690	.68997	.68271	.67507	
					1.00310	1.00624	1.01272	1.01946	1.02293	1.02649	1.03013	1.03386	
.738	.7484	.9728	.93469	0.18093	.73247	.72678	.71489	.70216	.69542	.68838	.68100	.67322	
					1.00311	1.00627	1.01278	1.01956	1.02306	1.02663	1.03030	1.03405	
.737	.7501	.9726	.93559	0.17781	.73142	.72568	.71365	.70076	.69392	.68678	.67927	.67135	
					1.00312	1.00630	1.01284	1.01966	1.02318	1.02678	1.03047	1.03425	
.736	.7518	.9725	.93648	0.17473	.73036	.72456	.71240	.69935	.69242	.68516	.67753	.66945	
					1.00314	1.00633	1.01291	1.01976	1.02330	1.02692	1.03064	1.03445	
.735	.7535	.9724	.93737	0.17170	.72931	.72345	.71116	.69794	.69090	.68353	.67576	.66753	
					1.00315	1.00636	1.01297	1.01986	1.02342	1.02707	1.03081	1.03465	
.734	.7551	.9723	.93825	0.16870	.72826	.72234	.70991	.69651	.68938	.68189	.67398	.66558	
					1.00317	1.00639	1.01303	1.01996	1.02355	1.02722	1.03098	1.03485	
.733	.7568	.9722	.93912	0.16574	.72720	.72122	.70865	.69508	.68784	.68023	.67218	.66360	
					1.00318	1.00642	1.01310	1.02007	1.02367	1.02736	1.03115	1.03505	
.732	.7585	.9720	.93999	0.16283	.72615	.72011	.70739	.69365	.68629	.67855	.67035	.66159	
					1.00320	1.00645	1.01316	1.02017	1.02379	1.02751	1.03133	1.03526	
.731	.7602	.9719	.94085	0.15995	.72509	.71899	.70613	.69220	.68473	.67686	.66850	.65954	
					1.00321	1.00648	1.01323	1.02027	1.02392	1.02766	1.03151	1.03546	

TABLE I. Specific Heat Ratio = 1.10

R₂ and TPR versus Loss Coefficient K

p/pₜ	MACH	T/Tₜ	GAMMA	K_crit	.01	.02	.04	.05	.06	.07	.08	.09	.1
.730	.7619	.9718	.94170	0.15711	.72403	.71787	.70486	.69796	.69074	.68316	.67515	.66663	.65747
					1.00323	1.00651	1.01329	1.01679	1.02037	1.02405	1.02781	1.03168	1.03567
.729	.7636	.9717	.94254	0.15431	.72297	.71674	.70358	.69659	.68928	.68158	.67342	.66473	.65535
					1.00324	1.00654	1.01335	1.01687	1.02048	1.02417	1.02796	1.03186	1.03588
.728	.7652	.9716	.94338	0.15155	.72191	.71562	.70231	.69522	.68780	.67998	.67168	.66280	.65320
					1.00325	1.00657	1.01342	1.01696	1.02058	1.02430	1.02812	1.03204	1.03609
.727	.7669	.9714	.94421	0.14882	.72085	.71449	.70102	.69385	.68631	.67836	.66991	.66084	.65099
					1.00327	1.00660	1.01348	1.01704	1.02069	1.02443	1.02827	1.03223	1.03631
.726	.7686	.9713	.94503	0.14614	.71979	.71337	.69974	.69246	.68482	.67673	.66812	.65886	.64875
					1.00328	1.00663	1.01355	1.01713	1.02079	1.02456	1.02843	1.03241	1.03653
.725	.7703	.9712	.94585	0.14348	.71873	.71224	.69844	.69107	.68331	.67509	.66631	.65684	.64645
					1.00330	1.00666	1.01361	1.01721	1.02090	1.02469	1.02858	1.03260	1.03675
.724	.7720	.9711	.94666	0.14087	.71767	.71110	.69714	.68967	.68179	.67343	.66447	.65478	.64410
					1.00331	1.00669	1.01368	1.01730	1.02101	1.02482	1.02874	1.03278	1.03697
.723	.7736	.9709	.94746	0.13829	.71661	.70997	.69584	.68826	.68026	.67175	.66261	.65268	.64169
					1.00333	1.00672	1.01375	1.01738	1.02112	1.02495	1.02890	1.03298	1.03720
.722	.7753	.9708	.94826	0.13574	.71554	.70883	.69453	.68684	.67871	.67005	.66072	.65054	.63922
					1.00334	1.00675	1.01381	1.01747	1.02122	1.02508	1.02906	1.03317	1.03743
.721	.7770	.9707	.94905	0.13323	.71448	.70770	.69321	.68542	.67716	.66833	.65881	.64836	.63667
					1.00336	1.00678	1.01388	1.01756	1.02133	1.02522	1.02922	1.03337	1.03766
.720	.7786	.9706	.94983	0.13076	.71341	.70656	.69189	.68398	.67558	.66660	.65686	.64613	.63404
					1.00337	1.00681	1.01394	1.01764	1.02144	1.02535	1.02938	1.03356	1.03790
.719	.7803	.9705	.95060	0.12832	.71234	.70541	.69056	.68254	.67400	.66484	.65487	.64385	.63133
					1.00339	1.00684	1.01401	1.01773	1.02155	1.02549	1.02955	1.03376	1.03814
.718	.7820	.9703	.95137	0.12591	.71128	.70427	.68923	.68108	.67240	.66305	.65285	.64152	.62851
					1.00340	1.00687	1.01408	1.01782	1.02166	1.02562	1.02972	1.03396	1.03838
.717	.7837	.9702	.95213	0.12353	.71021	.70312	.68788	.67961	.67078	.66125	.65080	.63912	.62557
					1.00342	1.00691	1.01415	1.01791	1.02178	1.02576	1.02989	1.03417	1.03863
.716	.7853	.9701	.95289	0.12119	.70914	.70197	.68654	.67814	.66915	.65942	.64870	.63665	.62251
					1.00343	1.00694	1.01421	1.01800	1.02189	1.02590	1.03006	1.03437	1.03888
.715	.7870	.9700	.95364	0.11888	.70806	.70082	.68518	.67685	.66750	.65755	.64657	.63410	.61928
					1.00345	1.00697	1.01428	1.01809	1.02200	1.02605	1.03023	1.03458	1.03914
.714	.7887	.9698	.95438	0.11660	.70699	.69966	.68381	.67515	.66583	.65566	.64438	.63147	.61588
					1.00346	1.00700	1.01435	1.01818	1.02212	1.02619	1.03041	1.03480	1.03940
.713	.7903	.9697	.95511	0.11435	.70592	.69850	.68244	.67363	.66414	.65374	.64214	.62874	.61225
					1.00348	1.00703	1.01442	1.01827	1.02223	1.02633	1.03058	1.03501	1.03967
.712	.7920	.9696	.95584	0.11214	.70484	.69734	.68106	.67211	.66243	.65179	.63984	.62591	.60835
					1.00349	1.00706	1.01449	1.01836	1.02235	1.02648	1.03076	1.03523	1.03994
.711	.7936	.9695	.95656	0.10995	.70377	.69618	.67967	.67057	.66068	.64980	.63749	.62294	.60410
					1.00351	1.00709	1.01456	1.01845	1.02247	1.02662	1.03094	1.03546	1.04023
.710	.7953	.9693	.95728	0.10780	.70269	.69501	.67827	.66901	.65893	.64777	.63506	.61984	.59938
					1.00352	1.00712	1.01463	1.01854	1.02258	1.02677	1.03112	1.03569	1.04052
.709	.7970	.9692	.95799	0.10568	.70161	.69384	.67686	.66744	.65715	.64570	.63255	.61655	.59394
					1.00354	1.00716	1.01470	1.01863	1.02270	1.02692	1.03131	1.03592	1.04083
.708	.7986	.9691	.95869	0.10358	.70053	.69267	.67544	.66586	.65534	.64358	.62996	.61306	.58739
					1.00355	1.00719	1.01477	1.01873	1.02282	1.02707	1.03150	1.03616	1.04115
.707	.8003	.9690	.95939	0.10152	.69945	.69149	.67402	.66425	.65350	.64142	.62727	.60932	.57841
					1.00357	1.00722	1.01484	1.01882	1.02294	1.02722	1.03169	1.03640	1.04149
.706	.8019	.9688	.96008	0.09949	.69837	.69031	.67258	.66262	.65164	.63920	.62447	.60524	
					1.00358	1.00725	1.01491	1.01892	1.02307	1.02738	1.03189	1.03665	
.705	.8036	.9687	.96076	0.09748	.69728	.68913	.67113	.66098	.64974	.63692	.62154	.60072	
					1.00360	1.00728	1.01498	1.01901	1.02319	1.02753	1.03209	1.03691	
.704	.8053	.9686	.96144	0.09550	.69620	.68794	.66966	.65932	.64781	.63458	.61845	.59554	
					1.00361	1.00732	1.01505	1.01911	1.02331	1.02769	1.03229	1.03719	
.703	.8069	.9685	.96211	0.09355	.69511	.68675	.66819	.65764	.64584	.63216	.61519	.58936	
					1.00363	1.00735	1.01513	1.01921	1.02344	1.02785	1.03250	1.03747	
.702	.8086	.9683	.96277	0.09163	.69402	.68556	.66670	.65594	.64383	.62966	.61170	.58113	
					1.00364	1.00738	1.01520	1.01930	1.02357	1.02802	1.03271	1.03777	
.701	.8102	.9682	.96343	0.08974	.69293	.68436	.66520	.65421	.64177	.62706	.60794		
					1.00366	1.00741	1.01527	1.01940	1.02369	1.02818	1.03293		

TABLE I. Specific Heat Ratio = 1.10

p/p_t	MACH	T/T_t	GAMMA	K_crit	.01	.02	.03	.04	.05	.055	.06	.065	.07
.700	.8119	.9681	.96408	0.08787	.69184	.68315	.67383	.66368	.65245	.64630	.63967	.63243	.62437
					1.00367	1.00745	1.01133	1.01535	1.01950	1.02164	1.02382	1.02606	1.02835
.699	.8135	.9680	.96472	0.08603	.69074	.68195	.67248	.66215	.65067	.64435	.63751	.63000	.62153
					1.00369	1.00748	1.01138	1.01542	1.01960	1.02175	1.02395	1.02621	1.02852
.698	.8152	.9678	.96536	0.08422	.68965	.68074	.67113	.66060	.64886	.64236	.63530	.62748	.61856
					1.00371	1.00751	1.01144	1.01550	1.01970	1.02187	1.02409	1.02636	1.02870
.697	.8168	.9677	.96599	0.08243	.68855	.67952	.66976	.65904	.64702	.64033	.63302	.62486	.61542
					1.00372	1.00754	1.01149	1.01557	1.01981	1.02199	1.02422	1.02651	1.02888
.696	.8185	.9676	.96662	0.08068	.68745	.67830	.66838	.65746	.64514	.63825	.63066	.62210	.61206
					1.00374	1.00758	1.01154	1.01565	1.01991	1.02211	1.02436	1.02668	1.02906
.695	.8201	.9675	.96724	0.07894	.68635	.67707	.66699	.65586	.64323	.63611	.62823	.61922	.60844
					1.00375	1.00761	1.01160	1.01572	1.02001	1.02223	1.02450	1.02684	1.02925
.694	.8218	.9673	.96785	0.07723	.68525	.67584	.66559	.65424	.64127	.63392	.62570	.61618	.60448
					1.00377	1.00764	1.01165	1.01580	1.02012	1.02235	1.02464	1.02700	1.02945
.693	.8234	.9672	.96845	0.07555	.68414	.67461	.66419	.65259	.63928	.63166	.62305	.61295	.60004
					1.00378	1.00768	1.01170	1.01588	1.02022	1.02247	1.02479	1.02717	1.02965
.692	.8251	.9671	.96905	0.07389	.68304	.67336	.66277	.65093	.63723	.62933	.62029	.60947	.59489
					1.00380	1.00771	1.01176	1.01596	1.02033	1.02260	1.02493	1.02735	1.02986
.691	.8267	.9670	.96965	0.07226	.68193	.67212	.66134	.64924	.63513	.62691	.61738	.60567	.58847
					1.00381	1.00774	1.01181	1.01603	1.02044	1.02272	1.02508	1.02752	1.03008
.690	.8284	.9668	.97024	0.07065	.68081	.67085	.65989	.64752	.63297	.62438	.61430	.60145	.57887
					1.00383	1.00778	1.01187	1.01611	1.02055	1.02286	1.02524	1.02771	1.03031
.689	.8300	.9667	.97082	0.06906	.67970	.66959	.65843	.64578	.63075	.62176	.61099	.59660	
					1.00385	1.00781	1.01192	1.01619	1.02066	1.02299	1.02539	1.02790	
.688	.8316	.9666	.97139	0.06750	.67858	.66832	.65696	.64400	.62846	.61900	.60742	.59072	
					1.00386	1.00785	1.01198	1.01628	1.02078	1.02313	1.02556	1.02810	
.687	.8333	.9664	.97196	0.06597	.67747	.66705	.65547	.64220	.62608	.61609	.60347	.58241	
					1.00388	1.00788	1.01203	1.01636	1.02089	1.02326	1.02572	1.02831	
.686	.8349	.9663	.97253	0.06445	.67634	.66577	.65397	.64035	.62360	.61300	.59901		
					1.00389	1.00792	1.01209	1.01644	1.02101	1.02340	1.02589		
.685	.8366	.9662	.97308	0.06296	.67522	.66448	.65245	.63847	.62102	.60967	.59374		
					1.00391	1.00795	1.01215	1.01652	1.02113	1.02355	1.02607		
.684	.8382	.9661	.97363	0.06150	.67409	.66318	.65091	.63654	.61831	.60605	.58685		
					1.00392	1.00799	1.01220	1.01661	1.02125	1.02370	1.02627		
.683	.8399	.9659	.97418	0.06005	.67295	.66187	.64935	.63457	.61545	.60201	.57290		
					1.00395	1.00802	1.01226	1.01669	1.02138	1.02385	1.02648		
.682	.8415	.9658	.97472	0.05863	.67182	.66056	.64777	.63255	.61240	.59739			
					1.00396	1.00806	1.01232	1.01678	1.02150	1.02401			
.681	.8431	.9657	.97525	0.05723	.67068	.65924	.64617	.63047	.60911	.59173			
					1.00398	1.00809	1.01238	1.01687	1.02163	1.02418			
.680	.8448	.9655	.97578	0.05585	.66954	.65790	.64455	.62832	.60549	.58383			
					1.00399	1.00813	1.01244	1.01696	1.02177	1.02436			
.679	.8464	.9654	.97630	0.05450	.66840	.65656	.64290	.62610	.60147				
					1.00401	1.00816	1.01250	1.01705	1.02191				
.678	.8480	.9653	.97681	0.05316	.66725	.65521	.64122	.62380	.59682				
					1.00403	1.00820	1.01256	1.01714	1.02206				
.677	.8497	.9652	.97732	0.05185	.66610	.65384	.63951	.62141	.59109				
					1.00404	1.00824	1.01262	1.01723	1.02221				
.676	.8513	.9650	.97782	0.05056	.66494	.65246	.63777	.61890	.58251				
					1.00406	1.00827	1.01268	1.01733	1.02237				
.675	.8530	.9649	.97832	0.04929	.66378	.65107	.63600	.61626					
					1.00408	1.00831	1.01274	1.01742					
.674	.8546	9648	.97881	0.04804	.66262	.64967	.63418	.61347					
					1.00409	1.00835	1.01280	1.01752					
.673	.8562	.9646	.97929	0.04681	.66145	.64825	.63233	.61045					
					1.00411	1.00838	1.01287	1.01763					
.672	.8579	.9645	.97977	0.04560	.66028	.64682	.63042	.60719					
					1.00413	1.00842	1.01293	1.01773					
.671	.8595	.9644	.98024	0.04441	.65910	.64537	.62847	.60360					
					1.00414	1.00846	1.01299	1.01784					

TABLE I. Specific Heat Ratio = 1.10

p/p_t	MACH	T/T_t	GAMMA	K_{crit}	.006	.008	.01	.015	.02	.025	.03	.035	.04
.670	.8611	.9642	.98071	0.04324	.66294	.66047	.65793	.65122	.64391	.63578	.62645	.61516	.59953
					1.00247	1.00331	1.00416	1.00631	1.00849	1.01074	1.01306	1.01545	1.01795
.669	.8628	.9641	.98117	0.04209	.66184	.65932	.65674	.64990	.64242	.63406	.62436	.61241	.59463
					1.00248	1.00333	1.00418	1.00633	1.00853	1.01079	1.01313	1.01554	1.01807
.668	.8644	.9640	.98163	0.04096	.66073	.65818	.65555	.64857	.64092	.63229	.62220	.60945	.58814
					1.00249	1.00334	1.00419	1.00636	1.00857	1.01085	1.01319	1.01563	1.01820
.667	.8660	.9639	.98208	0.03985	.65963	.65703	.65435	.64724	.63939	.63049	.61995	.60623	
					1.00250	1.00335	1.00421	1.00639	1.00861	1.01090	1.01326	1.01572	
.666	.8677	.9637	.98252	0.03876	.65851	.65587	.65314	.64589	.63784	.62864	.61760	.60265	
					1.00251	1.00337	1.00423	1.00642	1.00865	1.01095	1.01333	1.01581	
.665	.8693	.9636	.98296	0.03768	.65740	.65471	.65193	.64452	.63627	.62674	.61513	.59856	
					1.00252	1.00338	1.00424	1.00644	1.00869	1.01101	1.01341	1.01591	
.664	.8709	.9635	.98339	0.03663	.65628	.65354	.65072	.64314	.63467	.62479	.61250	.59351	
					1.00253	1.00339	1.00426	1.00647	1.00872	1.01106	1.01348	1.01602	
.663	.8726	.9633	.98381	0.03559	.65516	.65237	.64949	.64175	.63302	.62277	.60969	.58647	
					1.00254	1.00341	1.00428	1.00650	1.00877	1.01112	1.01355	1.01613	
.662	.8742	.9632	.98423	0.03458	.65403	.65120	.64826	.64034	.63135	.62067	.60664		
					1.00255	1.00342	1.00430	1.00653	1.00881	1.01117	1.01363		
.661	.8758	.9631	.98465	0.03358	.65290	.65002	.64702	.63892	.62965	.61849	.60328		
					1.00256	1.00343	1.00431	1.00655	1.00885	1.01123	1.01371		
.660	.8775	.9629	.98506	0.03260	.65177	.64883	.64577	.63747	.62790	.61621	.59945		
					1.00257	1.00345	1.00433	1.00658	1.00889	1.01129	1.01379		
.659	.8791	.9628	.98546	0.03163	.65064	.64764	.64450	.63601	.62611	.61380	.59479		
					1.00258	1.00346	1.00435	1.00661	1.00893	1.01135	1.01388		
.658	.8807	.9627	.98585	0.03069	.64949	.64644	.64323	.63451	.62428	.61125	.58853		
					1.00259	1.00347	1.00437	1.00664	1.00898	1.01141	1.01397		
.657	.8823	.9625	.98625	0.02976	.64835	.64521	.64195	.63300	.62238	.60851			
					1.00260	1.00349	1.00439	1.00667	1.00902	1.01147			
.656	.8840	.9624	.98663	0.02885	.64720	.64400	.64066	.63146	.62042	.60553			
					1.00261	1.00351	1.00441	1.00670	1.00906	1.01153			
.655	.8856	.9623	.98701	0.02796	.64603	.64277	.63936	.62989	.61838	.60221			
					1.00263	1.00352	1.00442	1.00673	1.00911	1.01160			
.654	.8872	.9621	.98738	0.02708	.64486	.64154	.63804	.62830	.61626	.59835			
					1.00264	1.00353	1.00444	1.00676	1.00915	1.01167			
.653	.8889	.9620	.98775	0.02622	.64370	.64030	.63672	.62667	.61403	.59364			
					1.00265	1.00355	1.00446	1.00679	1.00920	1.01174			
.652	.8905	.9619	.98812	0.02538	.64252	.63905	.63538	.62500	.61168	.58669			
					1.00266	1.00356	1.00448	1.00683	1.00925	1.01182			
.651	.8921	.9617	.98847	0.02455	.64134	.63779	.63402	.62329	.60917				
					1.00267	1.00358	1.00450	1.00686	1.00930				
.650	.8937	.9616	.98882	0.02374	.64015	.63652	.63264	.62153	.60646				
					1.00268	1.00359	1.00452	1.00689	1.00935				
.649	.8954	.9615	.98917	0.02295	.63896	.63523	.63125	.61971	.60349				
					1.00269	1.00361	1.00454	1.00692	1.00940				
.648	.8970	.9613	.98951	0.02217	.63775	.63393	.62984	.61783	.60009				
					1.00270	1.00362	1.00456	1.00695	1.00945				
.647	.8986	.9612	.98984	0.02141	.63654	.63262	.62841	.61588	.59609				
					1.00271	1.00364	1.00458	1.00699	1.00951				
.646	.9003	.9611	.99017	0.02066	.63532	.63130	.62695	.61384	.59087				
					1.00272	1.00365	1.00459	1.00702	1.00956				
.645	.9019	.9609	.99050	0.01993	.63409	.62996	.62546	.61170					
					1.00273	1.00367	1.00461	1.00705					
.644	.9035	.9608	.99081	0.01922	.63285	.62860	.62395	.60943					
					1.00274	1.00368	1.00463	1.00709					
.643	.9051	.9606	.99112	0.01852	.63160	.62722	.62241	.60700					
					1.00275	1.00370	1.00465	1.00712					
.642	.9068	.9605	.99143	0.01783	.63034	.62582	.62082	.60437					
					1.00277	1.00371	1.00467	1.00716					
.641	.9084	.9604	.99173	0.01716	.62906	.62440	.61920	.60140					
					1.00278	1.00373	1.00469	1.00720					

TABLE I. Specific Heat Ratio = 1.10

R₂ and TPR versus Loss Coefficient K

p/p,	MACH	T/T,	GAMMA	K_crit	.001	.002	.003	.004	.005	.006	.008	.01	.015
.640	.9100	.9602	.99203	0.01651	.63813	.63621	.63421	.63216	.63001	.62777	.62295	.61753	.598C1
					1.00046	1.00092	1.00138	1.00185	1.00231	1.00279	1.00374	1.00472	1.00724
.639	.9116	.9601	.99232	0.01587	.63710	.63513	.63309	.63098	.62877	.62647	.62147	.61581	.59385
					1.00046	1.00092	1.00138	1.00185	1.00232	1.00280	1.00376	1.00474	1.00728
.638	.9133	.9600	.99260	0.01524	.63606	.63406	.63196	.62979	.62752	.62514	.61997	.61403	.58758
					1.00046	1.00092	1.00139	1.00186	1.00233	1.00281	1.00377	1.00476	1.00733
.637	.9149	.9598	.99288	0.01463	.63500	.63296	.63082	.62859	.62626	.62380	.61842	.61217	
					1.00047	1.00093	1.00139	1.00187	1.00234	1.00282	1.00379	1.00478	
.636	.9165	.9597	.99315	0.01403	.63397	.63187	.62968	.62739	.62498	.62244	.61684	.61023	
					1.00046	1.00093	1.00140	1.00187	1.00235	1.00283	1.00381	1.00480	
.635	.9181	.9596	.99342	0.01345	.63293	.63078	.62853	.62617	.62369	.62106	.61521	.60818	
					1.00046	1.00093	1.00141	1.00188	1.00236	1.00284	1.00382	1.00482	
.634	.9198	.9594	.99368	0.01288	.63188	.62968	.62737	.62494	.62238	.61965	.61352	.60594	
					1.00047	1.00094	1.00141	1.00189	1.00237	1.00285	1.00384	1.00485	
.633	.9214	.9593	.99394	0.01233	.63083	.62857	.62620	.62370	.62105	.61822	.61177	.60356	
					1.00047	1.00094	1.00142	1.00189	1.00238	1.00286	1.00386	1.00487	
.632	.9230	.9591	.99419	0.01179	.62978	.62747	.62503	.62245	.61970	.61675	.60995	.60093	
					1.00047	1.00094	1.00142	1.00190	1.00239	1.00288	1.00387	1.00490	
.631	.9246	.9590	.99444	0.01126	.62873	.62635	.62384	.62118	.61833	.61522	.60800	.59794	
					1.00047	1.00095	1.00143	1.00191	1.00240	1.00289	1.00389	1.00492	
.630	.9263	.9589	.99468	0.01075	.62768	.62523	.62265	.61989	.61693	.61367	.60596	.59432	
					1.00047	1.00095	1.00143	1.00192	1.00240	1.00290	1.00391	1.00495	
.629	.9279	.9587	.99491	0.01025	.62662	.62411	.62144	.61859	.61547	.61207	.60376	.58907	
					1.00047	1.00095	1.00144	1.00192	1.00242	1.00292	1.00393	1.00498	
.628	.9295	.9586	.99514	0.00976	.62556	.62298	.62022	.61726	.61400	.61041	.60137		
					1.00048	1.00096	1.00144	1.00193	1.00243	1.00293	1.00395		
.627	.9311	.9585	.99536	0.00929	.62450	.62184	.61899	.61588	.61249	.60868	.59868		
					1.00048	1.00096	1.00145	1.00194	1.00244	1.00294	1.00397		
.626	.9328	.9583	.99558	0.00883	.62343	.62069	.61774	.61450	.61094	.60687	.59552		
					1.00048	1.00096	1.00145	1.00195	1.00245	1.00296	1.00399		
.625	.9344	.9582	.99580	0.00838	.62236	.61953	.61648	.61309	.60932	.60496	.59133		
					1.00048	1.00097	1.00146	1.00196	1.00246	1.00297	1.00402		
.624	.9360	.9580	.99600	0.00795	.62129	.61837	.61516	.61164	.60765	.60291			
					1.00048	1.00097	1.00147	1.00197	1.00247	1.00298			
.623	.9376	.9579	.99621	0.00753	.62021	.61719	.61385	.61015	.60589	.60068			
					1.00048	1.00097	1.00147	1.00197	1.00248	1.00300			
.622	.9393	.9578	.99640	0.00712	.61913	.61597	.61252	.60861	.60403	.59820			
					1.00048	1.00098	1.00148	1.00198	1.00249	1.00301			
.621	.9409	.9576	.99660	0.00672	.61804	.61477	.61115	.60702	.60204	.59521			
					1.00049	1.00098	1.00148	1.00199	1.00250	1.00303			
.620	.9425	.9575	.99678	0.00634	.61695	.61354	.60975	.60535	.59988	.59141			
					1.00049	1.00099	1.00149	1.00200	1.00251	1.00304			
.619	.9441	.9573	.99696	0.00597	.61582	.61230	.60832	.60359	.59746				
					1.00049	1.00099	1.00150	1.00201	1.00252				
.618	.9458	.9572	.99714	0.00561	.61471	.61105	.60683	.60172	.59451				
					1.00049	1.00099	1.00150	1.00201	1.00254				
.617	.9474	.9571	.99731	0.00526	.61359	.60976	.60530	.59970	.59072				
					1.00050	1.00100	1.00151	1.00202	1.00255				
.616	.9490	.9569	.99748	0.00493	.61247	.60845	.60369	.59747					
					1.00050	1.00100	1.00151	1.00203					
.615	.9506	.9568	.99763	0.00461	.61133	.60712	.60200	.59476					
					1.00050	1.00101	1.00152	1.00204					
.614	.9523	9566	.99779	0.00430	.61018	.60574	.60019	.59143					
					1.00050	1.00101	1.00152	1.00205					
.613	.9539	.9565	.99794	0.00400	.60903	.60432	.59824						
					1.00050	1.00101	1.00153						
.612	.9555	.9563	.99808	0.00371	.60785	.60285	.59596						
					1.00051	1.00102	1.00154						
.611	.9571	.9562	.99822	0.00343	.60666	.60131	.59334						
					1.00051	1.00102	1.00155						

TABLE I. Specific Heat Ratio = 1.10

p/p_t	MACH	T/T_t	GAMMA	K_{crit}	.001	.002	.003	.004	.005	.006	.007	.008	.009
								R_t and TPR versus Loss Coefficient K					
.610	.9588	.9561	.99835	0.00317	.60545 1.00051	.59969 1.00102	.58988 1.00155						
.609	.9604	.9559	.99848	0.00291	.60422 1.00051	.59788 1.00103							
.608	.9620	.9558	.99861	0.00267	.60296 1.00051	.59595 1.00104							
.607	.9636	.9556	.99872	0.00244	.60167 1.00051	.59373 1.00104							
.606	.9653	.9555	.99884	0.00222	.60034 1.00051	.59100 1.00104							
.605	.9669	.9553	.99894	0.00201	.59896 1.00051	.58620 1.00105							
.604	.9685	.9552	.99904	0.00181	.59742 1.00052								
.603	.9701	.9551	.99914	0.00163	.59585 1.00052								
.602	.9718	.9549	.99923	0.00145	.59413 1.00052								
.601	.9734	.9548	.99932	0.00128	.59218 1.00053								
.600	.9750	.9546	.99940	0.00113	.58962 1.00053								

TABLE II. Specific Heat Ratio = 1.20

R₂ and TPR versus Loss Coefficient K

p/p,	MACH	T/T,	GAMMA	K_crit	.1	.2	.4	.6	1.0	2.0	4.0	6.0	10.0
.999	.0408	.9998	.06891	493.11995	.99890	.99880	.99860	.99840	.99800	.99699	.99499	.99297	.98893
					1.00010	1.00020	1.00040	1.00060	1.00100	1.00201	1.00403	1.00606	1.01016
.998	.0578	.9997	.09740	243.73775	.99780	.99760	.99720	.99680	.99599	.99398	.98994	.98589	.97773
					1.00020	1.00040	1.00080	1.00120	1.00201	1.00403	1.00811	1.01224	1.02065
.997	.0708	.9995	.11921	160.77549	.99670	.99640	.99579	.99519	.99398	.99095	.98487	.97875	.96638
					1.00030	1.00060	1.00120	1.00181	1.00302	1.00607	1.01224	1.01854	1.03148
.996	.0817	.9993	.13757	119.37062	.99560	.99519	.99439	.99358	.99197	.98792	.97976	.97154	.95488
					1.00040	1.00080	1.00161	1.00241	1.00403	1.00812	1.01644	1.02497	1.04269
.995	.0914	.9992	.15371	94.57540	.99450	.99399	.99298	.99197	.98995	.98487	.97463	.96428	.94323
					1.00050	1.00100	1.00201	1.00302	1.00505	1.01018	1.02069	1.03153	1.05429
.994	.1002	.9990	.16828	78.07666	.99339	.99279	.99157	.99036	.98792	.98181	.96946	.95695	.93141
					1.00060	1.00121	1.00242	1.00363	1.00608	1.01227	1.02500	1.03823	1.06631
.993	.1082	.9988	.18165	66.31324	.99229	.99158	.99016	.98874	.98590	.97874	.96426	.94955	.91941
					1.00070	1.00141	1.00282	1.00424	1.00710	1.01436	1.02937	1.04507	1.07878
.992	.1157	.9987	.19407	57.50756	.99119	.99038	.98875	.98713	.98386	.97566	.95902	.94209	.90724
					1.00080	1.00161	1.00323	1.00486	1.00814	1.01648	1.03381	1.05206	1.09172
.991	.1228	.9985	.20571	50.67121	.99009	.98917	.98734	.98551	.98183	.97256	.95375	.93455	.89487
					1.00091	1.00181	1.00364	1.00547	1.00917	1.01861	1.03831	1.05921	1.10517
.990	.1295	.9983	.21670	45.21265	.98898	.98797	.98593	.98388	.97978	.96945	.94845	.92694	.88229
					1.00101	1.00202	1.00405	1.00609	1.01022	1.02075	1.04288	1.06652	1.11917
.989	.1358	.9982	.22713	40.75477	.98788	.98676	.98451	.98226	.97774	.96634	.94310	.91926	.86951
					1.00111	1.00222	1.00446	1.00671	1.01126	1.02292	1.04751	1.07400	1.13375
.988	.1419	.9980	.23708	37.04706	.98678	.98555	.98309	.98063	.97569	.96321	.93773	.91150	.85649
					1.00121	1.00243	1.00487	1.00733	1.01231	1.02510	1.05222	1.08165	1.14897
.987	.1478	.9978	.24661	33.91583	.98567	.98434	.98167	.97900	.97363	.96006	.93231	.90365	.84324
					1.00131	1.00263	1.00528	1.00796	1.01337	1.02729	1.05700	1.08949	1.16486
.986	.1534	.9977	.25575	31.23667	.98457	.98313	.98025	.97737	.97157	.95691	.92685	.89572	.82972
					1.00141	1.00284	1.00570	1.00858	1.01443	1.02951	1.06185	1.09752	1.18150
.985	.1588	.9975	.26456	28.91947	.98346	.98192	.97883	.97573	.96950	.95374	.92135	.88770	.81593
					1.00152	1.00304	1.00611	1.00921	1.01549	1.03174	1.06679	1.10575	1.19894
.984	.1641	.9973	.27306	26.89552	.98236	.98071	.97741	.97409	.96743	.95056	.91581	.87959	.80185
					1.00162	1.00325	1.00652	1.00984	1.01656	1.03399	1.07180	1.11420	1.21725
.983	.1692	.9971	.28129	25.11320	.98125	.97950	.97598	.97245	.96536	.94736	.91023	.87139	.78745
					1.00172	1.00345	1.00694	1.01047	1.01764	1.03626	1.07689	1.12286	1.23653
.982	.1741	.9970	.28926	23.53204	.98014	.97829	.97456	.97081	.96327	.94415	.90460	.86308	.77270
					1.00182	1.00366	1.00736	1.01110	1.01872	1.03855	1.08207	1.13175	1.25684
.981	.1789	.9968	.29700	22.11986	.97904	.97707	.97313	.96917	.96119	.94093	.89892	.85467	.75759
					1.00193	1.00387	1.00778	1.01174	1.01980	1.04086	1.08734	1.14089	1.27832
.980	.1837	.9966	.30452	20.85143	.97793	.97586	.97170	.96752	.95910	.93769	.89320	.84615	.74208
					1.00203	1.00407	1.00820	1.01237	1.02089	1.04318	1.09269	1.15028	1.30108
.979	.1882	.9965	.31184	19.70608	.97682	.97464	.97026	.96587	.95700	.93444	.88743	.83752	.72613
					1.00213	1.00428	1.00862	1.01301	1.02198	1.04553	1.09814	1.15994	1.32527
.978	.1927	.9963	.31897	18.66684	.97572	.97343	.96883	.96421	.95490	.93117	.88161	.82877	.70970
					1.00224	1.00449	1.00904	1.01365	1.02308	1.04790	1.10369	1.16988	1.35104
.977	.1971	.9961	.32593	17.71979	.97461	.97221	.96740	.96255	.95279	.92789	.87574	.81989	.69274
					1.00234	1.00470	1.00946	1.01430	1.02419	1.05028	1.10934	1.18013	1.37862
.976	.2014	.9960	.33273	16.85333	.97350	.97099	.96596	.96089	.95068	.92459	.86982	.81089	.67519
					1.00244	1.00491	1.00989	1.01494	1.02530	1.05269	1.11509	1.19069	1.40823
.975	.2056	.9958	.33937	16.05779	.97239	.96978	.96452	.95923	.94856	.92128	.86384	.80174	.65698
					1.00255	1.00511	1.01031	1.01559	1.02641	1.05512	1.12095	1.20158	1.44017
.974	.2098	.9956	.34587	15.32481	.97128	.96856	.96308	.95757	.94643	.91795	.85780	.79245	.63802
					1.00265	1.00532	1.01074	1.01624	1.02754	1.05757	1.12691	1.21284	1.47480
.973	.2138	.9954	.35223	14.64750	.97017	.96734	.96164	.95590	.94430	.91461	.85170	.78301	.61821
					1.00276	1.00553	1.01116	1.01689	1.02866	1.06004	1.13300	1.22447	1.51256
.972	.2178	.9953	.35847	14.01985	.96906	.96612	.96019	.95423	.94217	.91124	.84555	.77340	.59741
					1.00286	1.00574	1.01159	1.01755	1.02979	1.06254	1.13920	1.23650	1.55400
.971	.2217	.9951	.36458	13.43659	.96795	.96489	.95874	.95255	.94003	.90787	.83933	.76362	.57545
					1.00296	1.00595	1.01202	1.01820	1.03093	1.06506	1.14553	1.24896	1.59988

TABLE II. Specific Heat Ratio = 1.20

p/p_t	MACH	T/T_t	GAMMA	K_crit	.1	.2	.4	.6	1.0	2.0	4.0	6.0	10.0
.970	.2256	.9949	.37057	12.89327	.96684	.96367	.95730	.95087	.93788	.90447	.83304	.75366	.55209
					1.00307	1.00617	1.01245	1.01886	1.03207	1.06760	1.15199	1.26188	1.65114
.969	.2294	.9948	.37645	12.38606	.96573	.96245	.95585	.94919	.93572	.90106	.82668	.74351	.52702
					1.00317	1.00638	1.01288	1.01952	1.03322	1.07016	1.15858	1.27530	1.70912
.968	.2331	.9946	.38223	11.91149	.96462	.96122	.95439	.94751	.93356	.89762	.82026	.73314	.49978
					1.00328	1.00659	1.01332	1.02019	1.03438	1.07276	1.16531	1.28924	1.77572
.967	.2368	.9944	.38790	11.46657	.96351	.96000	.95294	.94582	.93140	.89417	.81376	.72256	.46963
					1.00338	1.00680	1.01375	1.02085	1.03554	1.07537	1.17218	1.30375	1.85378
.966	.2405	.9943	.39348	11.04865	.96239	.95877	.95148	.94413	.92923	.89070	.80718	.71174	.43533
					1.00349	1.00701	1.01419	1.02152	1.03670	1.07801	1.17921	1.31888	1.94796
.965	.2440	.9941	.39897	10.65540	.96128	.95755	.95003	.94244	.92705	.88722	.80052	.70066	.39437
					1.00359	1.00723	1.01462	1.02219	1.03788	1.08068	1.18640	1.33467	2.06675
.964	.2476	.9939	.40436	10.28474	.96017	.95632	.94857	.94074	.92486	.88371	.79378	.68930	.33964
					1.00370	1.00744	1.01506	1.02286	1.03905	1.08337	1.19375	1.35119	2.22961
.963	.2511	.9937	.40967	9.93483	.95905	.95509	.94710	.93904	.92267	.88018	.78696	.67764	
					1.00380	1.00765	1.01550	1.02354	1.04024	1.08609	1.20128	1.36849	
.962	.2545	.9936	.41490	9.60403	.95794	.95386	.94564	.93734	.92047	.87663	.78004	.66566	
					1.00391	1.00787	1.01594	1.02421	1.04143	1.08883	1.20898	1.38666	
.961	.2579	.9934	.42005	9.29081	.95683	.95263	.94418	.93563	.91826	.87306	.77302	.65331	
					1.00402	1.00808	1.01638	1.02489	1.04262	1.09161	1.21688	1.40578	
.960	.2613	.9932	.42513	8.99385	.95571	.95140	.94271	.93392	.91605	.86947	.76591	.64057	
					1.00412	1.00830	1.01682	1.02558	1.04383	1.09441	1.22498	1.42595	
.959	.2646	.9930	.43013	8.71196	.95460	.95017	.94124	.93221	.91383	.86586	.75869	.62739	
					1.00423	1.00851	1.01727	1.02626	1.04504	1.09724	1.23329	1.44729	
.958	.2679	.9929	.43506	8.44404	.95348	.94893	.93977	.93049	.91161	.86222	.75136	.61372	
					1.00434	1.00873	1.01771	1.02695	1.04625	1.10010	1.24181	1.46994	
.957	.2711	.9927	.43992	8.18911	.95236	.94770	.93829	.92877	.90937	.85856	.74392	.59950	
					1.00444	1.00895	1.01816	1.02764	1.04747	1.10300	1.25057	1.49407	
.956	.2744	.9925	.44471	7.94627	.95125	.94646	.93681	.92705	.90713	.85488	.73635	.58464	
					1.00455	1.00916	1.01860	1.02833	1.04870	1.10592	1.25958	1.51987	
.955	.2776	.9924	.44944	7.71470	.95013	.94523	.93534	.92532	.90488	.85117	.72865	.56906	
					1.00466	1.00938	1.01905	1.02902	1.04994	1.10887	1.26884	1.54759	
.954	.2807	.9922	.45411	7.49365	.94901	.94399	.93386	.92359	.90262	.84744	.72082	.55263	
					1.00476	1.00960	1.01950	1.02972	1.05118	1.11186	1.27838	1.57754	
.953	.2838	.9920	.45872	7.28246	.94789	.94275	.93237	.92185	.90036	.84369	.71284	.53518	
					1.00487	1.00982	1.01995	1.03042	1.05243	1.11489	1.28821	1.61011	
.952	.2869	.9918	.46327	7.08048	.94677	.94151	.93089	.92011	.89809	.83990	.70470	.51650	
					1.00498	1.01004	1.02040	1.03112	1.05369	1.11794	1.29835	1.64581	
.951	.2900	.9917	.46776	6.88715	.94566	.94027	.92940	.91837	.89581	.83609	.69640	.49628	
					1.00509	1.01026	1.02086	1.03183	1.05495	1.12103	1.30882	1.68533	
.950	.2930	.9915	.47220	6.70194	.94454	.93903	.92791	.91662	.89352	.83225	.68793	.47403	
					1.00519	1.01048	1.02131	1.03253	1.05622	1.12416	1.31964	1.72966	
.949	.2960	.9913	.47658	6.52436	.94342	.93779	.92642	.91487	.89122	.82839	.67927	.44898	
					1.00530	1.01070	1.02177	1.03324	1.05750	1.12733	1.33083	1.78025	
.948	.2990	.9911	.48091	6.35399	.94230	.93655	.92492	.91312	.88892	.82449	.67040	.41966	
					1.00541	1.01092	1.02223	1.03396	1.05878	1.13053	1.34243	1.83947	
.947	.3019	.9910	.48519	6.19039	.94117	.93530	.92343	.91136	.88661	.82057	.66132	.38257	
					1.00552	1.01114	1.02269	1.03467	1.06008	1.13377	1.35446	1.91185	
.946	.3049	.9908	.48943	6.03317	.94005	.93406	.92193	.90960	.88429	.81661	.65199	.32136	
					1.00563	1.01136	1.02315	1.03539	1.06138	1.13705	1.36695	2.00978	
.945	.3078	.9906	.49361	5.88200	.93893	.93281	.92043	.90783	.88196	.81262	.64242		
					1.00574	1.01158	1.02361	1.03611	1.06269	1.14038	1.37996		
.944	.3107	.9904	.49774	5.73653	.93781	.93156	.91892	.90606	.87962	.80860	.63256		
					1.00585	1.01180	1.02407	1.03683	1.06400	1.14374	1.39350		
.943	.3135	.9903	.50184	5.59648	.93668	.93032	.91742	.90428	.87727	.80455	.62239		
					1.00595	1.01202	1.02453	1.03756	1.06533	1.14715	1.40765		
.942	.3164	.9901	.50588	5.46153	.93556	.92907	.91591	.90250	.87491	.80046	.61187		
					1.00606	1.01225	1.02500	1.03829	1.06666	1.15060	1.42246		
.941	.3192	.9899	.50988	5.33145	.93444	.92782	.91439	.90072	.87254	.79633	.60098		
					1.00617	1.01247	1.02547	1.03902	1.06800	1.15410	1.43798		

TABLE II. Specific Heat Ratio = 1.20

R₂ and TPR versus Loss Coefficient K

p/p,	MACH	T/T,	GAMMA	K_crit	.1	.2	.4	.6	.8	1.0	2.0	3.0	4.0
.940	.3220	.9897	.51384	5.20597	.93331	.92656	.91288	.89893	.88470	.87017	.79217	.70191	.58967
					1.00628	1.01270	1.02594	1.03976	1.05421	1.06935	1.15765	1.27673	1.45431
.939	.3247	.9896	.51776	5.08487	.93219	.92531	.91136	.89714	.88262	.86778	.78797	.69505	.57787
					1.00639	1.01292	1.02641	1.04050	1.05524	1.07070	1.16124	1.28446	1.47152
.938	.3275	.9894	.52164	4.96793	.93106	.92406	.90984	.89534	.88053	.86539	.78373	.68807	.56552
					1.00650	1.01315	1.02688	1.04124	1.05628	1.07207	1.16489	1.29239	1.48973
.937	.3302	.9892	.52548	4.85494	.92994	.92280	.90832	.89354	.87843	.86298	.77945	.68094	.55253
					1.00661	1.01337	1.02735	1.04198	1.05732	1.07344	1.16858	1.30055	1.50908
.936	.3329	.9890	.52928	4.74573	.92881	.92155	.90680	.89173	.87633	.86056	.77513	.67367	.53878
					1.00672	1.01360	1.02782	1.04273	1.05837	1.07482	1.17233	1.30895	1.52972
.935	.3356	.9889	.53304	4.64011	.92768	.92029	.90527	.88992	.87422	.85814	.77077	.66624	.52414
					1.00683	1.01382	1.02830	1.04348	1.05943	1.07621	1.17613	1.31760	1.55187
.934	.3383	.9887	.53676	4.53792	.92655	.91903	.90374	.88810	.87210	.85570	.76635	.65863	.50838
					1.00695	1.01405	1.02878	1.04423	1.06049	1.07761	1.17999	1.32651	1.57579
.933	.3410	.9885	.54045	4.43900	.92543	.91777	.90221	.88628	.86998	.85325	.76190	.65085	.49120
					1.00706	1.01428	1.02925	1.04499	1.06155	1.07902	1.18391	1.33571	1.60183
.932	.3436	.9883	.54410	4.34319	.92430	.91651	.90067	.88446	.86784	.85079	.75739	.64286	.47214
					1.00717	1.01451	1.02973	1.04575	1.06262	1.08044	1.18789	1.34522	1.63051
.931	.3462	.9882	.54772	4.25037	.92317	.91525	.89913	.88263	.86570	.84832	.75283	.63466	.45039
					1.00728	1.01473	1.03022	1.04651	1.06370	1.08187	1.19193	1.35505	1.66257
.930	.3488	.9880	.55130	4.16040	.92204	.91398	.89759	.88079	.86355	.84584	.74822	.62622	.42432
					1.00739	1.01496	1.03070	1.04728	1.06478	1.08331	1.19604	1.36524	1.69924
.929	.3514	.9878	.55485	4.07316	.92091	.91272	.89605	.87895	.86140	.84334	.74356	.61753	.38952
					1.00750	1.01519	1.03118	1.04805	1.06587	1.08476	1.20021	1.37581	1.74296
.928	.3540	.9876	.55837	3.98854	.91977	.91145	.89450	.87710	.85923	.84084	.73884	.60855	
					1.00761	1.01542	1.03167	1.04882	1.06696	1.08621	1.20445	1.38679	
.927	.3566	.9874	.56185	3.90641	.91864	.91018	.89295	.87525	.85706	.83832	.73406	.59926	
					1.00773	1.01565	1.03216	1.04960	1.06806	1.08768	1.20877	1.39823	
.926	.3591	.9873	.56530	3.82668	.91751	.90892	.89139	.87340	.85488	.83579	.72921	.58961	
					1.00784	1.01588	1.03265	1.05037	1.06917	1.08916	1.21315	1.41016	
.925	.3616	.9871	.56872	3.74926	.91638	.90765	.88984	.87153	.85269	.83325	.72431	.57957	
					1.00795	1.01612	1.03314	1.05116	1.07028	1.09065	1.21762	1.42264	
.924	.3642	.9869	.57211	3.67405	.91524	.90638	.88828	.86967	.85049	.83069	.71933	.56908	
					1.00806	1.01635	1.03363	1.05194	1.07140	1.09215	1.22217	1.43572	
.923	.3667	.9867	.57547	3.60095	.91411	.90510	.88672	.86779	.84828	.82812	.71429	.55808	
					1.00818	1.01658	1.03412	1.05273	1.07253	1.09366	1.22680	1.44947	
.922	.3691	.9866	.57880	3.52990	.91298	.90383	.88515	.86591	.84607	.82554	.70917	.54647	
					1.00829	1.01681	1.03462	1.05353	1.07366	1.09518	1.23152	1.46399	
.921	.3716	.9864	.58210	3.46079	.91184	.90255	.88358	.86403	.84384	.82294	.70397	.53415	
					1.00840	1.01705	1.03512	1.05432	1.07480	1.09671	1.23632	1.47937	
.920	.3741	.9862	.58538	3.39358	.91070	.90128	.88201	.86214	.84160	.82033	.69869	.52097	
					1.00852	1.01728	1.03562	1.05512	1.07595	1.09825	1.24123	1.49575	
.919	.3765	.9860	.58862	3.32817	.90957	.90000	.88043	.86024	.83936	.81771	.69332	.50670	
					1.00863	1.01752	1.03612	1.05593	1.07710	1.09981	1.24623	1.51330	
.918	.3790	.9858	.59184	3.26450	.90843	.89872	.87885	.85834	.83711	.81507	.68786	.49103	
					1.00874	1.01775	1.03662	1.05674	1.07826	1.10137	1.25134	1.53225	
.917	.3814	.9857	.59503	3.20251	.90729	.89744	.87727	.85643	.83484	.81241	.68231	.47347	
					1.00886	1.01799	1.03712	1.05755	1.07942	1.10295	1.25656	1.55293	
.916	.3838	.9855	.59819	3.14214	.90615	.89616	.87568	.85451	.83257	.80974	.67665	.45306	
					1.00897	1.01822	1.03763	1.05836	1.08060	1.10454	1.26189	1.57583	
.915	.3862	.9853	.60133	3.08333	.90501	.89487	.87409	.85259	.83028	.80705	.67089	.42780	
					1.00909	1.01846	1.03814	1.05918	1.08178	1.10615	1.26734	1.60179	
.914	.3886	9851	.60444	3.02602	.90387	.89359	.87250	.85067	.82799	.80435	.66501	.39045	
					1.00920	1.01870	1.03865	1.06001	1.08297	1.10776	1.27292	1.63270	
.913	.3910	.9849	.60753	2.97015	.90273	.89230	.87091	.84873	.82568	.80163	.65901		
					1.00932	1.01893	1.03916	1.06083	1.08416	1.10939	1.27863		
.912	.3933	.9848	.61059	2.91569	.90159	.89101	.86931	.84679	.82336	.79889	.65287		
					1.00943	1.01917	1.03967	1.06166	1.08536	1.11104	1.28449		
.911	.3957	.9846	.61362	2.86257	.90045	.88972	.86770	.84484	.82103	.79614	.64660		
					1.00955	1.01941	1.04019	1.06250	1.08658	1.11269	1.29049		

TABLE II. Specific Heat Ratio = 1.20

p/p,	MACH	T/T,	GAMMA	K_{crit}	.1	.2	.4	.6	.8	.9	1.0	1.5	2.0
.910	.3980	.9844	.61664	2.81076	.89931	.88843	.86609	.84289	.81869	.80618	.79337	.72370	.64017
					1.00966	1.01965	1.04070	1.06334	1.08779	1.10079	1.11436	1.19270	1.29665
.909	.4004	.9842	.61962	2.76021	.89816	.88714	.86448	.84092	.81634	.80362	.79058	.71949	.63358
					1.00978	1.01989	1.04122	1.06418	1.08902	1.10224	1.11605	1.19601	1.30298
.908	.4027	.9840	.62259	2.71087	.89702	.88585	.86287	.83896	.81398	.80104	.78777	.71523	.62681
					1.00989	1.02013	1.04174	1.06503	1.09026	1.10369	1.11774	1.19938	1.30949
.907	.4050	.9839	.62553	2.66271	.89588	.88455	.86125	.83698	.81160	.79845	.78494	.71091	.61985
					1.01001	1.02037	1.04226	1.06588	1.09150	1.10516	1.11946	1.20280	1.31619
.906	.4073	.9837	.62845	2.61569	.89473	.88325	.85962	.83500	.80922	.79584	.78209	.70653	.61268
					1.01013	1.02061	1.04279	1.06674	1.09275	1.10664	1.12118	1.20627	1.32310
.905	.4096	.9835	.63134	2.56976	.89358	.88195	.85800	.83300	.80681	.79321	.77923	.70208	.60527
					1.01024	1.02086	1.04331	1.06760	1.09401	1.10813	1.12293	1.20980	1.33023
.904	.4119	.9833	.63421	2.52491	.89244	.88065	.85637	.83100	.80440	.79057	.77634	.69757	.59761
					1.01036	1.02110	1.04384	1.06846	1.09528	1.10963	1.12469	1.21339	1.33760
.903	.4141	.9831	.63706	2.48108	.89129	.87935	.85473	.82900	.80197	.78791	.77343	.69300	.58966
					1.01048	1.02134	1.04437	1.06933	1.09656	1.11114	1.12646	1.21704	1.34523
.902	.4164	.9830	.63989	2.43826	.89014	.87805	.85309	.82698	.79953	.78523	.77049	.68835	.58139
					1.01059	1.02159	1.04490	1.07021	1.09784	1.11267	1.12825	1.22076	1.35314
.901	.4187	.9828	.64270	2.39640	.88899	.87674	.85145	.82496	.79708	.78254	.76754	.68362	.57276
					1.01071	1.02183	1.04544	1.07108	1.09914	1.11421	1.13006	1.22454	1.36136
.900	.4209	.9826	.64549	2.35548	.88784	.87544	.84980	.82293	.79461	.77982	.76456	.67882	.56370
					1.01083	1.02208	1.04597	1.07197	1.10044	1.11576	1.13188	1.22839	1.36992
.899	.4231	.9824	.64825	2.31547	.88669	.87413	.84815	.82089	.79213	.77709	.76156	.67393	.55417
					1.01095	1.02232	1.04651	1.07285	1.10176	1.11732	1.13373	1.23231	1.37887
.898	.4254	.9822	.65100	2.27634	.88554	.87282	.84649	.81884	.78963	.77434	.75853	.66895	.54405
					1.01106	1.02257	1.04705	1.07375	1.10308	1.11890	1.13559	1.23631	1.38825
.897	.4276	.9820	.65372	2.23806	.88439	.87150	.84483	.81679	.78712	.77157	.75548	.66387	.53325
					1.01118	1.02282	1.04759	1.07465	1.10441	1.12049	1.13747	1.24039	1.39810
.896	.4298	.9819	.65643	2.20062	.88324	.87019	.84316	.81472	.78459	.76878	.75240	.65870	.52159
					1.01130	1.02306	1.04813	1.07555	1.10576	1.12209	1.13936	1.24455	1.40852
.895	.4320	.9817	.65911	2.16397	.88208	.86887	.84149	.81264	.78204	.76597	.74929	.65341	.50884
					1.01142	1.02331	1.04868	1.07646	1.10711	1.12371	1.14128	1.24880	1.41959
.894	.4342	.9815	.66178	2.12811	.88093	.86756	.83982	.81056	.77948	.76313	.74616	.64801	.49462
					1.01154	1.02356	1.04923	1.07737	1.10848	1.12534	1.14322	1.25314	1.43145
.893	.4364	.9813	.66442	2.09300	.87977	.86624	.83814	.80847	.77690	.76028	.74299	.64249	.47830
					1.01166	1.02381	1.04978	1.07829	1.10985	1.12699	1.14517	1.25758	1.44428
.892	.4385	.9811	.66705	2.05864	.87862	.86492	.83645	.80636	.77431	.75740	.73980	.63683	.45854
					1.01178	1.02406	1.05033	1.07921	1.11124	1.12865	1.14715	1.26213	1.45840
.891	.4407	.9809	.66966	2.02499	.87746	.86359	.83476	.80425	.77169	.75449	.73657	.63103	.43148
					1.01190	1.02431	1.05089	1.08014	1.11264	1.13033	1.14915	1.26678	1.47440
.890	.4429	.9808	.67225	1.99203	.87630	.86227	.83307	.80213	.76906	.75157	.73332	.62508	
					1.01202	1.02457	1.05144	1.08107	1.11404	1.13203	1.15117	1.27155	
.889	.4450	.9806	.67482	1.95975	.87514	.86094	.83137	.79999	.76641	.74862	.73003	.61895	
					1.01214	1.02482	1.05200	1.08201	1.11546	1.13374	1.15322	1.27644	
.888	.4472	.9804	.67737	1.92813	.87398	.85961	.82966	.79785	.76374	.74564	.72670	.61264	
					1.01226	1.02507	1.05256	1.08296	1.11690	1.13546	1.15528	1.28146	
.887	.4493	.9802	.67991	1.89714	.87282	.85828	.82795	.79570	.76105	.74264	.72334	.60613	
					1.01238	1.02532	1.05313	1.08391	1.11834	1.13721	1.15737	1.28662	
.886	.4514	.9800	.68242	1.86678	.87166	.85694	.82624	.79353	.75834	.73960	.71994	.59939	
					1.01250	1.02558	1.05369	1.08486	1.11980	1.13897	1.15949	1.29194	
.885	.4535	.9798	.68492	1.83702	.87050	.85561	.82451	.79135	.75561	.73654	.71651	.59240	
					1.01262	1.02583	1.05426	1.08583	1.12127	1.14075	1.16163	1.29742	
.884	.4557	.9797	.68740	1.80785	.86934	.85427	.82279	.78917	.75286	.73345	.71303	.58514	
					1.01274	1.02609	1.05483	1.08680	1.12275	1.14255	1.16380	1.30308	
.883	.4578	.9795	.68987	1.77926	.86817	.85293	.82106	.78697	.75008	.73034	.70951	.57755	
					1.01286	1.02634	1.05541	1.08777	1.12424	1.14436	1.16599	1.30893	
.882	.4599	.9793	.69232	1.75122	.86701	.85159	.81932	.78475	.74729	.72718	.70595	.56960	
					1.01299	1.02660	1.05598	1.08875	1.12575	1.14620	1.16822	1.31499	
.881	.4620	.9791	.69475	1.72372	.86584	.85025	.81757	.78253	.74447	.72400	.70234	.56123	
					1.01311	1.02686	1.05656	1.08974	1.12727	1.14806	1.17047	1.32129	

TABLE II. Specific Heat Ratio = 1.20

p/p,	MACH	T/T,	GAMMA	K_{crit}	.1	.2	.4	.6	.7	.8	.9	1.0	1.5
.880	.4640	.9789	.69716	1.69676	.86467	.84890	.81582	.78029	.76141	.74162	.72078	.69869	.55236
					1.01323	1.02712	1.05714	1.09073	1.10914	1.12881	1.14994	1.17275	1.32785
.879	.4661	.9787	.69956	1.67031	.86351	.84755	.81407	.77805	.75887	.73875	.71753	.69498	.54291
					1.01335	1.02738	1.05773	1.09173	1.11039	1.13036	1.15183	1.17506	1.33470
.878	.4682	.9785	.70194	1.64436	.86234	.84620	.81230	.77578	.75631	.73585	.71424	.69123	.53272
					1.01348	1.02764	1.05831	1.09274	1.11166	1.13193	1.15375	1.17741	1.34189
.877	.4703	.9784	.70430	1.61890	.86117	.84485	.81054	.77351	.75373	.73293	.71091	.68742	.52160
					1.01360	1.02790	1.05890	1.09375	1.11293	1.13351	1.15570	1.17978	1.34947
.876	.4723	.9782	.70665	1.59392	.86000	.84350	.80876	.77122	.75113	.72998	.70755	.68355	.50925
					1.01372	1.02816	1.05949	1.09477	1.11422	1.13510	1.15766	1.18220	1.35750
.875	.4744	.9780	.70899	1.56940	.85883	.84214	.80698	.76891	.74852	.72700	.70414	.67962	.49514
					1.01384	1.02842	1.06009	1.09580	1.11551	1.13672	1.15965	1.18464	1.36610
.874	.4764	.9778	.71130	1.54534	.85765	.84078	.80519	.76660	.74588	.72399	.70069	.67563	.47827
					1.01397	1.02868	1.06068	1.09683	1.11682	1.13835	1.16167	1.18713	1.37541
.873	.4785	.9776	.71360	1.52172	.85648	.83942	.80340	.76426	.74322	.72095	.69719	.67157	.45590
					1.01409	1.02894	1.06128	1.09787	1.11814	1.13999	1.16371	1.18965	1.38575
.872	.4805	.9774	.71589	1.49853	.85531	.83805	.80159	.76191	.74054	.71788	.69365	.66744	
					1.01422	1.02921	1.06189	1.09892	1.11947	1.14166	1.16578	1.19222	
.871	.4826	.9772	.71816	1.47576	.85413	.83669	.79979	.75955	.73783	.71477	.69005	.66324	
					1.01434	1.02947	1.06249	1.09998	1.12081	1.14334	1.16787	1.19482	
.870	.4846	.9771	.72042	1.45341	.85295	.83532	.79797	.75717	.73510	.71163	.68641	.65895	
					1.01447	1.02974	1.06310	1.10105	1.12217	1.14504	1.16999	1.19747	
.869	.4866	.9769	.72266	1.43145	.85178	.83394	.79615	.75477	.73235	.70845	.68271	.65459	
					1.01459	1.03000	1.06371	1.10212	1.12353	1.14676	1.17214	1.20017	
.868	.4886	.9767	.72489	1.40989	.85060	.83257	.79432	.75236	.72958	.70524	.67896	.65013	
					1.01472	1.03027	1.06432	1.10320	1.12491	1.14850	1.17433	1.20291	
.867	.4906	.9765	.72710	1.38871	.84942	.83119	.79248	.74993	.72677	.70199	.67514	.64557	
					1.01484	1.03054	1.06494	1.10429	1.12630	1.15026	1.17654	1.20571	
.866	.4926	.9763	.72930	1.36790	.84824	.82981	.79063	.74748	.72394	.69869	.67127	.64092	
					1.01497	1.03081	1.06556	1.10538	1.12771	1.15204	1.17879	1.20855	
.865	.4946	.9761	.73148	1.34746	.84705	.82843	.78878	.74502	.72108	.69536	.66732	.63615	
					1.01509	1.03108	1.06618	1.10649	1.12913	1.15384	1.18107	1.21146	
.864	.4966	.9759	.73365	1.32737	.84587	.82705	.78692	.74253	.71820	.69198	.66331	.63127	
					1.01522	1.03135	1.06681	1.10760	1.13056	1.15566	1.18338	1.21442	
.863	.4986	.9757	.73580	1.30764	.84469	.82566	.78505	.74002	.71528	.68855	.65922	.62625	
					1.01535	1.03162	1.06744	1.10873	1.13201	1.15751	1.18574	1.21744	
.862	.5006	.9756	.73794	1.28824	.84350	.82427	.78317	.73750	.71233	.68508	.65505	.62110	
					1.01547	1.03189	1.06807	1.10986	1.13347	1.15938	1.18813	1.22053	
.861	.5026	.9754	.74007	1.26917	.84231	.82288	.78128	.73495	.70935	.68155	.65080	.61579	
					1.01560	1.03216	1.06871	1.11100	1.13495	1.16128	1.19056	1.22369	
.860	.5045	.9752	.74218	1.25044	.84113	.82148	.77939	.73238	.70634	.67798	.64646	.61032	
					1.01573	1.03243	1.06935	1.11215	1.13644	1.16320	1.19304	1.22692	
.859	.5065	.9750	.74428	1.23202	.83994	.82008	.77748	.72979	.70329	.67434	.64202	.60466	
					1.01586	1.03271	1.06999	1.11331	1.13795	1.16514	1.19556	1.23024	
.858	.5085	.9748	.74637	1.21391	.83875	.81868	.77557	.72718	.70021	.67065	.63748	.59880	
					1.01598	1.03298	1.07064	1.11449	1.13948	1.16712	1.19813	1.23364	
.857	.5104	.9746	.74844	1.19610	.83756	.81727	.77365	.72454	.69709	.66689	.63283	.59270	
					1.01611	1.03326	1.07129	1.11567	1.14102	1.16912	1.20075	1.23713	
.856	.5124	.9744	.75050	1.17859	.83636	.81587	.77171	.72188	.69393	.66307	.62805	.58635	
					1.01624	1.03354	1.07194	1.11686	1.14258	1.17116	1.20342	1.24072	
.855	.5143	.9742	.75254	1.16138	.83517	.81445	.76977	.71920	.69072	.65917	.62315	.57970	
					1.01637	1.03381	1.07259	1.11806	1.14416	1.17322	1.20614	1.24442	
.854	.5163	9740	.75458	1.14444	.83397	.81304	.76782	.71648	.68748	.65521	.61811	.57271	
					1.01650	1.03409	1.07326	1.11928	1.14576	1.17532	1.20892	1.24824	
.853	.5182	.9738	.75660	1.12779	.83278	.81162	.76586	.71374	.68419	.65116	.61290	.56532	
					1.01663	1.03437	1.07392	1.12050	1.14737	1.17745	1.21177	1.25220	
.852	.5201	.9737	.75860	1.11141	.83158	.81020	.76388	.71098	.68085	.64703	.60753	.55745	
					1.01676	1.03465	1.07459	1.12174	1.14901	1.17962	1.21468	1.25629	
.851	.5221	.9735	.76060	1.09529	.83038	.80878	.76190	.70818	.67746	.64281	.60197	.54900	
					1.01689	1.03493	1.07526	1.12299	1.15067	1.18182	1.21767	1.26056	

TABLE II. Specific Heat Ratio = 1.20

R₂ and TPR versus Loss Coefficient K

p/p_t	MACH	T/T_t	GAMMA	K_crit	.1	.2	.4	.5	.6	.7	.8	.9	1.0
.850	.5240	.9733	.76258	1.07944	.82918	.80735	.75991	.73373	.70535	.67402	.63849	.59619	.53983
					1.01702	1.03521	1.07593	1.09896	1.12425	1.15235	1.18406	1.22073	1.26501
.849	.5259	.9731	.76455	1.06385	.82798	.80592	.75790	.73135	.70249	.67053	.63407	.59017	.52970
					1.01715	1.03550	1.07661	1.09991	1.12553	1.15405	1.18635	1.22387	1.26968
.848	.5278	.9729	.76650	1.04850	.82678	.80449	.75588	.72895	.69960	.66698	.62953	.58388	.51826
					1.01728	1.03578	1.07730	1.10086	1.12681	1.15578	1.18868	1.22711	1.27462
.847	.5297	.9727	.76845	1.03340	.82557	.80305	.75385	.72653	.69667	.66336	.62488	.57729	.50481
					1.01741	1.03607	1.07798	1.10182	1.12812	1.15753	1.19105	1.23044	1.27988
.846	.5316	.9725	.77038	1.01854	.82436	.80161	.75181	.72408	.69371	.65968	.62009	.57032	.48777
					1.01754	1.03635	1.07868	1.10278	1.12943	1.15930	1.19347	1.23388	1.28558
.845	.5335	.9723	.77230	1.00392	.82316	.80016	.74976	.72162	.69071	.65594	.61515	.56293	.46022
					1.01767	1.03664	1.07937	1.10376	1.13076	1.16111	1.19594	1.23744	1.29198
.844	.5354	.9721	.77421	0.98952	.82195	.79871	.74769	.71913	.68767	.65211	.61006	.55502	
					1.01780	1.03693	1.08007	1.10474	1.13211	1.16294	1.19847	1.24113	
.843	.5373	.9719	.77610	0.97535	.82074	.79726	.74562	.71662	.68459	.64822	.60479	.54645	
					1.01794	1.03721	1.08078	1.10573	1.13347	1.16479	1.20106	1.24498	
.842	.5392	.9717	.77799	0.96141	.81953	.79581	.74352	.71409	.68147	.64423	.59931	.53706	
					1.01807	1.03750	1.08149	1.10673	1.13484	1.16668	1.20371	1.24900	
.841	.5411	.9716	.77986	0.94768	.81831	.79435	.74142	.71153	.67831	.64016	.59362	.52654	
					1.01820	1.03779	1.08220	1.10774	1.13623	1.16860	1.20643	1.25323	
.840	.5430	.9714	.78172	0.93416	.81710	.79288	.73930	.70895	.67509	.63600	.58768	.51436	
					1.01833	1.03809	1.08292	1.10875	1.13764	1.17056	1.20922	1.25772	
.839	.5449	.9712	.78357	0.92086	.81588	.79142	.73716	.70634	.67183	.63173	.58145	.49942	
					1.01847	1.03838	1.08364	1.10978	1.13907	1.17255	1.21210	1.26253	
.838	.5468	.9710	.78541	0.90775	.81466	.78994	.73501	.70370	.66852	.62734	.57489	.47812	
					1.01860	1.03867	1.08437	1.11082	1.14051	1.17457	1.21506	1.26784	
.837	.5486	.9708	.78723	0.89485	.81344	.78847	.73285	.70104	.66515	.62284	.56794		
					1.01874	1.03897	1.08511	1.11186	1.14198	1.17663	1.21811		
.836	.5505	.9706	.78905	0.88215	.81222	.78699	.73066	.69834	.66172	.61820	.56051		
					1.01887	1.03926	1.08584	1.11292	1.14346	1.17874	1.22128		
.835	.5524	.9704	.79085	0.86963	.81100	.78550	.72847	.69562	.65823	.61342	.55250		
					1.01900	1.03956	1.08659	1.11398	1.14496	1.18089	1.22457		
.834	.5542	.9702	.79264	0.85731	.80978	.78401	.72625	.69286	.65468	.60847	.54374		
					1.01914	1.03986	1.08734	1.11506	1.14649	1.18309	1.22800		
.833	.5561	.9700	.79442	0.84517	.80855	.78252	.72402	.69007	.65106	.60335	.53400		
					1.01927	1.04016	1.08809	1.11615	1.14804	1.18534	1.23159		
.832	.5579	.9698	.79619	0.83322	.80732	.78102	.72177	.68725	.64736	.59803	.52284		
					1.01941	1.04046	1.08885	1.11725	1.14961	1.18764	1.23538		
.831	.5598	.9696	.79795	0.82144	.80609	.77952	.71950	.68438	.64358	.59248	.50940		
					1.01955	1.04076	1.08962	1.11836	1.15120	1.19000	1.23943		
.830	.5616	.9694	.79970	0.80984	.80486	.77801	.71721	.68148	.63972	.58667	.49134		
					1.01968	1.04106	1.09039	1.11948	1.15282	1.19242	1.24383		
.829	.5635	.9692	.80144	0.79841	.80363	.77650	.71490	.67854	.63577	.58057			
					1.01982	1.04137	1.09117	1.12061	1.15447	1.19491			
.828	.5653	.9690	.80316	0.78715	.80239	.77498	.71258	.67556	.63172	.57412			
					1.01996	1.04167	1.09195	1.12176	1.15615	1.19748			
.827	.5671	.9688	.80488	0.77606	.80116	.77346	.71023	.67254	.62757	.56725			
					1.02009	1.04198	1.09274	1.12292	1.15785	1.20013			
.826	.5690	.9686	.80658	0.76513	.79992	.77194	.70785	.66947	.62329	.55988			
					1.02023	1.04228	1.09354	1.12410	1.15959	1.20288			
.825	.5708	.9684	.80828	0.75436	.79868	.77040	.70546	.66635	.61889	.55187			
					1.02037	1.04259	1.09434	1.12529	1.16136	1.20573			
.824	5726	.9683	.80996	0.74374	.79744	.76887	.70304	.66317	.61435	.54304			
					1.02051	1.04290	1.09515	1.12649	1.16317	1.20871			
.823	.5745	.9681	.81164	0.73328	.79619	.76732	.70060	.65995	.60966	.53305			
					1.02065	1.04321	1.09597	1.12771	1.16501	1.21184			
.822	.5763	.9679	.81330	0.72298	.79495	.76578	.69813	.65666	.60480	.52133			
					1.02079	1.04352	1.09679	1.12895	1.16690	1.21515			
.821	.5781	.9677	.81495	0.71282	.79370	.76422	.69564	.65332	.59975	.50651			
					1.02093	1.04384	1.09762	1.13020	1.16882	1.21871			

TABLE II. Specific Heat Ratio = 1.20

p/p_t	MACH	T/T_t	GAMMA	K_crit	R_2 and TPR versus Loss Coefficient K								
					.08	.09	.1	.2	.3	.4	.5	.6	.7
.820	.5799	.9675	.81660	0.70280	.79812 1.01670	.79530 1.01887	.79245 1.02106	.76266 1.04415	.72995 1.06972	.69312 1.09846	.64991 1.13147	.59448 1.17080	.48268 1.22264
.819	.5817	.9673	.81823	0.69293	.79692 1.01681	.79407 1.01900	.79120 1.02120	.76110 1.04447	.72797 1.07026	.69057 1.09931	.64643 1.13275	.58897 1.17282	
.818	.5835	.9671	.81985	0.68321	.79572 1.01692	.79284 1.01912	.78994 1.02134	.75953 1.04478	.72598 1.07081	.68799 1.10016	.64287 1.13406	.58317 1.17491	
.817	.5853	.9669	.82147	0.67362	.79452 1.01703	.79162 1.01925	.78869 1.02149	.75795 1.04510	.72398 1.07136	.68538 1.10103	.63924 1.13539	.57704 1.17705	
.816	.5871	.9667	.82307	0.66416	.79332 1.01714	.79039 1.01937	.78743 1.02163	.75636 1.04542	.72197 1.07191	.68274 1.10190	.63551 1.13674	.57051 1.17926	
.815	.5889	.9665	.82466	0.65484	.79211 1.01725	.78915 1.01950	.78617 1.02177	.75477 1.04574	.71994 1.07247	.68006 1.10278	.63170 1.13811	.56350 1.18154	
.814	.5907	.9663	.82624	0.64565	.79091 1.01737	.78792 1.01963	.78490 1.02191	.75318 1.04606	.71789 1.07302	.67735 1.10366	.62779 1.13950	.55587 1.18391	
.813	.5925	.9661	.82782	0.63659	.78970 1.01748	.78668 1.01975	.78363 1.02205	.75158 1.04639	.71583 1.07359	.67460 1.10456	.62377 1.14092	.54743 1.18639	
.812	.5943	.9659	.82938	0.62766	.78849 1.01759	.78545 1.01988	.78237 1.02220	.74997 1.04671	.71375 1.07416	.67181 1.10547	.61962 1.14236	.53788 1.18898	
.811	.5961	.9657	.83094	0.61886	.78728 1.01770	.78420 1.02001	.78110 1.02234	.74835 1.04704	.71166 1.07473	.66898 1.10639	.61535 1.14383	.52660 1.19172	
.810	.5979	.9655	.83248	0.61017	.78607 1.01781	.78296 1.02014	.77982 1.02248	.74672 1.04737	.70955 1.07530	.66610 1.10732	.61093 1.14533	.51219 1.19465	
.809	.5996	.9653	.83401	0.60161	.78485 1.01792	.78172 1.02026	.77855 1.02263	.74509 1.04770	.70742 1.07588	.66318 1.10825	.60635 1.14686	.48755 1.19791	
.808	.6014	.9651	.83554	0.59316	.78364 1.01804	.78047 1.02039	.77727 1.02277	.74345 1.04803	.70527 1.07646	.66022 1.10920	.60159 1.14843		
.807	.6032	.9649	.83706	0.58483	.78242 1.01815	.77922 1.02052	.77599 1.02292	.74181 1.04836	.70311 1.07705	.65720 1.11017	.59663 1.15003		
.806	.6050	.9647	.83856	0.57662	.78120 1.01826	.77797 1.02065	.77471 1.02306	.74015 1.04870	.70092 1.07764	.65412 1.11114	.59143 1.15167		
.805	.6067	.9645	.84006	0.56852	.77997 1.01838	.77672 1.02078	.77342 1.02321	.73849 1.04903	.69871 1.07824	.65099 1.11213	.58596 1.15335		
.804	.6085	.9643	.84155	0.56053	.77875 1.01849	.77546 1.02091	.77214 1.02335	.73682 1.04937	.69649 1.07884	.64780 1.11313	.58017 1.15509		
.803	.6103	.9641	.84303	0.55265	.77752 1.01860	.77420 1.02104	.77085 1.02350	.73514 1.04971	.69423 1.07945	.64454 1.11414	.57400 1.15687		
.802	.6120	.9639	.84450	0.54488	.77630 1.01872	.77294 1.02117	.76955 1.02365	.73346 1.05005	.69196 1.08006	.64121 1.11517	.56737 1.15871		
.801	.6138	.9637	.84596	0.53722	.77507 1.01883	.77168 1.02130	.76826 1.02380	.73176 1.05039	.68966 1.08068	.63780 1.11621	.56013 1.16063		
.800	.6156	.9635	.84741	0.52966	.77383 1.01895	.77041 1.02143	.76696 1.02394	.73005 1.05074	.68734 1.08130	.63432 1.11727	.55211 1.16261		
.799	.6173	.9633	.84885	0.52220	.77260 1.01906	.76915 1.02156	.76566 1.02409	.72834 1.05108	.68499 1.08192	.63075 1.11834	.54300 1.16470		
.798	.6191	.9631	.85028	0.51484	.77136 1.01918	.76788 1.02170	.76435 1.02424	.72661 1.05143	.68262 1.08256	.62708 1.11943	.53217 1.16689		
.797	.6208	.9629	.85171	0.50758	.77012 1.01930	.76660 1.02183	.76304 1.02439	.72488 1.05178	.68022 1.08320	.62330 1.12054	.51806 1.16925		
.796	.6226	.9627	.85312	0.50042	.76888 1.01941	.76533 1.02196	.76173 1.02454	.72314 1.05213	.67778 1.08384	.61942 1.12167	.49030 1.17188		
.795	.6243	.9625	.85453	0.49335	.76764 1.01953	.76405 1.02209	.76042 1.02469	.72138 1.05249	.67532 1.08449	.61540 1.12282			
.794	.6261	.9623	.85593	0.48638	.76639 1.01965	.76277 1.02223	.75910 1.02484	.71962 1.05284	.67282 1.08515	.61125 1.12400			
.793	.6278	.9621	.85732	0.47950	.76515 1.01976	.76149 1.02236	.75778 1.02499	.71784 1.05320	.67029 1.08581	.60694 1.12519			
.792	.6295	.9619	.85870	0.47272	.76390 1.01988	.76020 1.02250	.75646 1.02514	.71605 1.05356	.66773 1.08648	.60246 1.12642			
.791	.6313	.9617	.86007	0.46602	.76264 1.02000	.75891 1.02263	.75513 1.02529	.71425 1.05392	.66512 1.08716	.59778 1.12767			

TABLE II. Specific Heat Ratio = 1.20

R₁ and TPR versus Loss Coefficient K

p/p_t	MACH	T/T_t	GAMMA	K_crit	.01	.02	.04	.06	.08	.09	.1	.2	.3
.790	.6330	.9615	.86143	0.45941	.78656	.78308	.77602	.76879	.76139	.75762	.75380	.71244	.66247
					1.00242	1.00487	1.00983	1.01491	1.02012	1.02276	1.02545	1.05428	1.08784
.789	.6347	.9613	.86279	0.45289	.78553	.78202	.77490	.76760	.76013	.75632	.75247	.71062	.65979
					1.00243	1.00489	1.00989	1.01500	1.02023	1.02290	1.02560	1.05465	1.08853
.788	.6365	.9611	.86414	0.44646	.78450	.78096	.77377	.76641	.75887	.75503	.75113	.70878	.65706
					1.00245	1.00492	1.00994	1.01509	1.02035	1.02304	1.02575	1.05502	1.08923
.787	.6382	.9609	.86547	0.44011	.78347	.77991	.77265	.76522	.75761	.75373	.74979	.70693	.65428
					1.00246	1.00495	1.01000	1.01517	1.02047	1.02317	1.02591	1.05539	1.08994
.786	.6399	.9607	.86680	0.43385	.78244	.77885	.77153	.76403	.75634	.75242	.74845	.70507	.65145
					1.00247	1.00497	1.01005	1.01526	1.02059	1.02331	1.02606	1.05576	1.09065
.785	.6416	.9605	.86812	0.42767	.78141	.77778	.77040	·76284	.75507	.75111	.74710	.70319	.64857
					1.00249	1.00500	1.01011	1.01535	1.02071	1.02345	1.02622	1.05613	1.09138
.784	.6434	.9603	.86944	0.42157	.78038	.77672	.76928	.76164	.75380	.74980	.74575	.70130	.64564
					1.00250	1.00503	1.01017	1.01543	1.02083	1.02358	1.02637	1.05651	1.09211
.783	.6451	.9600	.87074	0.41555	.77935	.77566	.76815	.76044	.75253	.74849	.74439	.69939	.64264
					1.00251	1.00506	1.01022	1.01552	1.02095	1.02372	1.02653	1.05689	1.09285
.782	.6468	.9598	.87204	0.40961	.77832	.77460	.76702	.75924	.75125	.74717	.74303	.69747	.63958
					1.00253	1.00508	1.01028	1.01561	1.02107	1.02386	1.02668	1.05727	1.09360
.781	.6485	.9596	.87333	0.40374	.77729	.77354	.76589	.75804	·74997	.74585	.74167	.69553	.63644
					1.00254	1.00511	1.01034	1.01570	1.02119	1.02400	1.02684	1.05766	1.09437
.780	.6502	.9594	.87460	0.39795	.77626	.77247	.76476	.75684	.74869	.74453	.74030	.69358	.63324
					1.00255	1.00514	1.01039	1.01578	1.02132	1.02414	1.02700	1.05805	1.09514
.779	.6519	.9592	.87588	0.39224	.77523	.77141	.76363	.75563	.74740	.74320	.73892	.69161	.62995
					1.00257	1.00517	1.01045	1.01587	1.02144	1.02428	1.02716	1.05843	1.09592
.778	.6536	.9590	.87714	0.38660	.77420	.77035	.76249	·75442	.74612	.74187	.73755	.68962	.62657
					1.00258	1.00519	1.01051	1.01596	1.02156	1.02442	1.02732	1.05883	1.09672
.777	.6554	.9588	.87840	0.38104	.77316	.76928	.76136	.75321	.74482	.74053	.73617	·68761	.62310
					1.00259	1.00522	1.01056	1.01605	1.02168	1.02456	1.02747	1.05922	1.09753
.776	.6571	.9586	.87964	0.37555	.77213	.76822	.76022	.75200	.74353	.73919	.73478	.68558	.61952
					1.00261	1.00525	1.01062	1.01614	1.02181	1.02470	1.02763	1.05962	1.09836
.775	.6588	.9584	.88088	0.37013	.77110	.76715	.75909	.75079	.74223	.73785	.73339	.68353	.61582
					1.00262	1.00528	1.01068	1.01623	1.02193	1.02484	1.02779	1.06002	1.09919
.774	.6605	.9582	.88211	0.36477	.77007	.76608	.75795	.74957	.74093	.73650	.73199	.68145	.61199
					1.00263	1.00531	1.01074	1.01632	1.02205	1.02498	1.02796	1.06043	1.10005
.773	.6622	.9580	.88334	0.35949	.76903	.76502	.75681	.74835	.73962	.73515	.73059	.67936	.60801
					1.00265	1.00533	1.01079	1.01641	1.02218	1.02513	1.02812	1.06084	1.10092
.772	.6639	.9578	.88455	0.35428	.76800	.76395	.75567	·74713	.73831	.73379	.72919	.67725	.60387
					1.00266	1.00536	1.01085	1.01650	1.02230	1.02527	1.02828	1.06125	1.10180
.771	.6656	.9576	.88576	0.34913	.76697	.76288	.75452	.74591	.73700	.73243	.72778	.67511	.59954
					1.00268	1.00539	1.01091	1.01659	1.02243	1.02541	1.02844	1.06166	1.10271
.770	.6673	.9574	.88696	0.34405	.76593	.76181	.75338	.74468	.73569	.73107	.72636	.67294	.59500
					1.00269	1.00542	1.01097	1.01668	1.02255	1.02556	1.02861	1.06208	1.10364
.769	.6690	.9572	.88815	0.33904	.76490	.76074	.75223	·74345	.73437	.72970	.72494	.67075	.59020
					1.00270	1.00545	1.01103	1.01677	1.02268	1.02570	1.02877	1.06250	1.10459
.768	.6706	.9570	.88934	0.33408	.76387	.75967	.75109	.74222	.73304	.72832	.72351	.66854	.58508
					1.00272	1.00547	1.01108	1.01686	1.02281	1.02585	1.02894	1.06293	1.10557
.767	.6723	.9568	.89051	0.32920	.76283	.75860	.74994	·74099	.73171	.72694	.72208	.66629	.57959
					1.00273	1.00550	1.01114	1.01695	1.02293	1.02599	1.02910	1.06336	1.10657
.766	.6740	.9565	.89168	0.32437	.76180	.75753	.74879	·73975	.73038	.72556	.72064	.66401	.57364
					1.00275	1.00552	1.01120	1.01704	1.02306	1.02614	1.02927	1.06379	1.10761
.765	.6757	.9563	.89284	0.31961	.76076	.75645	.74763	.73851	.72905	.72417	.71920	.66171	·56707
					1.00276	1.00556	1.01126	1.01714	1.02319	1.02629	1.02943	1.06423	1.10868
.764	·6774	.9561	.89400	0.31491	.75972	.75538	·74648	.73727	.72771	.72278	.71775	.65936	.55966
					1.00277	1.00559	1.01132	1.01723	1.02332	1.02643	1.02960	1.06467	1.10980
.763	.6791	.9559	.89514	0.31026	.75869	.75431	.74533	.73602	.72636	.72138	.71629	.65699	.55096
					1.00279	1.00562	1.01138	1.01732	1.02344	1.02658	1.02977	1.06512	1.11097
.762	.6808	.9557	.89628	0.30568	.75765	.75323	.74417	.73478	.72501	.71998	.71483	.65457	.53988
					1.00280	1.00564	1.01144	1.01741	1.02357	1.02673	1.02994	1.06557	1.11221
.761	.6824	.9555	.89741	0.30116	.75661	.75216	.74301	.73353	.72366	.71857	.71336	.65212	.52188
					1.00282	1.00567	1.01150	1.01751	1.02370	1.02688	1.03011	1.06602	1.11357

TABLE II. Specific Heat Ratio = 1.20

R₂ and TPR versus Loss Coefficient K

p/p_t	MACH	T/T_t	GAMMA	K_{crit}	.01	.02	.04	.06	.07	.08	.09	.1	.2
.760	.6841	.9553	.89854	0.29669	.75558	.75108	.74185	.73227	.72734	.72230	.71715	.71188	.64962
					1.00283	1.00570	1.01156	1.01760	1.02069	1.02383	1.02703	1.03028	1.06648
.759	.6858	.9551	.89965	0.29228	.75454	.75000	.74069	.73102	.72603	.72094	.71573	.71040	.64708
					1.00284	1.00572	1.01162	1.01769	1.02080	1.02396	1.02718	1.03045	1.06695
.758	.6875	.9549	.90076	0.28793	.75350	.74893	.73953	.72976	.72472	.71957	.71430	.70891	.64449
					1.00286	1.00575	1.01168	1.01779	1.02092	1.02410	1.02733	1.03062	1.06742
.757	.6891	.9547	.90186	0.28363	.75246	.74785	.73836	.72850	.72340	.71820	.71287	.70741	.64185
					1.00287	1.00578	1.01174	1.01788	1.02103	1.02423	1.02748	1.03079	1.06789
.756	.6908	.9545	.90296	0.27938	.75142	.74677	.73719	.72723	.72208	.71682	.71143	.70590	.63915
					1.00289	1.00581	1.01180	1.01797	1.02114	1.02436	1.02763	1.03097	1.06837
.755	.6925	.9542	.90404	0.27519	.75039	.74569	.73603	.72596	.72076	.71544	.70999	.70439	.63640
					1.00290	1.00584	1.01186	1.01807	1.02125	1.02449	1.02779	1.03114	1.06886
.754	.6942	.9540	.90512	0.27106	.74935	.74461	.73485	.72469	.71943	.71405	.70853	.70287	.63358
					1.00291	1.00587	1.01192	1.01816	1.02137	1.02462	1.02794	1.03132	1.06935
.753	.6958	.9538	.90620	0.26697	.74831	.74353	.73368	.72341	.71810	.71266	.70707	.70134	.63070
					1.00293	1.00590	1.01198	1.01826	1.02148	1.02476	1.02809	1.03149	1.06985
.752	.6975	.9536	.90726	0.26294	.74727	.74244	.73251	.72213	.71676	.71126	.70561	.69980	.62774
					1.00294	1.00593	1.01204	1.01835	1.02160	1.02489	1.02825	1.03167	1.07036
.751	.6992	.9534	.90832	0.25896	.74623	.74136	.73133	.72085	.71542	.70985	.70414	.69825	.62469
					1.00296	1.00596	1.01210	1.01845	1.02171	1.02503	1.02840	1.03185	1.07088
.750	.7008	.9532	.90937	0.25502	.74518	.74027	.73015	.71956	.71407	.70844	.70266	.69670	.62156
					1.00297	1.00598	1.01216	1.01855	1.02183	1.02516	1.02856	1.03203	1.07140
.749	.7025	.9530	.91041	0.25114	.74414	.73919	.72897	.71827	.71272	.70703	.70117	.69513	.61833
					1.00299	1.00601	1.01222	1.01864	1.02194	1.02530	1.02872	1.03221	1.07193
.748	.7041	.9528	.91145	0.24731	.74310	.73810	.72779	.71698	.71137	.70560	.69967	.69356	.61499
					1.00300	1.00604	1.01228	1.01874	1.02206	1.02543	1.02888	1.03239	1.07247
.747	.7058	.9525	.91248	0.24353	.74206	.73702	.72660	.71568	.71001	.70417	.69817	.69197	.61153
					1.00301	1.00607	1.01234	1.01884	1.02217	1.02557	1.02903	1.03257	1.07302
.746	.7075	.9523	.91350	0.23979	.74102	.73593	.72541	.71438	.70864	.70274	.69666	.69038	.60793
					1.00303	1.00610	1.01241	1.01893	1.02229	1.02571	1.02919	1.03275	1.07358
.745	.7091	.9521	.91451	0.23610	.73997	.73484	.72422	.71307	.70727	.70130	.69514	.68877	.60416
					1.00304	1.00613	1.01247	1.01903	1.02241	1.02585	1.02935	1.03293	1.07415
.744	.7108	.9519	.91552	0.23246	.73893	.73375	.72303	.71176	.70589	.69985	.69361	.68716	.60022
					1.00306	1.00616	1.01253	1.01913	1.02253	1.02598	1.02951	1.03312	1.07474
.743	.7124	.9517	.91652	0.22886	.73788	.73266	.72183	.71045	.70451	.69839	.69207	.68553	.59606
					1.00307	1.00619	1.01259	1.01923	1.02264	1.02612	1.02968	1.03330	1.07533
.742	.7141	.9515	.91752	0.22531	.73684	.73157	.72064	.70913	.70312	.69692	.69052	.68389	.59166
					1.00309	1.00622	1.01266	1.01933	1.02276	1.02626	1.02984	1.03349	1.07595
.741	.7157	.9513	.91850	0.22181	.73580	.73047	.71944	.70780	.70172	.69545	.68896	.68224	.58694
					1.00310	1.00625	1.01272	1.01943	1.02288	1.02640	1.03000	1.03368	1.07657
.740	.7174	.9511	.91948	0.21835	.73475	.72938	.71823	.70647	.70032	.69397	.68740	.68057	.58185
					1.00312	1.00628	1.01278	1.01953	1.02300	1.02655	1.03017	1.03387	1.07722
.739	.7190	.9508	.92046	0.21493	.73370	.72828	.71703	.70513	.69891	.69248	.68582	.67889	.57626
					1.00313	1.00631	1.01284	1.01963	1.02312	1.02669	1.03033	1.03406	1.07788
.738	.7207	.9506	.92142	0.21155	.73266	.72719	.71582	.70380	.69750	.69099	.68423	.67720	.57001
					1.00314	1.00634	1.01291	1.01973	1.02324	1.02683	1.03050	1.03425	1.07857
.737	.7223	.9504	.92238	0.20822	.73161	.72609	.71461	.70245	.69608	.68948	.68263	.67550	.56274
					1.00316	1.00637	1.01297	1.01983	1.02336	1.02697	1.03066	1.03444	1.07930
.736	.7240	.9502	.92333	0.20493	.73056	.72499	.71340	.70110	.69465	.68797	.68102	.67378	.55379
					1.00317	1.00640	1.01303	1.01993	1.02349	1.02712	1.03083	1.03463	1.08006
.735	.7256	.9500	.92428	0.20169	.72951	.72389	.71218	.69975	.69322	.68644	.67940	.67204	.54077
					1.00319	1.00643	1.01310	1.02003	1.02361	1.02726	1.03100	1.03483	1.08088
.734	.7273	.9498	.92522	0.19848	.72846	.72279	.71096	.69839	.69177	.68491	.67776	.67029	
					1.00320	1.00646	1.01316	1.02014	1.02373	1.02741	1.03117	1.03502	
.733	.7289	.9495	.92615	0.19531	.72742	.72169	.70973	.69702	.69033	.68337	.67611	.66852	
					1.00322	1.00649	1.01323	1.02024	1.02386	1.02755	1.03134	1.03522	
.732	.7306	.9493	.92708	0.19219	.72637	.72058	.70851	.69565	.68887	.68181	.67445	.66673	
					1.00323	1.00652	1.01329	1.02034	1.02398	1.02770	1.03151	1.03542	
.731	.7322	.9491	.92800	0.18910	.72532	.71948	.70728	.69427	.68740	.68025	.67278	.66493	
					1.00325	1.00655	1.01336	1.02044	1.02410	1.02785	1.03168	1.03562	

TABLE II. Specific Heat Ratio = 1.20

p/p_t	MACH	T/T_t	GAMMA	K_{crit}	.01	.02	.04	.05	.06	.07	.08	.09	.1
.730	.7338	.9489	.92891	0.18605	.72426	.71837	.70605	.69958	.69289	.68593	.67868	.67109	.66310
					1.00326	1.00658	1.01342	1.01695	1.02055	1.02423	1.02800	1.03186	1.03582
.729	.7355	.9487	.92981	0.18305	.72321	.71726	.70481	.69827	.69149	.68445	.67709	.66938	.66126
					1.00328	1.00661	1.01349	1.01703	1.02065	1.02436	1.02815	1.03203	1.03602
.728	.7371	.9485	.93071	0.18008	.72216	.71615	.70357	.69696	.69010	.68295	.67549	.66766	.65939
					1.00329	1.00664	1.01355	1.01711	1.02076	1.02448	1.02830	1.03221	1.03623
.727	.7388	.9482	.93160	0.17714	.72111	.71504	.70233	.69564	.68869	.68145	.67388	.66592	.65751
					1.00331	1.00667	1.01362	1.01720	1.02086	1.02461	1.02845	1.03239	1.03643
.726	.7404	.9480	.93249	0.17425	.72005	.71393	.70108	.69432	.68728	.67994	.67226	.66417	.65560
					1.00332	1.00670	1.01368	1.01728	1.02097	1.02474	1.02860	1.03257	1.03664
.725	.7420	.9478	.93337	0.17139	.71900	.71282	.69983	.69299	.68586	.67842	.67062	.66239	.65366
					1.00334	1.00674	1.01375	1.01737	1.02107	1.02487	1.02875	1.03275	1.03685
.724	.7437	.9476	.93424	0.16857	.71794	.71170	.69858	.69165	.68444	.67689	.66897	.66060	.65170
					1.00335	1.00677	1.01381	1.01745	1.02118	1.02499	1.02891	1.03293	1.03706
.723	.7453	.9474	.93511	0.16578	.71689	.71059	.69732	.69031	.68300	.67535	.66730	.65879	.64971
					1.00337	1.00680	1.01388	1.01754	1.02129	1.02512	1.02906	1.03311	1.03728
.722	.7469	.9472	.93597	0.16303	.71583	.70947	.69606	.68897	.68156	.67380	.66562	.65696	.64769
					1.00338	1.00683	1.01395	1.01763	1.02139	1.02526	1.02922	1.03329	1.03749
.721	.7486	.9469	.93682	0.16032	.71477	.70835	.69479	.68761	.68011	.67223	.66393	.65510	.64564
					1.00340	1.00686	1.01401	1.01771	1.02150	1.02539	1.02938	1.03348	1.03771
.720	.7502	.9467	.93766	0.15764	.71372	.70723	.69352	.68625	.67865	.67066	.66221	.65322	.64356
					1.00341	1.00689	1.01408	1.01780	1.02161	1.02552	1.02953	1.03367	1.03793
.719	.7518	.9465	.93850	0.15499	.71266	.70610	.69225	.68489	.67718	.66907	.66048	.65132	.64145
					1.00343	1.00692	1.01415	1.01789	1.02172	1.02565	1.02969	1.03385	1.03815
.718	.7534	.9463	.93934	0.15238	.71160	.70498	.69097	.68352	.67570	.66747	.65874	.64940	.63930
					1.00344	1.00695	1.01422	1.01797	1.02183	1.02579	1.02985	1.03404	1.03837
.717	.7551	.9461	.94017	0.14980	.71054	.70385	.68969	.68214	.67422	.66585	.65697	.64744	.63709
					1.00346	1.00699	1.01428	1.01806	1.02194	1.02592	1.03001	1.03423	1.03860
.716	.7567	.9458	.94099	0.14726	.70948	.70272	.68840	.68075	.67272	.66423	.65518	.64546	.63486
					1.00347	1.00702	1.01435	1.01815	1.02205	1.02605	1.03018	1.03443	1.03883
.715	.7583	.9456	.94180	0.14474	.70842	.70159	.68711	.67936	.67121	.66258	.65337	.64344	.63257
					1.00349	1.00705	1.01442	1.01824	1.02216	1.02619	1.03034	1.03462	1.03906
.714	.7599	.9454	.94261	0.14226	.70735	.70046	.68581	.67796	.66969	.66092	.65154	.64140	.63024
					1.00350	1.00708	1.01449	1.01833	1.02227	1.02633	1.03051	1.03482	1.03929
.713	.7616	.9452	.94341	0.13981	.70629	.69933	.68450	.67656	.66816	.65925	.64969	.63931	.62785
					1.00352	1.00711	1.01456	1.01842	1.02239	1.02647	1.03067	1.03502	1.03953
.712	.7632	.9450	.94421	0.13740	.70523	.69819	.68319	.67514	.66662	.65756	.64781	.63719	.62541
					1.00353	1.00714	1.01463	1.01851	1.02250	1.02661	1.03084	1.03522	1.03977
.711	.7648	.9447	.94499	0.13501	.70416	.69705	.68188	.67372	.66507	.65585	.64591	.63503	.62290
					1.00355	1.00718	1.01470	1.01860	1.02261	1.02675	1.03101	1.03543	1.04001
.710	.7664	.9445	.94578	0.13266	.70310	.69591	.68056	.67228	.66350	.65412	.64397	.63282	.62032
					1.00356	1.00721	1.01477	1.01869	1.02273	1.02689	1.03118	1.03563	1.04026
.709	.7681	.9443	.94655	0.13033	.70203	.69477	.67923	.67084	.66193	.65237	.64200	.63058	.61766
					1.00358	1.00724	1.01484	1.01878	1.02284	1.02703	1.03136	1.03584	1.04051
.708	.7697	.9441	.94732	0.12804	.70096	.69363	.67790	.66939	.66033	.65060	.64001	.62828	.61492
					1.00360	1.00727	1.01491	1.01887	1.02296	1.02717	1.03153	1.03605	1.04076
.707	.7713	.9439	.94809	0.12578	.69989	.69248	.67656	.66793	.65873	.64881	.63798	.62592	.61207
					1.00361	1.00730	1.01498	1.01897	1.02308	1.02731	1.03171	1.03626	1.04102
.706	.7729	.9436	.94885	0.12354	.69882	.69133	.67522	.66646	.65710	.64700	.63592	.62351	.60912
					1.00363	1.00734	1.01505	1.01906	1.02319	1.02746	1.03188	1.03648	1.04128
.705	.7745	.9434	.94960	0.12134	.69775	.69018	.67386	.66498	.65547	.64515	.63381	.62103	.60604
					1.00364	1.00737	1.01512	1.01915	1.02331	1.02761	1.03206	1.03670	1.04155
.704	.7762	.9432	.95034	0.11916	.69668	.68903	.67250	.66349	.65381	.64329	.63167	.61848	.60282
					1.00366	1.00740	1.01519	1.01925	1.02343	1.02776	1.03225	1.03692	1.04182
.703	.7778	.9430	.95108	0.11702	.69561	.68787	.67114	.66199	.65214	.64140	.62948	.61585	.59943
					1.00367	1.00743	1.01526	1.01934	1.02355	1.02791	1.03243	1.03714	1.04210
.702	.7794	.9427	.95182	0.11490	.69453	.68671	.66976	.66047	.65045	.63948	.62724	.61314	.59583
					1.00369	1.00747	1.01533	1.01944	1.02367	1.02806	1.03261	1.03737	1.04239
.701	.7810	.9425	.95254	0.11281	.69346	.68555	.66838	.65894	.64874	.63753	.62496	.61032	.59199
					1.00371	1.00750	1.01541	1.01953	1.02379	1.02821	1.03280	1.03761	1.04268

TABLE II. Specific Heat Ratio = 1.20

p/p_t	MACH	T/T_t	GAMMA	K_{crit}	.01	.02	.04	R₂ and TPR versus Loss Coefficient K .05	.06	.07	.08	.09	.1
.700	.7826	.9423	.95326	0.11074	.69238 1.00372	.68439 1.00753	.66699 1.01548	.65740 1.01963	.64700 1.02392	.63554 1.02836	.62261 1.03299	.60738 1.03784	.58784 1.04298
.699	.7842	.9421	.95398	0.10871	.69130 1.00374	.68322 1.00757	.66559 1.01555	.65585 1.01972	.64525 1.02404	.63352 1.02852	.62020 1.03319	.60432 1.03808	.58330 1.04328
.698	.7858	.9418	.95469	0.10670	.69022 1.00375	.68205 1.00760	.66419 1.01562	.65428 1.01982	.64347 1.02417	.63146 1.02868	.61772 1.03338	.60110 1.03833	.57818 1.04361
.697	.7875	.9416	.95539	0.10472	.68914 1.00377	.68088 1.00763	.66277 1.01570	.65270 1.01992	.64167 1.02429	.62936 1.02884	.61516 1.03358	.59771 1.03858	.57226 1.04394
.696	.7891	.9414	.95609	0.10276	.68806 1.00378	.67970 1.00767	.66134 1.01577	.65110 1.02002	.63984 1.02442	.62722 1.02900	.61251 1.03378	.59410 1.03884	.56491 1.04429
.695	.7907	.9412	.95678	0.10083	.68698 1.00380	.67852 1.00770	.65991 1.01585	.64948 1.02012	.63799 1.02455	.62503 1.02916	.60977 1.03399	.59022 1.03910	.55391 1.04467
.694	.7923	.9409	.95746	0.09893	.68590 1.00382	.67734 1.00773	.65846 1.01592	.64785 1.02022	.63611 1.02468	.62278 1.02932	.60692 1.03420	.58600 1.03937	
.693	.7939	.9407	.95814	0.09705	.68481 1.00383	.67615 1.00777	.65701 1.01600	.64620 1.02032	.63420 1.02481	.62048 1.02949	.60394 1.03441	.58133 1.03965	
.692	.7955	.9405	.95882	0.09520	.68373 1.00385	.67496 1.00780	.65554 1.01607	.64453 1.02042	.63225 1.02494	.61811 1.02966	.60081 1.03462	.57596 1.03994	
.691	.7971	.9403	.95948	0.09337	.68264 1.00386	.67377 1.00783	.65406 1.01615	.64284 1.02052	.63027 1.02507	.61567 1.02983	.59750 1.03485	.56957 1.04025	
.690	.7987	.9400	.96014	0.09157	.68155 1.00388	.67257 1.00787	.65256 1.01623	↖64113 1.02063	.62824 1.02521	.61315 1.03000	.59398 1.03507	.56103 1.04057	
.689	.8003	.9398	.96080	0.08979	.68046 1.00390	.67137 1.00790	.65106 1.01630	.63939 1.02073	.62618 1.02534	.61054 1.03018	.59020 1.03530		
.688	.8019	.9396	.96145	0.08803	.67936 1.00391	.67017 1.00794	.64954 1.01638	.63763 1.02084	.62407 1.02548	.60784 1.03036	.58608 1.03554		
.687	.8035	.9393	.96209	0.08630	.67827 1.00393	.66896 1.00797	.64801 1.01646	.63585 1.02094	.62191 1.02562	.60501 1.03054	.58148 1.03579		
.686	.8052	.9391	.96273	0.08460	.67717 1.00395	.66775 1.00800	.64646 1.01654	.63404 1.02105	.61970 1.02576	.60206 1.03073	.57623 1.03605		
.685	.8068	.9389	.96336	0.08291	.67608 1.00396	.66653 1.00804	.64490 1.01662	.63220 1.02116	.61742 1.02590	.59895 1.03092	.56991 1.03632		
.684	.8084	.9387	.96399	0.08126	.67498 1.00398	.66531 1.00807	.64332 1.01670	.63033 1.02127	.61508 1.02605	.59565 1.03111	.56127 1.03661		
.683	.8100	.9384	.96461	0.07962	.67388 1.00399	.66409 1.00811	.64172 1.01678	.62842 1.02138	.61267 1.02620	.59212 1.03131			
.682	.8116	.9382	.96522	0.07801	.67277 1.00401	.66286 1.00814	.64010 1.01686	.62648 1.02149	.61017 1.02634	.58830 1.03151			
.681	.8132	.9380	.96583	0.07641	.67167 1.00403	.66162 1.00818	.63847 1.01694	.62450 1.02160	.60758 1.02650	.58407 1.03173			
.680	.8148	.9377	.96643	0.07485	.67056 1.00404	.66038 1.00821	.63681 1.01702	.62247 1.02171	.60488 1.02665	.57932 1.03194			
.679	.8164	.9375	.96703	0.07330	.66945 1.00406	.65913 1.00825	.63513 1.01711	.62040 1.02183	.60205 1.02681	.57375 1.03217			
.678	.8180	.9373	.96762	0.07177	.66834 1.00408	.65787 1.00829	.63343 1.01719	.61827 1.02194	.59908 1.02697	.56668 1.03241			
.677	.8196	.9371	.96820	0.07027	.66723 1.00409	.65661 1.00832	.63170 1.01727	.61609 1.02206	.59593 1.02713	.55452 1.03267			
.676	.8212	.9368	.96878	0.06879	.66611 1.00411	.65535 1.00836	.62995 1.01736	.61385 1.02218	.59257 1.02730				
.675	.8228	.9366	.96936	0.06733	.66499 1.00413	.65408 1.00840	.62816 1.01745	.61153 1.02230	.58893 1.02747				
.674	.8244	.9364	.96992	0.06589	.66387 1.00414	.65280 1.00843	.62635 1.01753	.60914 1.02243	.58491 1.02765				
.673	.8260	.9361	.97049	0.06447	.66275 1.00416	.65152 1.00847	.62450 1.01762	.60665 1.02255	.58041 1.02784				
.672	.8276	.9359	.97104	0.06307	.66163 1.00418	.65023 1.00851	.62261 1.01771	.60406 1.02268	.57515 1.02803				
.671	.8292	.9357	.97160	0.06169	.66049 1.00420	.64893 1.00854	.62069 1.01780	.60136 1.02281	.56855 1.02823				

53

TABLE II. Specific Heat Ratio = 1.20

R₂ and TPR versus Loss Coefficient K

p/p_t	MACH	T/T_t	GAMMA	K_{crit}	.01	.015	.02	.025	.03	.035	.04	.045	.05
.670	.8308	.9354	.97214	0.06034	.65936	.65365	.64762	.64122	.63436	.62692	.61872	.60943	.59850
					1.00421	1.00638	1.00858	1.01083	1.01312	1.01548	1.01789	1.02037	1.02294
.669	.8324	.9352	.97268	0.05900	.65822	.65243	.64631	.63979	.63279	.62516	.61670	.60704	.59548
					1.00423	1.00641	1.00862	1.01088	1.01319	1.01555	1.01798	1.02048	1.02308
.668	.8340	.9350	.97321	0.05768	.65709	.65121	.64499	.63835	.63119	.62336	.61464	.60456	.59225
					1.00425	1.00643	1.00865	1.01092	1.01325	1.01563	1.01807	1.02060	1.02321
.667	.8356	.9347	.97374	0.05638	.65595	.64998	.64366	.63689	.62957	.62153	.61249	.60196	.58876
					1.00427	1.00646	1.00869	1.01097	1.01331	1.01570	1.01817	1.02071	1.02336
.666	.8372	.9345	.97427	0.05510	.65480	.64875	.64232	.63541	.62793	.61967	.61030	.59924	.58492
					1.00428	1.00649	1.00873	1.01102	1.01337	1.01578	1.01827	1.02083	1.02350
.665	.8388	.9343	.97478	0.05384	.65366	.64751	.64096	.63393	.62626	.61776	.60803	.59636	.58060
					1.00430	1.00652	1.00877	1.01107	1.01343	1.01586	1.01836	1.02095	1.02365
.664	.8404	.9340	.97530	0.05260	.65251	.64627	.63960	.63242	.62457	.61580	.60568	.59330	.57553
					1.00432	1.00654	1.00880	1.01112	1.01350	1.01594	1.01846	1.02107	1.02381
.663	.8420	.9338	.97580	0.05138	.65135	.64501	.63823	.63090	.62284	.61378	.60323	.59000	.56899
					1.00434	1.00657	1.00884	1.01117	1.01356	1.01602	1.01856	1.02120	1.02398
.662	.8436	.9336	.97630	0.05018	.65020	.64376	.63685	.62935	.62109	.61172	.60067	.58639	.55765
					1.00435	1.00660	1.00888	1.01122	1.01363	1.01611	1.01866	1.02133	1.02416
.661	.8452	.9333	.97680	0.04899	.64903	.64249	.63545	.62779	.61930	.60960	.59798	.58238	
					1.00437	1.00662	1.00892	1.01127	1.01369	1.01619	1.01877	1.02146	
.660	.8468	.9331	.97729	0.04782	.64787	.64122	.63404	.62621	.61748	.60740	.59513	.57775	
					1.00439	1.00665	1.00896	1.01133	1.01376	1.01627	1.01887	1.02160	
.659	.8484	.9329	.97777	0.04667	.64670	.63993	.63262	.62460	.61559	.60513	.59209	.57200	
					1.00441	1.00668	1.00900	1.01138	1.01383	1.01636	1.01898	1.02175	
.658	.8500	.9326	.97825	0.04554	.64553	.63865	.63119	.62297	.61368	.60276	.58880	.56372	
					1.00442	1.00671	1.00904	1.01143	1.01390	1.01644	1.01910	1.02190	
.657	.8516	.9324	.97872	0.04443	.64435	.63735	.62973	.62131	.61171	.60029	.58519		
					1.00444	1.00674	1.00908	1.01148	1.01397	1.01653	1.01921		
.656	.8532	.9321	.97919	0.04333	.64316	.63604	.62827	.61962	.60969	.59770	.58113		
					1.00446	1.00676	1.00912	1.01154	1.01404	1.01662	1.01933		
.655	.8548	.9319	.97965	0.04226	.64198	.63472	.62678	.61790	.60761	.59495	.57633		
					1.00448	1.00679	1.00916	1.01159	1.01411	1.01672	1.01946		
.654	.8564	.9317	.98011	0.04119	.64078	.63340	.62528	.61613	.60546	.59203	.57024		
					1.00450	1.00682	1.00920	1.01165	1.01418	1.01681	1.01959		
.653	.8580	.9314	.98056	0.04015	.63959	.63206	.62375	.61434	.60323	.58887	.55963		
					1.00451	1.00685	1.00924	1.01170	1.01425	1.01691	1.01973		
.652	.8596	.9312	.98101	0.03912	.63838	.63071	.62221	.61250	.60090	.58540			
					1.00453	1.00688	1.00928	1.01176	1.01433	1.01701			
.651	.8612	.9310	.98145	0.03811	.63717	.62935	.62064	.61062	.59847	.58151			
					1.00455	1.00691	1.00932	1.01182	1.01440	1.01711			
.650	.8628	.9307	.98188	0.03712	.63595	.62798	.61905	.60869	.59591	.57692			
					1.00457	1.00694	1.00936	1.01188	1.01448	1.01722			
.649	.8644	.9305	.98231	0.03614	.63473	.62659	.61741	.60670	.59319	.57114			
					1.00459	1.00697	1.00941	1.01193	1.01456	1.01733			
.648	.8660	.9302	.98274	0.03518	.63350	.62519	.61576	.60464	.59027	.56140			
					1.00461	1.00700	1.00945	1.01199	1.01464	1.01746			
.647	.8676	.9300	.98315	0.03423	.63226	.62377	.61408	.60251	.58711				
					1.00462	1.00703	1.00949	1.01205	1.01472				
.646	.8692	.9298	.98357	0.03330	.63102	.62234	.61236	.60029	.58361				
					1.00464	1.00706	1.00954	1.01211	1.01481				
.645	.8708	.9295	.98398	0.03239	.62977	.62089	.61060	.59797	.57957				
					1.00466	1.00709	1.00958	1.01218	1.01490				
.644	.8723	.9293	.98438	0.03149	.62850	.61940	.60880	.59554	.57474				
					1.00468	1.00712	1.00963	1.01224	1.01500				
.643	.8739	.9290	.98478	0.03061	.62723	.61791	.60695	.59296	.56815				
					1.00470	1.00715	1.00967	1.01231	1.01509				
.642	.8755	.9288	.98517	0.02974	.62595	.61639	.60505	.59020					
					1.00472	1.00718	1.00972	1.01237					
.641	.8771	.9286	.98555	0.02889	.62466	.61485	.60308	.58721					
					1.00474	1.00722	1.00977	1.01244					

TABLE II. Specific Heat Ratio = 1.20

R_2 and TPR versus Loss Coefficient K

p/p_t	MACH	T/T_t	GAMMA	K_{crit}	.001	.002	.003	.004	.005	.006	.01	.015	.02
.640	.8787	.9283	.98594	0.02806	.63848	.63694	.63537	.63376	.63212	.63045	.62337	.61329	.60105.
					1.00046	1.00093	1.00140	1.00187	1.00235	1.00282	1.00476	1.00725	1.00981
.639	.8803	.9281	.98631	0.02723	.63746	.63589	.63429	.63266	.63099	.62929	.62205	.61169	.59894
					1.00047	1.00094	1.00141	1.00188	1.00236	1.00283	1.00478	1.00728	1.00986
.638	.8819	.9278	.98668	0.02643	.63644	.63484	.63322	.63156	.62986	.62812	.62073	.61007	.59673
					1.00047	1.00094	1.00141	1.00189	1.00237	1.00285	1.00480	1.00731	1.00991
.637	.8835	.9276	.98705	0.02563	.63541	.63379	.63214	.63045	.62872	.62695	.61939	.60841	.59441
					1.00047	1.00094	1.00142	1.00189	1.00237	1.00286	1.00481	1.00734	1.00996
.636	.8851	.9273	.98741	0.02486	.63439	.63274	.63106	.62934	.62758	.62577	.61804	.60671	.59195
					1.00047	1.00095	1.00142	1.00190	1.00238	1.00287	1.00483	1.00738	1.01001
.635	.8867	.9271	.98776	0.02409	.63336	.63169	.62997	.62822	.62643	.62458	.61667	.60497	.58932
					1.00047	1.00095	1.00143	1.00191	1.00239	1.00288	1.00485	1.00741	1.01007
.634	.8883	.9269	.98811	0.02334	.63233	.63063	.62889	.62710	.62527	.62339	.61529	.60317	.58648
					1.00048	1.00095	1.00143	1.00192	1.00240	1.00289	1.00487	1.00745	1.01012
.633	.8899	.9266	.98846	0.02261	.63131	.62957	.62780	.62598	.62411	.62219	.61389	.60133	.58334
					1.00048	1.00096	1.00144	1.00192	1.00241	1.00290	1.00489	1.00748	1.01017
.632	.8915	.9264	.98880	0.02189	.63028	.62851	.62671	.62485	.62295	.62099	.61247	.59942	.57971
					1.00048	1.00096	1.00144	1.00193	1.00242	1.00291	1.00491	1.00751	1.01024
.631	.8931	.9261	.98913	0.02118	.62925	.62745	.62561	.62372	.62178	.61978	.61100	.59744	.57540
					1.00048	1.00096	1.00145	1.00194	1.00243	1.00292	1.00494	1.00755	1.01030
.630	.8947	.9259	.98946	0.02049	.62822	.62639	.62452	.62259	.62059	.61855	.60954	.59537	.56948
					1.00048	1.00097	1.00145	1.00194	1.00244	1.00294	1.00496	1.00759	1.01036
.629	.8963	.9256	.98978	0.01981	.62719	.62533	.62341	.62145	.61941	.61731	.60805	.59320	
					1.00048	1.00097	1.00146	1.00195	1.00245	1.00295	1.00498	1.00762	
.628	.8979	.9254	.99010	0.01914	.62615	.62426	.62231	.62031	.61822	.61607	.60653	.59091	
					1.00049	1.00097	1.00146	1.00196	1.00246	1.00296	1.00500	1.00766	
.627	.8995	.9251	.99042	0.01849	.62512	.62319	.62119	.61915	.61702	.61482	.60498	.58846	
					1.00049	1.00098	1.00147	1.00197	1.00247	1.00297	1.00502	1.00770	
.626	.9011	.9249	.99072	0.01785	.62409	.62212	.62009	.61799	.61582	.61356	.60339	.58582	
					1.00049	1.00098	1.00148	1.00197	1.00248	1.00298	1.00505	1.00774	
.625	.9027	.9247	.99103	0.01722	.62305	.62105	.61897	.61683	.61460	.61228	.60177	.58286	
					1.00049	1.00098	1.00148	1.00198	1.00249	1.00299	1.00507	1.00778	
.624	.9043	.9244	.99132	0.01661	.62200	.61997	.61785	.61566	.61337	.61099	.60010	.57954	
					1.00050	1.00099	1.00149	1.00199	1.00249	1.00300	1.00509	1.00783	
.623	.9059	.9242	.99162	0.01601	.62098	.61889	.61672	.61448	.61214	.60969	.59839	.57556	
					1.00049	1.00099	1.00149	1.00200	1.00250	1.00302	1.00511	1.00787	
.622	.9075	.9239	.99190	0.01542	.61994	.61781	.61559	.61329	.61089	.60837	.59661	.57013	
					1.00050	1.00099	1.00150	1.00200	1.00251	1.00303	1.00514	1.00792	
.621	.9091	.9237	.99219	0.01485	.61890	.61672	.61446	.61210	.60963	.60703	.59478		
					1.00050	1.00100	1.00150	1.00201	1.00252	1.00304	1.00516		
.620	.9107	.9234	.99246	0.01429	.61786	.61563	.61331	.61090	.60836	.60568	.59286		
					1.00050	1.00100	1.00151	1.00202	1.00253	1.00305	1.00518		
.619	.9123	.9232	.99274	0.01374	.61681	.61454	.61217	.60968	.60707	.60431	.59085		
					1.00050	1.00101	1.00151	1.00203	1.00254	1.00306	1.00521		
.618	.9139	.9229	.99300	0.01320	.61577	.61344	.61101	.60846	.60577	.60291	.58873		
					1.00050	1.00101	1.00152	1.00203	1.00255	1.00308	1.00523		
.617	.9155	.9227	.99326	0.01267	.61472	.61234	.60985	.60723	.60445	.60149	.58647		
					1.00050	1.00101	1.00152	1.00204	1.00256	1.00309	1.00526		
.616	.9171	.9224	.99352	0.01216	.61367	.61124	.60868	.60598	.60312	.60005	.58402		
					1.00051	1.00102	1.00153	1.00205	1.00257	1.00310	1.00528		
.615	.9187	.9222	.99377	0.01166	.61262	.61013	.60750	.60472	.60176	.59854	.58127		
					1.00051	1.00102	1.00154	1.00206	1.00258	1.00312	1.00531		
.614	.9203	.9219	.99402	0.01117	.61157	.60901	.60631	.60345	.60038	.59702	.57816		
					1.00051	1.00102	1.00154	1.00206	1.00259	1.00313	1.00534		
.613	.9219	.9217	.99426	0.01069	.61051	.60789	.60512	.60216	.59894	.59546	.57439		
					1.00051	1.00103	1.00155	1.00207	1.00261	1.00314	1.00537		
.612	.9235	.9214	.99450	0.01023	.60946	.60677	.60391	.60085	.59751	.59385	.56894		
					1.00051	1.00103	1.00155	1.00208	1.00262	1.00316	1.00540		
.611	.9251	.9212	.99473	0.00978	.60839	.60563	.60269	.59953	.59604	.59220			
					1.00051	1.00103	1.00156	1.00209	1.00263	1.00317			

TABLE II. Specific Heat Ratio = 1.20

R₂ and TPR versus Loss Coefficient K

p/P_t	MACH	T/T_t	GAMMA	K_{crit}	.001	.002	.003	.004	.005	.006	.007	.008	.009
.610	.9267	.9209	.99495	0.00933	.60733	.60449	.60146	.59814	.59453	.59047	.58575	.57989	.57077
					1.00052	1.00104	1.00156	1.00210	1.00264	1.00318	1.00374	1.00430	1.00488
.609	.9283	.9207	.99517	0.00890	.60627	.60335	.60021	.59676	.59298	.58867	.58355	.57675	
					1.00052	1.00104	1.00157	1.00211	1.00265	1.00320	1.00375	1.00432	
.608	.9299	.9204	.99539	0.00848	.60520	.60219	.59892	.59536	.59138	.58678	.58113	.57286	
					1.00052	1.00104	1.00158	1.00212	1.00266	1.00321	1.00377	1.00435	
.607	.9315	.9202	.99560	0.00808	.60413	.60103	.59763	.59392	.58972	.58476	.57843	.56642	
					1.00052	1.00105	1.00159	1.00213	1.00267	1.00323	1.00379	1.00437	
.606	.9331	.9199	.99581	0.00768	.60305	.59983	.59633	.59244	.58798	.58260	.57512		
					1.00052	1.00106	1.00159	1.00213	1.00268	1.00324	1.00381		
.605	.9347	.9197	.99601	0.00729	.60197	.59864	.59501	.59092	.58617	.58023	.57083		
					1.00052	1.00106	1.00160	1.00214	1.00269	1.00326	1.00383		
.604	.9363	.9194	.99620	0.00692	.60089	.59744	.59365	.58935	.58424	.57747			
					1.00052	1.00106	1.00160	1.00215	1.00271	1.00328			
.603	.9379	.9191	.99640	0.00656	.59976	.59622	.59227	.58772	.58218	.57422			
					1.00053	1.00107	1.00161	1.00216	1.00272	1.00329			
.602	.9395	.9189	.99658	0.00620	.59867	.59500	.59086	.58602	.57992	.56976			
					1.00053	1.00107	1.00162	1.00217	1.00273	1.00331			
.601	.9411	.9186	.99676	0.00586	.59756	.59375	.58941	.58423	.57730				
					1.00053	1.00107	1.00162	1.00218	1.00275				
.600	.9427	.9184	.99694	0.00553	.59645	.59248	.58791	.58232	.57424				
					1.00054	1.00108	1.00163	1.00219	1.00276				
.599	.9443	.9181	.99711	0.00521	.59533	.59120	.58636	.58027	.57014				
					1.00054	1.00108	1.00164	1.00220	1.00277				
.598	.9459	.9179	.99728	0.00490	.59420	.58988	.58474	.57791					
					1.00054	1.00109	1.00164	1.00221					
.597	.9475	.9176	.99744	0.00460	.59307	.58854	.58304	.57525					
					1.00054	1.00109	1.00165	1.00222					
.596	.9491	.9174	.99759	0.00431	.59192	.58717	.58124	.57192					
					1.00054	1.00110	1.00165	1.00223					
.595	.9507	.9171	.99774	0.00403	.59076	.58575	.57930	.56610					
					1.00055	1.00110	1.00166	1.00224					
.594	.9523	.9168	.99789	0.00376	.58959	.58429	.57705						
					1.00055	1.00110	1.00167						
.593	.9539	.9166	.99803	0.00350	.58840	.58276	.57453						
					1.00055	1.00111	1.00168						
.592	.9555	.9163	.99817	0.00325	.58720	.58117	.57139						
					1.00055	1.00111	1.00169						
.591	.9571	.9161	.99830	0.00301	.58597	.57948	.56520						
					1.00055	1.00111	1.00170						
.590	.9588	.9158	.99843	0.00278	.58473	.57755							
					1.00055	1.00112							
.589	.9604	.9156	.99855	0.00256	.58345	.57549							
					1.00056	1.00113							
.588	.9620	.9153	.99866	0.00235	.58214	.57308							
					1.00056	1.00113							
.587	.9636	.9150	.99877	0.00215	.58080	.57003							
					1.00056	1.00113							
.586	.9652	.9148	.99888	0.00196	.57932								
					1.00056								
.585	.9668	.9145	.99898	0.00177	.57783								
					1.00057								
.584	.9684	.9143	.99908	0.00160	.57623								
					1.00057								
.583	.9700	.9140	.99917	0.00144	.57449								
					1.00057								
.582	.9716	.9137	.99926	0.00128	.57253								
					1.00057								
.581	.9733	.9135	.99934	0.00114	.57018								
					1.00057								

TABLE III. Specific Heat Ratio = 1.30

p/p_t	MACH	T/T_t	GAMMA	K_{crit}	.1	.2	.4	.6	1.0	2.0	4.0	6.0	10.0
								R₂ and TPR versus Loss Coefficient K					
.999	.0392	.9998	.06698	493.33739	.99890	.99880	.99860	.99840	.99800	.99699	.99499	.99297	.98893
					1.00010	1.00020	1.00040	1.00060	1.00100	1.00201	1.00403	1.00606	1.01016
.998	.0555	.9995	.09467	243.93227	.99780	.99760	.99720	.99680	.99599	.99398	.98994	.98589	.97773
					1.00020	1.00040	1.00080	1.00120	1.00201	1.00403	1.00811	1.01224	1.02065
.997	.0680	.9993	.11588	160.95893	.99670	.99640	.99579	.99519	.99398	.99095	.98487	.97874	.96638
					1.00030	1.00060	1.00120	1.00181	1.00302	1.00607	1.01224	1.01854	1.03149
.996	.0785	.9991	.13373	119.54497	.99560	.99519	.99439	.99358	.99197	.98791	.97976	.97154	.95488
					1.00040	1.00080	1.00161	1.00241	1.00403	1.00812	1.01644	1.02497	1.04270
.995	.0878	.9988	.14943	94.74316	.99450	.99399	.99298	.99197	.98995	.98487	.97462	.96427	.94322
					1.00050	1.00100	1.00201	1.00302	1.00505	1.01019	1.02069	1.03153	1.05430
.994	.0963	.9986	.16360	78.23738	.99339	.99279	.99157	.99036	.98792	.98181	.96945	.95694	.93140
					1.00060	1.00121	1.00242	1.00363	1.00608	1.01227	1.02500	1.03823	1.06633
.993	.1040	.9984	.17661	66.46895	.99229	.99158	.99016	.98874	.98589	.97873	.96425	.94954	.91940
					1.00070	1.00141	1.00282	1.00425	1.00711	1.01437	1.02938	1.04508	1.07880
.992	.1112	.9981	.18869	57.65916	.99119	.99038	.98875	.98712	.98386	.97565	.95902	.94208	.90722
					1.00080	1.00161	1.00323	1.00486	1.00814	1.01648	1.03381	1.05208	1.09174
.991	.1180	.9979	.20002	50.81905	.99009	.98917	.98734	.98550	.98182	.97256	.95374	.93454	.89485
					1.00091	1.00181	1.00364	1.00548	1.00918	1.01861	1.03832	1.05923	1.10520
.990	.1244	.9977	.21072	45.35694	.98898	.98796	.98593	.98388	.97978	.96945	.94844	.92693	.88227
					1.00101	1.00202	1.00405	1.00609	1.01022	1.02076	1.04289	1.06654	1.11920
.989	.1305	.9975	.22087	40.89609	.98788	.98676	.98451	.98226	.97773	.96633	.94309	.91924	.86948
					1.00111	1.00222	1.00446	1.00671	1.01127	1.02292	1.04753	1.07402	1.13379
.988	.1364	.9972	.23056	37.18552	.98678	.98555	.98309	.98063	.97568	.96320	.93771	.91147	.85646
					1.00121	1.00243	1.00487	1.00733	1.01232	1.02511	1.05224	1.08168	1.14902
.987	.1420	.9970	.23983	34.05155	.98567	.98434	.98167	.97900	.97362	.96005	.93229	.90362	.84320
					1.00131	1.00263	1.00528	1.00796	1.01337	1.02731	1.05702	1.08953	1.16492
.986	.1474	.9968	.24874	31.37007	.98457	.98313	.98025	.97736	.97156	.95690	.92683	.89569	.82968
					1.00142	1.00284	1.00570	1.00858	1.01443	1.02952	1.06188	1.09756	1.18157
.985	.1526	.9965	.25732	29.05055	.98346	.98192	.97883	.97573	.96950	.95373	.92133	.88767	.81589
					1.00152	1.00304	1.00611	1.00921	1.01550	1.03176	1.06682	1.10580	1.19902
.984	.1577	.9963	.26560	27.02460	.98236	.98071	.97741	.97409	.96742	.95054	.91578	.87956	.80181
					1.00162	1.00325	1.00653	1.00984	1.01657	1.03401	1.07183	1.11425	1.21734
.983	.1626	.9961	.27362	25.24031	.98125	.97950	.97598	.97245	.96535	.94734	.91020	.87135	.78740
					1.00172	1.00345	1.00695	1.01047	1.01765	1.03628	1.07693	1.12292	1.23662
.982	.1673	.9958	.28139	23.65713	.98014	.97828	.97455	.97081	.96327	.94413	.90457	.86304	.77266
					1.00183	1.00366	1.00736	1.01111	1.01873	1.03857	1.08212	1.13182	1.25694
.981	.1720	.9956	.28893	22.24335	.97904	.97707	.97312	.96916	.96118	.94091	.89889	.85463	.75756
					1.00193	1.00387	1.00778	1.01174	1.01981	1.04088	1.08739	1.14096	1.27843
.980	.1765	.9953	.29626	20.97316	.97793	.97586	.97169	.96751	.95909	.93767	.89316	.84610	.74205
					1.00203	1.00408	1.00820	1.01238	1.02090	1.04321	1.09275	1.15036	1.30119
.979	.1809	.9951	.30339	19.82618	.97682	.97464	.97026	.96586	.95699	.93441	.88739	.83747	.72612
					1.00213	1.00428	1.00862	1.01302	1.02200	1.04556	1.09820	1.16003	1.32537
.978	.1852	.9949	.31035	18.78538	.97572	.97342	.96883	.96420	.95488	.93115	.88157	.82872	.70970
					1.00224	1.00449	1.00905	1.01366	1.02310	1.04793	1.10376	1.16998	1.35114
.977	.1895	.9946	.31714	17.83688	.97461	.97221	.96739	.96255	.95278	.92786	.87570	.81984	.69277
					1.00234	1.00470	1.00947	1.01431	1.02421	1.05032	1.10941	1.18023	1.37870
.976	.1936	.9944	.32377	16.96901	.97350	.97099	.96595	.96088	.95066	.92456	.86977	.81083	.67525
					1.00245	1.00491	1.00989	1.01496	1.02532	1.05273	1.11517	1.19080	1.40829
.975	.1976	.9942	.33025	16.17210	.97239	.96977	.96451	.95922	.94854	.92125	.86379	.80169	.65708
					1.00255	1.00512	1.01032	1.01560	1.02644	1.05516	1.12103	1.20170	1.44019
.974	.2016	.9939	.33659	15.43790	.97128	.96855	.96307	.95755	.94642	.91792	.85775	.79240	.63818
					1.00265	1.00533	1.01075	1.01625	1.02756	1.05762	1.12701	1.21296	1.47475
.973	.2055	.9937	.34280	14.75934	.97017	.96733	.96163	.95589	.94429	.91457	.85165	.78296	.61845
					1.00276	1.00554	1.01117	1.01691	1.02869	1.06009	1.13310	1.22460	1.51241
.972	.2094	.9935	.34888	14.13046	.96906	.96611	.96018	.95421	.94215	.91121	.84549	.77336	.59776
					1.00286	1.00575	1.01160	1.01756	1.02982	1.06259	1.13931	1.23664	1.55372
.971	.2131	.9932	.35484	13.54604	.96795	.96489	.95874	.95254	.94000	.90783	.83927	.76359	.57593
					1.00297	1.00596	1.01203	1.01822	1.03096	1.06512	1.14564	1.24911	1.59940

TABLE III. Specific Heat Ratio = 1.30

p/p_t	MACH	T/T_t	GAMMA	K_{crit}	.1	.2	.4	.6	1.0	2.0	4.0	6.0	10.0
								R_2 and TPR versus Loss Coefficient K					
.970	.2169	.9930	.36069	13.00161	.96684	.96367	.95729	.95086	.93786	.90443	.83298	.75364	.55277
					1.00307	1.00617	1.01246	1.01888	1.03210	1.06766	1.15211	1.26203	1.65036
.969	.2205	.9928	.36644	12.49330	.96573	.96244	.95584	.94918	.93570	.90101	.82663	.74350	.52797
					1.00318	1.00638	1.01290	1.01954	1.03326	1.07023	1.15870	1.27545	1.70789
.968	.2241	.9925	.37208	12.01765	.96462	.96122	.95438	.94749	.93354	.89758	.82020	.73315	.50110
					1.00328	1.00660	1.01333	1.02021	1.03441	1.07282	1.16544	1.28939	1.77379
.967	.2277	.9923	.37762	11.57170	.96350	.95999	.95293	.94580	.93137	.89413	.81370	.72259	.47153
					1.00339	1.00681	1.01377	1.02087	1.03557	1.07544	1.17233	1.30390	1.85074
.966	.2312	.9920	.38307	11.15279	.96239	.95877	.95147	.94411	.92920	.89066	.80713	.71179	.43817
					1.00349	1.00702	1.01420	1.02154	1.03674	1.07809	1.17936	1.31901	1.94298
.965	.2346	.9918	.38843	10.75856	.96128	.95754	.95001	.94242	.92702	.88717	.80048	.70074	.39898
					1.00360	1.00724	1.01464	1.02222	1.03792	1.08076	1.18656	1.33479	2.05807
.964	.2380	.9916	.39371	10.38698	.96016	.95631	.94855	.94072	.92483	.88366	.79374	.68942	.34889
					1.00370	1.00745	1.01508	1.02289	1.03910	1.08346	1.19392	1.35129	2.21195
.963	.2414	.9913	.39890	10.03615	.95905	.95508	.94708	.93902	.92264	.88013	.78692	.67781	.25845
					1.00381	1.00766	1.01552	1.02357	1.04028	1.08618	1.20145	1.36857	2.45943
.962	.2447	.9911	.40401	9.70443	.95794	.95385	.94563	.93732	.92044	.87658	.78001	.66588	
					1.00392	1.00788	1.01596	1.02424	1.04148	1.08893	1.20916	1.38670	
.961	.2480	.9909	.40904	9.39032	.95682	.95262	.94416	.93561	.91823	.87301	.77300	.65360	
					1.00402	1.00809	1.01640	1.02493	1.04268	1.09171	1.21706	1.40578	
.960	.2512	.9906	.41401	9.09252	.95571	.95139	.94269	.93390	.91602	.86941	.76590	.64094	
					1.00413	1.00831	1.01684	1.02561	1.04388	1.09452	1.22516	1.42589	
.959	.2544	.9904	.41890	8.80980	.95459	.95016	.94122	.93218	.91379	.86580	.75869	.62785	
					1.00423	1.00853	1.01729	1.02629	1.04509	1.09736	1.23347	1.44715	
.958	.2576	.9901	.42372	8.54107	.95347	.94892	.93975	.93047	.91157	.86216	.75138	.61429	
					1.00434	1.00874	1.01773	1.02698	1.04631	1.10022	1.24200	1.46969	
.957	.2607	.9899	.42847	8.28534	.95236	.94769	.93827	.92874	.90933	.85851	.74395	.60021	
					1.00445	1.00896	1.01818	1.02767	1.04754	1.10312	1.25076	1.49368	
.956	.2638	.9897	.43316	8.04173	.95124	.94646	.93680	.92702	.90709	.85482	.73640	.58553	
					1.00456	1.00918	1.01863	1.02837	1.04877	1.10605	1.25977	1.51929	
.955	.2669	.9894	.43779	7.80939	.95012	.94522	.93532	.92529	.90484	.85112	.72873	.57016	
					1.00466	1.00940	1.01908	1.02906	1.05001	1.10901	1.26903	1.54678	
.954	.2699	.9892	.44236	7.58759	.94901	.94398	.93384	.92356	.90258	.84739	.72092	.55400	
					1.00477	1.00961	1.01953	1.02976	1.05125	1.11201	1.27856	1.57642	
.953	.2729	.9890	.44687	7.37566	.94789	.94274	.93235	.92182	.90032	.84363	.71297	.53691	
					1.00488	1.00983	1.01998	1.03046	1.05251	1.11504	1.28839	1.60857	
.952	.2759	.9887	.45133	7.17298	.94677	.94150	.93087	.92008	.89804	.83985	.70488	.51869	
					1.00499	1.01005	1.02044	1.03117	1.05377	1.11810	1.29851	1.64371	
.951	.2788	.9885	.45573	6.97894	.94565	.94026	.92938	.91834	.89576	.83604	.69662	.49909	
					1.00509	1.01027	1.02089	1.03188	1.05503	1.12120	1.30896	1.68246	
.950	.2817	.9882	.46008	6.79304	.94453	.93902	.92789	.91659	.89347	.83220	.68819	.47772	
					1.00520	1.01049	1.02135	1.03258	1.05631	1.12433	1.31976	1.72567	
.949	.2846	.9880	.46437	6.61479	.94341	.93778	.92639	.91484	.89118	.82834	.67959	.45400	
					1.00531	1.01071	1.02180	1.03330	1.05759	1.12750	1.33093	1.77459	
.948	.2875	.9878	.46861	6.44375	.94229	.93654	.92490	.91308	.88887	.82445	.67078	.42688	
					1.00542	1.01093	1.02226	1.03401	1.05888	1.13071	1.34250	1.83113	
.947	.2904	.9875	.47281	6.27949	.94117	.93529	.92340	.91132	.88656	.82053	.66177	.39429	
					1.00553	1.01116	1.02272	1.03473	1.06018	1.13396	1.35449	1.89864	
.946	.2932	.9873	.47696	6.12163	.94005	.93405	.92190	.90956	.88424	.81657	.65254	.35029	
					1.00564	1.01138	1.02319	1.03545	1.06148	1.13725	1.36693	1.98418	
.945	.2960	.9870	.48106	5.96983	.93892	.93280	.92040	.90779	.88191	.81259	.64306		
					1.00575	1.01160	1.02365	1.03617	1.06279	1.14058	1.37987		
.944	.2988	.9868	.48512	5.82375	.93780	.93155	.91890	.90602	.87957	.80858	.63331		
					1.00586	1.01182	1.02411	1.03690	1.06411	1.14395	1.39335		
.943	.3015	.9865	.48913	5.68308	.93668	.93030	.91739	.90425	.87722	.80453	.62328		
					1.00597	1.01205	1.02458	1.03763	1.06544	1.14736	1.40740		
.942	.3042	.9863	.49310	5.54754	.93555	.92905	.91588	.90247	.87486	.80045	.61292		
					1.00608	1.01227	1.02505	1.03836	1.06677	1.15082	1.42210		
.941	.3069	.9861	.49702	5.41686	.93443	.92780	.91437	.90068	.87249	.79633	.60221		
					1.00619	1.01250	1.02552	1.03910	1.06812	1.15432	1.43749		

TABLE III. Specific Heat Ratio = 1.30

p/p_t	MACH	T/T_t	GAMMA	K_crit	.1	.2	.4	R₂ and TPR versus Loss Coefficient K .6	.8	1.0	2.0	3.0	4.0
.940	.3096	.9858	.50091	5.29080	.93330	.92655	.91285	.89889	.88465	.87012	.79218	.70221	.59111
					1.00630	1.01272	1.02599	1.03983	1.05431	1.06947	1.15787	1.27691	1.45364
.939	.3123	.9856	.50475	5.16912	.93218	.92530	.91133	.89710	.88257	.86773	.78799	.69541	.57957
					1.00641	1.01295	1.02646	1.04057	1.05535	1.07083	1.16147	1.28462	1.47065
.938	.3150	.9853	.50856	5.05162	.93105	.92404	.90982	.89530	.88048	.86534	.78376	.68848	.56753
					1.00652	1.01317	1.02693	1.04132	1.05639	1.07220	1.16512	1.29254	1.48861
.937	.3176	.9851	.51233	4.93808	.92993	.92279	.90829	.89350	.87839	.86293	.77949	.68142	.55492
					1.00663	1.01340	1.02740	1.04206	1.05743	1.07358	1.16882	1.30067	1.50764
.936	.3202	.9849	.51606	4.82833	.92880	.92153	.90677	.89169	.87628	.86052	.77518	.67422	.54165
					1.00674	1.01363	1.02788	1.04281	1.05849	1.07497	1.17258	1.30904	1.52789
.935	.3228	.9846	.51975	4.72216	.92767	.92027	.90524	.88988	.87417	.85809	.77083	.66687	.52759
					1.00685	1.01385	1.02836	1.04357	1.05954	1.07636	1.17638	1.31765	1.54953
.934	.3254	.9844	.52341	4.61944	.92655	.91901	.90371	.88807	.87206	.85566	.76644	.65936	.51260
					1.00696	1.01408	1.02884	1.04432	1.06061	1.07777	1.18025	1.32653	1.57280
.933	.3279	.9841	.52703	4.51998	.92542	.91775	.90218	.88624	.86993	.85321	.76200	.65168	.49644
					1.00707	1.01431	1.02932	1.04508	1.06168	1.07918	1.18417	1.33568	1.59798
.932	.3305	.9839	.53062	4.42366	.92429	.91649	.90064	.88442	.86780	.85075	.75751	.64381	.47881
					1.00718	1.01454	1.02980	1.04585	1.06275	1.08060	1.18815	1.34512	1.62548
.931	.3330	.9836	.53418	4.33033	.92316	.91523	.89910	.88259	.86566	.84828	.75298	.63574	.45920
					1.00730	1.01477	1.03028	1.04661	1.06383	1.08204	1.19219	1.35488	1.65585
.930	.3355	.9834	.53770	4.23986	.92203	.91397	.89756	.88075	.86351	.84580	.74839	.62745	.43676
					1.00741	1.01500	1.03077	1.04738	1.06492	1.08348	1.19630	1.36499	1.68993
.929	.3380	.9831	.54119	4.15212	.92090	.91270	.89602	.87891	.86136	.84331	.74375	.61893	.40974
					1.00752	1.01523	1.03126	1.04815	1.06601	1.08493	1.20047	1.37546	1.72914
.928	.3405	.9829	.54465	4.06700	.91977	.91143	.89447	.87707	.85919	.84081	.73906	.61015	.37322
					1.00763	1.01546	1.03174	1.04893	1.06711	1.08639	1.20471	1.38633	1.77629
.927	.3430	.9827	.54807	3.98439	.91863	.91017	.89292	.87522	.85702	.83829	.73431	.60109	
					1.00774	1.01569	1.03223	1.04971	1.06821	1.08787	1.20902	1.39762	
.926	.3454	.9824	.55147	3.90419	.91750	.90890	.89136	.87336	.85484	.83577	.72951	.59172	
					1.00786	1.01592	1.03272	1.05049	1.06932	1.08935	1.21341	1.40939	
.925	.3479	.9822	.55483	3.82629	.91637	.90763	.88981	.87150	.85266	.83323	.72464	.58199	
					1.00797	1.01616	1.03322	1.05128	1.07044	1.09084	1.21787	1.42167	
.924	.3503	.9819	.55817	3.75061	.91523	.90636	.88825	.86963	.85046	.83068	.71970	.57187	
					1.00808	1.01639	1.03371	1.05207	1.07156	1.09235	1.22241	1.43451	
.923	.3527	.9817	.56148	3.67705	.91410	.90508	.88669	.86776	.84825	.82811	.71470	.56131	
					1.00820	1.01662	1.03421	1.05286	1.07269	1.09386	1.22703	1.44799	
.922	.3551	.9814	.56476	3.60554	.91297	.90381	.88512	.86588	.84604	.82553	.70963	.55024	
					1.00831	1.01686	1.03471	1.05366	1.07383	1.09539	1.23174	1.46215	
.921	.3575	.9812	.56801	3.53599	.91183	.90254	.88355	.86400	.84382	.82294	.70449	.53858	
					1.00842	1.01709	1.03521	1.05446	1.07497	1.09692	1.23654	1.47711	
.920	.3599	.9809	.57123	3.46832	.91069	.90126	.88198	.86211	.84159	.82034	.69927	.52622	
					1.00854	1.01733	1.03571	1.05526	1.07612	1.09847	1.24143	1.49296	
.919	.3623	.9807	.57443	3.40247	.90956	.89998	.88040	.86021	.83934	.81772	.69397	.51301	
					1.00865	1.01756	1.03621	1.05607	1.07728	1.10003	1.24642	1.50983	
.918	.3646	.9804	.57760	3.33836	.90842	.89870	.87883	.85831	.83709	.81508	.68858	.49877	
					1.00877	1.01780	1.03672	1.05688	1.07844	1.10160	1.25151	1.52791	
.917	.3669	.9802	.58074	3.27595	.90728	.89742	.87724	.85641	.83483	.81244	.68311	.48321	
					1.00888	1.01804	1.03722	1.05769	1.07961	1.10318	1.25670	1.54742	
.916	.3693	.9800	.58386	3.21515	.90614	.89614	.87566	.85449	.83256	.80977	.67754	.46586	
					1.00900	1.01827	1.03773	1.05851	1.08079	1.10478	1.26201	1.56867	
.915	.3716	.9797	.58695	3.15591	.90500	.89486	.87407	.85257	.83028	.80710	.67187	.44595	
					1.00911	1.01851	1.03824	1.05933	1.08198	1.10639	1.26743	1.59216	
.914	.3739	.9795	.59002	3.09818	.90386	.89357	.87248	.85065	.82799	.80440	.66609	.42187	
					1.00923	1.01875	1.03875	1.06016	1.08317	1.10801	1.27297	1.61866	
.913	.3762	.9792	.59306	3.04190	.90272	.89228	.87088	.84872	.82569	.80169	.66021	.38894	
					1.00934	1.01899	1.03927	1.06099	1.08437	1.10964	1.27865	1.64973	
.912	.3785	.9790	.59608	2.98703	.90158	.89100	.86928	.84678	.82338	.79897	.65420		
					1.00946	1.01923	1.03978	1.06183	1.08558	1.11129	1.28445		
.911	.3807	.9787	.59908	2.93350	.90044	.88971	.86768	.84483	.82106	.79622	.64807		
					1.00958	1.01947	1.04030	1.06267	1.08679	1.11295	1.29040		

TABLE III. Specific Heat Ratio = 1.30

R_t and TPR versus Loss Coefficient K

p/p_t	MACH	T/T_t	GAMMA	K_{crit}	.1	.2	.4	.6	.8	.9	1.0	1.5	2.0
.910	.3830	.9785	.60205	2.88129	.89930	.88842	.86607	.84288	.81872	.80624	.79347	.72419	.64180
					1.00969	1.01971	1.04082	1.06351	1.08802	1.10104	1.11462	1.19296	1.29651
.909	.3852	.9782	.60500	2.83034	.89816	.88712	.86446	.84092	.81638	.80369	.79069	.72003	.63538
					1.00981	1.01995	1.04134	1.06436	1.08925	1.10248	1.11631	1.19627	1.30277
.908	.3875	.9780	.60793	2.78061	.89701	.88583	.86285	.83896	.81402	.80112	.78789	.71581	.62881
					1.00992	1.02019	1.04186	1.06521	1.09049	1.10394	1.11801	1.19963	1.30920
.907	.3897	.9777	.61083	2.73206	.89587	.88454	.86123	.83698	.81166	.79854	.78508	.71155	.62206
					1.01004	1.02044	1.04239	1.06606	1.09173	1.10542	1.11973	1.20304	1.31581
.906	.3919	.9775	.61371	2.68465	.89472	.88324	.85961	.83500	.80928	.79594	.78225	.70723	.61514
					1.01016	1.02068	1.04291	1.06692	1.09299	1.10690	1.12146	1.20651	1.32262
.905	.3941	.9772	.61657	2.63834	.89358	.88194	.85799	.83302	.80689	.79332	.77940	.70285	.60801
					1.01028	1.02092	1.04344	1.06779	1.09425	1.10839	1.12321	1.21003	1.32963
.904	.3963	.9770	.61941	2.59310	.89243	.88064	.85636	.83102	.80448	.79069	.77652	.69841	.60066
					1.01039	1.02117	1.04397	1.06865	1.09553	1.10990	1.12497	1.21360	1.33686
.903	.3985	.9767	.62223	2.54890	.89128	.87934	.85472	.82902	.80206	.78805	.77363	.69390	.59308
					1.01051	1.02141	1.04450	1.06953	1.09681	1.11142	1.12675	1.21724	1.34433
.902	.4007	.9765	.62503	2.50570	.89013	.87804	.85309	.82701	.79963	.78539	.77072	.68934	.58522
					1.01063	1.02166	1.04504	1.07041	1.09810	1.11294	1.12854	1.22094	1.35206
.901	.4029	.9762	.62780	2.46347	.88898	.87673	.85144	.82499	.79719	.78271	.76779	.68470	.57707
					1.01075	1.02190	1.04558	1.07129	1.09940	1.11449	1.13035	1.22470	1.36007
.900	.4051	.9760	.63056	2.42219	.88784	.87542	.84980	.82297	.79474	.78001	.76483	.67999	.56859
					1.01087	1.02215	1.04611	1.07218	1.10071	1.11604	1.13218	1.22852	1.36838
.899	.4072	.9757	.63329	2.38182	.88668	.87412	.84815	.82094	.79227	.77730	.76185	.67520	.55972
					1.01098	1.02240	1.04665	1.07307	1.10202	1.11761	1.13403	1.23242	1.37703
.898	.4094	.9755	.63601	2.34232	.88553	.87281	.84650	.81890	.78978	.77457	.75885	.67034	.55042
					1.01110	1.02265	1.04720	1.07396	1.10335	1.11919	1.13589	1.23639	1.38605
.897	.4115	.9752	.63871	2.30369	.88438	.87150	.84484	.81685	.78728	.77181	.75583	.66538	.54062
					1.01122	1.02289	1.04774	1.07487	1.10469	1.12078	1.13777	1.24044	1.39548
.896	.4136	.9750	.64138	2.26589	.88323	.87018	.84317	.81479	.78477	.76904	.75277	.66034	.53021
					1.01134	1.02314	1.04829	1.07577	1.10604	1.12239	1.13967	1.24456	1.40538
.895	.4158	.9747	.64404	2.22889	.88208	.86887	.84151	.81272	.78224	.76625	.74970	.65520	.51909
					1.01146	1.02339	1.04884	1.07668	1.10739	1.12401	1.14159	1.24877	1.41581
.894	.4179	.9745	.64668	2.19268	.88092	.86755	.83984	.81065	.77970	.76344	.74660	.64997	.50708
					1.01158	1.02364	1.04939	1.07760	1.10876	1.12564	1.14352	1.25306	1.42685
.893	.4200	.9742	.64930	2.15723	.87977	.86623	.83816	.80856	.77714	.76061	.74347	.64462	.49392
					1.01170	1.02389	1.04994	1.07852	1.11014	1.12729	1.14548	1.25745	1.43860
.892	.4221	.9740	.65190	2.12252	.87861	.86491	.83648	.80647	.77456	.75776	.74031	.63916	.47923
					1.01182	1.02415	1.05050	1.07945	1.11153	1.12896	1.14746	1.26193	1.45122
.891	.4242	.9737	.65449	2.08853	.87745	.86359	.83479	.80436	.77197	.75489	.73713	.63357	.46231
					1.01194	1.02440	1.05106	1.08038	1.11293	1.13064	1.14946	1.26651	1.46491
.890	.4263	.9735	.65705	2.05523	.87630	.86227	.83310	.80225	.76936	.75199	.73391	.62785	.44172
					1.01206	1.02465	1.05162	1.08132	1.11434	1.13233	1.15147	1.27121	1.48003
.889	.4284	.9732	.65960	2.02262	.87514	.86094	.83141	.80013	.76673	.74907	.73066	.62199	.41320
					1.01218	1.02491	1.05218	1.08226	1.11576	1.13405	1.15352	1.27601	1.49728
.888	.4304	.9730	.66213	1.99066	.87398	.85961	.82971	.79800	.76408	.74613	.72739	.61598	
					1.01230	1.02516	1.05274	1.08321	1.11720	1.13577	1.15558	1.28094	
.887	.4325	.9727	.66465	1.95935	.87282	.85828	.82800	.79585	.76142	.74316	.72408	.60980	
					1.01242	1.02542	1.05331	1.08417	1.11864	1.13752	1.15767	1.28599	
.886	.4346	.9725	.66714	1.92866	.87166	.85695	.82629	.79370	.75873	.74017	.72073	.60344	
					1.01255	1.02567	1.05388	1.08513	1.12010	1.13928	1.15978	1.29118	
.885	.4366	.9722	.66962	1.89858	.87050	.85562	.82457	.79154	.75603	.73715	.71735	.59687	
					1.01267	1.02593	1.05445	1.08609	1.12157	1.14106	1.16192	1.29652	
.884	.4387	.9719	.67209	1.86908	.86933	.85428	.82285	.78936	.75331	.73410	.71394	.59009	
					1.01279	1.02619	1.05503	1.08706	1.12306	1.14286	1.16408	1.30201	
.883	.4407	.9717	.67453	1.84017	.86817	.85294	.82113	.78718	.75057	.73102	.71048	.58307	
					1.01291	1.02644	1.05560	1.08804	1.12455	1.14467	1.16627	1.30767	
.882	.4427	.9714	.67696	1.81181	.86701	.85160	.81939	.78498	.74780	.72792	.70699	.57577	
					1.01304	1.02670	1.05618	1.08903	1.12606	1.14650	1.16848	1.31353	
.881	.4447	.9712	.67938	1.78400	.86584	.85026	.81766	.78278	.74502	.72479	.70346	.56816	
					1.01316	1.02696	1.05676	1.09002	1.12759	1.14836	1.17072	1.31958	

TABLE III. Specific Heat Ratio = 1.30

p/p_t	MACH	T/T_t	GAMMA	K_{crit}	R₁ and TPR versus Loss Coefficient K								
					.1	.2	.4	.6	.7	.8	.9	1.0	1.5
.880	.4468	.9709	.68177	1.75672	.86468	.84892	.81591	.78056	.76181	.74221	.72162	.69988	.56020
					1.01328	1.02722	1.05735	1.09101	1.10944	1.12912	1.15023	1.17299	1.32584
.879	.4488	.9707	.68416	1.72996	.86351	.84757	.81417	.77833	.75929	.73938	.71843	.69626	.55185
					1.01341	1.02748	1.05793	1.09202	1.11070	1.13067	1.15213	1.17529	1.33234
.878	.4508	.9704	.68652	1.70370	.86234	.84622	.81241	.77608	.75676	.73652	.71520	.69259	.54302
					1.01353	1.02774	1.05852	1.09302	1.11197	1.13224	1.15404	1.17762	1.33912
.877	.4528	.9702	.68887	1.67793	.86117	.84487	.81065	.77383	.75422	.73364	.71194	.68888	.53364
					1.01365	1.02801	1.05911	1.09404	1.11324	1.13382	1.15598	1.17999	1.34619
.876	.4548	.9699	.69121	1.65265	.86000	.84352	.80888	.77156	.75165	.73074	.70864	.68511	.52358
					1.01378	1.02827	1.05971	1.09506	1.11453	1.13541	1.15794	1.18238	1.35359
.875	.4568	.9697	.69353	1.62783	.85883	.84217	.80711	.76928	.74906	.72781	.70530	.68129	.51268
					1.01390	1.02853	1.06031	1.09609	1.11583	1.13702	1.15992	1.18481	1.36139
.874	.4587	.9694	.69583	1.60346	.85766	.84081	.80533	.76698	.74646	.72485	.70193	.67742	.50069
					1.01403	1.02880	1.06091	1.09713	1.11714	1.13865	1.16192	1.18727	1.36964
.873	.4607	.9691	.69812	1.57954	.85649	.83945	.80355	.76467	.74384	.72186	.69851	.67349	.48722
					1.01415	1.02906	1.06151	1.09818	1.11846	1.14029	1.16395	1.18977	1.37843
.872	.4627	.9689	.70040	1.55606	.85532	.83809	.80176	.76235	.74119	.71885	.69506	.66949	.47153
					1.01428	1.02933	1.06211	1.09923	1.11979	1.14195	1.16601	1.19231	1.38792
.871	.4647	.9686	.70266	1.53299	.85414	.83672	.79996	.76001	.73853	.71580	.69156	.66544	.45204
					1.01440	1.02960	1.06272	1.10029	1.12113	1.14363	1.16809	1.19489	1.39832
.870	.4666	.9684	.70491	1.51035	.85297	.83536	.79815	.75766	.73585	.71273	.68802	.66131	.42309
					1.01453	1.02986	1.06333	1.10135	1.12248	1.14533	1.17020	1.19751	1.41012
.869	.4686	.9681	.70714	1.48810	.85179	.83399	.79634	.75529	.73314	.70962	.68443	.65712	
					1.01465	1.03013	1.06395	1.10243	1.12385	1.14704	1.17233	1.20017	
.868	.4705	.9679	.70936	1.46625	.85061	.83262	.79452	.75291	.73041	.70648	.68079	.65285	
					1.01478	1.03040	1.06456	1.10351	1.12522	1.14877	1.17450	1.20287	
.867	.4725	.9676	.71156	1.44478	.84943	.83125	.79270	.75051	.72765	.70330	.67710	.64850	
					1.01491	1.03067	1.06518	1.10460	1.12661	1.15052	1.17669	1.20562	
.866	.4744	.9673	.71375	1.42368	.84825	.82987	.79086	.74809	.72488	.70009	.67335	.64406	
					1.01503	1.03094	1.06581	1.10570	1.12802	1.15229	1.17891	1.20842	
.865	.4763	.9671	.71593	1.40296	.84707	.82849	.78902	.74566	.72207	.69685	.66955	.63954	
					1.01516	1.03121	1.06643	1.10680	1.12943	1.15408	1.18117	1.21127	
.864	.4783	.9668	.71809	1.38259	.84589	.82711	.78718	.74321	.71925	.69356	.66569	.63492	
					1.01529	1.03148	1.06706	1.10792	1.13086	1.15589	1.18346	1.21417	
.863	.4802	.9666	.72024	1.36257	.84471	.82573	.78532	.74075	.71639	.69023	.66177	.63020	
					1.01542	1.03176	1.06770	1.10904	1.13230	1.15773	1.18578	1.21713	
.862	.4821	.9663	.72237	1.34289	.84352	.82434	.78346	.73826	.71351	.68687	.65778	.62536	
					1.01554	1.03203	1.06833	1.11017	1.13376	1.15958	1.18814	1.22014	
.861	.4840	.9661	.72450	1.32355	.84234	.82295	.78159	.73576	.71060	.68345	.65372	.62041	
					1.01567	1.03230	1.06897	1.11132	1.13523	1.16146	1.19054	1.22322	
.860	.4859	.9658	.72660	1.30454	.84115	.82156	.77971	.73324	.70766	.68000	.64959	.61533	
					1.01580	1.03258	1.06961	1.11247	1.13672	1.16337	1.19297	1.22636	
.859	.4878	.9655	.72870	1.28584	.83997	.82017	.77782	.73069	.70469	.67649	.64537	.61011	
					1.01593	1.03286	1.07026	1.11363	1.13822	1.16529	1.19545	1.22958	
.858	.4897	.9653	.73078	1.26746	.83878	.81877	.77593	.72813	.70169	.67293	.64108	.60474	
					1.01606	1.03313	1.07091	1.11480	1.13974	1.16725	1.19796	1.23286	
.857	.4916	.9650	.73285	1.24938	.83759	.81737	.77402	.72555	.69865	.66933	.63670	.59920	
					1.01619	1.03341	1.07156	1.11598	1.14127	1.16923	1.20053	1.23623	
.856	.4935	.9648	.73491	1.23161	.83640	.81597	.77211	.72294	.69558	.66566	.63222	.59349	
					1.01632	1.03369	1.07221	1.11717	1.14282	1.17124	1.20313	1.23967	
.855	.4954	.9645	.73696	1.21412	.83521	.81456	.77019	.72031	.69248	.66194	.62764	.58757	
					1.01645	1.03397	1.07287	1.11837	1.14439	1.17327	1.20579	1.24321	
.854	.4973	.9642	.73899	1.19692	.83401	.81315	.76826	.71766	.68934	.65816	.62295	.58142	
					1.01658	1.03425	1.07354	1.11958	1.14598	1.17534	1.20850	1.24684	
.853	.4992	.9640	.74101	1.18000	.83282	.81174	.76632	.71499	.68616	.65431	.61815	.57503	
					1.01671	1.03453	1.07420	1.12080	1.14758	1.17743	1.21126	1.25058	
.852	.5010	.9637	.74302	1.16336	.83162	.81033	.76437	.71229	.68294	.65040	.61322	.56834	
					1.01684	1.03481	1.07487	1.12203	1.14920	1.17956	1.21408	1.25443	
.851	.5029	.9635	.74501	1.14698	.83043	.80891	.76241	.70956	.67968	.64641	.60815	.56133	
					1.01697	1.03510	1.07555	1.12328	1.15085	1.18172	1.21696	1.25840	

61

TABLE III. Specific Heat Ratio = 1.30

R₁ and TPR versus Loss Coefficient K

p/p_t	MACH	T/T_t	GAMMA	K_{crit}	.1	.2	.4	.5	.6	.7	.8	.9	1.0
.850	.5048	.9632	.74700	1.13087	.82923	.80749	.76044	.73463	.70681	.67637	.64235	.60293	.55394
					1.01710	1.03538	1.07622	1.09927	1.12453	1.15251	1.18392	1.21991	1.26251
.849	.5066	.9629	.74897	1.11501	.82803	.80607	.75846	.73229	.70403	.67302	.63820	.59754	.54610
					1.01723	1.03567	1.07690	1.10022	1.12580	1.15419	1.18615	1.22292	1.26677
.848	.5085	.9627	.75093	1.09941	.82683	.80464	.75647	.72994	.70122	.66962	.63397	.59198	.53772
					1.01736	1.03595	1.07759	1.10117	1.12708	1.15589	1.18842	1.22600	1.27119
.847	.5103	.9624	.75287	1.08405	.82563	.80321	.75447	.72757	.69838	.66617	.62964	.58621	.52866
					1.01750	1.03624	1.07828	1.10212	1.12838	1.15762	1.19073	1.22917	1.27581
.846	.5122	.9621	.75481	1.06893	.82442	.80178	.75246	.72517	.69552	.66267	.62522	.58021	.51875
					1.01763	1.03653	1.07897	1.10309	1.12968	1.15937	1.19309	1.23242	1.28065
.845	.5140	.9619	.75673	1.05405	.82322	.80034	.75043	.72277	.69262	.65911	.62069	.57396	.50770
					1.01776	1.03682	1.07967	1.10406	1.13100	1.16114	1.19549	1.23576	1.28575
.844	.5158	.9616	.75865	1.03940	.82201	.79890	.74840	.72034	.68969	.65550	.61604	.56743	.49499
					1.01790	1.03710	1.08037	1.10504	1.13234	1.16294	1.19793	1.23920	1.29117
.843	.5177	.9614	.76055	1.02499	.82081	.79746	.74635	.71789	.68672	.65182	.61127	.56056	.47958
					1.01803	1.03740	1.08108	1.10603	1.13368	1.16476	1.20042	1.24275	1.29700
.842	.5195	.9611	.76244	1.01079	.81960	.79601	.74430	.71542	.68372	.64807	.60636	.55330	.45851
					1.01816	1.03769	1.08179	1.10703	1.13505	1.16661	1.20297	1.24642	1.30341
.841	.5213	.9608	.76432	0.99681	.81839	.79456	.74223	.71293	.68068	.64426	.60130	.54557	
					1.01830	1.03798	1.08250	1.10803	1.13642	1.16849	1.20557	1.25022	
.840	.5232	.9606	.76618	0.98305	.81718	.79310	.74014	.71042	.67760	.64037	.59608	.53728	
					1.01843	1.03827	1.08322	1.10905	1.13782	1.17040	1.20823	1.25417	
.839	.5250	.9603	.76804	0.96949	.81596	.79165	.73805	.70788	.67448	.63641	.59067	.52829	
					1.01857	1.03857	1.08395	1.11007	1.13922	1.17234	1.21096	1.25829	
.838	.5268	.9600	.76989	0.95614	.81475	.79018	.73594	.70533	.67132	.63236	.58506	.51838	
					1.01870	1.03886	1.08468	1.11110	1.14065	1.17431	1.21375	1.26261	
.837	.5286	.9598	.77172	0.94300	.81353	.78872	.73382	.70274	.66812	.62822	.57923	.50722	
					1.01884	1.03916	1.08541	1.11214	1.14209	1.17631	1.21661	1.26716	
.836	.5304	.9595	.77354	0.93005	.81232	.78725	.73168	.70014	.66487	.62399	.57315	.49416	
					1.01897	1.03946	1.08615	1.11319	1.14355	1.17835	1.21955	1.27202	
.835	.5322	.9592	.77536	0.91729	.81110	.78578	.72953	.69751	.66157	.61965	.56676	.47780	
					1.01911	1.03976	1.08689	1.11425	1.14503	1.18042	1.22258	1.27725	
.834	.5340	.9590	.77716	0.90473	.80988	.78430	.72737	.69485	.65821	.61520	.56004	.45270	
					1.01924	1.04006	1.08764	1.11532	1.14653	1.18254	1.22571	1.28308	
.833	.5358	.9587	.77895	0.89235	.80865	.78282	.72519	.69216	.65481	.61063	.55293		
					1.01938	1.04036	1.08839	1.11639	1.14805	1.18469	1.22893		
.832	.5376	.9584	.78073	0.88016	.80743	.78133	.72299	.68944	.65135	.60593	.54533		
					1.01952	1.04066	1.08915	1.11748	1.14959	1.18689	1.23228		
.831	.5394	.9582	.78250	0.86814	.80621	.77984	.72078	.68669	.64782	.60108	.53714		
					1.01966	1.04096	1.08992	1.11858	1.15114	1.18914	1.23575		
.830	.5412	.9579	.78426	0.85630	.80498	.77835	.71855	.68392	.64424	.59608	.52821		
					1.01979	1.04127	1.09069	1.11969	1.15273	1.19143	1.23938		
.829	.5430	.9576	.78601	0.84464	.80375	.77685	.71631	.68111	.64058	.59090	.51829		
					1.01993	1.04157	1.09146	1.12082	1.15433	1.19378	1.24318		
.828	.5448	.9574	.78775	0.83314	.80252	.77535	.71404	.67826	.63685	.58552	.50695		
					1.02007	1.04188	1.09224	1.12195	1.15596	1.19618	1.24720		
.827	.5466	.9571	.78948	0.82182	.80129	.77384	.71176	.67538	.63305	.57992	.49340		
					1.02021	1.04219	1.09303	1.12310	1.15761	1.19865	1.25148		
.826	.5483	.9568	.79120	0.81065	.80006	.77233	.70946	.67246	.62916	.57408	.47553		
					1.02035	1.04250	1.09383	1.12425	1.15929	1.20118	1.25612		
.825	.5501	.9566	.79290	0.79965	.79882	.77081	.70714	.66951	.62519	.56794			
					1.02049	1.04281	1.09463	1.12543	1.16100	1.20378			
.824	.5519	.9563	.79460	0.78881	.79758	.76929	.70480	.66651	.62112	.56148			
					1.02063	1.04312	1.09543	1.12661	1.16274	1.20646			
.823	.5537	.9560	.79629	0.77812	.79635	.76777	.70244	.66347	.61695	.55462			
					1.02077	1.04343	1.09625	1.12781	1.16451	1.20922			
.822	.5554	.9558	.79797	0.76758	.79511	.76624	.70006	.66039	.61266	.54729			
					1.02091	1.04374	1.09706	1.12902	1.16631	1.21209			
.821	.5572	.9555	.79964	0.75719	.79386	.76470	.69766	.65726	.60826	.53938			
					1.02105	1.04406	1.09789	1.13025	1.16815	1.21506			

TABLE III. Specific Heat Ratio = 1.30

R₁ and TPR versus Loss Coefficient K

p/p_t	MACH	T/T_t	GAMMA	K_{crit}	.08	.09	.1	.2	.3	.4	.5	.6	.7
.820	.5589	.9552	.80130	0.74695	.79825	.79545	.79262	.76316	.73102	.69524	.65408	.60373	.53072
					1.01680	1.01899	1.02119	1.04438	1.07001	1.09872	1.13149	1.17002	1.21816
.819	.5607	.9550	.80295	0.73686	.79705	.79423	.79138	.76161	.72909	.69279	.65085	.59905	.52105
					1.01691	1.01911	1.02133	1.04469	1.07055	1.09956	1.13275	1.17193	1.22141
.818	.5624	.9547	.80459	0.72690	.79586	.79301	.79013	.76006	.72715	.69032	.64756	.59421	.50994
					1.01703	1.01924	1.02147	1.04501	1.07110	1.10041	1.13402	1.17388	1.22483
.817	.5642	.9544	.80622	0.71709	.79466	.79178	.78888	.75850	.72520	.68782	.64422	.58919	.49648
					1.01714	1.01936	1.02161	1.04533	1.07165	1.10126	1.13531	1.17587	1.22849
.816	.5659	.9542	.80784	0.70741	.79347	.79056	.78763	.75694	.72323	.68530	.64081	.58398	.47808
					1.01725	1.01949	1.02176	1.04565	1.07220	1.10212	1.13662	1.17791	1.23246
.815	.5677	.9539	.80945	0.69787	.79227	.78933	.78637	.75537	.72126	.68274	.63734	.57854	
					1.01736	1.01962	1.02190	1.04598	1.07276	1.10299	1.13794	1.18001	
.814	.5694	.9536	.81105	0.68846	.79107	.78810	.78512	.75380	.71927	.68016	.63380	.57285	
					1.01747	1.01975	1.02204	1.04630	1.07331	1.10387	1.13929	1.18216	
.813	.5712	.9533	.81265	0.67918	.78986	.78688	.78386	.75222	.71726	.67756	.63019	.56685	
					1.01759	1.01987	1.02219	1.04663	1.07388	1.10476	1.14066	1.18436	
.812	.5729	.9531	.81423	0.67003	.78866	.78564	.78260	.75064	.71525	.67492	.62648	.56051	
					1.01770	1.02000	1.02233	1.04695	1.07444	1.10565	1.14205	1.18664	
.811	.5746	.9528	.81580	0.66100	.78746	.78441	.78134	.74904	.71322	.67224	.62270	.55376	
					1.01781	1.02013	1.02248	1.04728	1.07501	1.10656	1.14346	1.18899	
.810	.5764	.9525	.81737	0.65210	.78625	.78318	.78007	.74745	.71117	.66954	.61883	.54650	
					1.01792	1.02026	1.02262	1.04762	1.07559	1.10747	1.14489	1.19142	
.809	.5781	.9523	.81892	0.64332	.78504	.78194	.77880	.74584	.70911	.66680	.61486	.53859	
					1.01804	1.02039	1.02277	1.04794	1.07616	1.10839	1.14635	1.19394	
.808	.5798	.9520	.82047	0.63466	.78383	.78070	.77754	.74423	.70704	.66402	.61077	.52983	
					1.01815	1.02052	1.02291	1.04828	1.07674	1.10932	1.14783	1.19658	
.807	.5815	.9517	.82201	0.62611	.78262	.77946	.77626	.74261	.70495	.66121	.60657	.51990	
					1.01827	1.02065	1.02306	1.04861	1.07733	1.11027	1.14934	1.19935	
.806	.5833	.9514	.82354	0.61769	.78141	.77821	.77499	.74099	.70284	.65835	.60224	.50815	
					1.01838	1.02078	1.02321	1.04895	1.07792	1.11122	1.15088	1.20228	
.805	.5850	.9512	.82506	0.60937	.78019	.77697	.77372	.73936	.70071	.65545	.59777	.49302	
					1.01850	1.02091	1.02335	1.04929	1.07851	1.11218	1.15245	1.20543	
.804	.5867	.9509	.82657	0.60117	.77897	.77572	.77244	.73772	.69857	.65251	.59314	.46620	
					1.01861	1.02104	1.02350	1.04963	1.07911	1.11316	1.15406	1.20893	
.803	.5884	.9506	.82807	0.59308	.77776	.77447	.77116	.73608	.69641	.64952	.58834		
					1.01873	1.02118	1.02365	1.04997	1.07971	1.11414	1.15569		
.802	.5901	.9504	.82956	0.58509	.77653	.77322	.76988	.73442	.69424	.64648	.58333		
					1.01884	1.02131	1.02380	1.05031	1.08032	1.11514	1.15737		
.801	.5918	.9501	.83105	0.57722	.77531	.77197	.76859	.73276	.69204	.64339	.57810		
					1.01896	1.02144	1.02395	1.05065	1.08093	1.11615	1.15909		
.800	.5935	.9498	.83253	0.56945	.77409	.77071	.76730	.73110	.68982	.64024	.57261		
					1.01907	1.02157	1.02410	1.05100	1.08155	1.11718	1.16085		
.799	.5952	.9495	.83399	0.56178	.77286	.76946	.76601	.72942	.68758	.63703	.56682		
					1.01919	1.02170	1.02425	1.05134	1.08217	1.11821	1.16266		
.798	.5969	.9493	.83545	0.55421	.77163	.76820	.76472	.72774	.68532	.63376	.56066		
					1.01931	1.02184	1.02440	1.05169	1.08279	1.11926	1.16453		
.797	.5986	.9490	.83690	0.54674	.77040	.76693	.76343	.72604	.68304	.63042	.55408		
					1.01942	1.02197	1.02455	1.05204	1.08342	1.12033	1.16645		
.796	.6003	.9487	.83834	0.53937	.76917	.76567	.76213	.72434	.68074	.62702	.54696		
					1.01954	1.02211	1.02470	1.05239	1.08406	1.12141	1.16844		
.795	.6020	.9484	.83978	0.53210	.76794	.76440	.76083	.72263	.67841	.62353	.53912		
					1.01966	1.02224	1.02485	1.05275	1.08470	1.12251	1.17052		
.794	.6037	.9482	.84120	0.52492	.76670	.76313	.75952	.72092	.67606	.61996	.53035		
					1.01978	1.02237	1.02500	1.05310	1.08535	1.12362	1.17268		
.793	.6054	.9479	.84262	0.51784	.76547	.76186	.75822	.71919	.67368	.61630	.52019		
					1.01990	1.02251	1.02515	1.05346	1.08600	1.12475	1.17496		
.792	.6071	.9476	.84403	0.51085	.76423	.76059	.75691	.71745	.67128	.61255	.50772		
					1.02002	1.02265	1.02531	1.05382	1.08666	1.12590	1.17737		
.791	.6088	.9473	.84543	0.50395	.76298	.75931	.75559	.71570	.66884	.60869	.48980		
					1.02013	1.02278	1.02546	1.05418	1.08732	1.12707	1.18000		

TABLE III. Specific Heat Ratio = 1.30

R₂ and TPR versus Loss Coefficient K

p/p_t	MACH	T/T_t	GAMMA	K_{crit}	.01	.02	.04	.06	.08	.09	.1	.2	.3
.790	.6105	.9471	.84682	0.49714	.78659	.78315	.77616	.76903	.76174	.75803	.75428	.71395	.66638
					1.00244	1.00490	1.00990	1.01502	1.02025	1.02292	1.02561	1.05455	1.08799
.789	.6122	.9468	.84820	0.49042	.78556	.78209	.77505	.76785	.76049	.75675	.75296	.71218	.66389
					1.00245	1.00493	1.00996	1.01511	1.02037	1.02305	1.02577	1.05491	1.08867
.788	.6139	.9465	.84958	0.48379	.78454	.78104	.77393	.76667	.75924	.75546	.75164	.71040	.66136
					1.00246	1.00496	1.01001	1.01519	1.02049	1.02319	1.02592	1.05528	1.08935
.787	.6155	.9462	.85095	0.47724	.78351	.77998	.77281	.76549	.75799	.75418	.75032	.70862	.65881
					1.00248	1.00498	1.01007	1.01528	1.02061	1.02333	1.02608	1.05565	1.09004
.786	.6172	.9459	.85231	0.47077	.78248	.77892	.77169	.76430	.75674	.75289	.74899	.70682	.65621
					1.00249	1.00501	1.01013	1.01537	1.02073	1.02347	1.02623	1.05602	1.09074
.785	.6189	.9457	.85366	0.46439	.78145	.77786	.77057	.76312	.75548	.75159	.74766	.70501	.65358
					1.00250	1.00504	1.01018	1.01546	1.02086	1.02361	1.02639	1.05639	1.09145
.784	.6206	.9454	.85500	0.45809	.78042	.77680	.76945	.76193	.75422	.75030	.74632	.70319	.65091
					1.00252	1.00507	1.01024	1.01554	1.02098	1.02374	1.02655	1.05677	1.09216
.783	.6222	.9451	.85634	0.45187	.77939	.77575	.76833	.76074	.75296	.74900	.74499	.70135	.64820
					1.00253	1.00509	1.01030	1.01563	1.02110	1.02388	1.02670	1.05715	1.09288
.782	.6239	.9448	.85766	0.44573	.77836	.77469	.76721	.75955	.75170	.74770	.74364	.69950	.64544
					1.00255	1.00512	1.01036	1.01572	1.02122	1.02402	1.02686	1.05753	1.09361
.781	.6256	.9446	.85898	0.43967	.77733	.77363	.76608	.75836	.75044	.74640	.74230	.69765	.64264
					1.00256	1.00515	1.01041	1.01581	1.02134	1.02416	1.02702	1.05791	1.09434
.780	.6272	.9443	.86030	0.43369	.77630	.77256	.76496	.75716	.74917	.74509	.74095	.69577	.63979
					1.00257	1.00518	1.01047	1.01590	1.02147	1.02430	1.02718	1.05830	1.09509
.779	.6289	.9440	.86160	0.42778	.77527	.77150	.76383	.75597	.74790	.74378	.73960	.69389	.63690
					1.00259	1.00521	1.01053	1.01599	1.02159	1.02445	1.02734	1.05868	1.09584
.778	.6306	.9437	.86290	0.42195	.77424	.77044	.76270	.75477	.74662	.74246	.73824	.69199	.63394
					1.00260	1.00523	1.01059	1.01608	1.02171	1.02459	1.02750	1.05907	1.09661
.777	.6322	.9434	.86419	0.41619	.77321	.76938	.76157	.75357	.74535	.74115	.73689	.69007	.63093
					1.00261	1.00526	1.01065	1.01617	1.02184	1.02473	1.02766	1.05947	1.09738
.776	.6339	.9432	.86547	0.41050	.77218	.76832	.76045	.75237	.74407	.73983	.73552	.68814	.62786
					1.00263	1.00529	1.01070	1.01626	1.02196	1.02487	1.02782	1.05986	1.09816
.775	.6355	.9429	.86674	0.40489	.77115	.76725	.75931	.75116	.74279	.73850	.73416	.68620	.62472
					1.00264	1.00532	1.01076	1.01635	1.02209	1.02501	1.02798	1.06026	1.09896
.774	.6372	.9426	.86801	0.39934	.77012	.76619	.75818	.74996	.74150	.73718	.73278	.68424	.62150
					1.00266	1.00535	1.01082	1.01644	1.02221	1.02516	1.02814	1.06066	1.09976
.773	.6388	.9423	.86926	0.39387	.76909	.76513	.75705	.74875	.74021	.73585	.73141	.68226	.61822
					1.00267	1.00538	1.01088	1.01653	1.02234	1.02530	1.02831	1.06106	1.10058
.772	.6405	.9420	.87052	0.38846	.76806	.76406	.75592	.74754	.73892	.73451	.73003	.68027	.61484
					1.00268	1.00540	1.01094	1.01662	1.02246	1.02545	1.02847	1.06147	1.10141
.771	.6421	.9417	.87176	0.38312	.76702	.76300	.75478	.74633	.73763	.73317	.72865	.67826	.61138
					1.00270	1.00543	1.01100	1.01671	1.02259	1.02559	1.02863	1.06188	1.10225
.770	.6438	.9415	.87299	0.37785	.76599	.76193	.75364	.74512	.73633	.73183	.72726	.67623	.60782
					1.00271	1.00546	1.01106	1.01680	1.02272	1.02574	1.02880	1.06229	1.10310
.769	.6454	.9412	.87422	0.37264	.76496	.76086	.75250	.74390	.73503	.73049	.72587	.67418	.60415
					1.00273	1.00549	1.01111	1.01690	1.02284	1.02588	1.02896	1.06271	1.10398
.768	.6471	.9409	.87544	0.36750	.76393	.75980	.75137	.74268	.73373	.72914	.72447	.67210	.60036
					1.00274	1.00552	1.01117	1.01699	1.02297	1.02603	1.02913	1.06313	1.10486
.767	.6487	.9406	.87666	0.36243	.76289	.75873	.75023	.74146	.73242	.72778	.72307	.67001	.59643
					1.00275	1.00555	1.01123	1.01708	1.02310	1.02617	1.02930	1.06355	1.10576
.766	.6504	.9403	.87786	0.35741	.76186	.75766	.74908	.74024	.73111	.72643	.72166	.66790	.59236
					1.00277	1.00558	1.01129	1.01717	1.02323	1.02632	1.02946	1.06397	1.10668
.765	.6520	.9401	.87906	0.35246	.76083	.75659	.74794	.73902	.72980	.72507	.72025	.66576	.58812
					1.00278	1.00561	1.01135	1.01727	1.02336	1.02647	1.02963	1.06440	1.10762
.764	.6537	.9398	.88025	0.34757	.75979	.75552	.74680	.73779	.72848	.72370	.71883	.66360	.58369
					1.00280	1.00563	1.01141	1.01736	1.02348	1.02662	1.02980	1.06483	1.10858
.763	.6553	.9395	.88144	0.34274	.75876	.75445	.74565	.73656	.72716	.72233	.71741	.66142	.57904
					1.00281	1.00566	1.01147	1.01745	1.02361	1.02677	1.02997	1.06527	1.10956
.762	.6569	.9392	.88262	0.33797	.75772	.75338	.74450	.73533	.72584	.72096	.71598	.65921	.57412
					1.00283	1.00569	1.01153	1.01755	1.02374	1.02692	1.03014	1.06571	1.11057
.761	.6586	.9389	.88379	0.33326	.75669	.75231	.74335	.73410	.72451	.71958	.71455	.65698	.56891
					1.00284	1.00572	1.01159	1.01764	1.02387	1.02707	1.03031	1.06615	1.11161

TABLE III. Specific Heat Ratio = 1.30

					R₂ and TPR versus Loss Coefficient K								
p/p,	MACH	T/T,	GAMMA	K_crit	.01	.02	.04	.06	.08	.09	.1	.2	.3
.760	.6602	.9386	.88495	0.32861	.75565	.75124	.74220	.73286	.72318	.71820	.71311	.65472	.56332
					1.00285	1.00575	1.01165	1.01773	1.02401	1.02722	1.03048	1.06660	1.11267
.759	.6618	.9383	.88610	0.32401	.75462	.75017	.74105	.73162	.72184	.71681	.71167	.65243	.55727
					1.00287	1.00578	1.01171	1.01783	1.02414	1.02737	1.03065	1.06705	1.11377
.758	.6635	.9381	.88725	0.31948	.75358	.74909	.73990	.73038	.72051	.71542	.71022	.65010	.55061
					1.00288	1.00581	1.01177	1.01792	1.02427	1.02752	1.03082	1.06751	1.11490
.757	.6651	.9378	.88839	0.31499	.75254	.74802	.73874	.72914	.71916	.71402	.70877	.64775	.54313
					1.00290	1.00584	1.01183	1.01802	1.02440	1.02767	1.03100	1.06797	1.11609
.756	.6667	.9375	.88953	0.31056	.75151	.74694	.73759	.72789	.71782	.71262	.70731	.64536	.53442
					1.00291	1.00587	1.01189	1.01811	1.02453	1.02783	1.03117	1.06843	1.11732
.755	.6684	.9372	.89066	0.30619	.75047	.74587	.73643	.72664	.71646	.71121	.70584	.64293	.52355
					1.00293	1.00590	1.01196	1.01821	1.02467	1.02798	1.03135	1.06890	1.11864
.754	.6700	.9369	.89178	0.30187	.74943	.74479	.73527	.72539	.71511	.70980	.70437	.64047	.50728
					1.00294	1.00593	1.01202	1.01831	1.02480	1.02813	1.03152	1.06937	1.12006
.753	.6716	.9366	.89289	0.29760	.74840	.74372	.73411	.72413	.71375	.70838	.70289	.63797	
					1.00295	1.00596	1.01208	1.01840	1.02494	1.02829	1.03170	1.06985	
.752	.6732	.9363	.89400	0.29339	.74736	.74264	.73295	.72288	.71238	.70696	.70140	.63542	
					1.00297	1.00598	1.01214	1.01850	1.02507	1.02844	1.03188	1.07034	
.751	.6748	.9361	.89510	0.28923	.74632	.74156	.73178	.72162	.71101	.70553	.69991	.63283	
					1.00298	1.00601	1.01220	1.01860	1.02521	1.02860	1.03205	1.07083	
.750	.6765	.9358	.89619	0.28511	.74528	.74048	.73061	.72035	.70964	.70409	.69841	.63020	
					1.00300	1.00604	1.01226	1.01869	1.02534	1.02876	1.03223	1.07132	
.749	.6781	.9355	.89728	0.28105	.74424	.73940	.72945	.71908	.70826	.70265	.69690	.62751	
					1.00301	1.00607	1.01233	1.01879	1.02548	1.02891	1.03241	1.07182	
.748	.6797	.9352	.89836	0.27704	.74320	.73832	.72828	.71781	.70688	.70120	.69539	.62476	
					1.00303	1.00610	1.01239	1.01889	1.02561	1.02907	1.03259	1.07233	
.747	.6813	.9349	.89943	0.27308	.74216	.73724	.72711	.71654	.70549	.69975	.69387	.62196	
					1.00304	1.00613	1.01245	1.01899	1.02575	1.02923	1.03277	1.07284	
.746	.6829	.9346	.90049	0.26916	.74112	.73616	.72593	.71527	.70410	.69829	.69234	.61909	
					1.00306	1.00616	1.01251	1.01908	1.02589	1.02939	1.03295	1.07336	
.745	.6846	.9343	.90155	0.26529	.74008	.73508	.72476	.71399	.70270	.69683	.69080	.61616	
					1.00307	1.00619	1.01258	1.01918	1.02603	1.02955	1.03314	1.07389	
.744	.6862	.9340	.90260	0.26147	.73904	.73399	.72358	.71270	.70129	.69535	.68925	.61315	
					1.00309	1.00622	1.01264	1.01928	1.02617	1.02971	1.03332	1.07442	
.743	.6878	.9337	.90365	0.25770	.73800	.73291	.72240	.71142	.69988	.69387	.68770	.61006	
					1.00310	1.00625	1.01270	1.01938	1.02631	1.02987	1.03351	1.07497	
.742	.6894	.9335	.90469	0.25397	.73696	.73182	.72122	.71012	.69847	.69239	.68613	.60688	
					1.00312	1.00628	1.01276	1.01948	1.02645	1.03003	1.03369	1.07552	
.741	.6910	.9332	.90572	0.25029	.73592	.73074	.72004	.70883	.69704	.69089	.68456	.60360	
					1.00313	1.00631	1.01283	1.01958	1.02659	1.03020	1.03388	1.07608	
.740	.6926	.9329	.90675	0.24665	.73488	.72965	.71885	.70753	.69562	.68939	.68298	.60022	
					1.00314	1.00634	1.01289	1.01968	1.02673	1.03036	1.03407	1.07665	
.739	.6942	.9326	.90777	0.24306	.73383	.72856	.71766	.70623	.69418	.68788	.68139	.59671	
					1.00316	1.00637	1.01295	1.01978	1.02687	1.03053	1.03425	1.07722	
.738	.6958	.9323	.90878	0.23951	.73279	.72747	.71647	.70493	.69274	.68637	.67978	.59307	
					1.00317	1.00641	1.01302	1.01988	1.02701	1.03069	1.03444	1.07781	
.737	.6974	.9320	.90978	0.23600	.73175	.72638	.71528	.70362	.69129	.68484	.67817	.58927	
					1.00319	1.00644	1.01308	1.01998	1.02716	1.03086	1.03463	1.07842	
.736	.6990	.9317	.91078	0.23254	.73070	.72529	.71409	.70230	.68983	.68331	.67655	.58530	
					1.00320	1.00647	1.01315	1.02009	1.02730	1.03102	1.03483	1.07904	
.735	.7006	.9314	.91178	0.22911	.72966	.72420	.71289	.70099	.68837	.68176	.67492	.58113	
					1.00322	1.00650	1.01321	1.02019	1.02745	1.03119	1.03502	1.07966	
.734	.7022	.9311	.91276	0.22573	.72861	.72310	.71169	.69966	.68691	.68021	.67327	.57672	
					1.00323	1.00653	1.01328	1.02029	1.02759	1.03136	1.03521	1.08031	
.733	.7038	.9308	.91374	0.22239	.72757	.72201	.71049	.69834	.68543	.67865	.67162	.57202	
					1.00325	1.00655	1.01334	1.02039	1.02774	1.03153	1.03541	1.08097	
.732	.7055	.9305	.91472	0.21909	.72652	.72092	.70929	.69700	.68395	.67708	.66995	.56697	
					1.00326	1.00658	1.01341	1.02050	1.02789	1.03170	1.03560	1.08164	
.731	.7071	.9302	.91568	0.21583	.72548	.71982	.70808	.69566	.68246	.67550	.66827	.56148	
					1.00328	1.00661	1.01347	1.02060	1.02803	1.03187	1.03580	1.08234	

TABLE III. Specific Heat Ratio = 1.30

p/p_t	MACH	T/T_t	GAMMA	K_{crit}	.01	.02	.04	.06	.07	.08	.09	.1	.2
								R₂ and TPR versus Loss Coefficient K					
.730	.7087	.9299	.91664	0.21261	.72443	.71872	.70687	.69432	.68776	.68096	.67391	.66658	.55540
					1.00329	1.00665	1.01354	1.02071	1.02440	1.02818	1.03204	1.03600	1.08307
.729	.7102	.9297	.91760	0.20943	.72338	.71762	.70566	.69298	.68633	.67945	.67231	.66487	.54848
					1.00331	1.00668	1.01360	1.02081	1.02453	1.02833	1.03222	1.03620	1.08382
.728	.7118	.9294	.91855	0.20629	.72233	.71653	.70444	.69163	.68490	.67794	.67070	.66315	.54026
					1.00332	1.00671	1.01367	1.02092	1.02466	1.02848	1.03239	1.03640	1.08461
.727	.7134	.9291	.91949	0.20319	.72128	.71542	.70323	.69027	.68347	.67641	.66907	.66141	.52957
					1.00334	1.00674	1.01373	1.02102	1.02478	1.02863	1.03257	1.03660	1.08545
.726	.7150	.9288	.92042	0.20012	.72024	.71432	.70201	.68891	.68203	.67488	.66744	.65966	.50794
					1.00336	1.00677	1.01380	1.02113	1.02491	1.02878	1.03274	1.03681	1.08637
.725	.7166	.9285	.92135	0.19709	.71919	.71322	.70078	.68754	.68058	.67334	.66579	.65789	
					1.00337	1.00680	1.01387	1.02123	1.02504	1.02893	1.03292	1.03701	
.724	.7182	.9282	.92227	0.19410	.71814	.71212	.69955	.68617	.67912	.67179	.66413	.65611	
					1.00339	1.00683	1.01394	1.02134	1.02517	1.02909	1.03310	1.03722	
.723	.7198	.9279	.92319	0.19114	.71709	.71101	.69832	.68480	.67766	.67022	.66246	.65431	
					1.00340	1.00686	1.01400	1.02145	1.02530	1.02924	1.03328	1.03743	
.722	.7214	.9276	.92410	0.18822	.71603	.70991	.69709	.68341	.67618	.66865	.66077	.65249	
					1.00342	1.00690	1.01407	1.02156	1.02543	1.02939	1.03346	1.03764	
.721	.7230	.9273	.92501	0.18534	.71498	.70880	.69586	.68202	.67470	.66707	.65907	.65066	
					1.00343	1.00693	1.01414	1.02166	1.02556	1.02955	1.03364	1.03785	
.720	.7246	.9270	.92590	0.18249	.71393	.70769	.69462	.68063	.67322	.66548	.65736	.64880	
					1.00345	1.00696	1.01420	1.02177	1.02569	1.02970	1.03382	1.03806	
.719	.7262	.9267	.92680	0.17968	.71288	.70658	.69338	.67922	.67172	.66387	.65563	.64693	
					1.00346	1.00699	1.01427	1.02188	1.02582	1.02986	1.03401	1.03827	
.718	.7278	.9264	.92768	0.17690	.71182	.70547	.69213	.67782	.67021	.66225	.65388	.64503	
					1.00348	1.00702	1.01434	1.02199	1.02595	1.03002	1.03419	1.03849	
.717	.7294	.9261	.92856	0.17415	.71077	.70435	.69088	.67640	.66870	.66062	.65212	.64311	
					1.00349	1.00705	1.01441	1.02210	1.02609	1.03018	1.03438	1.03871	
.716	.7310	.9258	.92943	0.17144	.70972	.70324	.68963	.67498	.66717	.65898	.65034	.64117	
					1.00351	1.00709	1.01448	1.02221	1.02622	1.03034	1.03457	1.03893	
.715	.7325	.9255	.93030	0.16876	.70866	.70212	.68838	.67355	.66564	.65733	.64854	.63920	
					1.00352	1.00712	1.01455	1.02232	1.02636	1.03050	1.03476	1.03915	
.714	.7341	.9252	.93116	0.16611	.70760	.70101	.68712	.67211	.66409	.65566	.64673	.63720	
					1.00354	1.00715	1.01462	1.02243	1.02649	1.03066	1.03495	1.03937	
.713	.7357	.9249	.93202	0.16349	.70655	.69989	.68585	.67067	.66254	.65397	.64489	.63518	
					1.00356	1.00718	1.01468	1.02255	1.02663	1.03082	1.03514	1.03959	
.712	.7373	.9246	.93287	0.16091	.70549	.69877	.68459	.66921	.66097	.65227	.64303	.63313	
					1.00357	1.00722	1.01475	1.02266	1.02676	1.03099	1.03533	1.03982	
.711	.7389	.9243	.93371	0.15836	.70443	.69764	.68331	.66775	.65939	.65056	.64116	.63105	
					1.00359	1.00725	1.01482	1.02277	1.02690	1.03115	1.03553	1.04005	
.710	.7405	.9240	.93455	0.15584	.70337	.69652	.68204	.66628	.65781	.64883	.63925	.62894	
					1.00360	1.00728	1.01489	1.02289	1.02704	1.03132	1.03573	1.04028	
.709	.7420	.9237	.93538	0.15335	.70231	.69540	.68076	.66481	.65620	.64708	.63733	.62679	
					1.00362	1.00731	1.01496	1.02300	1.02718	1.03148	1.03592	1.04051	
.708	.7436	.9234	.93620	0.15089	.70125	.69427	.67948	.66332	.65459	.64532	.63538	.62459	
					1.00363	1.00735	1.01503	1.02311	1.02732	1.03165	1.03612	1.04076	
.707	.7452	.9231	.93702	0.14846	.70019	.69314	.67819	.66182	.65296	.64353	.63340	.62237	
					1.00365	1.00738	1.01510	1.02323	1.02746	1.03182	1.03633	1.04099	
.706	.7468	.9228	.93784	0.14607	.59913	.69201	.67689	.66031	.65132	.64173	.63139	.62010	
					1.00367	1.00741	1.01518	1.02335	1.02760	1.03199	1.03653	1.04124	
.705	.7484	.9225	.93864	0.14370	.69807	.69088	.67560	.65880	.64967	.63990	.62936	.61778	
					1.00368	1.00744	1.01525	1.02346	1.02775	1.03217	1.03673	1.04148	
.704	.7500	.9222	.93945	0.14136	.69700	.68974	.67429	.65727	.64800	.63806	.62728	.61542	
					1.00370	1.00748	1.01532	1.02358	1.02789	1.03234	1.03695	1.04173	
.703	.7515	.9219	.94024	0.13905	.69594	.68861	.67298	.65573	.64631	.63619	.62518	.61300	
					1.00371	1.00751	1.01539	1.02370	1.02803	1.03251	1.03716	1.04198	
.702	.7531	.9216	.94103	0.13676	.69488	.68747	.67167	.65418	.64461	.63430	.62304	.61053	
					1.00373	1.00754	1.01546	1.02382	1.02818	1.03269	1.03737	1.04223	
.701	.7547	.9213	.94181	0.13451	.69381	.68633	.67035	.65262	.64289	.63238	.62086	.60799	
					1.00375	1.00758	1.01553	1.02394	1.02833	1.03287	1.03758	1.04249	

TABLE III. Specific Heat Ratio = 1.30

					R₁ and TPR versus Loss Coefficient K								
p/p,	MACH	T/T,	GAMMA	K_crit	.01	.02	.04	.05	.06	.07	.08	.09	.1
.700	.7563	.9210	.94259	0.13228	.69274	.68519	.66903	.66031	.65106	.64115	.63042	.61864	.60539
					1.00376	1.00761	1.01561	1.01977	1.02405	1.02848	1.03305	1.03780	1.04275
.699	.7578	.9207	.94336	0.13009	.69168	.68405	.66770	.65887	.64947	.63940	.62845	.61638	.60270
					1.00378	1.00764	1.01568	1.01986	1.02417	1.02862	1.03323	1.03802	1.04301
.698	.7594	.9204	.94413	0.12791	.69061	.68290	.66636	.65741	.64787	.63762	.62645	.61406	.59994
					1.00379	1.00768	1.01575	1.01996	1.02430	1.02877	1.03342	1.03824	1.04328
.697	.7610	.9201	.94489	0.12577	.68954	.68175	.66502	.65595	.64626	.63582	.62441	.61170	.59708
					1.00381	1.00771	1.01583	1.02006	1.02442	1.02893	1.03360	1.03846	1.04355
.696	.7626	.9198	.94565	0.12365	.68847	.68060	.66367	.65447	.64463	.63401	.62234	.60928	.59412
					1.00383	1.00774	1.01590	1.02015	1.02454	1.02908	1.03379	1.03869	1.04383
.695	.7642	.9195	.94640	0.12156	.68740	.67945	.66232	.65299	.64299	.63215	.62023	.60679	.59103
					1.00384	1.00778	1.01598	1.02025	1.02467	1.02924	1.03398	1.03892	1.04411
.694	.7657	.9192	.94714	0.11950	.68632	.67830	.66096	.65150	.64134	.63029	.61809	.60424	.58781
					1.00386	1.00781	1.01605	1.02035	1.02479	1.02939	1.03417	1.03915	1.04440
.693	.7673	.9189	.94788	0.11746	.68525	.67714	.65959	.64999	.63966	.62839	.61590	.60162	.58444
					1.00388	1.00785	1.01613	1.02045	1.02492	1.02955	1.03436	1.03939	1.04469
.692	.7689	.9185	.94861	0.11545	.68418	.67598	.65822	.64848	.63797	.62647	.61366	.59891	.58088
					1.00389	1.00788	1.01620	1.02055	1.02504	1.02971	1.03456	1.03963	1.04499
.691	.7705	.9182	.94934	0.11346	.68310	.67482	.65683	.64695	.63626	.62452	.61138	.59611	.57709
					1.00391	1.00791	1.01628	1.02065	1.02517	1.02987	1.03475	1.03987	1.04529
.690	.7720	.9179	.95006	0.11149	.68202	.67365	.65544	.64541	.63452	.62254	.60904	.59320	.57304
					1.00392	1.00795	1.01635	1.02075	1.02530	1.03003	1.03495	1.04012	1.04561
.689	.7736	.9176	.95077	0.10956	.68095	.67249	.65405	.64386	.63277	.62053	.60665	.59018	.56864
					1.00394	1.00798	1.01643	1.02085	1.02543	1.03019	1.03515	1.04038	1.04593
.688	.7752	.9173	.95148	0.10764	.67987	.67132	.65264	.64229	.63100	.61848	.60419	.58702	.56379
					1.00396	1.00802	1.01651	1.02095	1.02556	1.03035	1.03536	1.04063	1.04626
.687	.7767	.9170	.95218	0.10575	.67879	.67014	.65123	.64071	.62921	.61640	.60166	.58370	.55827
					1.00397	1.00805	1.01658	1.02105	1.02569	1.03052	1.03557	1.04089	1.04661
.686	.7783	.9167	.95288	0.10389	.67771	.66897	.64980	.63912	.62739	.61427	.59905	.58019	.55179
					1.00399	1.00809	1.01666	1.02116	1.02583	1.03069	1.03578	1.04116	1.04697
.685	.7799	.9164	.95358	0.10205	.67662	.66779	.64837	.63750	.62555	.61210	.59636	.57647	.54348
					1.00401	1.00812	1.01674	1.02126	1.02596	1.03086	1.03599	1.04144	1.04735
.684	.7815	.9161	.95426	0.10023	.67554	.66661	.64693	.63588	.62368	.60988	.59357	.57246	.52850
					1.00402	1.00816	1.01682	1.02137	1.02610	1.03103	1.03621	1.04172	1.04777
.683	.7830	.9158	.95494	0.09843	.67446	.66542	.64547	.63423	.62178	.60761	.59067	.56810	
					1.00404	1.00819	1.01690	1.02147	1.02623	1.03120	1.03643	1.04201	
.682	.7846	.9155	.95562	0.09666	.67337	.66424	.64401	.63257	.61986	.60528	.58764	.56325	
					1.00406	1.00823	1.01697	1.02158	1.02637	1.03138	1.03666	1.04231	
.681	.7862	.9152	.95629	0.09491	.67228	.66304	.64254	.63090	.61790	.60289	.58447	.55773	
					1.00407	1.00826	1.01705	1.02169	1.02651	1.03156	1.03689	1.04262	
.680	.7877	.9148	.95695	0.09319	.67119	.66185	.64105	.62920	.61590	.60043	.58112	.55113	
					1.00409	1.00830	1.01713	1.02179	1.02665	1.03174	1.03712	1.04295	
.679	.7893	.9145	.95761	0.09148	.67010	.66065	.63956	.62748	.61387	.59790	.57757	.54230	
					1.00411	1.00833	1.01722	1.02190	1.02679	1.03192	1.03736	1.04330	
.678	.7909	.9142	.95827	0.08980	.66901	.65945	.63805	.62574	.61180	.59528	.57377		
					1.00412	1.00837	1.01730	1.02201	1.02693	1.03211	1.03761		
.677	.7924	.9139	.95892	0.08814	.66792	.65825	.63653	.62398	.60969	.59257	.56964		
					1.00414	1.00841	1.01738	1.02212	1.02708	1.03230	1.03786		
.676	.7940	.9136	.95956	0.08650	.66682	.65704	.63499	.62220	.60752	.58975	.56506		
					1.00416	1.00844	1.01746	1.02223	1.02723	1.03249	1.03813		
.675	.7956	.9133	.96020	0.08489	.66573	.65582	.63344	.62038	.60531	.58680	.55991		
					1.00418	1.00848	1.01754	1.02235	1.02737	1.03269	1.03840		
.674	.7972	.9130	.96083	0.08329	.66463	.65461	.63188	.61854	.60304	.58371	.55382		
					1.00419	1.00851	1.01763	1.02246	1.02752	1.03289	1.03868		
.673	.7987	.9127	.96145	0.08172	.66353	.65339	.63030	.61667	.60071	.58044	.54602		
					1.00421	1.00855	1.01771	1.02257	1.02768	1.03309	1.03898		
.672	.8003	.9124	.96208	0.08016	.66243	.65216	.62870	.61477	.59831	.57698	.53162		
					1.00423	1.00859	1.01779	1.02269	1.02783	1.03330	1.03931		
.671	.8019	.9120	.96269	0.07863	.66133	.65093	.62709	.61284	.59583	.57326			
					1.00424	1.00862	1.01788	1.02281	1.02799	1.03351			

TABLE III. Specific Heat Ratio = 1.30

R₂ and TPR versus Loss Coefficient K

p/p,	MACH	T/T,	GAMMA	K_crit	.01	.02	.03	.04	.05	.055	.06	.065	.07
.670	.8034	.9117	.96330	0.07712	.66022	.64969	.63821	.62545	.61087	.60257	.59327	.58250	.56919
					1.00426	1.00866	1.01322	1.01797	1.02292	1.02550	1.02815	1.03089	1.03374
.669	.8050	.9114	.96391	0.07562	.65911	.64845	.63679	.62380	.60885	.60030	.59061	.57924	.56470
					1.00428	1.00870	1.01328	1.01805	1.02304	1.02563	1.02831	1.03108	1.03396
.668	.8066	.9111	.96451	0.07415	.65801	.64721	.63537	.62213	.60680	.59796	.58784	.57575	.55959
					1.00430	1.00873	1.01334	1.01814	1.02316	1.02577	1.02848	1.03127	1.03420
.667	.8081	.9108	.96510	0.07270	.65689	.64595	.63394	.62044	.60470	.59554	.58494	.57200	.55351
					1.00431	1.00877	1.01340	1.01823	1.02329	1.02592	1.02864	1.03147	1.03445
.666	.8097	.9105	.96569	0.07126	.65578	.64469	.63249	.61872	.60255	.59302	.58190	.56790	.54535
					1.00433	1.00881	1.01346	1.01832	1.02341	1.02607	1.02882	1.03168	1.03471
.665	.8113	.9101	.96627	0.06985	.65467	.64343	.63103	.61698	.60034	.59043	.57867	.56332	
					1.00435	1.00885	1.01353	1.01841	1.02353	1.02621	1.02899	1.03189	
.664	.8128	.9098	.96685	0.06845	.65355	.64216	.62956	.61521	.59807	.58773	.57523	.55804	
					1.00436	1.00889	1.01359	1.01850	1.02366	1.02636	1.02917	1.03211	
.663	.8144	.9095	.96742	0.06708	.65243	.64089	.62808	.61342	.59573	.58490	.57151	.55152	
					1.00438	1.00893	1.01365	1.01859	1.02379	1.02652	1.02936	1.03235	
.662	.8160	.9092	.96799	0.06572	.65131	.63961	.62658	.61159	.59332	.58194	.56743	.54219	
					1.00440	1.00896	1.01371	1.01868	1.02392	1.02667	1.02955	1.03260	
.661	.8175	.9089	.96855	0.06438	.65019	.63833	.62507	.60974	.59082	.57880	.56286		
					1.00442	1.00900	1.01377	1.01877	1.02405	1.02683	1.02974		
.660	.8191	.9086	.96911	0.06306	.64905	.63703	.62354	.60784	.58822	.57545	.55756		
					1.00444	1.00904	1.01384	1.01886	1.02419	1.02700	1.02995		
.659	.8207	.9082	.96966	0.06176	.64792	.63573	.62200	.60591	.58550	.57183	.55091		
					1.00446	1.00908	1.01390	1.01896	1.02432	1.02717	1.03017		
.658	.8222	.9079	.97020	0.06048	.64679	.63442	.62044	.60394	.58266	.56788	.54078		
					1.00448	1.00912	1.01396	1.01905	1.02446	1.02734	1.03040		
.657	.8238	.9076	.97075	0.05921	.64565	.63311	.61886	.60193	.57966	.56346			
					1.00449	1.00916	1.01403	1.01915	1.02461	1.02752			
.656	.8254	.9073	.97128	0.05796	.64452	.63179	.61727	.59986	.57647	.55835			
					1.00451	1.00920	1.01409	1.01925	1.02475	1.02770			
.655	.8269	.9070	.97181	0.05673	.64338	.63045	.61565	.59774	.57301	.55196			
					1.00453	1.00924	1.01416	1.01935	1.02491	1.02790			
.654	.8285	.9067	.97234	0.05552	.64223	.62911	.61401	.59556	.56927	.54246			
					1.00455	1.00928	1.01423	1.01945	1.02506	1.02811			
.653	.8301	.9063	.97286	0.05432	.64109	.62777	.61235	.59331	.56512				
					1.00457	1.00932	1.01429	1.01955	1.02522				
.652	.8316	.9060	.97337	0.05314	.63994	.62641	.61066	.59099	.56038				
					1.00458	1.00936	1.01436	1.01966	1.02539				
.651	.8332	.9057	.97388	0.05198	.63879	.62504	.60895	.58859	.55468				
					1.00460	1.00940	1.01443	1.01976	1.02556				
.650	.8347	.9054	.97439	0.05084	.63763	.62366	.60720	.58608	.54681				
					1.00462	1.00944	1.01450	1.01987	1.02575				
.649	.8363	.9050	.97488	0.04971	.63647	.62227	.60543	.58347					
					1.00464	1.00948	1.01456	1.01998					
.648	.8379	.9047	.97538	0.04860	.63531	.62086	.60362	.58072					
					1.00466	1.00952	1.01463	1.02009					
.647	.8394	.9044	.97587	0.04750	.63414	.61945	.60178	.57781					
					1.00468	1.00956	1.01471	1.02020					
.646	.8410	.9041	.97635	0.04642	.63297	.61802	.59989	.57468					
					1.00470	1.00960	1.01478	1.02032					
.645	.8426	.9038	.97683	0.04536	.63179	.61658	.59797	.57132					
					1.00471	1.00964	1.01485	1.02044					
.644	.8441	.9034	.97730	0.04431	.63061	.61512	.59599	.56765					
					1.00473	1.00968	1.01492	1.02057					
.643	.8457	.9031	.97777	0.04328	.62943	.61365	.59397	.56352					
					1.00475	1.00973	1.01500	1.02069					
.642	.8473	.9028	.97823	0.04226	.62824	.61216	.59188	.55872					
					1.00477	1.00977	1.01507	1.02083					
.641	.8488	.9025	.97869	0.04126	.62705	.61065	.58973	.55264					
					1.00479	1.00981	1.01515	1.02097					

TABLE III. Specific Heat Ratio = 1.30

R₁ and TPR versus Loss Coefficient K

p/p_t	MACH	T/T_t	GAMMA	K_{crit}	.006	.008	.01	.015	.02	.025	.03	.035	.04
.640	.8504	.9021	.97915	0.04027	.63176	.62885	.62586	.61791	.60914	.59920	.58749	.57238	.54299
					1.00286	1.00383	1.00481	1.00730	1.00985	1.01249	1.01523	1.01808	1.02112
.639	.8520	.9018	.97959	0.03930	.63064	.62770	.62465	.61656	.60760	.59737	.58517	.56895	
					1.00287	1.00384	1.00482	1.00733	1.00990	1.01255	1.01531	1.01818	
.638	.8535	.9015	.98004	0.03834	.62953	.62654	.62344	.61520	.60603	.59549	.58275	.56515	
					1.00288	1.00386	1.00484	1.00736	1.00994	1.01261	1.01539	1.01829	
.637	.8551	.9012	.98047	0.03740	.62841	.62537	.62223	.61382	.60444	.59357	.58022	.56076	
					1.00289	1.00387	1.00486	1.00739	1.00998	1.01267	1.01547	1.01841	
.636	.8567	.9008	.98091	0.03648	.62729	.62421	.62101	.61244	.60281	.59159	.57755	.55545	
					1.00290	1.00389	1.00488	1.00742	1.01003	1.01274	1.01555	1.01853	
.635	.8582	.9005	.98133	0.03556	.62617	.62303	.61978	.61104	.60116	.58956	.57470	.54808	
					1.00291	1.00390	1.00490	1.00746	1.01008	1.01280	1.01564	1.01865	
.634	.8598	.9002	.98176	0.03467	.62504	.62186	.61853	.60963	.59949	.58746	.57165		
					1.00292	1.00392	1.00493	1.00749	1.01013	1.01286	1.01572		
.633	.8614	.8998	.98217	0.03378	.62391	.62067	.61729	.60821	.59779	.58528	.56832		
					1.00293	1.00393	1.00495	1.00752	1.01017	1.01293	1.01581		
.632	.8629	.8995	.98259	0.03291	.62278	.61947	.61604	.60676	.59605	.58301	.56461		
					1.00294	1.00395	1.00497	1.00755	1.01022	1.01299	1.01591		
.631	.8645	.8992	.98299	0.03206	.62165	.61828	.61478	.60528	.59427	.58065	.56028		
					1.00296	1.00397	1.00499	1.00759	1.01027	1.01306	1.01601		
.630	.8661	.8989	.98340	0.03122	.62050	.61708	.61351	.60380	.59245	.57816	.55501		
					1.00297	1.00398	1.00501	1.00762	1.01031	1.01313	1.01611		
.629	.8677	.8985	.98379	0.03039	.61936	.61587	.61224	.60230	.59059	.57552	.54727		
					1.00298	1.00400	1.00503	1.00765	1.01036	1.01320	1.01622		
.628	.8692	.8982	.98419	0.02958	.61821	.61466	.61096	.60078	.58867	.57271			
					1.00299	1.00401	1.00505	1.00769	1.01041	1.01327			
.627	.8708	.8979	.98457	0.02878	.61704	.61344	.60966	.59923	.58669	.56967			
					1.00301	1.00403	1.00507	1.00772	1.01046	1.01334			
.626	.8724	.8975	.98496	0.02799	.61588	.61221	.60836	.59766	.58464	.56634			
					1.00302	1.00405	1.00509	1.00775	1.01051	1.01342			
.625	.8739	.8972	.98533	0.02722	.61472	.61098	.60704	.59607	.58252	.56251			
					1.00303	1.00406	1.00511	1.00779	1.01056	1.01350			
.624	.8755	.8969	.98571	0.02646	.61355	.60974	.60572	.59444	.58031	.55802			
					1.00304	1.00408	1.00513	1.00782	1.01062	1.01358			
.623	.8771	.8965	.98608	0.02571	.61238	.60849	.60438	.59278	.57799	.55220			
					1.00305	1.00409	1.00515	1.00786	1.01067	1.01366			
.622	.8786	.8962	.98644	0.02497	.61120	.60723	.60302	.59108	.57554				
					1.00306	1.00411	1.00517	1.00789	1.01072				
.621	.8802	.8959	.98680	0.02425	.61001	.60596	.60166	.58934	.57294				
					1.00308	1.00413	1.00519	1.00793	1.01078				
.620	.8818	.8956	.98715	0.02354	.60882	.60468	.60027	.58756	.57015				
					1.00309	1.00414	1.00521	1.00796	1.01084				
.619	.8834	.8952	.98750	0.02285	.60762	.60340	.59888	.58573	.56711				
					1.00310	1.00416	1.00523	1.00800	1.01089				
.618	.8849	.8949	.98784	0.02216	.60642	.60210	.59746	.58384	.56365				
					1.00311	1.00418	1.00526	1.00804	1.01096				
.617	.8865	.8945	.98818	0.02149	.60521	.60078	.59602	.58188	.55968				
					1.00312	1.00419	1.00528	1.00808	1.01102				
.616	.8881	.8942	.98851	0.02083	.60399	.59946	.59457	.57986	.55474				
					1.00314	1.00421	1.00530	1.00811	1.01109				
.615	.8896	.8939	.98884	0.02019	.60276	.59812	.59309	.57774	.54694				
					1.00315	1.00423	1.00532	1.00815	1.01116				
.614	.8912	.8935	.98916	0.01955	.60152	.59677	.59158	.57552					
					1.00316	1.00424	1.00535	1.00819					
.613	.8928	.8932	.98948	0.01893	.60027	.59540	.59005	.57317					
					1.00317	1.00426	1.00537	1.00823					
.612	.8944	.8929	.98980	0.01832	.59901	.59401	.58849	.57067					
					1.00318	1.00428	1.00539	1.00827					
.611	.8959	.8925	.99010	0.01772	.59775	.59260	.58690	.56797					
					1.00320	1.00430	1.00541	1.00832					

TABLE III. Specific Heat Ratio = 1.30

p/p₁	MACH	T/T₁	GAMMA	K_crit	R₁ and TPR versus Loss Coefficient K								
					.001	.002	.003	.004	.005	.006	.008	.01	.015
.610	.8975	.8922	.99041	0.01713	.60793	.60580	.60359	.60131	.59894	.59647	.59117	.58527	.56494
					1.00053	1.00105	1.00159	1.00212	1.00266	1.00321	1.00431	1.00544	1.00836
.609	.8991	.8919	.99071	0.01656	.60689	.60472	.60247	.60014	.59771	.59518	.58972	.58359	.56155
					1.00053	1.00106	1.00159	1.00213	1.00267	1.00322	1.00433	1.00546	1.00841
.608	.9007	.8915	.99100	0.01599	.60586	.60364	.60134	.59896	.59647	.59388	.58825	.58188	.55752
					1.00053	1.00106	1.00160	1.00214	1.00268	1.00323	1.00435	1.00548	1.00846
.607	.9022	.8912	.99129	0.01544	.60482	.60256	.60021	.59777	.59523	.59256	.58675	.58010	.55209
					1.00053	1.00107	1.00160	1.00215	1.00269	1.00325	1.00437	1.00551	1.00851
.606	.9038	.8908	.99158	0.01490	.60378	.60147	.59908	.59658	.59397	.59122	.58521	.57827	
					1.00053	1.00107	1.00161	1.00215	1.00270	1.00326	1.00438	1.00553	
.605	.9054	.8905	.99185	0.01437	.60273	.60038	.59793	.59538	.59270	.58987	.58365	.57637	
					1.00053	1.00107	1.00162	1.00216	1.00271	1.00327	1.00440	1.00556	
.604	.9070	.8902	.99213	0.01385	.60169	.59929	.59679	.59417	.59141	.58850	.58204	.57438	
					1.00054	1.00108	1.00162	1.00217	1.00272	1.00328	1.00442	1.00558	
.603	.9085	.8898	.99240	0.01334	.60065	.59819	.59563	.59295	.59011	.58711	.58039	.57225	
					1.00054	1.00108	1.00163	1.00218	1.00273	1.00330	1.00444	1.00561	
.602	.9101	.8895	.99267	0.01284	.59960	.59709	.59447	.59172	.58880	.58570	.57870	.57002	
					1.00054	1.00108	1.00163	1.00219	1.00275	1.00331	1.00446	1.00564	
.601	.9117	.8891	.99293	0.01236	.59855	.59599	.59331	.59047	.58747	.58426	.57695	.56763	
					1.00054	1.00109	1.00164	1.00219	1.00276	1.00332	1.00448	1.00567	
.600	.9133	.8888	.99318	0.01188	.59750	.59488	.59213	.58922	.58612	.58276	.57513	.56503	
					1.00054	1.00109	1.00164	1.00220	1.00277	1.00334	1.00450	1.00570	
.599	.9149	.8885	.99343	0.01142	.59645	.59377	.59095	.58795	.58475	.58126	.57319	.56213	
					1.00054	1.00110	1.00165	1.00221	1.00278	1.00335	1.00452	1.00573	
.598	.9164	.8881	.99368	0.01096	.59540	.59265	.58976	.58667	.58333	.57972	.57119	.55877	
					1.00055	1.00110	1.00166	1.00222	1.00279	1.00337	1.00454	1.00576	
.597	.9180	.8878	.99392	0.01052	.59434	.59153	.58855	.58538	.58191	.57814	.56907	.55450	
					1.00055	1.00110	1.00166	1.00223	1.00280	1.00338	1.00457	1.00579	
.596	.9196	.8874	.99416	0.01008	.59328	.59040	.58734	.58403	.58046	.57652	.56680	.54746	
					1.00055	1.00111	1.00167	1.00224	1.00282	1.00340	1.00459	1.00583	
.595	.9212	.8871	.99439	0.00966	.59222	.58927	.58612	.58269	.57897	.57484	.56432		
					1.00055	1.00111	1.00167	1.00225	1.00283	1.00341	1.00461		
.594	.9228	.8867	.99462	0.00925	.59116	.58813	.58489	.58133	.57746	.57310	.56157		
					1.00055	1.00111	1.00168	1.00226	1.00284	1.00343	1.00463		
.593	.9243	.8864	.99484	0.00885	.59009	.58698	.58360	.57995	.57590	.57128	.55840		
					1.00055	1.00112	1.00169	1.00227	1.00285	1.00344	1.00465		
.592	.9259	.8861	.99505	0.00845	.58902	.58583	.58233	.57854	.57429	.56937	.55433		
					1.00056	1.00112	1.00170	1.00228	1.00286	1.00346	1.00468		
.591	.9275	.8857	.99527	0.00807	.58795	.58463	.58105	.57710	.57262	.56735	.54762		
					1.00056	1.00113	1.00170	1.00229	1.00287	1.00347	1.00471		
.590	.9291	.8854	.99548	0.00770	.58687	.58345	.57974	.57562	.57089	.56519			
					1.00056	1.00113	1.00171	1.00229	1.00289	1.00349			
.589	.9307	.8850	.99568	0.00733	.58580	.58226	.57842	.57410	.56908	.56283			
					1.00056	1.00114	1.00172	1.00230	1.00290	1.00350			
.588	.9323	.8847	.99588	0.00698	.58468	.58107	.57707	.57254	.56717	.56012			
					1.00057	1.00114	1.00172	1.00231	1.00291	1.00352			
.587	.9339	.8843	.99607	0.00664	.58359	.57985	.57570	.57092	.56513	.55700			
					1.00057	1.00115	1.00173	1.00232	1.00292	1.00354			
.586	.9354	.8840	.99626	0.00630	.58249	.57863	.57429	.56924	.56293	.55297			
					1.00057	1.00115	1.00174	1.00233	1.00294	1.00356			
.585	.9370	.8836	.99644	0.00598	.58139	.57739	.57285	.56748	.56042				
					1.00057	1.00115	1.00174	1.00234	1.00295				
.584	.9386	.8833	.99662	0.00566	.58028	.57613	.57137	.56561	.55758				
					1.00058	1.00116	1.00175	1.00235	1.00297				
.583	.9402	.8829	.99680	0.00536	.57916	.57485	.56984	.56363	.55405				
					1.00058	1.00116	1.00176	1.00236	1.00298				
.582	.9418	.8826	.99697	0.00506	.57804	.57355	.56826	.56147	.54826				
					1.00058	1.00117	1.00176	1.00237	1.00300				
.581	.9434	.8822	.99713	0.00477	.57691	.57223	.56661	.55895					
					1.00058	1.00117	1.00177	1.00239					

TABLE III. Specific Heat Ratio = 1.30

R₁ and TPR versus Loss Coefficient K

p/p,	MACH	T/T,	GAMMA	K,ₚₕ	.001	.002	.003	.004	.005	.006	.007	.008	.009
.580	.9450	.8819	.99730	0.00449	.57577	.57087	.56487	.55609					
					1.00058	1.00118	1.00178	1.00240					
.579	.9466	.8815	.99745	0.00422	.57461	.56948	.56303	.55240					
					1.00059	1.00118	1.00178	1.00241					
.578	.9482	.8812	.99760	0.00396	.57345	.56805	.56096						
					1.00059	1.00118	1.00180						
.577	.9498	.8808	.99775	0.00371	.57228	.56658	.55872						
					1.00059	1.00119	1.00181						
.576	.9514	.8805	.99789	0.00347	.57109	.56505	.55614						
					1.00059	1.00119	1.00181						
.575	.9529	.8801	.99803	0.00324	.56988	.56344	.55292						
					1.00059	1.00120	1.00182						
.574	.9545	.8798	.99816	0.00301	.56866	.56175	.54714						
					1.00060	1.00120	1.00183						
.573	.9561	.8794	.99829	0.00279	.56741	.55982							
					1.00060	1.00121							
.572	.9577	.8791	.99841	0.00259	.56614	.55778							
					1.00060	1.00122							
.571	.9593	.8787	.99853	0.00239	.56484	.55545							
					1.00060	1.00122							
.570	.9609	.8783	.99865	0.00220	.56350	.55259							
					1.00060	1.00122							
.569	.9625	.8780	.99876	0.00201	.56212	.54763							
					1.00060	1.00123							
.568	.9641	.8776	.99886	0.00184	.56057								
					1.00061								
.567	.9657	.8773	.99896	0.00167	.55902								
					1.00061								
.566	.9673	.8769	.99906	0.00152	.55736								
					1.00062								
.565	.9689	.8766	.99915	0.00137	.55552								
					1.00062								
.564	.9705	.8762	.99923	0.00123	.55343								
					1.00062								
.563	.9721	.8758	.99932	0.00109	.55055								
					1.00062								

TABLE IV. Specific Heat Ratio = 1.40

R₂ and TPR versus Loss Coefficient K

p/p_t	MACH	T/T_t	GAMMA	K_crit	.1	.2	.4	.6	1.0	2.0	4.0	6.0	10.0
.999	.0378	.9997	.06528	493.52676	.99890	.99880	.99860	.99840	.99800	.99699	.99499	.99297	.98893
					1.00010	1.00020	1.00040	1.00060	1.00100	1.00201	1.00403	1.00606	1.01016
.998	.0535	.9994	.09227	244.10224	.99780	.99760	.99720	.99680	.99599	.99398	.98994	.98589	.97773
					1.00020	1.00040	1.00080	1.00120	1.00201	1.00403	1.00811	1.01224	1.02065
.997	.0655	.9991	.11294	161.11626	.99670	.99640	.99579	.99519	.99398	.99095	.98487	.97874	.96638
					1.00030	1.00060	1.00120	1.00181	1.00302	1.00607	1.01225	1.01854	1.03149
.996	.0757	.9989	.13034	119.69574	.99560	.99519	.99439	.99358	.99197	.98791	.97976	.97154	.95487
					1.00040	1.00080	1.00161	1.00241	1.00403	1.00812	1.01644	1.02497	1.04270
.995	.0847	.9986	.14565	94.88672	.99450	.99399	.99298	.99197	.98995	.98486	.97462	.96427	.94321
					1.00050	1.00100	1.00201	1.00302	1.00505	1.01019	1.02069	1.03154	1.05431
.994	.0928	.9983	.15947	78.37618	.99339	.99279	.99157	.99036	.98792	.98180	.96945	.95694	.93139
					1.00060	1.00121	1.00242	1.00363	1.00608	1.01227	1.02501	1.03824	1.06634
.993	.1002	.9980	.17215	66.60353	.99229	.99158	.99016	.98874	.98589	.97873	.96425	.94954	.91939
					1.00070	1.00141	1.00282	1.00425	1.00711	1.01437	1.02938	1.04509	1.07881
.992	.1072	.9977	.18394	57.78987	.99119	.99038	.98875	.98712	.98386	.97565	.95901	.94207	.90720
					1.00081	1.00161	1.00323	1.00486	1.00814	1.01648	1.03382	1.05209	1.09176
.991	.1137	.9974	.19499	50.94672	.99009	.98917	.98734	.98550	.98182	.97255	.95374	.93453	.89483
					1.00091	1.00182	1.00364	1.00548	1.00918	1.01862	1.03833	1.05924	1.10522
.990	.1199	.9971	.20543	45.48177	.98898	.98796	.98592	.98388	.97978	.96944	.94843	.92691	.88225
					1.00101	1.00202	1.00405	1.00609	1.01022	1.02077	1.04290	1.06656	1.11923
.989	.1258	.9968	.21534	41.01807	.98788	.98676	.98451	.98225	.97773	.96632	.94308	.91922	.86945
					1.00111	1.00222	1.00446	1.00671	1.01127	1.02293	1.04754	1.07405	1.13383
.988	.1314	.9966	.22479	37.30507	.98677	.98555	.98309	.98063	.97568	.96319	.93770	.91145	.85643
					1.00121	1.00243	1.00487	1.00734	1.01232	1.02511	1.05225	1.08171	1.14906
.987	.1369	.9963	.23384	34.16883	.98567	.98434	.98167	.97899	.97362	.96004	.93227	.90360	.84317
					1.00131	1.00263	1.00529	1.00796	1.01338	1.02732	1.05704	1.08956	1.16497
.986	.1421	.9960	.24254	31.48539	.98457	.98313	.98025	.97736	.97156	.95689	.92681	.89567	.82965
					1.00142	1.00284	1.00570	1.00859	1.01444	1.02953	1.06190	1.09760	1.18162
.985	.1471	.9957	.25091	29.16386	.98346	.98192	.97883	.97573	.96949	.95371	.92131	.88764	.81586
					1.00152	1.00304	1.00612	1.00922	1.01551	1.03177	1.06684	1.10584	1.19908
.984	.1520	.9954	.25900	27.13613	.98235	.98071	.97740	.97409	.96742	.95053	.91576	.87952	.80177
					1.00162	1.00325	1.00653	1.00985	1.01658	1.03402	1.07186	1.11429	1.21741
.983	.1567	.9951	.26683	25.35015	.98125	.97949	.97598	.97245	.96534	.94733	.91017	.87131	.78737
					1.00172	1.00346	1.00695	1.01048	1.01765	1.03630	1.07697	1.12297	1.23669
.982	.1613	.9948	.27441	23.76545	.98014	.97828	.97455	.97080	.96326	.94412	.90454	.86300	.77263
					1.00183	1.00366	1.00737	1.01111	1.01874	1.03859	1.08215	1.13187	1.25703
.981	.1658	.9945	.28178	22.35004	.97904	.97707	.97312	.96915	.96117	.94089	.89886	.85459	.75753
					1.00193	1.00387	1.00779	1.01175	1.01982	1.04090	1.08743	1.14102	1.27852
.980	.1701	.9942	.28894	21.07846	.97793	.97585	.97169	.96750	.95908	.93765	.89313	.84606	.74203
					1.00203	1.00408	1.00821	1.01239	1.02091	1.04323	1.09280	1.15043	1.30128
.979	.1744	.9940	.29591	19.93013	.97682	.97464	.97026	.96585	.95698	.93439	.88736	.83743	.72611
					1.00214	1.00429	1.00863	1.01303	1.02201	1.04559	1.09826	1.16010	1.32546
.978	.1786	.9937	.30271	18.88802	.97571	.97342	.96882	.96420	.95487	.93112	.88153	.82867	.70971
					1.00224	1.00449	1.00905	1.01367	1.02311	1.04796	1.10381	1.17006	1.35123
.977	.1826	.9934	.30934	17.93821	.97461	.97221	.96738	.96254	.95276	.92784	.87566	.81980	.69279
					1.00234	1.00470	1.00948	1.01432	1.02422	1.05035	1.10947	1.18032	1.37878
.976	.1866	.9931	.31582	17.06919	.97350	.97099	.96595	.96088	.95065	.92454	.86973	.81079	.67530
					1.00245	1.00491	1.00990	1.01497	1.02533	1.05276	1.11524	1.19090	1.40834
.975	.1905	.9928	.32216	16.27111	.97239	.96977	.96451	.95921	.94853	.92122	.86374	.80165	.65717
					1.00255	1.00512	1.01033	1.01561	1.02645	1.05520	1.12111	1.20181	1.44021
.974	.1944	.9925	.32836	15.53578	.97128	.96855	.96306	.95755	.94640	.91789	.85770	.79236	.63832
					1.00266	1.00533	1.01075	1.01627	1.02758	1.05766	1.12709	1.21307	1.47472
.973	.1981	.9922	.33443	14.85616	.97017	.96733	.96162	.95588	.94427	.91454	.85160	.78292	.61865
					1.00276	1.00554	1.01118	1.01692	1.02871	1.06014	1.13318	1.22472	1.51231
.972	.2018	.9919	.34038	14.22622	.96906	.96611	.96018	.95420	.94213	.91117	.84544	.77333	.59804
					1.00286	1.00575	1.01161	1.01758	1.02984	1.06264	1.13940	1.23676	1.55350
.971	.2055	.9916	.34621	13.64082	.96795	.96489	.95873	.95253	.93999	.90779	.83922	.76356	.57634
					1.00297	1.00597	1.01204	1.01824	1.03098	1.06517	1.14574	1.24923	1.59901

TABLE IV. Specific Heat Ratio = 1.40

p/p_t	MACH	T/T_t	GAMMA	K_crit	.1	.2	.4	.6	1.0	2.0	4.0	6.0	10.0
.970	.2091	.9913	.35193	13.09542	.96684	.96366	.95728	.95085	.93784	.90439	.83293	.75362	.55333
					1.00307	1.00618	1.01247	1.01890	1.03213	1.06771	1.15221	1.26216	1.64973
.969	.2126	.9910	.35755	12.58617	.96573	.96244	.95583	.94916	.93568	.90097	.82658	.74349	.52874
					1.00318	1.00639	1.01291	1.01956	1.03328	1.07029	1.15882	1.27558	1.70690
.968	.2161	.9908	.36307	12.10963	.96461	.96121	.95437	.94748	.93352	.89754	.82015	.73316	.50218
					1.00328	1.00660	1.01334	1.02023	1.03444	1.07289	1.16556	1.28952	1.77223
.967	.2195	.9905	.36850	11.66280	.96350	.95999	.95292	.94579	.93135	.89409	.81366	.72261	.47306
					1.00339	1.00682	1.01378	1.02089	1.03561	1.07551	1.17245	1.30402	1.84827
.966	.2228	.9902	.37383	11.24304	.96239	.95876	.95146	.94410	.92917	.89061	.80708	.71183	.44043
					1.00350	1.00703	1.01421	1.02156	1.03678	1.07816	1.17949	1.31914	1.93899
.965	.2262	.9899	.37908	10.84799	.96128	.95753	.95000	.94240	.92699	.88712	.80043	.70081	.40255
					1.00360	1.00724	1.01465	1.02224	1.03795	1.08083	1.18669	1.33491	2.05124
.964	.2295	.9896	.38424	10.47558	.96016	.95631	.94854	.94070	.92480	.88361	.79370	.68952	.35547
					1.00371	1.00746	1.01509	1.02291	1.03914	1.08353	1.19406	1.35139	2.19880
.963	.2327	.9893	.38933	10.12396	.95905	.95508	.94708	.93900	.92261	.88008	.78688	.67794	.28376
					1.00381	1.00767	1.01553	1.02359	1.04032	1.08626	1.20159	1.36864	2.42082
.962	.2359	.9890	.39433	9.79149	.95793	.95385	.94561	.93730	.92041	.87653	.77998	.66606	
					1.00392	1.00789	1.01597	1.02427	1.04152	1.08902	1.20931	1.38675	
.961	.2391	.9887	.39926	9.47664	.95682	.95262	.94415	.93559	.91820	.87296	.77298	.65383	
					1.00403	1.00810	1.01642	1.02495	1.04272	1.09180	1.21722	1.40578	
.960	.2422	.9884	.40412	9.17810	.95570	.95138	.94268	.93388	.91598	.86936	.76589	.64123	
					1.00413	1.00832	1.01686	1.02564	1.04393	1.09461	1.22533	1.42584	
.959	.2453	.9881	.40891	8.89468	.95459	.95015	.94121	.93216	.91376	.86575	.75869	.62822	
					1.00424	1.00854	1.01731	1.02632	1.04514	1.09745	1.23364	1.44704	
.958	.2483	.9878	.41364	8.62524	.95347	.94892	.93973	.93044	.91153	.86211	.75139	.61476	
					1.00435	1.00875	1.01775	1.02701	1.04637	1.10033	1.24217	1.46949	
.957	.2514	.9875	.41830	8.36883	.95235	.94768	.93826	.92872	.90930	.85845	.74397	.60079	
					1.00445	1.00897	1.01820	1.02771	1.04759	1.10323	1.25093	1.49337	
.956	.2544	.9872	.42290	8.12454	.95124	.94645	.93678	.92700	.90705	.85477	.73644	.58625	
					1.00456	1.00919	1.01865	1.02840	1.04883	1.10616	1.25994	1.51884	
.955	.2573	.9869	.42744	7.89156	.95012	.94521	.93530	.92527	.90480	.85106	.72879	.57106	
					1.00467	1.00941	1.01910	1.02910	1.05007	1.10913	1.26920	1.54613	
.954	.2602	.9866	.43192	7.66913	.94900	.94397	.93382	.92353	.90254	.84733	.72101	.55512	
					1.00478	1.00963	1.01956	1.02980	1.05132	1.11213	1.27872	1.57552	
.953	.2631	.9863	.43634	7.45658	.94788	.94273	.93233	.92180	.90028	.84358	.71308	.53830	
					1.00488	1.00985	1.02001	1.03050	1.05257	1.11517	1.28854	1.60735	
.952	.2660	.9860	.44071	7.25326	.94676	.94149	.93085	.92006	.89800	.83980	.70502	.52044	
					1.00499	1.01007	1.02046	1.03121	1.05384	1.11823	1.29866	1.64205	
.951	.2689	.9857	.44502	7.05863	.94564	.94025	.92936	.91831	.89572	.83599	.69679	.50131	
					1.00510	1.01029	1.02092	1.03192	1.05511	1.12134	1.30910	1.68019	
.950	.2717	.9855	.44929	6.87215	.94452	.93901	.92787	.91656	.89343	.83216	.68841	.48060	
					1.00521	1.01051	1.02138	1.03263	1.05638	1.12448	1.31988	1.72254	
.949	.2745	.9852	.45350	6.69332	.94340	.93777	.92638	.91481	.89114	.82829	.67984	.45782	
					1.00532	1.01073	1.02184	1.03334	1.05767	1.12765	1.33103	1.77022	
.948	.2773	.9849	.45767	6.52171	.94228	.93653	.92488	.91305	.88883	.82441	.67110	.43219	
					1.00543	1.01095	1.02230	1.03406	1.05896	1.13087	1.34257	1.82484	
.947	.2800	.9846	.46178	6.35689	.94116	.93528	.92338	.91130	.88652	.82049	.66214	.40225	
					1.00554	1.01117	1.02276	1.03478	1.06026	1.13412	1.35453	1.88909	
.946	.2827	.9843	.46586	6.19849	.94004	.93403	.92188	.90953	.88419	.81654	.65298	.36466	
					1.00565	1.01140	1.02322	1.03550	1.06157	1.13741	1.36693	1.96803	
.945	.2854	.9840	.46988	6.04615	.93892	.93279	.92038	.90776	.88186	.81256	.64358	.30552	
					1.00576	1.01162	1.02368	1.03623	1.06288	1.14075	1.37982	2.07497	
.944	.2881	.9837	.47386	5.89954	.93779	.93154	.91887	.90599	.87952	.80855	.63392		
					1.00587	1.01184	1.02415	1.03696	1.06421	1.14412	1.39324		
.943	.2908	.9834	.47780	5.75835	.93667	.93029	.91737	.90421	.87718	.80451	.62399		
					1.00598	1.01207	1.02462	1.03769	1.06554	1.14754	1.40722		
.942	.2934	.9831	.48170	5.62229	.93555	.92904	.91586	.90243	.87482	.80043	.61376		
					1.00609	1.01229	1.02509	1.03842	1.06688	1.15101	1.42183		
.941	.2960	.9828	.48556	5.49111	.93442	.92779	.91434	.90065	.87245	.79632	.60320		
					1.00620	1.01252	1.02556	1.03916	1.06822	1.15451	1.43711		

TABLE IV. Specific Heat Ratio = 1.40

p/p,	MACH	T/T,	GAMMA	K_crit	R₁ and TPR versus Loss Coefficient K								
					.1	.2	.4	.6	.8	1.0	2.0	3.0	4.0
.940	.2986	.9825	.48938	5.36456	.93330	.92654	.91283	.89886	.88461	.87008	.79217	.70246	.59227
					1.00631	1.01274	1.02603	1.03990	1.05440	1.06958	1.15807	1.27707	1.45314
.939	.3012	.9822	.49316	5.24240	.93217	.92528	.91131	.89707	.88253	.86769	.78799	.69570	.58093
					1.00642	1.01297	1.02650	1.04064	1.05544	1.07094	1.16168	1.28477	1.46999
.938	.3038	.9819	.49690	5.12440	.93105	.92403	.90979	.89527	.88044	.86529	.78377	.68882	.56913
					1.00653	1.01320	1.02698	1.04139	1.05648	1.07232	1.16533	1.29267	1.48775
.937	.3063	.9816	.50060	5.01040	.92992	.92277	.90827	.89347	.87835	.86289	.77952	.68181	.55681
					1.00664	1.01342	1.02745	1.04214	1.05753	1.07370	1.16903	1.30079	1.50654
.936	.3088	.9813	.50427	4.90017	.92879	.92152	.90674	.89166	.87624	.86048	.77522	.67467	.54389
					1.00675	1.01365	1.02793	1.04289	1.05859	1.07509	1.17279	1.30914	1.52648
.935	.3114	.9810	.50790	4.79355	.92767	.92026	.90521	.88985	.87413	.85805	.77088	.66738	.53027
					1.00686	1.01388	1.02841	1.04364	1.05965	1.07649	1.17660	1.31772	1.54773
.934	.3138	.9807	.51150	4.69036	.92654	.91900	.90368	.88803	.87202	.85562	.76650	.65994	.51583
					1.00697	1.01411	1.02889	1.04440	1.06071	1.07790	1.18047	1.32656	1.57051
.933	.3163	.9804	.51506	4.59047	.92541	.91774	.90215	.88621	.86989	.85317	.76208	.65234	.50039
					1.00709	1.01434	1.02937	1.04516	1.06179	1.07931	1.18439	1.33567	1.59505
.932	.3188	.9801	.51859	4.49370	.92428	.91648	.90061	.88438	.86776	.85071	.75761	.64457	.48372
					1.00720	1.01457	1.02986	1.04593	1.06286	1.08074	1.18837	1.34507	1.62171
.931	.3212	.9798	.52209	4.39993	.92315	.91521	.89907	.88255	.86562	.84825	.75309	.63660	.46547
					1.00731	1.01480	1.03034	1.04670	1.06395	1.08218	1.19242	1.35478	1.65091
.930	.3237	.9795	.52555	4.30903	.92202	.91395	.89753	.88072	.86348	.84577	.74852	.62844	.44507
					1.00742	1.01503	1.03083	1.04747	1.06504	1.08363	1.19653	1.36482	1.68333
.929	.3261	.9792	.52899	4.22087	.92089	.91269	.89599	.87888	.86132	.84328	.74391	.62005	.42156
					1.00753	1.01526	1.03132	1.04824	1.06613	1.08508	1.20070	1.37521	1.71993
.928	.3285	.9789	.53239	4.13533	.91976	.91142	.89444	.87703	.85916	.84078	.73924	.61143	.39287
					1.00765	1.01549	1.03181	1.04902	1.06723	1.08655	1.20494	1.38599	1.76242
.927	.3309	.9786	.53576	4.05231	.91863	.91015	.89289	.87518	.85699	.83827	.73452	.60254	.35261
					1.00776	1.01572	1.03230	1.04980	1.06834	1.08802	1.20925	1.39718	1.81443
.926	.3332	.9783	.53911	3.97169	.91749	.90888	.89134	.87333	.85481	.83574	.72974	.59337	
					1.00787	1.01596	1.03279	1.05059	1.06946	1.08951	1.21363	1.40882	
.925	.3356	.9780	.54242	3.89339	.91636	.90761	.88978	.87147	.85263	.83321	.72490	.58388	
					1.00799	1.01619	1.03329	1.05138	1.07058	1.09101	1.21809	1.42095	
.924	.3379	.9777	.54571	3.81731	.91523	.90634	.88822	.86960	.85043	.83066	.72000	.57404	
					1.00810	1.01642	1.03378	1.05217	1.07170	1.09252	1.22263	1.43361	
.923	.3403	.9774	.54897	3.74336	.91409	.90507	.88666	.86773	.84823	.82810	.71504	.56381	
					1.00822	1.01666	1.03428	1.05297	1.07284	1.09403	1.22725	1.44686	
.922	.3426	.9771	.55220	3.67146	.91296	.90379	.88509	.86585	.84602	.82552	.71001	.55312	
					1.00833	1.01689	1.03478	1.05377	1.07398	1.09556	1.23195	1.46078	
.921	.3449	.9768	.55540	3.60152	.91182	.90252	.88352	.86397	.84379	.82294	.70491	.54193	
					1.00844	1.01713	1.03528	1.05457	1.07512	1.09711	1.23674	1.47542	
.920	.3472	.9765	.55858	3.53347	.91069	.90124	.88195	.86208	.84157	.82034	.69973	.53014	
					1.00856	1.01737	1.03579	1.05538	1.07628	1.09866	1.24162	1.49088	
.919	.3495	.9762	.56173	3.46724	.90955	.89997	.88038	.86019	.83933	.81772	.69448	.51766	
					1.00867	1.01760	1.03629	1.05619	1.07744	1.10022	1.24659	1.50727	
.918	.3518	.9759	.56485	3.40277	.90841	.89869	.87880	.85829	.83708	.81509	.68916	.50433	
					1.00879	1.01784	1.03680	1.05700	1.07861	1.10180	1.25167	1.52474	
.917	.3540	.9755	.56795	3.33998	.90727	.89741	.87722	.85638	.83482	.81245	.68374	.48998	
					1.00890	1.01808	1.03731	1.05782	1.07978	1.10339	1.25685	1.54346	
.916	.3563	.9752	.57103	3.27881	.90613	.89612	.87563	.85447	.83255	.80980	.67824	.47432	
					1.00902	1.01832	1.03782	1.05864	1.08096	1.10498	1.26213	1.56368	
.915	.3585	.9749	.57408	3.21922	.90500	.89484	.87405	.85255	.83028	.80713	.67265	.45691	
					1.00913	1.01856	1.03833	1.05947	1.08215	1.10660	1.26753	1.58572	
.914	.3607	.9746	.57711	3.16113	.90386	.89355	.87245	.85063	.82799	.80444	.66695	.43703	
					1.00925	1.01879	1.03884	1.06030	1.08335	1.10822	1.27305	1.61007	
.913	.3630	.9743	.58011	3.10450	.90271	.89227	.87086	.84870	.82569	.80174	.66116	.41321	
					1.00937	1.01904	1.03936	1.06113	1.08455	1.10986	1.27869	1.63752	
.912	.3652	.9740	.58309	3.04927	.90157	.89098	.86926	.84676	.82339	.79902	.65525	.38151	
					1.00948	1.01928	1.03988	1.06197	1.08576	1.11151	1.28447	1.66959	
.911	.3674	.9737	.58605	2.99540	.90043	.88969	.86766	.84482	.82107	.79629	.64922		
					1.00960	1.01952	1.04040	1.06281	1.08698	1.11318	1.29038		

TABLE IV. Specific Heat Ratio = 1.40

p/p_t	MACH	T/T_t	GAMMA	K_{crit}	.1	.2	.4	.6	.8	.9	1.0	1.5	2.0
.910	.3695	.9734	.58898	2.94285	.89929	.88840	.86605	.84287	.81874	.80629	.79354	.72458	.64307
					1.00972	1.01976	1.04092	1.06366	1.08821	1.10125	1.11485	1.19321	1.29643
.909	.3717	.9731	.59189	2.89156	.89815	.88711	.86445	.84091	.81640	.80374	.79077	.72045	.63679
					1.00983	1.02000	1.04144	1.06451	1.08944	1.10270	1.11654	1.19652	1.30264
.908	.3739	.9728	.59478	2.84149	.89700	.88582	.86283	.83895	.81405	.80118	.78799	.71628	.63036
					1.00995	1.02025	1.04197	1.06536	1.09068	1.10416	1.11825	1.19987	1.30901
.907	.3760	.9725	.59765	2.79260	.89586	.88452	.86122	.83698	.81169	.79860	.78518	.71205	.62378
					1.01007	1.02049	1.04249	1.06622	1.09194	1.10564	1.11997	1.20328	1.31556
.906	.3782	.9722	.60050	2.74486	.89471	.88322	.85960	.83501	.80932	.79602	.78237	.70778	.61704
					1.01019	1.02073	1.04302	1.06708	1.09320	1.10712	1.12171	1.20673	1.32229
.905	.3803	.9719	.60332	2.69823	.89357	.88193	.85797	.83302	.80694	.79341	.77953	.70345	.61012
					1.01030	1.02098	1.04355	1.06795	1.09446	1.10862	1.12346	1.21024	1.32921
.904	.3825	.9716	.60613	2.65267	.89242	.88063	.85634	.83103	.80454	.79079	.77667	.69906	.60300
					1.01042	1.02122	1.04409	1.06882	1.09574	1.11013	1.12522	1.21381	1.33634
.903	.3846	.9713	.60891	2.60815	.89127	.87933	.85471	.82904	.80213	.78816	.77379	.69462	.59568
					1.01054	1.02147	1.04462	1.06970	1.09703	1.11165	1.12700	1.21744	1.34369
.902	.3867	.9710	.61168	2.56463	.89013	.87802	.85308	.82703	.79971	.78551	.77090	.69011	.58812
					1.01066	1.02172	1.04516	1.07058	1.09832	1.11319	1.12880	1.22112	1.35129
.901	.3888	.9707	.61442	2.52209	.88898	.87672	.85144	.82502	.79728	.78284	.76798	.68554	.58031
					1.01078	1.02196	1.04570	1.07147	1.09962	1.11473	1.13061	1.22487	1.35914
.900	.3909	.9703	.61715	2.48050	.88783	.87541	.84980	.82300	.79483	.78016	.76504	.68091	.57222
					1.01090	1.02221	1.04624	1.07236	1.10093	1.11629	1.13244	1.22868	1.36727
.899	.3930	.9700	.61985	2.43981	.88668	.87411	.84815	.82097	.79237	.77746	.76208	.67620	.56381
					1.01102	1.02246	1.04678	1.07325	1.10226	1.11786	1.13429	1.23255	1.37572
.898	.3951	.9697	.62254	2.40001	.88553	.87280	.84650	.81893	.78990	.77474	.75910	.67142	.55505
					1.01113	1.02271	1.04733	1.07415	1.10359	1.11944	1.13615	1.23650	1.38449
.897	.3971	.9694	.62521	2.36107	.88438	.87149	.84484	.81689	.78741	.77201	.75610	.66656	.54588
					1.01125	1.02296	1.04787	1.07506	1.10493	1.12104	1.13804	1.24052	1.39363
.896	.3992	.9691	.62786	2.32297	.88322	.87017	.84318	.81484	.78491	.76925	.75307	.66162	.53625
					1.01137	1.02321	1.04842	1.07597	1.10628	1.12265	1.13994	1.24462	1.40318
.895	.4013	.9688	.63049	2.28568	.88207	.86886	.84152	.81278	.78239	.76648	.75002	.65660	.52608
					1.01149	1.02346	1.04897	1.07688	1.10764	1.12427	1.14186	1.24880	1.41319
.894	.4033	.9685	.63310	2.24917	.88092	.86754	.83985	.81071	.77986	.76369	.74694	.65148	.51526
					1.01162	1.02371	1.04953	1.07780	1.10901	1.12591	1.14379	1.25306	1.42372
.893	.4054	.9682	.63570	2.21342	.87976	.86623	.83817	.80863	.77732	.76088	.74384	.64627	.50366
					1.01174	1.02397	1.05008	1.07873	1.11039	1.12756	1.14575	1.25740	1.43483
.892	.4074	.9679	.63827	2.17842	.87861	.86491	.83649	.80654	.77475	.75805	.74071	.64095	.49100
					1.01186	1.02422	1.05064	1.07966	1.11179	1.12923	1.14773	1.26184	1.44662
.891	.4094	.9676	.64083	2.14414	.87745	.86358	.83481	.80445	.77218	.75519	.73756	.63552	.47725
					1.01198	1.02447	1.05120	1.08059	1.11319	1.13091	1.14973	1.26637	1.45923
.890	.4114	.9673	.64338	2.11056	.87629	.86226	.83313	.80234	.76958	.75232	.73437	.62998	.46169
					1.01210	1.02473	1.05177	1.08153	1.11460	1.13261	1.15175	1.27101	1.47282
.889	.4134	.9669	.64590	2.07766	.87514	.86094	.83143	.80023	.76697	.74943	.73116	.62431	.44355
					1.01222	1.02498	1.05233	1.08248	1.11603	1.13432	1.15379	1.27575	1.48766
.888	.4155	.9666	.54841	2.04542	.87398	.85961	.82974	.79811	.76435	.74651	.72792	.61851	.42097
					1.01234	1.02524	1.05290	1.08343	1.11747	1.13605	1.15585	1.28061	1.50420
.887	.4175	.9663	.65090	2.01382	.87282	.85828	.82804	.79598	.76170	.74357	.72465	.61256	.38719
					1.01247	1.02550	1.05347	1.08439	1.11891	1.13779	1.15794	1.28558	1.52343
.886	.4194	.9660	.65338	1.98285	.87166	.85695	.82633	.79383	.75904	.74061	.72135	.60646	
					1.01259	1.02575	1.05404	1.08535	1.12037	1.13956	1.16005	1.29068	
.885	.4214	.9657	.65584	1.95250	.87049	.85562	.82462	.79168	.75636	.73762	.71801	.60020	
					1.01271	1.02601	1.05461	1.08632	1.12185	1.14134	1.16218	1.29592	
.884	.4234	.9654	.65828	1.92273	.86933	.85428	.82290	.78952	.75366	.73460	.71464	.59374	
					1.01283	1.02627	1.05519	1.08730	1.12333	1.14313	1.16434	1.30129	
.883	.4254	.9651	.66071	1.89354	.86817	.85295	.82118	.78734	.75094	.73156	.71124	.58709	
					1.01296	1.02653	1.05577	1.08828	1.12483	1.14495	1.16652	1.30683	
.882	.4274	.9648	.66312	1.86491	.86701	.85161	.81945	.78516	.74820	.72850	.70780	.58022	
					1.01308	1.02679	1.05635	1.08927	1.12634	1.14678	1.16873	1.31252	
.881	.4293	.9644	.66551	1.83683	.86584	.85027	.81772	.78297	.74545	.72540	.70432	.57310	
					1.01320	1.02705	1.05694	1.09026	1.12786	1.14863	1.17097	1.31840	

75

TABLE IV. Specific Heat Ratio = 1.40

R₂ and TPR versus Loss Coefficient K

p/p,	MACH	T/T,	GAMMA	K_crit	.1	.2	.4	.6	.7	.8	.9	1.0	1.5
.880	.4313	.9641	.66789	1.80929	.86468	.84893	.81598	.78076	.76212	.74267	.72228	.70080	.56571
					1.01333	1.02731	1.05752	1.09126	1.10971	1.12940	1.15050	1.17323	1.32446
.879	.4332	.9638	.67026	1.78226	.86351	.84758	.81424	.77854	.75963	.73986	.71913	.69725	.55801
					1.01345	1.02758	1.05811	1.09226	1.11097	1.13095	1.15239	1.17552	1.33074
.878	.4352	.9635	.67261	1.75574	.86234	.84624	.81249	.77631	.75712	.73704	.71595	.69365	.54997
					1.01358	1.02784	1.05870	1.09328	1.11224	1.13252	1.15431	1.17784	1.33725
.877	.4371	.9632	.67494	1.72971	.86118	.84489	.81074	.77407	.75459	.73420	.71273	.69000	.54153
					1.01370	1.02810	1.05930	1.09430	1.11352	1.13410	1.15624	1.18019	1.34400
.876	.4390	.9629	.67727	1.70416	.86001	.84354	.80898	.77182	.75205	.73133	.70949	.68632	.53262
					1.01383	1.02837	1.05990	1.09532	1.11481	1.13569	1.15819	1.18258	1.35104
.875	.4410	.9626	.67957	1.67909	.85884	.84219	.80721	.76956	.74949	.72843	.70621	.68258	.52317
					1.01395	1.02863	1.06050	1.09635	1.11611	1.13730	1.16017	1.18499	1.35839
.874	.4429	.9623	.68186	1.65447	.85767	.84083	.80544	.76728	.74691	.72551	.70289	.67879	.51306
					1.01408	1.02890	1.06110	1.09739	1.11742	1.13893	1.16217	1.18744	1.36608
.873	.4448	.9619	.68414	1.63029	.85649	.83947	.80366	.76499	.74432	.72257	.69954	.67496	.50214
					1.01420	1.02917	1.06170	1.09844	1.11874	1.14057	1.16419	1.18992	1.37419
.872	.4467	.9616	.68640	1.60656	.85532	.83811	.80188	.76268	.74171	.71960	.69615	.67107	.49017
					1.01433	1.02943	1.06231	1.09949	1.12007	1.14223	1.16624	1.19244	1.38276
.871	.4486	.9613	.68865	1.58324	.85415	.83675	.80009	.76037	.73907	.71660	.69273	.66712	.47680
					1.01446	1.02970	1.06292	1.10056	1.12141	1.14390	1.16831	1.19500	1.39189
.870	.4505	.9610	.69088	1.56034	.85297	.83539	.79829	.75804	.73642	.71358	.68926	.66311	.46138
					1.01458	1.02997	1.06354	1.10162	1.12276	1.14559	1.17041	1.19759	1.40173
.869	.4524	.9607	.69310	1.53785	.85180	.83402	.79649	.75569	.73375	.71052	.68575	.65904	.44265
					1.01471	1.03024	1.06416	1.10270	1.12413	1.14730	1.17253	1.20023	1.41248
.868	.4543	.9604	.69531	1.51575	.85062	.83266	.79468	.75333	.73105	.70744	.68219	.65491	.41674
					1.01484	1.03051	1.06477	1.10378	1.12551	1.14903	1.17468	1.20290	1.42457
.867	.4562	.9600	.69751	1.49404	.84944	.83129	.79286	.75096	.72834	.70432	.67859	.65070	
					1.01496	1.03078	1.06540	1.10488	1.12690	1.15077	1.17686	1.20562	
.866	.4581	.9597	.69969	1.47270	.84827	.82991	.79104	.74857	.72560	.70117	.67495	.64643	
					1.01509	1.03105	1.06602	1.10598	1.12830	1.15254	1.17907	1.20839	
.865	.4599	.9594	.70185	1.45173	.84709	.82854	.78921	.74617	.72284	.69799	.67125	.64207	
					1.01522	1.03133	1.06665	1.10708	1.12971	1.15432	1.18131	1.21120	
.864	.4618	.9591	.70401	1.43112	.84591	.82716	.78738	.74375	.72006	.69478	.66750	.63764	
					1.01535	1.03160	1.06728	1.10820	1.13114	1.15613	1.18358	1.21406	
.863	.4637	.9588	.70615	1.41087	.84472	.82578	.78553	.74131	.71725	.69152	.66370	.63311	
					1.01548	1.03188	1.06792	1.10932	1.13258	1.15795	1.18588	1.21697	
.862	.4655	.9585	.70827	1.39095	.84354	.82440	.78368	.73885	.71442	.68823	.65983	.62850	
					1.01561	1.03215	1.06856	1.11046	1.13403	1.15980	1.18821	1.21994	
.861	.4674	.9581	.71039	1.37138	.84236	.82301	.78182	.73638	.71156	.68490	.65591	.62379	
					1.01574	1.03243	1.06920	1.11160	1.13550	1.16167	1.19058	1.22296	
.860	.4692	.9578	.71249	1.35213	.84117	.82162	.77996	.73390	.70868	.68154	.65193	.61897	
					1.01586	1.03271	1.06984	1.11275	1.13699	1.16356	1.19299	1.22604	
.859	.4711	.9575	.71458	1.33320	.83999	.82023	.77808	.73139	.70576	.67813	.64788	.61404	
					1.01599	1.03298	1.07049	1.11391	1.13848	1.16547	1.19543	1.22918	
.858	.4729	.9572	.71666	1.31458	.83880	.81884	.77620	.72886	.70282	.67467	.64375	.60899	
					1.01612	1.03326	1.07114	1.11508	1.14000	1.16741	1.19792	1.23239	
.857	.4747	.9569	.71872	1.29628	.83761	.81745	.77431	.72632	.69985	.67117	.63956	.60381	
					1.01625	1.03354	1.07180	1.11626	1.14152	1.16937	1.20044	1.23567	
.856	.4766	.9565	.72078	1.27827	.83642	.81605	.77241	.72376	.69685	.66762	.63528	.59849	
					1.01639	1.03382	1.07246	1.11745	1.14307	1.17136	1.20301	1.23902	
.855	.4784	.9562	.72282	1.26055	.83523	.81465	.77051	.72117	.69382	.66402	.63092	.59301	
					1.01652	1.03410	1.07312	1.11865	1.14463	1.17338	1.20562	1.24245	
.854	.4802	.9559	.72485	1.24313	.83404	.81324	.76859	.71857	.69076	.66037	.62647	.58737	
					1.01665	1.03439	1.07378	1.11986	1.14621	1.17542	1.20827	1.24596	
.853	.4821	.9556	.72686	1.22598	.83285	.81184	.76667	.71594	.68766	.65666	.62193	.58154	
					1.01678	1.03467	1.07445	1.12108	1.14780	1.17749	1.21098	1.24956	
.852	.4839	.9553	.72887	1.20912	.83166	.81043	.76474	.71329	.68452	.65290	.61729	.57550	
					1.01691	1.03495	1.07512	1.12231	1.14941	1.17960	1.21374	1.25326	
.851	.4857	.9549	.73086	1.19252	.83046	.80901	.76280	.71062	.68135	.64908	.61253	.56924	
					1.01704	1.03524	1.07580	1.12355	1.15104	1.18173	1.21655	1.25705	

TABLE IV. Specific Heat Ratio = 1.40

p/p_t	MACH	T/T_t	GAMMA	K_{crit}	.1	.2	.4	.5	.6	.7	.8	.9	1.0
.850	.4875	.9546	.73284	1.17618	.82926	.80760	.76085	.73532	.70792	.67815	.64519	.60766	.56273
					1.01718	1.03552	1.07648	1.09955	1.12480	1.15269	1.18389	1.21942	1.26096
.849	.4893	.9543	.73481	1.16010	.82807	.80618	.75889	.73302	.70520	.67490	.64123	.60266	.55591
					1.01731	1.03581	1.07716	1.10050	1.12607	1.15436	1.18609	1.22234	1.26499
.848	.4911	.9540	.73677	1.14428	.82687	.80476	.75692	.73070	.70246	.67161	.63720	.59753	.54878
					1.01744	1.03610	1.07785	1.10145	1.12734	1.15605	1.18832	1.22534	1.26915
.847	.4929	.9537	.73872	1.12870	.82567	.80333	.75494	.72836	.69969	.66828	.63310	.59224	.54125
					1.01757	1.03639	1.07854	1.10241	1.12863	1.15776	1.19059	1.22839	1.27346
.846	.4947	.9533	.74065	1.11337	.82447	.80191	.75295	.72601	.69689	.66490	.62891	.58679	.53328
					1.01771	1.03668	1.07924	1.10337	1.12993	1.15949	1.19290	1.23152	1.27792
.845	.4965	.9530	.74258	1.09828	.82327	.80048	.75095	.72365	.69406	.66147	.62464	.58117	.52476
					1.01784	1.03697	1.07994	1.10434	1.13124	1.16124	1.19525	1.23473	1.28256
.844	.4983	.9527	.74449	1.08341	.82206	.79904	.74894	.72126	.69121	.65800	.62028	.57534	.51558
					1.01798	1.03726	1.08064	1.10532	1.13257	1.16302	1.19764	1.23802	1.28740
.843	.5000	.9524	.74639	1.06878	.82086	.79761	.74692	.71886	.68832	.65447	.61582	.56929	.50555
					1.01811	1.03755	1.08135	1.10631	1.13391	1.16482	1.20007	1.24140	1.29248
.842	.5018	.9521	.74828	1.05437	.81965	.79617	.74489	.71644	.68541	.65089	.61126	.56299	.49439
					1.01825	1.03785	1.08206	1.10731	1.13526	1.16664	1.20254	1.24487	1.29783
.841	.5036	.9517	.75016	1.04018	.81844	.79472	.74285	.71400	.68246	.64725	.60658	.55640	.48162
					1.01838	1.03814	1.08278	1.10831	1.13663	1.16849	1.20507	1.24845	1.30352
.840	.5054	.9514	.75203	1.02621	.81724	.79328	.74080	.71154	.67948	.64355	.60178	.54950	.46629
					1.01852	1.03844	1.08350	1.10932	1.13801	1.17037	1.20765	1.25214	1.30964
.839	.5071	.9511	.75389	1.01244	.81603	.79182	.73873	.70906	.67647	.63979	.59685	.54222	.44589
					1.01865	1.03873	1.08422	1.11034	1.13941	1.17227	1.21028	1.25595	1.31634
.838	.5089	.9508	.75574	0.99888	.81481	.79037	.73665	.70656	.67341	.63596	.59177	.53449	
					1.01879	1.03903	1.08495	1.11137	1.14082	1.17421	1.21296	1.25990	
.837	.5106	.9504	.75758	0.98553	.81360	.78891	.73456	.70404	.67032	.63206	.58653	.52623	
					1.01892	1.03933	1.08569	1.11240	1.14225	1.17617	1.21571	1.26400	
.836	.5124	.9501	.75941	0.97237	.81239	.78745	.73246	.70150	.66719	.62808	.58112	.51730	
					1.01906	1.03963	1.08643	1.11345	1.14370	1.17816	1.21853	1.26828	
.835	.5142	.9498	.76122	0.95941	.81117	.78599	.73035	.69893	.66402	.62403	.57552	.50754	
					1.01920	1.03993	1.08717	1.11450	1.14516	1.18019	1.22141	1.27276	
.834	.5159	.9495	.76303	0.94665	.80995	.78452	.72822	.69634	.66081	.61988	.56970	.49664	
					1.01934	1.04023	1.08792	1.11557	1.14664	1.18225	1.22437	1.27747	
.833	.5177	.9491	.76483	0.93406	.80874	.78305	.72608	.69373	.65755	.61565	.56364	.48410	
					1.01947	1.04054	1.08867	1.11664	1.14814	1.18435	1.22741	1.28248	
.832	.5194	.9488	.76662	0.92167	.80752	.78157	.72392	.69109	.65425	.61132	.55731	.46893	
					1.01961	1.04084	1.08943	1.11772	1.14965	1.18648	1.23053	1.28786	
.831	.5211	.9485	.76839	0.90945	.80629	.78009	.72175	.68842	.65090	.60688	.55066	.44821	
					1.01975	1.04114	1.09020	1.11882	1.15119	1.18865	1.23375	1.29376	
.830	.5229	.9482	.77016	0.89741	.80507	.77861	.71957	.68573	.64749	.60232	.54365		
					1.01989	1.04145	1.09096	1.11992	1.15275	1.19087	1.23708		
.829	.5246	.9478	.77191	0.88555	.80385	.77712	.71737	.68301	.64403	.59764	.53620		
					1.02003	1.04176	1.09174	1.12103	1.15432	1.19313	1.24052		
.828	.5263	.9475	.77366	0.87386	.80262	.77563	.71515	.68027	.64051	.59283	.52822		
					1.02017	1.04207	1.09252	1.12216	1.15592	1.19543	1.24409		
.827	.5281	.9472	.77540	0.86233	.80139	.77413	.71292	.67749	.63693	.58786	.51960		
					1.02031	1.04238	1.09331	1.12330	1.15754	1.19779	1.24781		
.826	.5298	.9468	.77713	0.85097	.80016	.77263	.71068	.67468	.63329	.58273	.51015		
					1.02045	1.04269	1.09410	1.12444	1.15919	1.20020	1.25169		
.825	.5315	.9465	.77884	0.83977	.79893	.77113	.70841	.67184	.62958	.57743	.49958		
					1.02059	1.04300	1.09489	1.12560	1.16085	1.20267	1.25578		
.824	.5332	.9462	.78055	0.82873	.79770	.76962	.70613	.66896	.62579	.57192	.48736		
					1.02073	1.04331	1.09570	1.12677	1.16255	1.20519	1.26012		
.823	.5350	.9459	.78225	0.81785	.79647	.76811	.70383	.66605	.62193	.56619	.47244		
					1.02087	1.04363	1.09651	1.12796	1.16427	1.20778	1.26477		
.822	.5367	.9455	.78394	0.80711	.79523	.76659	.70151	.66310	.61799	.56020	.45156		
					1.02101	1.04394	1.09733	1.12916	1.16602	1.21045	1.26987		
.821	.5384	.9452	.78562	0.79653	.79399	.76507	.69918	.66012	.61396	.55392			
					1.02116	1.04426	1.09815	1.13037	1.16779	1.21318			

TABLE IV. Specific Heat Ratio = 1.40

								R_1 and TPR versus Loss Coefficient K				
p/p_t	MACH	T/T_t	GAMMA	K_{crit}	.08	.09	.1	.2	.3	.4	.5	.6	.7
.820	.5401	.9449	.78729	0.78610	.79835	.79556	.79275	.76354	.73184	.69682	.65709	.60984	.54730
					1.01689	1.01908	1.02130	1.04458	1.07027	1.09898	1.13159	1.16960	1.21601
.819	.5418	.9445	.78895	0.77582	.79716	.79435	.79151	.76201	.72994	.69444	.65402	.60561	.54028
					1.01700	1.01921	1.02144	1.04490	1.07082	1.09981	1.13283	1.17144	1.21892
.818	.5435	.9442	.79060	0.76567	.79597	.79313	.79027	.76047	.72804	.69205	.65091	.60127	.53278
					1.01712	1.01934	1.02158	1.04522	1.07136	1.10066	1.13408	1.17332	1.22194
.817	.5452	.9439	.79224	0.75567	.79477	.79191	.78903	.75893	.72612	.68963	.64775	.59681	.52468
					1.01723	1.01947	1.02173	1.04554	1.07191	1.10150	1.13535	1.17523	1.22508
.816	.5469	.9436	.79387	0.74580	.79358	.79069	.78778	.75739	.72419	.68719	.64455	.59222	.51582
					1.01734	1.01959	1.02187	1.04586	1.07247	1.10236	1.13663	1.17718	1.22835
.815	.5486	.9432	.79550	0.73607	.79239	.78947	.78653	.75583	.72226	.68473	.64129	.58749	.50594
					1.01745	1.01972	1.02201	1.04619	1.07302	1.10323	1.13793	1.17917	1.23178
.814	.5503	.9429	.79711	0.72647	.79119	.78825	.78528	.75428	.72031	.68224	.63798	.58261	.49462
					1.01757	1.01985	1.02216	1.04651	1.07358	1.10410	1.13925	1.18120	1.23540
.813	.5520	.9426	.79872	0.71701	.78999	.78702	.78403	.75272	.71835	.67973	.63461	.57755	.48095
					1.01768	1.01998	1.02230	1.04684	1.07414	1.10498	1.14059	1.18328	1.23927
.812	.5537	.9422	.80032	0.70767	.78879	.78580	.78278	.75115	.71637	.67719	.63118	.57229	.46245
					1.01779	1.02011	1.02245	1.04717	1.07471	1.10586	1.14194	1.18540	1.24347
.811	.5553	.9419	.80190	0.69846	.78759	.78457	.78152	.74958	.71439	.67463	.62769	.56682	
					1.01791	1.02024	1.02260	1.04750	1.07528	1.10676	1.14331	1.18758	
.810	.5570	.9416	.80348	0.68937	.78639	.78334	.78026	.74800	.71239	.67203	.62413	.56110	
					1.01802	1.02037	1.02274	1.04783	1.07585	1.10766	1.14470	1.18982	
.809	.5587	.9412	.80505	0.68041	.78519	.78211	.77900	.74641	.71038	.66941	.62050	.55510	
					1.01814	1.02050	1.02289	1.04816	1.07643	1.10857	1.14611	1.19212	
.808	.5604	.9409	.80662	0.67156	.78398	.78088	.77774	.74482	.70836	.66676	.61679	.54876	
					1.01825	1.02063	1.02304	1.04850	1.07701	1.10949	1.14755	1.19449	
.807	.5621	.9406	.80817	0.66284	.78277	.77964	.77647	.74323	.70632	.66408	.61300	.54203	
					1.01837	1.02076	1.02319	1.04883	1.07759	1.11042	1.14901	1.19693	
.806	.5637	.9402	.80971	0.65423	.78157	.77840	.77521	.74163	.70427	.66137	.60912	.53483	
					1.01848	1.02090	1.02333	1.04917	1.07818	1.11136	1.15049	1.19946	
.805	.5654	.9399	.81125	0.64573	.78036	.77716	.77394	.74002	.70220	.65862	.60514	.52704	
					1.01860	1.02103	1.02348	1.04951	1.07877	1.11231	1.15200	1.20208	
.804	.5671	.9396	.81278	0.63735	.77915	.77592	.77267	.73841	.70012	.65584	.60107	.51848	
					1.01872	1.02116	1.02363	1.04985	1.07937	1.11327	1.15353	1.20481	
.803	.5687	.9392	.81430	0.62908	.77793	.77468	.77140	.73678	.69803	.65302	.59688	.50890	
					1.01883	1.02129	1.02378	1.05019	1.07997	1.11424	1.15509	1.20768	
.802	.5704	.9389	.81581	0.62091	.77672	.77344	.77012	.73516	.69592	.65016	.59256	.49782	
					1.01895	1.02143	1.02393	1.05053	1.08057	1.11522	1.15668	1.21070	
.801	.5721	.9386	.81731	0.61286	.77550	.77219	.76885	.73352	.69379	.64727	.58812	.48425	
					1.01907	1.02156	1.02408	1.05088	1.08118	1.11621	1.15830	1.21392	
.800	.5737	.9382	.81880	0.60491	.77428	.77094	.76757	.73188	.69164	.64433	.58353	.46502	
					1.01918	1.02169	1.02423	1.05122	1.08179	1.11721	1.15995	1.21744	
.799	.5754	.9379	.82029	0.59706	.77306	.76969	.76629	.73024	.68948	.64134	.57877		
					1.01930	1.02183	1.02438	1.05157	1.08241	1.11823	1.16164		
.798	.5770	.9376	.82177	0.58932	.77184	.76844	.76500	.72858	.68730	.63831	.57383		
					1.01942	1.02196	1.02453	1.05192	1.08303	1.11925	1.16337		
.797	.5787	.9372	.82324	0.58168	.77062	.76718	.76372	.72692	.68511	.63523	.56869		
					1.01954	1.02210	1.02468	1.05227	1.08366	1.12029	1.16514		
.796	.5803	.9369	.82470	0.57413	.76939	.76593	.76243	.72525	.68289	.63210	.56331		
					1.01966	1.02223	1.02484	1.05263	1.08429	1.12134	1.16695		
.795	.5820	.9366	.82615	0.56669	.76817	.76467	.76114	.72357	.68066	.62890	.55767		
					1.01977	1.02237	1.02499	1.05298	1.08493	1.12241	1.16881		
.794	.5836	.9362	.82760	0.55934	.76694	.76341	.75984	.72189	.67840	.62565	.55171		
					1.01989	1.02250	1.02514	1.05334	1.08557	1.12349	1.17073		
.793	.5853	.9359	.82903	0.55208	.76571	.76215	.75855	.72020	.67612	.62234	.54539		
					1.02001	1.02264	1.02530	1.05370	1.08622	1.12458	1.17270		
.792	.5869	.9355	.83046	0.54492	.76448	.76088	.75725	.71850	.67382	.61896	.53862		
					1.02013	1.02278	1.02545	1.05406	1.08687	1.12569	1.17473		
.791	.5886	.9352	.83188	0.53785	.76324	.75961	.75595	.71679	.67150	.61552	.53128		
					1.02025	1.02291	1.02561	1.05442	1.08752	1.12682	1.17685		

TABLE IV. Specific Heat Ratio = 1.40

					R_z and TPR versus Loss Coefficient K								
p/p_t	MACH	T/T_t	GAMMA	K_{crit}	.04	.06	.08	.09	.1	.2	.3	.4	.5
.790	.5902	.9349	.83330	0.53086	.77628	.76921	.76201	.75834	.75464	.71507	.66916	.61199	.52323
					1.00996	1.01511	1.02037	1.02305	1.02576	1.05478	1.08819	1.12796	1.17904
.789	.5918	.9345	.83470	0.52397	.77517	.76804	.76077	.75707	.75334	.71334	.66679	.60839	.51420
					1.01002	1.01520	1.02049	1.02319	1.02592	1.05515	1.08886	1.12912	1.18134
.788	.5935	.9342	.83610	0.51717	.77405	.76686	.75953	.75580	.75203	.71161	.66440	.60469	.50375
					1.01008	1.01529	1.02062	1.02333	1.02607	1.05552	1.08953	1.13030	1.18375
.787	.5951	.9339	.83749	0.51045	.77293	.76569	.75829	.75452	.75071	.70987	.66198	.60090	.49091
					1.01014	1.01537	1.02074	1.02347	1.02623	1.05589	1.09021	1.13150	1.18632
.786	.5967	.9335	.83887	0.50381	.77182	.76451	.75704	.75324	.74940	.70811	.65953	.59701	.47248
					1.01019	1.01546	1.02086	1.02361	1.02639	1.05626	1.09090	1.13272	1.18912
.785	.5984	.9332	.84025	0.49726	.77070	.76333	.75579	.75196	.74808	.70635	.65705	.59301	
					1.01025	1.01555	1.02098	1.02375	1.02654	1.05663	1.09159	1.13396	
.784	.6000	.9328	.84161	0.49080	.76959	.76215	.75455	.75068	.74676	.70458	.65455	.58888	
					1.01031	1.01564	1.02110	1.02389	1.02670	1.05701	1.09229	1.13522	
.783	.6016	.9325	.84297	0.48441	.76847	.76097	.75329	.74939	.74544	.70279	.65201	.58461	
					1.01037	1.01573	1.02123	1.02403	1.02686	1.05738	1.09299	1.13651	
.782	.6033	.9322	.84433	0.47810	.76735	.75978	.75204	.74810	.74411	.70100	.64945	.58020	
					1.01042	1.01582	1.02135	1.02417	1.02702	1.05776	1.09371	1.13782	
.781	.6049	.9318	.84567	0.47188	.76623	.75860	.75079	.74681	.74278	.69919	.64684	.57561	
					1.01048	1.01591	1.02147	1.02431	1.02718	1.05815	1.09443	1.13916	
.780	.6065	.9315	.84701	0.46573	.76511	.75741	.74953	.74551	.74145	.69738	.64420	.57084	
					1.01054	1.01600	1.02160	1.02445	1.02734	1.05853	1.09516	1.14053	
.779	.6081	.9311	.84834	0.45965	.76399	.75622	.74827	.74422	.74011	.69555	.64152	.56585	
					1.01060	1.01609	1.02172	1.02459	1.02750	1.05892	1.09590	1.14194	
.778	.6097	.9308	.84966	0.45365	.76286	.75503	.74701	.74292	.73877	.69371	.63881	.56061	
					1.01066	1.01618	1.02185	1.02473	1.02766	1.05931	1.09664	1.14338	
.777	.6114	.9304	.85097	0.44773	.76174	.75384	.74574	.74161	.73743	.69186	.63605	.55509	
					1.01072	1.01627	1.02197	1.02488	1.02782	1.05970	1.09739	1.14486	
.776	.6130	.9301	.85228	0.44188	.76061	.75265	.74447	.74031	.73608	.69000	.63325	.54922	
					1.01078	1.01636	1.02210	1.02502	1.02798	1.06009	1.09815	1.14637	
.775	.6146	.9298	.85358	0.43610	.75949	.75145	.74320	.73900	.73473	.68813	.63041	.54292	
					1.01083	1.01646	1.02222	1.02517	1.02815	1.06049	1.09892	1.14795	
.774	.6162	.9294	.85487	0.43039	.75836	.75025	.74193	.73769	.73338	.68624	.62751	.53612	
					1.01089	1.01655	1.02235	1.02531	1.02831	1.06089	1.09970	1.14957	
.773	.6178	.9291	.85616	0.42476	.75723	.74906	.74066	.73637	.73202	.68434	.62457	.52864	
					1.01095	1.01664	1.02248	1.02546	1.02847	1.06129	1.10049	1.15125	
.772	.6194	.9287	.85743	0.41919	.75610	.74786	.73938	.73505	.73066	.68242	.62157	.52025	
					1.01101	1.01673	1.02260	1.02560	1.02864	1.06170	1.10128	1.15301	
.771	.6210	.9284	.85871	0.41369	.75497	.74665	.73810	.73373	.72930	.68049	.61851	.51053	
					1.01107	1.01682	1.02273	1.02575	1.02880	1.06210	1.10209	1.15485	
.770	.6226	.9280	.85997	0.40825	.75384	.74545	.73682	.73241	.72793	.67855	.61539	.49852	
					1.01113	1.01692	1.02286	1.02589	1.02897	1.06251	1.10291	1.15681	
.769	.6242	.9277	.86123	0.40289	.75271	.74424	.73553	.73108	.72656	.67659	.61221	.48118	
					1.01119	1.01701	1.02299	1.02604	1.02914	1.06292	1.10374	1.15893	
.768	.6258	.9274	.86247	0.39759	.75158	.74304	.73424	.72975	.72518	.67462	.60895		
					1.01125	1.01710	1.02312	1.02619	1.02930	1.06334	1.10458		
.767	.6274	.9270	.86372	0.39235	.75044	.74183	.73295	.72841	.72380	.67263	.60562		
					1.01131	1.01719	1.02324	1.02634	1.02947	1.06376	1.10544		
.766	.6290	.9267	.86495	0.38718	.74931	.74061	.73166	.72708	.72242	.67062	.60220		
					1.01137	1.01729	1.02337	1.02648	1.02964	1.06418	1.10630		
.765	.6306	.9263	.86618	0.38207	.74817	.73940	.73036	.72573	.72103	.66859	.59869		
					1.01143	1.01738	1.02350	1.02663	1.02981	1.06460	1.10719		
.764	.6322	.9260	.86740	0.37702	.74703	.73818	.72906	.72439	.71964	.66655	.59509		
					1.01149	1.01748	1.02363	1.02678	1.02998	1.06503	1.10808		
.763	.6338	.9256	.86862	0.37203	.74589	.73697	.72776	.72304	.71824	.66448	.59138		
					1.01155	1.01757	1.02376	1.02693	1.03015	1.06546	1.10899		
.762	.6354	.9253	.86982	0.36711	.74475	.73575	.72645	.72169	.71684	.66240	.58755		
					1.01161	1.01766	1.02390	1.02708	1.03032	1.06589	1.10992		
.761	.6370	.9249	.87102	0.36224	.74361	.73453	.72514	.72033	.71544	.66030	.58359		
					1.01167	1.01776	1.02403	1.02723	1.03049	1.06633	1.11086		

TABLE IV. Specific Heat Ratio = 1.40

R_2 and TPR versus Loss Coefficient K

p/p_t	MACH	T/T_t	GAMMA	K_crit	.01	.02	.04	.06	.08	.09	.1	.2	.3
.760	.6386	.9246	.87222	0.35743	.75571	.75136	.74247	.73330	.72383	.71897	.71403	.65818	.57949
					1.00288	1.00579	1.01173	1.01785	1.02416	1.02739	1.03066	1.06677	1.11182
.759	.6402	.9242	.87340	0.35268	.75468	.75029	.74132	.73208	.72252	.71761	.71261	.65603	.57523
					1.00289	1.00582	1.01180	1.01795	1.02429	1.02754	1.03083	1.06721	1.11281
.758	.6418	.9239	.87459	0.34798	.75364	.74922	.74018	.73085	.72120	.71624	.71119	.65387	.57078
					1.00290	1.00585	1.01186	1.01805	1.02442	1.02769	1.03101	1.06766	1.11381
.757	.6434	.9235	.87576	0.34334	.75261	.74815	.73903	.72962	.71987	.71487	.70977	.65168	.56612
					1.00292	1.00588	1.01192	1.01814	1.02456	1.02784	1.03118	1.06811	1.11483
.756	.6450	.9232	.87692	0.33876	.75157	.74708	.73788	.72838	.71855	.71349	.70834	.64946	.56121
					1.00293	1.00591	1.01198	1.01824	1.02469	1.02800	1.03136	1.06857	1.11589
.755	.6466	.9228	.87808	0.33423	.75054	.74601	.73673	.72715	.71722	.71211	.70690	.64722	.55602
					1.00295	1.00594	1.01204	1.01833	1.02483	1.02815	1.03153	1.06902	1.11697
.754	.6481	.9225	.87924	0.32976	.74950	.74493	.73558	.72591	.71589	.71073	.70546	.64496	.55048
					1.00296	1.00597	1.01210	1.01843	1.02496	1.02831	1.03171	1.06949	1.11808
.753	.6497	.9221	.88038	0.32534	.74847	.74386	.73443	.72467	.71455	.70934	.70402	.64266	.54452
					1.00298	1.00600	1.01217	1.01853	1.02510	1.02846	1.03189	1.06995	1.11922
.752	.6513	.9218	.88152	0.32097	.74743	.74279	.73327	.72343	.71321	.70794	.70257	.64034	.53803
					1.00299	1.00603	1.01223	1.01863	1.02523	1.02862	1.03206	1.07043	1.12041
.751	.6529	.9214	.88266	0.31666	.74639	.74171	.73212	.72218	.71186	.70654	.70111	.63798	.53082
					1.00301	1.00606	1.01229	1.01872	1.02537	1.02878	1.03224	1.07090	1.12163
.750	.6545	.9211	.88378	0.31239	.74536	.74064	.73096	.72093	.71051	.70514	.69965	.63560	.52260
					1.00302	1.00609	1.01235	1.01882	1.02550	1.02893	1.03242	1.07138	1.12291
.749	.6561	.9207	.88490	0.30818	.74432	.73956	.72980	.71968	.70916	.70373	.69818	.63318	.51276
					1.00304	1.00612	1.01242	1.01892	1.02564	1.02909	1.03260	1.07187	1.12427
.748	.6576	.9204	.88602	0.30401	.74328	.73849	.72864	.71843	.70781	.70232	.69671	.63072	.49983
					1.00305	1.00615	1.01248	1.01902	1.02578	1.02925	1.03278	1.07236	1.12571
.747	.6592	.9200	.88712	0.29990	.74224	.73741	.72748	.71718	.70644	.70090	.69523	.62823	
					1.00307	1.00618	1.01254	1.01912	1.02592	1.02941	1.03296	1.07285	
.746	.6608	.9197	.88823	0.29583	.74121	.73633	.72632	.71592	.70508	.69947	.69374	.62569	
					1.00308	1.00621	1.01261	1.01922	1.02606	1.02957	1.03314	1.07335	
.745	.6624	.9193	.88932	0.29182	.74017	.73525	.72515	.71466	.70371	.69805	.69225	.62312	
					1.00310	1.00624	1.01267	1.01932	1.02619	1.02973	1.03333	1.07386	
.744	.6639	.9190	.89041	0.28784	.73913	.73417	.72399	.71339	.70234	.69661	.69075	.62050	
					1.00311	1.00627	1.01273	1.01942	1.02633	1.02989	1.03351	1.07437	
.743	.6655	.9186	.89149	0.28392	.73809	.73309	.72282	.71212	.70096	.69517	.68924	.61783	
					1.00313	1.00630	1.01280	1.01952	1.02647	1.03005	1.03370	1.07489	
.742	.6671	.9183	.89256	0.28004	.73705	.73201	.72165	.71085	.69957	.69372	.68773	.61511	
					1.00314	1.00633	1.01286	1.01962	1.02662	1.03022	1.03388	1.07541	
.741	.6687	.9179	.89363	0.27621	.73601	.73093	.72048	.70958	.69819	.69227	.68621	.61234	
					1.00316	1.00636	1.01292	1.01972	1.02676	1.03038	1.03407	1.07594	
.740	.6702	.9176	.89469	0.27243	.73497	.72985	.71930	.70831	.69679	.69081	.68468	.60951	
					1.00317	1.00639	1.01299	1.01982	1.02690	1.03054	1.03426	1.07648	
.739	.6718	.9172	.89575	0.26868	.73393	.72877	.71813	.70703	.69540	.68935	.68314	.60661	
					1.00319	1.00643	1.01305	1.01992	1.02704	1.03071	1.03444	1.07702	
.738	.6734	.9169	.89680	0.26499	.73289	.72768	.71695	.70574	.69399	.68788	.68160	.60366	
					1.00320	1.00646	1.01312	1.02002	1.02718	1.03087	1.03463	1.07757	
.737	.6749	.9165	.89784	0.26133	.73185	.72660	.71577	.70446	.69258	.68640	.68005	.60062	
					1.00322	1.00649	1.01318	1.02012	1.02733	1.03104	1.03482	1.07813	
.736	.6765	.9161	.89888	0.25772	.73081	.72551	.71459	.70317	.69117	.68492	.67849	.59752	
					1.00323	1.00652	1.01325	1.02023	1.02747	1.03121	1.03502	1.07870	
.735	.6781	.9158	.89991	0.25415	.72977	.72443	.71341	.70188	.68975	.68343	.67692	.59432	
					1.00325	1.00655	1.01331	1.02033	1.02762	1.03137	1.03521	1.07928	
.734	.6796	.9154	.90093	0.25062	.72872	.72334	.71222	.70058	.68833	.68193	.67534	.59103	
					1.00326	1.00658	1.01338	1.02043	1.02776	1.03154	1.03540	1.07986	
.733	.6812	.9151	.90195	0.24714	.72768	.72225	.71104	.69928	.68690	.68043	.67376	.58764	
					1.00328	1.00661	1.01344	1.02054	1.02791	1.03171	1.03560	1.08046	
.732	.6827	.9147	.90296	0.24369	.72664	.72116	.70985	.69798	.68546	.67891	.67216	.58414	
					1.00329	1.00664	1.01351	1.02064	1.02806	1.03188	1.03579	1.08106	
.731	.6843	.9144	.90397	0.24029	.72559	.72007	.70866	.69667	.68401	.67739	.67056	.58051	
					1.00331	1.00667	1.01357	1.02074	1.02821	1.03205	1.03599	1.08167	

TABLE IV. Specific Heat Ratio = 1.40

					R₁ and TPR versus Loss Coefficient K								
p/p_t	MACH	T/T_t	GAMMA	K_{crit}	.01	.02	.04	.06	.07	.08	.09	.1	.2
.730	.6859	.9140	.90497	0.23692	.72455	.71898	.70746	.69536	.68905	.68256	.67587	.66894	.57672
					1.00332	1.00671	1.01364	1.02085	1.02456	1.02835	1.03223	1.03618	1.08231
.729	.6874	.9136	.90596	0.23360	.72351	.71789	.70627	.69404	.68767	.68111	.67433	.66732	.57278
					1.00334	1.00674	1.01371	1.02095	1.02469	1.02850	1.03240	1.03638	1.08295
.728	.6890	.9133	.90695	0.23031	.72246	.71680	.70507	.69273	.68628	.67964	.67279	.66569	.56865
					1.00335	1.00677	1.01377	1.02106	1.02482	1.02865	1.03257	1.03658	1.08360
.727	.6906	.9129	.90793	0.22706	.72142	.71571	.70387	.69140	.68489	.67817	.67123	.66404	.56431
					1.00337	1.00680	1.01384	1.02117	1.02494	1.02880	1.03275	1.03678	1.08427
.726	.6921	.9126	.90891	0.22385	.72037	.71461	.70267	.69008	.68349	.67670	.66967	.66238	.55971
					1.00338	1.00683	1.01391	1.02127	1.02507	1.02895	1.03292	1.03698	1.08496
.725	.6937	.9122	.90988	0.22068	.71932	.71352	.70147	.68874	.68209	.67522	.66810	.66072	.55481
					1.00340	1.00686	1.01397	1.02138	1.02520	1.02911	1.03310	1.03719	1.08566
.724	.6952	.9119	.91084	0.21754	.71828	.71242	.70026	.68740	.68068	.67372	.66652	.65904	.54952
					1.00342	1.00690	1.01404	1.02149	1.02533	1.02926	1.03328	1.03739	1.08639
.723	.6968	.9115	.91180	0.21444	.71723	.71132	.69905	.68606	.67926	.67223	.66493	.65734	.54375
					1.00343	1.00693	1.01411	1.02160	1.02546	1.02941	1.03345	1.03760	1.08714
.722	.6983	.9111	.91275	0.21138	.71618	.71022	.69784	.68471	.67784	.67072	.66333	.65564	.53733
					1.00345	1.00696	1.01418	1.02170	1.02559	1.02956	1.03363	1.03780	1.08791
.721	.6999	.9108	.91370	0.20836	.71514	.70913	.69662	.68336	.67641	.66921	.66172	.65392	.52995
					1.00346	1.00699	1.01424	1.02181	1.02572	1.02972	1.03381	1.03801	1.08872
.720	.7014	.9104	.91464	0.20536	.71409	.70802	.69541	.68201	.67498	.66769	.66010	.65219	.52108
					1.00348	1.00702	1.01431	1.02192	1.02585	1.02987	1.03399	1.03822	1.08957
.719	.7030	.9101	.91557	0.20241	.71304	.70692	.69418	.68065	.67354	.66615	.65847	.65045	.50908
					1.00349	1.00706	1.01438	1.02203	1.02598	1.03003	1.03418	1.03843	1.09048
.718	.7045	.9097	.91650	0.19949	.71199	.70582	.69296	.67929	.67209	.66462	.65683	.64868	
					1.00351	1.00709	1.01445	1.02214	1.02612	1.03019	1.03436	1.03864	
.717	.7061	.9093	.91742	0.19660	.71094	.70472	.69173	.67791	.67064	.66307	.65517	.64691	
					1.00352	1.00712	1.01452	1.02225	1.02625	1.03034	1.03454	1.03886	
.716	.7077	.9090	.91834	0.19374	.70989	.70361	.69051	.67654	.66917	.66151	.65351	.64512	
					1.00354	1.00715	1.01459	1.02236	1.02638	1.03050	1.03473	1.03907	
.715	.7092	.9086	.91925	0.19092	.70884	.70251	.68928	.67516	.66770	.65994	.65183	.64331	
					1.00356	1.00719	1.01466	1.02247	1.02652	1.03066	1.03492	1.03929	
.714	.7108	.9082	.92015	0.18814	.70779	.70140	.68804	.67377	.66623	.65837	.65014	.64149	
					1.00357	1.00722	1.01473	1.02258	1.02665	1.03082	1.03511	1.03951	
.713	.7123	.9079	.92105	0.18538	.70673	.70029	.68680	.67238	.66474	.65678	.64843	.63964	
					1.00359	1.00725	1.01480	1.02270	1.02679	1.03098	1.03529	1.03973	
.712	.7138	.9075	.92194	0.18266	.70568	.69918	.68556	.67098	.66325	.65518	.64671	.63778	
					1.00360	1.00729	1.01487	1.02281	1.02692	1.03115	1.03548	1.03995	
.711	.7154	.9071	.92283	0.17997	.70463	.69807	.68432	.66958	.66175	.65357	.64498	.63590	
					1.00362	1.00732	1.01494	1.02292	1.02706	1.03131	1.03568	1.04017	
.710	.7169	.9068	.92371	0.17731	.70357	.69696	.68307	.66816	.66024	.65195	.64323	.63400	
					1.00364	1.00735	1.01501	1.02304	1.02720	1.03147	1.03587	1.04039	
.709	.7185	.9064	.92459	0.17468	.70252	.69584	.68182	.66675	.65873	.65032	.64146	.63208	
					1.00365	1.00738	1.01508	1.02315	1.02734	1.03164	1.03606	1.04062	
.708	.7200	.9061	.92546	0.17208	.70146	.69473	.68057	.66532	.65720	.64867	.63968	.63013	
					1.00367	1.00741	1.01515	1.02326	1.02748	1.03180	1.03626	1.04085	
.707	.7216	.9057	.92632	0.16952	.70041	.69361	.67932	.66389	.65566	.64702	.63788	.62816	
					1.00368	1.00744	1.01522	1.02338	1.02762	1.03197	1.03646	1.04108	
.706	.7231	.9053	.92718	0.16698	.69935	.69249	.67805	.66245	.65412	.64535	.63607	.62617	
					1.00370	1.00748	1.01529	1.02350	1.02776	1.03214	1.03665	1.04131	
.705	.7247	.9050	.92803	0.16447	.69830	.69137	.67679	.66101	.65256	.64366	.63423	.62415	
					1.00372	1.00751	1.01537	1.02361	1.02790	1.03231	1.03685	1.04155	
.704	.7262	.9046	.92888	0.16199	.69724	.69025	.67552	.65956	.65100	.64197	.63238	.62211	
					1.00373	1.00754	1.01544	1.02373	1.02804	1.03248	1.03705	1.04178	
.703	.7278	.9042	.92972	0.15955	.69618	.68913	.67425	.65809	.64942	.64025	.63050	.62003	
					1.00375	1.00758	1.01551	1.02385	1.02818	1.03265	1.03726	1.04202	
.702	.7293	.9038	.93056	0.15713	.69512	.68801	.67298	.65663	.64783	.63853	.62860	.61793	
					1.00377	1.00761	1.01558	1.02396	1.02833	1.03282	1.03746	1.04226	
.701	.7309	.9035	.93139	0.15474	.69406	.68688	.67170	.65515	.64623	.63678	.62669	.61579	
					1.00378	1.00764	1.01565	1.02408	1.02847	1.03300	1.03767	1.04250	

TABLE IV. Specific Heat Ratio = 1.40

p/p_t	MACH	T/T_t	GAMMA	K_{crit}	.01	.02	.04	.05	.06	.07	.08	.09	.1
								R_t and TPR versus Loss Coefficient K					
.700	.7324	.9031	.93222	0.15237	.69300	.68576	.67041	.66224	.65367	.64462	.63502	.62474	.61360
					1.00380	1.00768	1.01573	1.01991	1.02420	1.02862	1.03317	1.03788	1.04275
.699	.7339	.9027	.93303	0.15004	.69194	.68463	.66913	.66086	.65217	.64300	.63324	.62278	.61139
					1.00381	1.00771	1.01580	1.02000	1.02432	1.02876	1.03335	1.03809	1.04300
.698	.7355	.9024	.93385	0.14773	.69088	.68350	.66783	.65947	.65067	.64136	.63145	.62078	.60915
					1.00383	1.00775	1.01587	1.02010	1.02444	1.02891	1.03353	1.03830	1.04325
.697	.7370	.9020	.93466	0.14545	.68982	.68237	.66654	.65807	.64916	.63971	.62963	.61876	.60686
					1.00385	1.00778	1.01595	1.02019	1.02456	1.02906	1.03370	1.03851	1.04351
.696	.7386	.9016	.93546	0.14320	.68875	.68123	.66524	.65667	.64763	.63805	.62780	.61670	.60452
					1.00386	1.00781	1.01602	1.02029	1.02468	1.02921	1.03388	1.03873	1.04376
.695	.7401	.9013	.93626	0.14097	.68769	.68010	.66393	.65526	.64610	.63637	.62594	.61462	.60214
					1.00388	1.00785	1.01610	1.02039	1.02480	1.02936	1.03407	1.03895	1.04402
.694	.7416	.9009	.93705	0.13877	.68662	.67896	.66262	.65384	.64456	.63468	.62406	.61250	.59970
					1.00390	1.00788	1.01617	1.02049	1.02493	1.02951	1.03425	1.03917	1.04428
.693	.7432	.9005	.93783	0.13660	.68556	.67782	.66130	.65242	.64301	.63297	.62216	.61035	.59721
					1.00391	1.00792	1.01625	1.02058	1.02505	1.02966	1.03443	1.03939	1.04455
.692	.7447	.9002	.93862	0.13445	.68449	.67668	.65998	.65098	.64144	.63125	.62024	.60816	.59466
					1.00393	1.00795	1.01632	1.02068	1.02518	1.02982	1.03462	1.03961	1.04482
.691	.7463	.8998	.93939	0.13233	.68342	.67554	.65866	.64954	.63986	.62950	.61827	.60593	.59205
					1.00395	1.00799	1.01640	1.02078	1.02530	1.02997	1.03481	1.03984	1.04509
.690	.7478	.8994	.94016	0.13023	.68236	.67439	.65732	.64809	.63828	.62774	.61629	.60366	.58936
					1.00396	1.00802	1.01647	1.02088	1.02543	1.03013	1.03500	1.04007	1.04537
.689	.7493	.8990	.94093	0.12816	.68129	.67325	.65599	.64664	.63667	.62596	.61429	.60135	.58659
					1.00398	1.00806	1.01655	1.02098	1.02555	1.03028	1.03519	1.04030	1.04565
.688	.7509	.8987	.94168	0.12611	.68022	.67210	.65464	.64517	.63506	.62416	.61225	.59898	.58373
					1.00400	1.00809	1.01663	1.02108	1.02568	1.03044	1.03539	1.04054	1.04594
.687	.7524	.8983	.94244	0.12409	.67915	.67095	.65329	.64370	.63343	.62233	.61018	.59656	.58077
					1.00401	1.00813	1.01670	1.02118	1.02581	1.03060	1.03558	1.04077	1.04623
.686	.7540	.8979	.94319	0.12209	.67808	.66980	.65194	.64221	.63179	.62049	.60807	.59408	.57771
					1.00403	1.00816	1.01678	1.02128	1.02594	1.03076	1.03578	1.04101	1.04652
.685	.7555	.8975	.94393	0.12012	.67700	.66864	.65058	.64072	.63013	.61863	.60593	.59154	.57451
					1.00405	1.00820	1.01686	1.02139	1.02607	1.03093	1.03598	1.04126	1.04682
.684	.7570	.8972	.94467	0.11817	.67593	.66748	.64921	.63921	.62846	.61674	.60375	.58893	.57118
					1.00406	1.00823	1.01694	1.02149	1.02620	1.03109	1.03618	1.04151	1.04713
.683	.7586	.8968	.94540	0.11624	.67486	.66632	.64783	.63770	.62677	.61482	.60152	.58624	.56767
					1.00408	1.00827	1.01702	1.02159	1.02633	1.03126	1.03638	1.04176	1.04744
.682	.7601	.8964	.94613	0.11433	.67378	.66516	.64645	.63617	.62506	.61288	.59925	.58347	.56397
					1.00410	1.00830	1.01709	1.02170	1.02646	1.03142	1.03659	1.04201	1.04776
.681	.7616	.8960	.94685	0.11245	.67270	.66400	.64506	.63463	.62333	.61091	.59693	.58060	.56003
					1.00412	1.00834	1.01717	1.02180	1.02660	1.03159	1.03680	1.04227	1.04809
.680	.7632	.8957	.94756	0.11059	.67163	.66283	.64367	.63309	.62158	.60890	.59456	.57763	.55580
					1.00413	1.00838	1.01725	1.02191	1.02674	1.03176	1.03701	1.04253	1.04842
.679	.7647	.8953	.94827	0.10876	.67055	.66166	.64226	.63152	.61982	.60687	.59213	.57453	.55120
					1.00415	1.00841	1.01733	1.02201	1.02687	1.03193	1.03722	1.04280	1.04877
.678	.7663	.8949	.94898	0.10695	.66947	.66049	.64085	.62995	.61804	.60480	.58963	.57130	.54608
					1.00417	1.00845	1.01741	1.02212	1.02701	1.03210	1.03744	1.04308	1.04913
.677	.7678	.8945	.94968	0.10515	.66839	.65931	.63943	.62836	.61623	.60269	.58707	.56791	.54027
					1.00418	1.00848	1.01749	1.02223	1.02715	1.03228	1.03766	1.04335	1.04950
.676	.7693	.8942	.95038	0.10338	.66730	.65814	.63800	.62675	.61440	.60055	.58443	.56433	.53334
					1.00420	1.00852	1.01757	1.02234	1.02729	1.03245	1.03788	1.04364	1.04989
.675	.7709	.8938	.95106	0.10164	.66622	.65696	.63656	.62513	.61255	.59836	.58170	.56052	.52425
					1.00422	1.00856	1.01766	1.02245	1.02743	1.03263	1.03811	1.04393	1.05030
.674	.7724	.8934	.95175	0.09991	.66514	.65577	.63512	.62350	.61067	.59612	.57888	.55643	
					1.00424	1.00859	1.01774	1.02256	1.02757	1.03281	1.03834	1.04423	
.673	.7739	.8930	.95243	0.09820	.66405	.65459	.63366	.62185	.60876	.59383	.57595	.55199	
					1.00425	1.00863	1.01782	1.02267	1.02771	1.03300	1.03857	1.04454	
.672	.7755	.8926	.95310	0.09652	.66296	.65340	.63219	.62019	.60683	.59149	.57289	.54705	
					1.00427	1.00867	1.01790	1.02278	1.02786	1.03318	1.03881	1.04486	
.671	.7770	.8923	.95377	0.09486	.66188	.65220	.63071	.61850	.60486	.58909	.56970	.54146	
					1.00429	1.00870	1.01799	1.02289	1.02800	1.03337	1.03905	1.04519	

TABLE IV. Specific Heat Ratio = 1.40

R_2 and TPR versus Loss Coefficient K

p/p_t	MACH	T/T_t	GAMMA	K_{crit}	.01	.02	.03	.04	.05	.06	.07	.08	.09
.670	.7785	.8919	.95444	0.09321	.66079	.65101	.64055	.62923	.61680	.60286	.58662	.56634	.53482
					1.00431	1.00874	1.01332	1.01807	1.02300	1.02815	1.03356	1.03930	1.04554
.669	.7801	.8915	.95509	0.09159	.65970	.64981	.63922	.62773	.61508	.60082	.58408	.56278	.52610
					1.00432	1.00878	1.01338	1.01816	1.02312	1.02830	1.03375	1.03955	1.04590
.668	.7816	.8911	.95575	0.08999	.65860	.64861	.63788	.62622	.61334	.59875	.58146	.55898	
					1.00434	1.00881	1.01344	1.01824	1.02323	1.02845	1.03395	1.03981	
.667	.7831	.8907	.95640	0.08841	.65751	.64740	.63654	.62469	.61158	.59663	.57875	.55489	
					1.00436	1.00885	1.01350	1.01833	1.02335	1.02860	1.03415	1.04008	
.666	.7847	.8904	.95704	0.08684	.65642	.64619	.63518	.62316	.60979	.59447	.57594	.55038	
					1.00438	1.00889	1.01356	1.01841	1.02346	1.02876	1.03435	1.04036	
.665	.7862	.8900	.95768	0.08530	.65532	.64498	.63382	.62161	.60798	.59226	.57301	.54537	
					1.00439	1.00893	1.01362	1.01850	1.02358	1.02891	1.03456	1.04064	
.664	.7878	.8896	.95831	0.08378	.65422	.64376	.63245	.62004	.60614	.59000	.56995	.53957	
					1.00441	1.00896	1.01368	1.01859	1.02370	1.02907	1.03477	1.04094	
.663	.7893	.8892	.95894	0.08227	.65312	.64254	.63107	.61847	.60428	.58768	.56674	.53246	
					1.00443	1.00900	1.01375	1.01867	1.02382	1.02923	1.03498	1.04125	
.662	.7908	.8888	.95956	0.08079	.65202	.64131	.62969	.61687	.60238	.58530	.56334	.52214	
					1.00445	1.00904	1.01381	1.01876	1.02394	1.02939	1.03520	1.04159	
.661	.7924	.8884	.96018	0.07932	.65092	.64009	.62830	.61526	.60046	.58285	.55972		
					1.00446	1.00908	1.01387	1.01885	1.02406	1.02956	1.03542		
.660	.7939	.8881	.96079	0.07787	.64981	.63885	.62689	.61364	.59850	.58033	.55583		
					1.00448	1.00912	1.01393	1.01894	1.02419	1.02972	1.03565		
.659	.7954	.8877	.96140	0.07644	.64871	.63761	.62548	.61199	.59650	.57771	.55154		
					1.00450	1.00915	1.01399	1.01903	1.02431	1.02989	1.03590		
.658	.7970	.8873	.96200	0.07503	.64760	.63637	.62406	.61033	.59447	.57500	.54680		
					1.00452	1.00919	1.01406	1.01912	1.02444	1.03007	1.03614		
.657	.7985	.8869	.96260	0.07364	.64649	.63511	.62263	.60865	.59239	.57218	.54135		
					1.00454	1.00924	1.01412	1.01921	1.02456	1.03024	1.03640		
.656	.8000	.8865	.96319	0.07226	.64538	.63386	.62119	.60694	.59027	.56922	.53475		
					1.00455	1.00928	1.01418	1.01931	1.02469	1.03042	1.03666		
.655	.8016	.8861	.96378	0.07090	.64426	.63260	.61974	.60521	.58810	.56612	.52552		
					1.00457	1.00932	1.01425	1.01940	1.02482	1.03060	1.03695		
.654	.8031	.8857	.96436	0.06956	.64315	.63133	.61828	.60346	.58587	.56284			
					1.00459	1.00935	1.01431	1.01949	1.02496	1.03079			
.653	.8046	.8854	.96494	0.06824	.64203	.63006	.61680	.60169	.58358	.55935			
					1.00461	1.00939	1.01437	1.01959	1.02509	1.03098			
.652	.8062	.8850	.96551	0.06693	.64091	.62879	.61532	.59988	.58123	.55555			
					1.00463	1.00943	1.01444	1.01968	1.02523	1.03118			
.651	.8077	.8846	.96608	0.06564	.63979	.62751	.61382	.59805	.57880	.55142			
					1.00465	1.00947	1.01451	1.01978	1.02536	1.03138			
.650	.8092	.8842	.96664	0.06437	.63865	.62622	.61230	.59619	.57629	.54682			
					1.00467	1.00951	1.01457	1.01988	1.02550	1.03159			
.649	.8108	.8838	.96720	0.06312	.63752	.62492	.61077	.59429	.57368	.54150			
					1.00469	1.00955	1.01464	1.01998	1.02565	1.03181			
.648	.8123	.8834	.96775	0.06188	.63639	.62362	.60923	.59236	.57097	.53501			
					1.00471	1.00960	1.01470	1.02008	1.02579	1.03203			
.647	.8139	.8830	.96830	0.06066	.63526	.62231	.60767	.59039	.56813	.52560			
					1.00473	1.00964	1.01477	1.02018	1.02594	1.03228			
.646	.8154	.8826	.96884	0.05945	.63413	.62100	.60609	.58837	.56515				
					1.00474	1.00968	1.01484	1.02028	1.02609				
.645	.8169	.8822	.96938	0.05826	.63299	.61968	.60449	.58631	.56199				
					1.00476	1.00972	1.01491	1.02038	1.02624				
.644	.8185	.8819	.96991	0.05709	.63185	.61834	.60288	.58420	.55859				
					1.00478	1.00976	1.01498	1.02049	1.02640				
.643	.8200	.8815	.97044	0.05593	.63071	.61700	.60124	.58204	.55495				
					1.00480	1.00980	1.01505	1.02059	1.02657				
.642	.8215	.8811	.97096	0.05479	.62956	.61566	.59959	.57981	.55095				
					1.00482	1.00984	1.01512	1.02070	1.02673				
.641	.8231	.8807	.97148	0.05366	.62841	.61430	.59791	.57752	.54647				
					1.00484	1.00988	1.01519	1.02081	1.02691				

TABLE IV: Specific Heat Ratio = 1.40

p/p₁	MACH	T/T₁	GAMMA	K_crit	.01	.015	.02	.025	.03	.035	.04	.045	.05
.640	.8246	.8803	.97199	0.05255	.62726	.62034	.61294	.60495	.59621	.58643	.57514	.56128	.54130
					1.00486	1.00737	1.00993	1.01255	1.01525	1.01804	1.02092	1.02392	1.02708
.639	.8261	.8799	.97250	0.05146	.62611	.61908	.61156	.60342	.59448	.58441	.57268	.55797	.53474
					1.00488	1.00740	1.00997	1.01261	1.01533	1.01813	1.02103	1.02406	1.02728
.638	.8277	.8795	.97300	0.05038	.62495	.61782	.61018	.60188	.59272	.58235	.57012	.55438	.52468
					1.00490	1.00743	1.01001	1.01267	1.01540	1.01822	1.02115	1.02421	1.02748
.637	.8292	.8791	.97350	0.04931	.62379	.61656	.60878	.60031	.59094	.58023	.56744	.55044	
					1.00492	1.00746	1.01005	1.01272	1.01547	1.01831	1.02126	1.02436	
.636	.8308	.8787	.97399	0.04826	.62262	.61528	.60737	.59873	.58911	.57804	.56463	.54601	
					1.00494	1.00749	1.01010	1.01278	1.01554	1.01841	1.02138	1.02451	
.635	.8323	.8783	.97448	0.04722	.62145	.61400	.60595	.59713	.58724	.57579	.56166	.54077	
					1.00496	1.00752	1.01014	1.01284	1.01562	1.01850	1.02151	1.02468	
.634	.8338	.8779	.97497	0.04620	.62028	.61271	.60452	.59551	.58534	.57347	.55849	.53416	
					1.00498	1.00755	1.01018	1.01290	1.01570	1.01860	1.02163	1.02485	
.633	.8354	.8775	.97544	0.04519	.61910	.61142	.60307	.59386	.58340	.57105	.55508	.52304	
					1.00500	1.00758	1.01023	1.01295	1.01578	1.01870	1.02176	1.02503	
.632	.8369	.8771	.97592	0.04420	.61792	.61011	.60161	.59219	.58141	.56854	.55135		
					1.00502	1.00761	1.01027	1.01301	1.01585	1.01880	1.02189		
.631	.8385	.8767	.97639	0.04322	.61673	.60880	.60014	.59049	.57938	.56591	.54717		
					1.00504	1.00764	1.01032	1.01307	1.01593	1.01890	1.02202		
.630	.8400	.8763	.97685	0.04226	.61554	.60748	.59865	.58874	.57728	.56314	.54229		
					1.00506	1.00768	1.01036	1.01314	1.01601	1.01901	1.02217		
.629	.8415	.8759	.97731	0.04131	.61435	.60615	.59714	.58698	.57512	.56021	.53627		
					1.00508	1.00771	1.01041	1.01320	1.01609	1.01911	1.02232		
.628	.8431	.8755	.97777	0.04037	.61315	.60481	.59561	.58518	.57289	.55707	.52716		
					1.00510	1.00774	1.01045	1.01326	1.01617	1.01922	1.02248		
.627	.8446	.8751	.97822	0.03945	.61195	.60347	.59407	.58335	.57058	.55368			
					1.00512	1.00777	1.01050	1.01333	1.01626	1.01933			
.626	.8462	.8747	.97866	0.03854	.61074	.60211	.59251	.58148	.56818	.54994			
					1.00514	1.00780	1.01054	1.01339	1.01634	1.01945			
.625	.8477	.8743	.97910	0.03764	.60952	.60074	.59092	.57956	.56567	.54573			
					1.00516	1.00784	1.01059	1.01345	1.01643	1.01956			
.624	.8492	.8739	.97954	0.03676	.60830	.59935	.58929	.57760	.56303	.54068			
					1.00518	1.00787	1.01064	1.01352	1.01652	1.01970			
.623	.8508	.8735	.97997	0.03589	.60707	.59796	.58765	.57558	.56024	.53422			
					1.00520	1.00790	1.01069	1.01358	1.01661	1.01983			
.622	.8523	.8731	.98040	0.03503	.60584	.59655	.58599	.57350	.55726	.52115			
					1.00522	1.00794	1.01074	1.01365	1.01670	1.01997			
.621	.8539	.8727	.98082	0.03419	.60460	.59513	.58430	.57135	.55405				
					1.00524	1.00797	1.01079	1.01372	1.01679				
.620	.8554	.8723	.98123	0.03335	.60335	.59369	.58257	.56913	.55053				
					1.00526	1.00800	1.01084	1.01379	1.01689				
.619	.8570	.8719	.98165	0.03254	.60210	.59222	.58082	.56681	.54660				
					1.00528	1.00804	1.01089	1.01386	1.01699				
.618	.8585	.8715	.98205	0.03173	.60083	.59075	.57902	.56439	.54190				
					1.00530	1.00808	1.01094	1.01393	1.01710				
.617	.8601	.8711	.98246	0.03094	.59956	.58926	.57718	.56185	.53605				
					1.00532	1.00811	1.01099	1.01400	1.01721				
.616	.8616	.8707	.98285	0.03016	.59829	.58775	.57529	.55916	.52631				
					1.00534	1.00814	1.01104	1.01408	1.01733				
.615	.8631	.8703	.98325	0.02939	.59700	.58622	.57336	.55629					
					1.00537	1.00818	1.01109	1.01415					
.614	.8647	.8699	.98364	0.02863	.59570	.58467	.57136	.55319					
					1.00539	1.00822	1.01115	1.01423					
.613	.8662	.8695	.98402	0.02789	.59439	.58309	.56930	.54979					
					1.00541	1.00825	1.01120	1.01431					
.612	.8678	.8691	.98440	0.02716	.59308	.58149	.56716	.54589					
					1.00543	1.00829	1.01125	1.01440					
.611	.8693	.8687	.98477	0.02644	.59175	.57986	.56494	.54135					
					1.00545	1.00832	1.01131	1.01449					

TABLE IV. Specific Heat Ratio = 1.40

R_2 and TPR versus Loss Coefficient K

p/p_t	MACH	T/T_t	GAMMA	K_{crit}	.001	.002	.003	.004	.005	.006	.01	.015	.02
.610	.8709	.8683	.98514	0.02573	.60823	.60643	.60460	.60272	.60080	.59883	.59042	.57819	.56262
					1.00053	1.00107	1.00161	1.00215	1.00269	1.00324	1.00547	1.00836	1.01137
.609	.8724	.8679	.98551	0.02503	.60721	.60538	.60352	.60161	.59965	.59765	.58907	.57650	.56018
					1.00053	1.00107	1.00161	1.00216	1.00270	1.00325	1.00549	1.00840	1.01142
.608	.8740	.8675	.98587	0.02435	.60618	.60433	.60243	.60049	.59849	.59646	.58770	.57476	.55759
					1.00054	1.00108	1.00162	1.00216	1.00272	1.00327	1.00552	1.00844	1.01148
.607	.8755	.8671	.98622	0.02367	.60516	.60327	.60135	.59937	.59735	.59527	.58632	.57298	.55483
					1.00054	1.00108	1.00162	1.00217	1.00272	1.00328	1.00554	1.00847	1.01154
.606	.8771	.8667	.98658	0.02301	.60413	.60222	.60026	.59826	.59619	.59407	.58493	.57116	.55184
					1.00054	1.00108	1.00163	1.00218	1.00273	1.00329	1.00556	1.00851	1.01160
.605	.8786	.8663	.98692	0.02236	.60310	.60116	.59916	.59713	.59503	.59287	.58352	.56928	.54848
					1.00054	1.00109	1.00164	1.00219	1.00275	1.00331	1.00558	1.00855	1.01167
.604	.8802	.8658	.98726	0.02172	.60207	.60010	.59808	.59600	.59386	.59166	.58206	.56735	.54471
					1.00054	1.00109	1.00164	1.00220	1.00276	1.00332	1.00561	1.00859	1.01174
.603	.8817	.8654	.98760	0.02109	.60104	.59904	.59698	.59486	.59269	.59044	.58061	.56534	.54019
					1.00055	1.00109	1.00165	1.00221	1.00277	1.00333	1.00564	1.00863	1.01181
.602	.8833	.8650	.98793	0.02047	.60000	.59797	.59588	.59373	.59151	.58922	.57914	.56326	.53407
					1.00055	1.00110	1.00165	1.00221	1.00278	1.00334	1.00566	1.00867	1.01189
.601	.8848	.8646	.98826	0.01987	.59900	.59691	.59478	.59259	.59032	.58798	.57765	.56109	
					1.00054	1.00110	1.00166	1.00222	1.00279	1.00336	1.00568	1.00872	
.600	.8864	.8642	.98858	0.01927	.59795	.59584	.59368	.59144	.58913	.58674	.57613	.55881	
					1.00055	1.00111	1.00167	1.00223	1.00280	1.00337	1.00571	1.00876	
.599	.8879	.8638	.98890	0.01869	.59692	.59477	.59257	.59029	.58793	.58549	.57458	.55640	
					1.00055	1.00111	1.00167	1.00224	1.00281	1.00338	1.00573	1.00880	
.598	.8895	.8634	.98922	0.01811	.59588	.59370	.59146	.58913	.58673	.58423	.57301	.55382	
					1.00056	1.00112	1.00168	1.00225	1.00282	1.00340	1.00576	1.00884	
.597	.8911	.8630	.98953	0.01755	.59485	.59263	.59034	.58797	.58551	.58296	.57140	.55105	
					1.00056	1.00112	1.00168	1.00225	1.00283	1.00341	1.00578	1.00889	
.596	.8926	.8626	.98983	0.01700	.59381	.59155	.58922	.58680	.58429	.58167	.56975	.54791	
					1.00056	1.00112	1.00169	1.00226	1.00284	1.00342	1.00580	1.00894	
.595	.8942	.8621	.99013	0.01645	.59278	.59048	.58810	.58563	.58306	.58038	.56807	.54441	
					1.00056	1.00113	1.00170	1.00227	1.00285	1.00343	1.00583	1.00899	
.594	.8957	.8617	.99043	0.01592	.59174	.58940	.58697	.58445	.58182	.57907	.56634	.54023	
					1.00056	1.00113	1.00170	1.00228	1.00286	1.00345	1.00586	1.00904	
.593	.8973	.8613	.99072	0.01540	.59070	.58831	.58584	.58326	.58057	.57775	.56455	.53455	
					1.00056	1.00113	1.00171	1.00229	1.00287	1.00346	1.00588	1.00910	
.592	.8988	.8609	.99101	0.01488	.58966	.58723	.58470	.58207	.57931	.57641	.56271		
					1.00057	1.00114	1.00172	1.00230	1.00288	1.00347	1.00591		
.591	.9004	.8605	.99129	0.01438	.58862	.58614	.58356	.58086	.57804	.57506	.56080		
					1.00057	1.00114	1.00172	1.00230	1.00289	1.00349	1.00594		
.590	.9020	.8601	.99157	0.01389	.58757	.58505	.58241	.57965	.57675	.57369	.55881		
					1.00057	1.00115	1.00173	1.00231	1.00290	1.00350	1.00596		
.589	.9035	.8596	.99184	0.01341	.58653	.58395	.58126	.57843	.57546	.57230	.55673		
					1.00057	1.00115	1.00173	1.00232	1.00292	1.00351	1.00599		
.588	.9051	.8592	.99211	0.01294	.58548	.58285	.58010	.57720	.57415	.57089	.55453		
					1.00057	1.00115	1.00174	1.00233	1.00293	1.00353	1.00602		
.587	.9067	.8588	.99237	0.01247	.58443	.58175	.57893	.57596	.57282	.56942	.55218		
					1.00058	1.00116	1.00175	1.00234	1.00294	1.00355	1.00605		
.586	.9082	.8584	.99263	0.01202	.58338	.58064	.57776	.57471	.57148	.56796	.54966		
					1.00058	1.00116	1.00175	1.00235	1.00295	1.00356	1.00608		
.585	.9098	.8580	.99288	0.01158	.58233	.57953	.57658	.57345	.57011	.56647	.54681		
					1.00058	1.00117	1.00176	1.00236	1.00296	1.00358	1.00611		
.584	.9113	.8576	.99313	0.01114	.58128	.57841	.57539	.57217	.56869	.56494	.54364		
					1.00058	1.00117	1.00176	1.00236	1.00298	1.00359	1.00614		
.583	.9129	.8571	.99338	0.01072	.58022	.57729	.57419	.57088	.56728	.56338	.53986		
					1.00058	1.00117	1.00177	1.00237	1.00299	1.00361	1.00618		
.582	.9145	.8567	.99362	0.01030	.57917	.57617	.57298	.56954	.56584	.56178	.53469		
					1.00059	1.00118	1.00178	1.00239	1.00300	1.00362	1.00621		
.581	.9160	.8563	.99386	0.00990	.57811	.57504	.57177	.56821	.56438	.56013			
					1.00059	1.00118	1.00178	1.00240	1.00301	1.00364			

TABLE IV. Specific Heat Ratio = 1.40

R₁ and TPR versus Loss Coefficient K

p/p_1	MACH	T/T_1	GAMMA	K_{crit}	.001	.002	.003	.004	.005	.006	.007	.008	.009
.580	.9176	.8559	.99409	0.00950	.57705	.57390	.57051	.56687	.56288	.55842	.55328	.54696	.53774
					1.00059	1.00119	1.00179	1.00241	1.00303	1.00365	1.00429	1.00494	1.00561
.579	.9192	.8554	.99432	0.00911	.57598	.57276	.56926	.56550	.56135	.55666	.55114	.54410	.53128
					1.00059	1.00119	1.00180	1.00242	1.00304	1.00367	1.00431	1.00496	1.00564
.578	.9207	.8550	.99454	0.00873	.57491	.57161	.56800	.56411	.55977	.55481	.54884	.54063	
					1.00059	1.00119	1.00181	1.00243	1.00305	1.00368	1.00433	1.00499	
.577	.9223	.8546	.99476	0.00836	.57384	.57041	.56673	.56269	.55815	.55288	.54634	.53629	
					1.00059	1.00120	1.00182	1.00243	1.00306	1.00370	1.00435	1.00502	
.576	.9239	.8542	.99497	0.00800	.57277	.56924	.56544	.56124	.55647	.55083	.54345	.52723	
					1.00060	1.00121	1.00182	1.00244	1.00308	1.00372	1.00438	1.00505	
.575	.9255	.8538	.99518	0.00765	.57170	.56807	.56413	.55976	.55473	.54864	.54011		
					1.00060	1.00121	1.00183	1.00245	1.00309	1.00373	1.00440		
.574	.9270	.8533	.99539	0.00731	.57058	.56688	.56281	.55823	.55291	.54626	.53580		
					1.00060	1.00122	1.00184	1.00246	1.00310	1.00375	1.00442		
.573	.9286	.8529	.99559	0.00698	.56949	.56568	.56146	.55667	.55099	.54353			
					1.00061	1.00122	1.00184	1.00247	1.00312	1.00377			
.572	.9302	.8525	.99578	0.00665	.56840	.56447	.56008	.55505	.54895	.54041			
					1.00061	1.00123	1.00185	1.00248	1.00313	1.00379			
.571	.9318	.8521	.99597	0.00633	.56730	.56324	.55868	.55337	.54677	.53648			
					1.00061	1.00123	1.00186	1.00249	1.00314	1.00381			
.570	.9333	.8516	.99616	0.00603	.56620	.56200	.55724	.55162	.54428	.52945			
					1.00061	1.00123	1.00186	1.00250	1.00316	1.00383			
.569	.9349	.8512	.99634	0.00573	.56509	.56075	.55577	.54978	.54152				
					1.00062	1.00124	1.00187	1.00252	1.00318				
.568	.9365	.8508	.99652	0.00543	.56398	.55947	.55425	.54782	.53820				
					1.00062	1.00124	1.00188	1.00253	1.00319				
.567	.9381	.8503	.99669	0.00515	.56286	.55818	.55269	.54572	.53367				
					1.00062	1.00125	1.00189	1.00254	1.00321				
.566	.9396	.8499	.99686	0.00488	.56173	.55686	.55106	.54330					
					1.00062	1.00125	1.00189	1.00255					
.565	.9412	.8495	.99703	0.00461	.56059	.55552	.54936	.54063					
					1.00062	1.00126	1.00190	1.00257					
.564	.9428	.8491	.99719	0.00435	.55945	.55415	.54756	.53739					
					1.00063	1.00126	1.00191	1.00258					
.563	.9444	.8486	.99734	0.00410	.55829	.55274	.54566	.53258					
					1.00063	1.00127	1.00192	1.00259					
.562	.9460	.8482	.99750	0.00386	.55712	.55129	.54348						
					1.00063	1.00127	1.00193						
.561	.9476	.8478	.99764	0.00362	.55594	.54979	.54113						
					1.00063	1.00128	1.00194						
.560	.9491	.8473	.99778	0.00340	.55475	.54824	.53838						
					1.00063	1.00128	1.00195						
.559	.9507	.8469	.99792	0.00318	.55353	.54662	.53489						
					1.00064	1.00128	1.00195						
.558	.9523	.8465	.99806	0.00297	.55231	.54481							
					1.00064	1.00129							
.557	.9539	.8460	.99818	0.00276	.55105	.54293							
					1.00064	1.00130							
.556	.9555	.8456	.99831	0.00257	.54978	.54086							
					1.00064	1.00131							
.555	.9571	.8452	.99843	0.00238	.54848	.53850							
					1.00064	1.00131							
.554	.9587	.8447	.99854	0.00220	.54714	.53565							
					1.00065	1.00131							
.553	.9603	.8443	.99866	0.00203	.54567	.53091							
					1.00065	1.00132							
.552	.9619	.8439	.99876	0.00186	.54421								
					1.00066								
.551	.9635	.8434	.99886	0.00171	.54268								
					1.00066								

TABLE IV. Specific Heat Ratio = 1.40

p/p,	MACH	T/T,	GAMMA	K_{crit}	.001	.002	.003	.004	.005	.006	.007	.008	.009
.550	.9651	.8430	.99896	0.00155	.54104 1.00066								
.549	.9667	.8425	.99906	0.00141	.53926 1.00066								
.548	.9682	.8421	.99915	0.00128	.53727 1.00066								
.547	.9698	.8417	.99923	0.00115	.53493 1.00066								
.546	.9714	.8412	.99931	0.00102	.53111 1.00067								

TABLE V. Specific Heat Ratio = 1.67

R₂ and TPR versus Loss Coefficient K

p/p,	MACH	T/T,	GAMMA	K_crit	.1	.2	.4	.6	1.0	2.0	4.0	6.0	10.0
.999	.0346	.9996	.06152	493.90381	.99890	.99880	.99860	.99840	.99800	.99699	.99499	.99297	.98893
					1.00010	1.00020	1.00040	1.00060	1.00100	1.00201	1.00403	1.00606	1.01016
.998	.0490	.9992	.08696	244.45897	.99780	.99760	.99720	.99680	.99599	.99398	.98994	.98589	.97772
					1.00020	1.00040	1.00080	1.00120	1.00201	1.00403	1.00811	1.01224	1.02065
.997	.0600	.9988	.10645	161.45030	.99670	.99640	.99579	.99519	.99398	.99095	.98486	.97874	.96637
					1.00030	1.00060	1.00120	1.00181	1.00302	1.00607	1.01225	1.01854	1.03150
.996	.0693	.9984	.12287	120.01288	.99560	.99519	.99439	.99358	.99196	.98791	.97976	.97153	.95487
					1.00040	1.00080	1.00161	1.00242	1.00404	1.00812	1.01644	1.02498	1.04271
.995	.0775	.9980	.13731	95.19130	.99450	.99399	.99298	.99197	.98994	.98486	.97462	.96426	.94320
					1.00050	1.00100	1.00201	1.00302	1.00506	1.01019	1.02070	1.03155	1.05432
.994	.0849	.9976	.15034	78.67066	.99339	.99279	.99157	.99036	.98792	.98180	.96944	.95692	.93137
					1.00060	1.00121	1.00242	1.00364	1.00608	1.01227	1.02502	1.03825	1.06636
.993	.0918	.9972	.16232	66.88948	.99229	.99158	.99016	.98874	.98589	.97873	.96424	.94952	.91936
					1.00070	1.00141	1.00283	1.00425	1.00711	1.01438	1.02939	1.04511	1.07884
.992	.0982	.9968	.17344	58.06774	.99119	.99038	.98875	.98712	.98386	.97564	.95899	.94205	.90717
					1.00081	1.00161	1.00323	1.00486	1.00814	1.01649	1.03384	1.05211	1.09180
.991	.1041	.9964	.18388	51.21767	.99009	.98917	.98734	.98550	.98182	.97254	.95372	.93450	.89479
					1.00091	1.00182	1.00364	1.00548	1.00918	1.01863	1.03835	1.05927	1.10527
.990	.1098	.9960	.19374	45.74662	.98898	.98796	.98592	.98388	.97977	.96943	.94840	.92688	.88220
					1.00101	1.00202	1.00405	1.00610	1.01023	1.02078	1.04292	1.06660	1.11929
.989	.1152	.9956	.20311	41.27752	.98788	.98676	.98451	.98225	.97772	.96631	.94305	.91919	.86940
					1.00111	1.00222	1.00446	1.00672	1.01128	1.02295	1.04757	1.07409	1.13390
.988	.1204	.9952	.21204	37.55943	.98677	.98555	.98309	₄98062	.97567	.96318	.93767	.91141	.85637
					1.00121	1.00243	1.00488	1.00734	1.01233	1.02513	1.05229	1.08176	1.14915
.987	.1253	.9948	.22060	34.41856	.98567	.98434	.98167	.97899	.97361	.96003	.93224	.90355	.84310
					1.00131	1.00263	1.00529	1.00797	1.01339	1.02734	1.05708	1.08962	1.16507
.986	.1301	.9944	.22882	31.73075	.98456	.98313	.98025	.97736	.97155	.95687	.92677	.89561	.82958
					1.00142	1.00284	1.00570	1.00859	1.01445	1.02956	1.06195	1.09767	1.18174
.985	.1347	.9940	.23674	29.40512	.98346	.98192	.97882	.97572	.96948	.95369	.92126	.88758	.81578
					1.00152	1.00305	1.00612	1.00922	1.01552	1.03180	1.06690	1.10592	1.19921
.984	.1392	.9935	.24440	27.37359	.98235	.98070	.97740	.97408	.96740	.95050	.91571	.87946	.80169
					1.00162	1.00325	1.00654	1.00986	1.01659	1.03406	1.07193	1.11439	1.21756
.983	.1435	.9931	.25180	25.58406	.98125	.97949	.97597	.97244	.96533	.94730	.91012	.87124	.78729
					1.00173	1.00346	1.00696	1.01049	1.01767	1.03633	1.07704	1.12308	1.23686
.982	.1478	.9927	.25898	23.99598	.98014	.97828	.97454	.97079	.96324	.94409	.90448	.86292	.77256
					1.00183	1.00367	1.00737	1.01112	1.01876	1.03863	1.08223	1.13199	1.25720
.981	.1519	.9923	.26596	22.57740	.97903	.97706	.97311	.96914	.96115	.94086	.89879	.85450	.75747
					1.00193	1.00387	1.00779	1.01176	1.01984	1.04095	1.08752	1.14116	1.27870
.980	.1559	.9919	.27274	21.30280	.97793	.97585	.97168	.96749	.95906	.93761	.89306	.84598	.74199
					1.00203	1.00408	1.00822	1.01240	1.02094	1.04328	1.09290	1.15058	1.30147
.979	.1598	.9915	.27935	20.15158	.97682	.97463	.97025	.96584	.95696	.93435	.88728	.83734	.72608
					1.00214	1.00429	1.00864	1.01305	1.02204	1.04564	1.09837	1.16027	1.32565
.978	.1636	.9911	.28579	19.10671	.97571	.97342	.96881	.96418	.95485	.93108	.88145	.82858	.70971
					1.00224	1.00450	1.00906	1.01369	1.02314	1.04802	1.10394	1.17024	1.35141
.977	.1673	.9907	.29208	18.15432	.97460	.97220	.96737	.96252	.95274	.92779	.87557	.81970	.69284
					1.00235	1.00471	1.00949	1.01434	1.02425	1.05042	1.10960	1.18051	1.37893
.976	.1710	.9903	.29823	17.28273	.97349	.97098	.96593	.96086	.95062	.92448	.86964	.81069	.67540
					1.00245	1.00492	1.00991	1.01499	1.02537	1.05284	1.11538	1.19110	1.40846
.975	.1746	.9899	.30424	16.48224	.97238	.96976	.96449	.95919	.94850	.92116	.86365	.80155	.65735
					1.00255	1.00513	1.01034	1.01564	1.02649	1.05528	1.12126	1.20203	1.44026
.974	.1781	.9895	.31012	15.74458	.97128	.96854	.96305	.95752	.94637	.91783	.85760	.79227	.63860
					1.00266	1.00534	1.01077	1.01629	1.02762	1.05774	1.12725	1.21331	1.47466
.973	.1816	.9891	.31588	15.06272	.97017	.96732	.96161	.95585	.94423	.91447	.85150	.78284	.61906
					1.00276	1.00555	1.01120	1.01695	1.02875	1.06023	1.13336	1.22496	1.51209
.972	.1849	.9887	.32153	14.43061	.96905	.96610	.96016	.95418	.94209	.91111	.84534	.77325	.59862
					1.00287	1.00576	1.01163	1.01761	1.02989	1.06274	1.13959	1.23702	1.55307
.971	.1883	.9883	.32707	13.84312	.96794	.96488	.95871	₄95250	.93995	.90772	.83911	.76350	.57714
					1.00297	1.00598	1.01206	1.01827	1.03104	1.06527	1.14595	1.24950	1.59824

TABLE V. Specific Heat Ratio = 1.67

p/p,	MACH	T/T,	GAMMA	K_crit	.1	.2	.4	.6	1.0	2.0	4.0	6.0	10.0
.970	.1916	.9879	.33250	13.29572	.96683	.96365	.95726	.95082	.93779	.90432	.83283	.75358	.55444
					1.00308	1.00619	1.01250	1.01893	1.03219	1.06783	1.15243	1.26244	1.64847
.969	.1948	.9874	.33784	12.78450	.96572	.96243	.95581	.94914	.93564	.90089	.82647	.74347	.53027
					1.00318	1.00640	1.01293	1.01960	1.03334	1.07041	1.15905	1.27586	1.70491
.968	.1980	.9870	.34309	12.30606	.96461	.96120	.95435	.94745	.93347	.89746	.82005	.73316	.50430
					1.00329	1.00661	1.01337	1.02026	1.03450	1.07301	1.16581	1.28980	1.76915
.967	.2011	.9866	.34825	11.85738	.96350	.95998	.95290	.94576	.93130	.89400	.81355	.72265	.47605
					1.00340	1.00683	1.01380	1.02093	1.03567	1.07564	1.17271	1.30430	1.84346
.966	.2042	.9862	.35332	11.43583	.96238	.95875	.95144	.94407	.92912	.89052	.80698	.71191	.44476
					1.00350	1.00704	1.01424	1.02161	1.03685	1.07830	1.17977	1.31940	1.93132
.965	.2073	.9858	.35831	11.03904	.96127	.95752	.94998	.94237	.92694	.88703	.80034	.70094	.40915
					1.00361	1.00726	1.01468	1.02228	1.03803	1.08098	1.18698	1.33515	2.03843
.964	.2103	.9854	.36322	10.66495	.96016	.95629	.94852	.94067	.92475	.88351	.79361	.68971	.36670
					1.00371	1.00747	1.01512	1.02296	1.03922	1.08369	1.19436	1.35160	2.17538
.963	.2133	.9850	.36806	10.31169	.95904	.95506	.94705	.93897	.92255	.87998	.78681	.67821	.31062
					1.00382	1.00769	1.01557	1.02364	1.04041	1.08643	1.20191	1.36882	2.36640
.962	.2162	.9846	.37283	9.97759	.95793	.95383	.94559	.93726	.92035	.87643	.77991	.66641	
					1.00393	1.00791	1.01601	1.02432	1.04161	1.08920	1.20964	1.38687	
.961	.2191	.9842	.37752	9.66119	.95681	.95260	.94412	.93555	.91814	.87285	.77293	.65429	
					1.00404	1.00812	1.01646	1.02501	1.04282	1.09199	1.21755	1.40583	
.960	.2220	.9838	.38215	9.36112	.95570	.95137	.94265	.93384	.91592	.86926	.76585	.64181	
					1.00414	1.00834	1.01690	1.02570	1.04403	1.09481	1.22567	1.42579	
.959	.2249	.9833	.38672	9.07620	.95458	.95014	.94118	.93212	.91369	.86564	.75868	.62896	
					1.00425	1.00856	1.01735	1.02639	1.04525	1.09767	1.23399	1.44685	
.958	.2277	.9829	.39122	8.80531	.95346	.94890	.93970	.93040	.91146	.86200	.75140	.61568	
					1.00436	1.00877	1.01780	1.02708	1.04648	1.10055	1.24253	1.46914	
.957	.2304	.9825	.39567	8.54748	.95235	.94767	.93823	.92867	.90922	.85834	.74401	.60193	
					1.00447	1.00899	1.01825	1.02778	1.04771	1.10346	1.25129	1.49280	
.956	.2332	.9821	.40005	8.30179	.95123	.94643	.93675	.92695	.90698	.85466	.73652	.58766	
					1.00457	1.00921	1.01870	1.02847	1.04895	1.10641	1.26030	1.51799	
.955	.2359	.9817	.40438	8.06744	.95011	.94519	.93527	.92522	.90473	.85095	.72890	.57280	
					1.00468	1.00943	1.01915	1.02918	1.05020	1.10938	1.26955	1.54492	
.954	.2386	.9813	.40866	7.84368	.94899	.94395	.93378	.92348	.90247	.84722	.72116	.55727	
					1.00479	1.00965	1.01961	1.02988	1.05145	1.11239	1.27908	1.57383	
.953	.2413	.9809	.41288	7.62982	.94787	.94271	.93230	.92174	.90020	.84347	.71329	.54097	
					1.00490	1.00987	1.02006	1.03059	1.05271	1.11544	1.28888	1.60503	
.952	.2439	.9805	.41705	7.42523	.94675	.94147	.93081	.92000	.89792	.83969	.70528	.52377	
					1.00501	1.01009	1.02052	1.03130	1.05398	1.11852	1.29899	1.63890	
.951	.2465	.9800	.42118	7.22934	.94563	.94023	.92932	.91825	.89564	.83588	.69713	.50550	
					1.00512	1.01032	1.02098	1.03201	1.05526	1.12163	1.30940	1.67592	
.950	.2491	.9796	.42525	7.04162	.94451	.93899	.92783	.91650	.89335	.83205	.68882	.48594	
					1.00523	1.01054	1.02144	1.03272	1.05654	1.12478	1.32016	1.71673	
.949	.2517	.9792	.42928	6.86158	.94339	.93775	.92633	.91475	.89105	.82820	.68034	.46475	
					1.00533	1.01076	1.02190	1.03344	1.05783	1.12797	1.33127	1.76220	
.948	.2543	.9788	.43326	6.68878	.94227	.93650	.92484	.91299	.88874	.82431	.67170	.44146	
					1.00544	1.01098	1.02236	1.03416	1.05913	1.13120	1.34276	1.81356	
.947	.2568	.9784	.43720	6.52280	.94115	.93526	.92334	.91123	.88643	.82040	.66286	.41530	
					1.00555	1.01121	1.02283	1.03489	1.06044	1.13446	1.35466	1.87265	
.946	.2593	.9780	.44109	6.36325	.94003	.93401	.92184	.90947	.88410	.81646	.65383	.38483	
					1.00566	1.01143	1.02329	1.03561	1.06175	1.13777	1.36699	1.94249	
.945	.2618	.9776	.44495	6.20979	.93891	.93276	.92033	.90770	.88177	.81248	.64458	.34691	
					1.00577	1.01166	1.02376	1.03634	1.06307	1.14111	1.37979	2.02873	
.944	.2642	.9771	.44876	6.06206	.93778	.93151	.91883	.90592	.87943	.80848	.63510	.28996	
					1.00588	1.01188	1.02423	1.03707	1.06440	1.14450	1.39309	2.14566	
.943	.2667	.9767	.45253	5.91979	.93666	.93027	.91732	.90415	.87708	.80445	.62538		
					1.00600	1.01211	1.02470	1.03781	1.06574	1.14793	1.40694		
.942	.2691	.9763	.45626	5.78266	.93553	.92901	.91581	.90236	.87472	.80039	.61538		
					1.00611	1.01233	1.02517	1.03855	1.06709	1.15140	1.42138		
.941	.2715	.9759	.45996	5.65043	.93441	.92776	.91429	.90058	.87236	.79629	.60509		
					1.00622	1.01256	1.02564	1.03929	1.06844	1.15492	1.43646		

TABLE V. Specific Heat Ratio = 1.67

R₁ and TPR versus Loss Coefficient K

p/p_t	MACH	T/T_t	GAMMA	K_{crit}	.1	.2	.4	.6	.8	1.0	2.0	3.0	4.0
.940	.2739	.9755	.46362	5.52283	.93328	.92651	.91278	.89879	.88453	.86998	.79216	.70293	.59448
					1.00633	1.01279	1.02612	1.04003	1.05458	1.06981	1.15849	1.27745	1.45224
.939	.2763	.9751	.46724	5.39965	.93216	.92526	.91126	.89699	.88245	.86760	.78799	.69624	.58351
					1.00644	1.01302	1.02660	1.04078	1.05563	1.07118	1.16210	1.28513	1.46879
.938	.2787	.9746	.47083	5.28066	.93103	.92400	.90974	.89520	.88036	.86520	.78379	.68945	.57215
					1.00655	1.01324	1.02707	1.04153	1.05668	1.07256	1.16577	1.29300	1.48619
.937	.2810	.9742	.47438	5.16565	.92991	.92274	.90821	.89339	.87826	.86280	.77955	.68255	.56035
					1.00666	1.01347	1.02755	1.04229	1.05773	1.07395	1.16948	1.30108	1.50453
.936	.2833	.9738	.47790	5.05445	.92878	.92149	.90669	.89159	.87616	.86038	.77528	.67551	.54806
					1.00678	1.01370	1.02803	1.04304	1.05879	1.07535	1.17325	1.30939	1.52391
.935	.2856	.9734	.48139	4.94687	.92765	.92023	.90516	.88977	.87405	.85796	.77096	.66835	.53520
					1.00689	1.01393	1.02852	1.04381	1.05986	1.07676	1.17707	1.31792	1.54448
.934	.2879	.9730	.48484	4.84274	.92652	.91897	.90363	.88796	.87193	.85553	.76661	.66106	.52170
					1.00700	1.01416	1.02900	1.04457	1.06094	1.07817	1.18094	1.32670	1.56639
.933	.2902	.9726	.48827	4.74190	.92539	.91771	.90209	.88614	.86981	.85308	.76221	.65361	.50744
					1.00711	1.01439	1.02949	1.04534	1.06201	1.07960	1.18487	1.33574	1.58983
.932	.2925	.9721	.49166	4.64422	.92426	.91645	.90056	.88431	.86768	.85063	.75778	.64601	.49229
					1.00723	1.01463	1.02997	1.04611	1.06310	1.08104	1.18886	1.34505	1.61506
.931	.2947	.9717	.49502	4.54954	.92313	.91518	.89902	.88248	.86554	.84817	.75329	.63824	.47605
					1.00734	1.01486	1.03046	1.04688	1.06419	1.08248	1.19291	1.35466	1.64238
.930	.2970	.9713	.49835	4.45774	.92200	.91392	.89748	.88084	.86339	.84569	.74877	.63030	.45846
					1.00745	1.01509	1.03095	1.04766	1.06529	1.08394	1.19702	1.36458	1.67221
.929	.2992	.9709	.50166	4.36869	.92087	.91265	.89593	.87880	.86124	.84321	.74419	.62217	.43912
					1.00757	1.01532	1.03145	1.04844	1.06639	1.08540	1.20120	1.37483	1.70512
.928	.3014	.9705	.50493	4.28227	.91974	.91139	.89438	.87696	.85908	.84071	.73957	.61383	.41739
					1.00768	1.01556	1.03194	1.04922	1.06750	1.08688	1.20544	1.38544	1.74196
.927	.3036	.9700	.50818	4.19839	.91861	.91012	.89283	.87511	.85691	.83820	.73490	.60526	.39211
					1.00779	1.01579	1.03243	1.05001	1.06861	1.08836	1.20975	1.39643	1.78401
.926	.3058	.9696	.51140	4.11692	.91748	.90885	.89128	.87325	.85474	.83568	.73017	.59646	.36076
					1.00791	1.01603	1.03293	1.05080	1.06974	1.08986	1.21413	1.40784	1.83362
.925	.3080	.9692	.51459	4.03778	.91634	.90758	.88972	.87139	.85255	.83315	.72539	.58740	.31449
					1.00802	1.01626	1.03343	1.05160	1.07086	1.09136	1.21859	1.41970	1.89618
.924	.3101	.9688	.51776	3.96086	.91521	.90631	.88816	.86953	.85036	.83061	.72056	.57805	
					1.00814	1.01650	1.03393	1.05239	1.07200	1.09288	1.22312	1.43204	
.923	.3123	.9684	.52090	3.88609	.91407	.90503	.88660	.86766	.84816	.82806	.71566	.56838	
					1.00825	1.01673	1.03443	1.05320	1.07314	1.09441	1.22773	1.44492	
.922	.3144	.9679	.52401	3.81337	.91294	.90376	.88504	.86578	.84595	.82549	.71071	.55836	
					1.00837	1.01697	1.03494	1.05400	1.07429	1.09595	1.23242	1.45837	
.921	.3165	.9675	.52710	3.74263	.91180	.90248	.88347	.86390	.84374	.82291	.70569	.54795	
					1.00848	1.01721	1.03544	1.05481	1.07544	1.09750	1.23720	1.47246	
.920	.3187	.9671	.53017	3.67379	.91067	.90121	.88190	.86202	.84151	.82032	.70061	.53709	
					1.00860	1.01745	1.03595	1.05562	1.07660	1.09906	1.24206	1.48727	
.919	.3208	.9667	.53321	3.60678	.90953	.89993	.88032	.86012	.83928	.81772	.69546	.52573	
					1.00872	1.01769	1.03646	1.05644	1.07777	1.10063	1.24702	1.50286	
.918	.3229	.9663	.53623	3.54152	.90839	.89865	.87874	.85823	.83704	.81510	.69023	.51380	
					1.00883	1.01793	1.03697	1.05726	1.07895	1.10222	1.25207	1.51933	
.917	.3250	.9658	.53922	3.47796	.90725	.89737	.87716	.85632	.83479	.81247	.68494	.50119	
					1.00895	1.01817	1.03748	1.05809	1.08013	1.10381	1.25721	1.53682	
.916	.3270	.9654	.54219	3.41604	.90612	.89609	.87558	.85442	.83253	.80982	.67956	.48778	
					1.00906	1.01841	1.03800	1.05891	1.08132	1.10542	1.26247	1.55545	
.915	.3291	.9650	.54514	3.35569	.90498	.89480	.87399	.85250	.83026	.80717	.67411	.47340	
					1.00918	1.01865	1.03852	1.05975	1.08252	1.10704	1.26783	1.57543	
.914	.3312	.9646	.54806	3.29686	.90384	.89352	.87240	.85058	.82798	.80449	.66856	.45781	
					1.00930	1.01889	1.03903	1.06058	1.08372	1.10868	1.27330	1.59700	
.913	.3332	.9641	.55097	3.23949	.90270	.89223	.87081	.84865	.82569	.80181	.66293	.44064	
					1.00942	1.01913	1.03956	1.06142	1.08493	1.11032	1.27889	1.62049	
.912	.3352	.9637	.55385	3.18354	.90155	.89095	.86921	.84672	.82339	.79911	.65720	.42133	
					1.00953	1.01938	1.04008	1.06227	1.08615	1.11198	1.28460	1.64638	
.911	.3373	.9633	.55671	3.12895	.90041	.88966	.86761	.84478	.82108	.79639	.65138	.39883	
					1.00965	1.01962	1.04060	1.06312	1.08738	1.11366	1.29044	1.67541	

TABLE V. Specific Heat Ratio = 1.67

p/p_t	MACH	T/T_t	GAMMA	K_{crit}	R₂ and TPR versus Loss Coefficient K								
					.1	.2	.4	.6	.8	.9	1.0	1.5	2.0
.910	.3393	.9629	.55955	3.07568	.89927	.88837	.86601	.84284	.81876	.80635	.79366	.72529	.64544
					1.00977	1.01987	1.04113	1.06397	1.08862	1.10170	1.11534	1.19376	1.29642
.909	.3413	.9624	.56237	3.02368	.89813	.88708	.86440	.84089	.81644	.80382	.79091	.72124	.63940
					1.00989	1.02011	1.04166	1.06483	1.08986	1.10316	1.11704	1.19706	1.30253
.908	.3433	.9620	.56517	2.97292	.89698	.88578	.86279	.83893	.81410	.80127	.78815	.71714	.63324
					1.01001	1.02036	1.04219	1.06569	1.09111	1.10463	1.11876	1.20041	1.30880
.907	.3453	.9616	.56795	2.92334	.89584	.88449	.86118	.83697	.81175	.79872	.78537	.71299	.62695
					1.01012	1.02060	1.04272	1.06655	1.09237	1.10612	1.12049	1.20381	1.31523
.906	.3473	.9612	.57072	2.87491	.89469	.88319	.85956	.83500	.80939	.79614	.78257	.70881	.62053
					1.01024	1.02085	1.04325	1.06742	1.09364	1.10761	1.12223	1.20726	1.32182
.905	.3492	.9607	.57346	2.82760	.89355	.88189	.85794	.83302	.80702	.79356	.77976	.70457	.61396
					1.01036	1.02110	1.04379	1.06830	1.09491	1.10912	1.12399	1.21076	1.32859
.904	.3512	.9603	.57618	2.78137	.89240	.88060	.85631	.83104	.80464	.79096	.77693	.70029	.60725
					1.01048	1.02134	1.04433	1.06918	1.09620	1.11064	1.12576	1.21431	1.33554
.903	.3532	.9599	.57888	2.73618	.89125	.87930	.85468	.82905	.80224	.78835	.77408	.69595	.60036
					1.01060	1.02159	1.04487	1.07006	1.09749	1.11217	1.12755	1.21791	1.34270
.902	.3551	.9595	.58157	2.69200	.89011	.87799	.85305	.82705	.79984	.78572	.77121	.69156	.59331
					1.01072	1.02184	1.04541	1.07095	1.09880	1.11371	1.12936	1.22158	1.35006
.901	.3571	.9590	.58423	2.64880	.88896	.87669	.85142	.82505	.79742	.78307	.76833	.68712	.58606
					1.01084	1.02209	1.04595	1.07184	1.10011	1.11526	1.13118	1.22530	1.35765
.900	.3590	.9586	.58688	2.60656	.88781	.87539	.84978	.82304	.79499	.78042	.76543	.68262	.57860
					1.01096	1.02234	1.04650	1.07274	1.10143	1.11682	1.13301	1.22908	1.36549
.899	.3609	.9582	.58951	2.56523	.88666	.87408	.84813	.82102	.79255	.77774	.76250	.67806	.57092
					1.01108	1.02260	1.04705	1.07364	1.10276	1.11840	1.13487	1.23293	1.37358
.898	.3629	.9578	.59212	2.52479	.88551	.87277	.84648	.81899	.79010	.77505	.75956	.67343	.56298
					1.01120	1.02285	1.04760	1.07455	1.10410	1.11999	1.13674	1.23684	1.38195
.897	.3648	.9573	.59472	2.48522	.88436	.87146	.84483	.81696	.78764	.77235	.75660	.66874	.55478
					1.01132	1.02310	1.04815	1.07546	1.10545	1.12160	1.13862	1.24082	1.39063
.896	.3667	.9569	.59730	2.44649	.88321	.87015	.84318	.81492	.78516	.76963	.75361	.66399	.54627
					1.01145	1.02335	1.04871	1.07638	1.10680	1.12321	1.14053	1.24487	1.39963
.895	.3686	.9565	.59986	2.40857	.88205	.86884	.84152	.81287	.78267	.76689	.75060	.65916	.53742
					1.01157	1.02361	1.04926	1.07730	1.10817	1.12484	1.14245	1.24899	1.40900
.894	.3705	.9560	.60240	2.37144	.88090	.86752	.83985	.81081	.78016	.76413	.74758	.65425	.52818
					1.01169	1.02386	1.04982	1.07823	1.10955	1.12649	1.14440	1.25319	1.41876
.893	.3724	.9556	.60493	2.33509	.87975	.86621	.83819	.80875	.77764	.76136	.74452	.64927	.51851
					1.01181	1.02412	1.05038	1.07916	1.11094	1.12815	1.14636	1.25747	1.42895
.892	.3742	.9552	.60744	2.29948	.87859	.86489	.83651	.80668	.77511	.75857	.74145	.64421	.50834
					1.01193	1.02438	1.05095	1.08010	1.11234	1.12982	1.14834	1.26183	1.43963
.891	.3761	.9548	.60994	2.26459	.87744	.86357	.83484	.80460	.77256	.75576	.73835	.63905	.49758
					1.01206	1.02463	1.05151	1.08104	1.11375	1.13151	1.15034	1.26628	1.45086
.890	.3780	.9543	.61242	2.23042	.87628	.86225	.83316	.80251	.77000	.75293	.73523	.63381	.48611
					1.01218	1.02489	1.05208	1.08199	1.11517	1.13321	1.15236	1.27083	1.46270
.889	.3799	.9539	.61488	2.19692	.87512	.86093	.83147	.80041	.76742	.75008	.73208	.62847	.47380
					1.01230	1.02515	1.05265	1.08295	1.11660	1.13493	1.15440	1.27546	1.47526
.888	.3817	.9535	.61733	2.16410	.87396	.85960	.82978	.79830	.76483	.74721	.72891	.62303	.46041
					1.01243	1.02541	1.05323	1.08391	1.11805	1.13666	1.15646	1.28020	1.48864
.887	.3836	.9530	.61976	2.13192	.87281	.85827	.82809	.79619	.76222	.74432	.72571	.61747	.44563
					1.01255	1.02567	1.05380	1.08487	1.11950	1.13841	1.15855	1.28504	1.50302
.886	.3854	.9526	.62218	2.10037	.87165	.85694	.82639	.79406	.75960	.74141	.72248	.61180	.42892
					1.01267	1.02593	1.05438	1.08584	1.12097	1.14017	1.16065	1.28999	1.51861
.885	.3872	.9522	.62458	2.06944	.87049	.85561	.82469	.79193	.75696	.73848	.71923	.60601	.40930
					1.01280	1.02619	1.05496	1.08682	1.12245	1.14195	1.16278	1.29506	1.53578
.884	.3891	.9517	.62697	2.03910	.86933	.85428	.82298	.78979	.75431	.73553	.71594	.60008	.38453
					1.01292	1.02645	1.05554	1.08780	1.12394	1.14375	1.16494	1.30025	1.55512
.883	.3909	.9513	.62935	2.00934	.86816	.85295	.82127	.78764	.75163	.73255	.71263	.59402	.34531
					1.01305	1.02671	1.05613	1.08879	1.12544	1.14557	1.16711	1.30557	1.57810
.882	.3927	.9509	.63170	1.98016	.86700	.85161	.81955	.78548	.74894	.72956	.70928	.58780	
					1.01318	1.02698	1.05672	1.08978	1.12695	1.14740	1.16932	1.31103	
.881	.3945	.9504	.63405	1.95152	.86584	.85028	.81783	.78331	.74623	.72653	.70591	.58141	
					1.01330	1.02724	1.05731	1.09078	1.12848	1.14925	1.17154	1.31663	

TABLE V. Specific Heat Ratio = 1.67

p/p_t	MACH	T/T_t	GAMMA	K_crit	.1	.2	.4	.6	.7	.8	.9	1.0	1.5
.880	.3963	.9500	.63638	1.92342	.86467	.84894	.81610	.78113	.76269	.74351	.72349	.70250	.57485
					1.01343	1.02751	1.05790	1.09179	1.11030	1.13002	1.15112	1.17380	1.32239
.879	.3981	.9496	.63870	1.89584	.86351	.84760	.81437	.77893	.76023	.74076	.72041	.69906	.56809
					1.01355	1.02777	1.05850	1.09280	1.11157	1.13158	1.15301	1.17608	1.32832
.878	.3999	.9491	.64100	1.86877	.86234	.84625	.81263	.77673	.75776	.73800	.71732	.69558	.56112
					1.01368	1.02804	1.05909	1.09382	1.11284	1.13315	1.15492	1.17838	1.33442
.877	.4017	.9487	.64329	1.84220	.86117	.84491	.81089	.77452	.75528	.73521	.71419	.69206	.55392
					1.01380	1.02831	1.05970	1.09485	1.11412	1.13473	1.15685	1.18072	1.34072
.876	.4035	.9483	.64556	1.81612	.86001	.84356	.80914	.77230	.75278	.73241	.71104	.68851	.54646
					1.01393	1.02857	1.06030	1.09588	1.11542	1.13632	1.15880	1.18308	1.34722
.875	.4053	.9478	.64783	1.79051	.85884	.84221	.80739	.77007	.75027	.72958	.70786	.68492	.53870
					1.01406	1.02884	1.06090	1.09692	1.11672	1.13793	1.16077	1.18548	1.35395
.874	.4071	.9474	.65007	1.76536	.85767	.84086	.80563	.76782	.74774	.72674	.70465	.68129	.53063
					1.01419	1.02911	1.06151	1.09797	1.11804	1.13956	1.16276	1.18790	1.36092
.873	.4089	.9470	.65231	1.74065	.85650	.83951	.80387	.76557	.74520	.72387	.70141	.67762	.52219
					1.01431	1.02938	1.06213	1.09902	1.11936	1.14120	1.16477	1.19036	1.36816
.872	.4106	.9465	.65453	1.71639	.85533	.83816	.80210	.76330	.74264	.72098	.69814	.67390	.51332
					1.01444	1.02965	1.06274	1.10008	1.12069	1.14285	1.16680	1.19285	1.37570
.871	.4124	.9461	.65674	1.69256	.85416	.83680	.80032	.76102	.74007	.71807	.69484	.67014	.50397
					1.01457	1.02993	1.06336	1.10115	1.12204	1.14453	1.16886	1.19537	1.38356
.870	.4142	.9457	.65894	1.66914	.85298	.83544	.79854	.75873	.73747	.71513	.69150	.66634	.49405
					1.01470	1.03020	1.06398	1.10222	1.12340	1.14621	1.17094	1.19793	1.39179
.869	.4159	.9452	.66112	1.64613	.85181	.83408	.79676	.75642	.73486	.71217	.68813	.66248	.48343
					1.01483	1.03047	1.06460	1.10330	1.12476	1.14792	1.17305	1.20052	1.40043
.868	.4177	.9448	.66330	1.62352	.85064	.83272	.79496	.75411	.73223	.70918	.68473	.65857	.47195
					1.01496	1.03075	1.06523	1.10439	1.12614	1.14964	1.17518	1.20316	1.40956
.867	.4194	.9444	.66545	1.60130	.84946	.83135	.79317	.75178	.72959	.70617	.68129	.65461	.45938
					1.01509	1.03102	1.06586	1.10549	1.12753	1.15138	1.17734	1.20582	1.41924
.866	.4212	.9439	.66760	1.57946	.84828	.82998	.79136	.74944	.72692	.70313	.67781	.65060	.44534
					1.01522	1.03130	1.06649	1.10659	1.12893	1.15313	1.17952	1.20853	1.42960
.865	.4229	.9435	.66974	1.55799	.84711	.82861	.78955	.74708	.72424	.70007	.67429	.64653	.42919
					1.01535	1.03158	1.06712	1.10770	1.13035	1.15491	1.18173	1.21129	1.44079
.864	.4246	.9430	.67186	1.53689	.84593	.82724	.78773	.74471	.72154	.69697	.67073	.64239	.40962
					1.01548	1.03185	1.06776	1.10882	1.13178	1.15670	1.18397	1.21408	1.45307
.863	.4264	.9426	.67397	1.51614	.84475	.82587	.78591	.74233	.71881	.69385	.66713	.63820	.38288
					1.01561	1.03213	1.06840	1.10995	1.13321	1.15851	1.18624	1.21692	1.46695
.862	.4281	.9422	.67607	1.49573	.84357	.82449	.78408	.73993	.71607	.69070	.66348	.63393	
					1.01574	1.03241	1.06905	1.11109	1.13467	1.16035	1.18854	1.21980	
.861	.4298	.9417	.67816	1.47566	.84239	.82311	.78225	.73752	.71330	.68751	.65979	.62960	
					1.01587	1.03269	1.06969	1.11223	1.13613	1.16220	1.19087	1.22274	
.860	.4315	.9413	.68023	1.45593	.84120	.82173	.78040	.73510	.71051	.68430	.65605	.62520	
					1.01600	1.03297	1.07034	1.11339	1.13761	1.16407	1.19323	1.22572	
.859	.4332	.9408	.68230	1.43652	.84002	.82035	.77856	.73266	.70770	.68105	.65226	.62071	
					1.01613	1.03325	1.07100	1.11455	1.13910	1.16596	1.19562	1.22875	
.858	.4349	.9404	.68435	1.41742	.83884	.81896	.77670	.73020	.70487	.67776	.64842	.61615	
					1.01626	1.03354	1.07166	1.11572	1.14061	1.16788	1.19805	1.23184	
.857	.4366	.9400	.68639	1.39864	.83765	.81757	.77484	.72772	.70202	.67445	.64452	.61149	
					1.01640	1.03382	1.07232	1.11690	1.14213	1.16982	1.20052	1.23499	
.856	.4383	.9395	.68842	1.38015	.83647	.81618	.77297	.72524	.69913	.67109	.64057	.60675	
					1.01653	1.03411	1.07298	1.11809	1.14366	1.17178	1.20302	1.23820	
.855	.4400	.9391	.69044	1.36197	.83528	.81479	.77109	.72273	.69623	.66770	.63656	.60191	
					1.01666	1.03439	1.07365	1.11929	1.14521	1.17377	1.20555	1.24146	
.854	.4417	.9386	.69245	1.34407	.83409	.81340	.76921	.72021	.69330	.66426	.63249	.59696	
					1.01680	1.03468	1.07432	1.12050	1.14678	1.17578	1.20813	1.24480	
.853	.4434	.9382	.69445	1.32646	.83290	.81200	.76731	.71766	.69034	.66079	.62835	.59190	
					1.01693	1.03497	1.07499	1.12172	1.14836	1.17781	1.21075	1.24820	
.852	.4451	.9378	.69643	1.30913	.83171	.81060	.76541	.71510	.68735	.65727	.62414	.58672	
					1.01706	1.03525	1.07567	1.12295	1.14996	1.17987	1.21341	1.25168	
.851	.4468	.9373	.69841	1.29207	.83052	.80919	.76351	.71253	.68434	.65371	.61985	.58141	
					1.01720	1.03554	1.07635	1.12418	1.15157	1.18196	1.21612	1.25523	

TABLE V. Specific Heat Ratio = 1.67

p/p,	MACH	T/T,	GAMMA	K_crit	.1	.2	.4	.5	.6	.7	.8	.9	1.0
								R₂ and TPR versus Loss Coefficient K					
.850	.4485	.9369	.70037	1.27527	.82933	.80779	.76159	.73657	.70993	.68129	.65010	.61550	.57597
					1.01733	1.03583	1.07704	1.10019	1.12543	1.15321	1.18408	1.21887	1.25886
.849	.4501	.9364	.70233	1.25874	.82813	.80638	.75967	.73433	.70731	.67822	.64644	.61106	.57037
					1.01747	1.03612	1.07773	1.10113	1.12669	1.15485	1.18622	1.22167	1.26258
.848	.4518	.9360	.70427	1.24246	.82694	.80497	.75774	.73208	.70468	.67511	.64274	.60653	.56461
					1.01760	1.03642	1.07842	1.10209	1.12796	1.15652	1.18839	1.22452	1.26639
.847	.4535	.9355	.70621	1.22643	.82574	.80356	.75580	.72981	.70202	.67197	.63898	.60191	.55868
					1.01774	1.03671	1.07912	1.10304	1.12925	1.15821	1.19060	1.22743	1.27030
.846	.4551	.9351	.70813	1.21065	.82455	.80214	.75385	.72753	.69934	.66880	.63516	.59720	.55254
					1.01787	1.03700	1.07982	1.10401	1.13054	1.15991	1.19284	1.23039	1.27431
.845	.4568	.9347	.71004	1.19511	.82335	.80072	.75189	.72524	.69664	.66559	.63129	.59238	.54620
					1.01801	1.03730	1.08052	1.10498	1.13185	1.16163	1.19511	1.23340	1.27843
.844	.4585	.9342	.71195	1.17980	.82215	.79930	.74993	.72293	.69392	.66235	.62735	.58745	.53961
					1.01815	1.03759	1.08123	1.10597	1.13316	1.16338	1.19741	1.23648	1.28267
.843	.4601	.9338	.71384	1.16473	.82095	.79787	.74795	.72060	.69117	.65906	.62335	.58240	.53274
					1.01828	1.03789	1.08194	1.10695	1.13449	1.16514	1.19975	1.23963	1.28705
.842	.4618	.9333	.71572	1.14988	.81975	.79645	.74597	.71827	.68840	.65574	.61929	.57722	.52558
					1.01842	1.03819	1.08266	1.10795	1.13583	1.16693	1.20213	1.24284	1.29157
.841	.4634	.9329	.71760	1.13525	.81854	.79502	.74398	.71591	.68561	.65238	.61515	.57190	.51806
					1.01856	1.03849	1.08338	1.10895	1.13719	1.16873	1.20454	1.24612	1.29623
.840	.4651	.9324	.71946	1.12084	.81734	.79358	.74197	.71355	.68279	.64898	.61094	.56644	.51014
					1.01870	1.03879	1.08410	1.10996	1.13855	1.17056	1.20700	1.24948	1.30107
.839	.4667	.9320	.72131	1.10664	.81613	.79215	.73996	.71116	.67994	.64553	.60664	.56080	.50175
					1.01883	1.03909	1.08483	1.11098	1.13994	1.17242	1.20949	1.25292	1.30610
.838	.4683	.9315	.72316	1.09265	.81493	.79071	.73794	.70876	.67707	.64204	.60227	.55499	.49279
					1.01897	1.03939	1.08557	1.11200	1.14133	1.17430	1.21204	1.25644	1.31134
.837	.4700	.9311	.72499	1.07887	.81372	.78926	.73591	.70634	.67417	.63849	.59780	.54898	.48313
					1.01911	1.03969	1.08630	1.11304	1.14274	1.17620	1.21462	1.26006	1.31682
.836	.4716	.9307	.72682	1.06529	.81251	.78782	.73387	.70391	.67124	.63490	.59324	.54275	.47259
					1.01925	1.04000	1.08705	1.11408	1.14416	1.17813	1.21726	1.26379	1.32258
.835	.4732	.9302	.72863	1.05190	.81130	.78637	.73181	.70146	.66828	.63126	.58857	.53627	.46088
					1.01939	1.04030	1.08779	1.11513	1.14560	1.18008	1.21994	1.26761	1.32866
.834	.4749	.9298	.73044	1.03871	.81009	.78492	.72975	.69899	.66529	.62755	.58380	.52952	.44752
					1.01953	1.04061	1.08854	1.11619	1.14705	1.18206	1.22268	1.27155	1.33514
.833	.4765	.9293	.73223	1.02571	.80888	.78346	.72768	.69650	.66226	.62380	.57891	.52245	.43158
					1.01967	1.04092	1.08930	1.11725	1.14852	1.18408	1.22547	1.27562	1.34213
.832	.4781	.9289	.73402	1.01289	.80766	.78200	.72559	.69399	.65921	.61998	.57389	.51502	.41068
					1.01981	1.04123	1.09006	1.11833	1.15001	1.18612	1.22833	1.27983	1.34981
.831	.4797	.9284	.73580	1.00026	.80645	.78054	.72349	.69146	.65612	.61609	.56874	.50716	.36579
					1.01995	1.04154	1.09083	1.11941	1.15151	1.18819	1.23124	1.28419	1.35877
.830	.4814	.9280	.73757	0.98781	.80523	.77908	.72139	.68892	.65299	.61214	.56343	.49881	
					1.02009	1.04185	1.09160	1.12051	1.15303	1.19029	1.23422	1.28871	
.829	.4830	.9275	.73933	0.97553	.80401	.77761	.71926	.68635	.64983	.60812	.55796	.48985	
					1.02024	1.04216	1.09237	1.12161	1.15456	1.19243	1.23728	1.29344	
.828	.4846	.9271	.74108	0.96343	.80280	.77613	.71713	.68374	.64663	.60402	.55231	.48013	
					1.02038	1.04247	1.09315	1.12272	1.15612	1.19460	1.24040	1.29837	
.827	.4862	.9266	.74282	0.95149	.80158	.77466	.71499	.68114	.64338	.59984	.54646	.46943	
					1.02052	1.04278	1.09394	1.12385	1.15769	1.19681	1.24361	1.30357	
.826	.4878	.9262	.74455	0.93972	.80035	.77318	.71283	.67851	.64010	.59558	.54039	.45738	
					1.02066	1.04310	1.09473	1.12498	1.15928	1.19906	1.24690	1.30907	
.825	.4894	.9257	.74628	0.92812	.79913	.77170	.71066	.67585	.63677	.59122	.53408	.44336	
					1.02081	1.04342	1.09553	1.12612	1.16089	1.20135	1.25029	1.31495	
.824	.4910	.9253	.74799	0.91668	.79790	.77021	.70847	.67317	.63340	.58677	.52749	.42594	
					1.02095	1.04373	1.09633	1.12728	1.16252	1.20368	1.25378	1.32132	
.823	.4926	.9248	.74970	0.90539	.79668	.76872	.70628	.67046	.62997	.58221	.52058	.40017	
					1.02110	1.04405	1.09714	1.12844	1.16418	1.20605	1.25737	1.32844	
.822	.4942	.9244	.75140	0.89426	.79545	.76723	.70406	.66773	.62650	.57754	.51329		
					1.02124	1.04437	1.09795	1.12962	1.16585	1.20847	1.26109		
.821	.4958	.9239	.75309	0.88328	.79422	.76573	.70184	.66497	.62298	.57275	.50558		
					1.02138	1.04469	1.09877	1.13081	1.16755	1.21094	1.26494		

TABLE V. Specific Heat Ratio = 1.67

R₂ and TPR versus Loss Coefficient K

p/p₁	MACH	T/T₁	GAMMA	K_{crit}	.08	.09	.1	.2	.3	.4	.5	.6	.7
.820	.4974	.9235	.75477	0.87245	.79852	.79577	.79299	.76423	.73330	.69959	.66218	.61940	.56783
					1.01708	1.01929	1.02153	1.04502	1.07086	1.09960	1.13200	1.16927	1.21346
.819	.4990	.9230	.75644	0.86177	.79734	.79456	.79176	.76272	.73146	.69734	.65936	.61576	.56276
					1.01719	1.01942	1.02167	1.04534	1.07141	1.10043	1.13322	1.17102	1.21604
.818	.5006	.9226	.75811	0.85123	.79616	.79335	.79053	.76121	.72961	.69507	.65651	.61206	.55754
					1.01731	1.01955	1.02182	1.04566	1.07196	1.10127	1.13444	1.17279	1.21867
.817	.5021	.9221	.75976	0.84083	.79497	.79214	.78929	.75970	.72776	.69278	.65364	.60829	.55215
					1.01742	1.01968	1.02197	1.04599	1.07251	1.10211	1.13568	1.17459	1.22136
.816	.5037	.9217	.76141	0.83058	.79378	.79093	.78806	.75818	.72590	.69047	.65073	.60446	.54657
					1.01754	1.01981	1.02211	1.04632	1.07306	1.10296	1.13693	1.17641	1.22413
.815	.5053	.9212	.76305	0.82046	.79260	.78972	.78682	.75666	.72402	.68815	.64779	.60056	.54078
					1.01765	1.01994	1.02226	1.04665	1.07362	1.10382	1.13819	1.17827	1.22696
.814	.5069	.9208	.76468	0.81048	.79141	.78850	.78558	.75513	.72214	.68581	.64481	.59658	.53476
					1.01777	1.02008	1.02241	1.04698	1.07418	1.10468	1.13946	1.18015	1.22987
.813	.5084	.9203	.76630	0.80062	.79022	.78729	.78434	.75360	.72026	.68346	.64180	.59251	.52847
					1.01788	1.02021	1.02256	1.04731	1.07475	1.10555	1.14076	1.18207	1.23286
.812	.5100	.9198	.76792	0.79090	.78903	.78607	.78309	.75207	.71836	.68108	.63875	.58836	.52187
					1.01800	1.02034	1.02270	1.04764	1.07532	1.10643	1.14206	1.18401	1.23594
.811	.5116	.9194	.76952	0.78131	.78784	.78485	.78185	.75053	.71645	.67869	.63566	.58412	.51493
					1.01812	1.02047	1.02285	1.04798	1.07589	1.10731	1.14338	1.18600	1.23912
.810	.5132	.9189	.77112	0.77184	.78664	.78363	.78060	.74898	.71453	.67628	.63253	.57977	.50756
					1.01823	1.02060	1.02300	1.04831	1.07646	1.10820	1.14472	1.18802	1.24241
.809	.5147	.9185	.77271	0.76250	.78545	.78241	.77935	.74743	.71261	.67385	.62936	.57532	.49971
					1.01835	1.02074	1.02315	1.04865	1.07704	1.10910	1.14607	1.19007	1.24582
.808	.5163	.9180	.77430	0.75328	.78425	.78119	.77810	.74588	.71067	.67139	.62614	.57075	.49125
					1.01847	1.02087	1.02330	1.04899	1.07762	1.11001	1.14744	1.19217	1.24937
.807	.5179	.9176	.77587	0.74417	.78305	.77997	.77685	.74432	.70872	.66892	.62288	.56605	.48202
					1.01858	1.02100	1.02345	1.04933	1.07820	1.11092	1.14882	1.19431	1.25308
.806	.5194	.9171	.77744	0.73519	.78185	.77874	.77560	.74276	.70676	.66642	.61957	.56122	.47177
					1.01870	1.02114	1.02360	1.04967	1.07879	1.11185	1.15023	1.19649	1.25697
.805	.5210	.9167	.77900	0.72632	.78065	.77751	.77435	.74119	.70479	.66390	.61620	.55624	.46009
					1.01882	1.02127	1.02375	1.05001	1.07938	1.11278	1.15165	1.19872	1.26109
.804	.5225	.9162	.78055	0.71757	.77945	.77628	.77309	.73962	.70281	.66136	.61279	.55110	.44619
					1.01894	1.02141	1.02390	1.05035	1.07998	1.11371	1.15309	1.20100	1.26548
.803	.5241	.9157	.78209	0.70893	.77825	.77505	.77183	.73804	.70082	.65879	.60931	.54577	.428C1
					1.01906	1.02154	1.02406	1.05070	1.08058	1.11466	1.15455	1.20334	1.27026
.802	.5256	.9153	.78363	0.70039	.77705	.77382	.77057	.73646	.69882	.65620	.60578	.54024	.39139
					1.01918	1.02168	1.02421	1.05105	1.08118	1.11562	1.15603	1.20573	1.27570
.801	.5272	.9148	.78516	0.69197	.77584	.77259	.76930	.73487	.69681	.65359	.60219	.53448	
					1.01930	1.02182	1.02437	1.05139	1.08179	1.11658	1.15753	1.20818	
.800	.5287	.9144	.78668	0.68366	.77463	.77135	.76804	.73328	.69478	.65094	.59852	.52846	
					1.01942	1.02195	1.02452	1.05174	1.08240	1.11755	1.15906	1.21070	
.799	.5303	.9139	.78819	0.67545	.77343	.77011	.76677	.73168	.69274	.64827	.59478	.52215	
					1.01954	1.02209	1.02467	1.05210	1.08301	1.11854	1.16061	1.21330	
.798	.5318	.9134	.78970	0.66734	.77222	.76888	.76551	.73008	.69069	.64557	.59097	.51549	
					1.01966	1.02223	1.02483	1.05245	1.08363	1.11953	1.16218	1.21597	
.797	.5334	.9130	.79120	0.65933	.77100	.76763	.76424	.72847	.68863	.64284	.58708	.50842	
					1.01978	1.02237	1.02498	1.05280	1.08425	1.12053	1.16378	1.21874	
.796	.5349	.9125	.79269	0.65143	.76979	.76639	.76296	.72685	.68655	.64008	.58311	.50088	
					1.01990	1.02250	1.02514	1.05316	1.08488	1.12155	1.16540	1.22160	
.795	.5365	.9121	.79417	0.64362	.76858	.76515	.76169	.72523	.68446	.63729	.57904	.49273	
					1.02002	1.02264	1.02529	1.05352	1.08551	1.12257	1.16706	1.22457	
.794	.5380	.9116	.79565	0.63592	.76736	.76390	.76041	.72360	.68235	.63446	.57487	.48379	
					1.02014	1.02278	1.02545	1.05388	1.08615	1.12360	1.16874	1.22768	
.793	.5395	.9111	.79712	0.62830	.76614	.76266	.75914	.72197	.68023	.63160	.57060	.47385	
					1.02026	1.02292	1.02561	1.05424	1.08679	1.12465	1.17045	1.23094	
.792	.5411	.9107	.79858	0.62078	.76493	.76141	.75786	.72033	.67810	.62870	.56621	.46243	
					1.02039	1.02306	1.02576	1.05460	1.08743	1.12571	1.17219	1.23438	
.791	.5426	.9102	.80004	0.61336	.76370	.76016	.75657	.71868	.67595	.62576	.56169	.44865	
					1.02051	1.02320	1.02592	1.05497	1.08808	1.12677	1.17397	1.23805	

TABLE V. Specific Heat Ratio=1.67

p/p_t	MACH	T/T_t	GAMMA	K_crit	R_2 and TPR versus Loss Coefficient K								
					.04	.06	.08	.09	.1	.2	.3	.4	.5
.790	.5441	.9098	.80148	0.60602	.77648	.76954	.76248	.75890	.75529	.71703	.67378	.62279	.55703
					1.01009	1.01531	1.02063	1.02334	1.02608	1.05533	1.08873	1.12785	1.17579
.789	.5457	.9093	.80293	0.59878	.77537	.76838	.76126	.75765	.75400	.71537	.67160	.61977	.55223
					1.01015	1.01540	1.02076	1.02348	1.02624	1.05570	1.08939	1.12895	1.17764
.788	.5472	.9088	.80436	0.59162	.77427	.76721	.76003	.75639	.75271	.71370	.66940	.61671	.54725
					1.01021	1.01549	1.02088	1.02362	1.02640	1.05607	1.09005	1.13005	1.17954
.787	.5487	.9084	.80579	0.58456	.77316	.76605	.75881	.75513	.75142	.71203	.66719	.61360	.54209
					1.01027	1.01558	1.02100	1.02376	1.02656	1.05644	1.09072	1.13117	1.18148
.786	.5502	.9079	.80721	0.57757	.77205	.76488	.75758	.75387	.75013	.71035	.66496	.61044	.53672
					1.01033	1.01567	1.02113	1.02391	1.02672	1.05681	1.09140	1.13231	1.18346
.785	.5518	.9074	.80862	0.57068	.77094	.76371	.75635	.75261	.74883	.70867	.66270	.60723	.53112
					1.01039	1.01576	1.02125	1.02405	1.02688	1.05719	1.09208	1.13346	1.18550
.784	.5533	.9070	.81002	0.56386	.76983	.76254	.75512	.75135	.74754	.70697	.66044	.60397	.52524
					1.01045	1.01585	1.02138	1.02419	1.02704	1.05757	1.09276	1.13462	1.18759
.783	.5548	.9065	.81142	0.55713	.76872	.76137	.75388	.75008	.74624	.70527	.65815	.60065	.51906
					1.01051	1.01594	1.02150	1.02433	1.02720	1.05795	1.09345	1.13580	1.18975
.782	.5563	.9061	.81281	0.55048	.76761	.76020	.75265	.74881	.74493	.70356	.65584	.59728	.51250
					1.01057	1.01603	1.02163	1.02448	1.02736	1.05833	1.09414	1.13700	1.19197
.781	.5579	.9056	.81420	0.54391	.76650	.75903	.75141	.74754	.74363	.70185	.65351	.59384	.50551
					1.01063	1.01613	1.02176	1.02462	1.02752	1.05871	1.09485	1.13821	1.19426
.780	.5594	.9051	.81558	0.53742	.76538	.75785	.75017	.74627	.74232	.70012	.65116	.59033	.49796
					1.01069	1.01622	1.02188	1.02477	1.02769	1.05910	1.09555	1.13944	1.19665
.779	.5609	.9047	.81695	0.53101	.76427	.75668	.74893	.74499	.74101	.69839	.64879	.58675	.48973
					1.01075	1.01631	1.02201	1.02491	1.02785	1.05948	1.09626	1.14069	1.19912
.778	.5624	.9042	.81831	0.52467	.76315	.75550	.74769	.74371	.73970	.69665	.64639	.58310	.48060
					1.01081	1.01640	1.02214	1.02506	1.02801	1.05987	1.09698	1.14196	1.20171
.777	.5639	.9037	.81967	0.51841	.76203	.75432	.74644	.74244	.73838	.69490	.64397	.57936	.47019
					1.01087	1.01650	1.02226	1.02520	1.02818	1.06026	1.09771	1.14325	1.20443
.776	.5654	.9033	.82102	0.51222	.76092	.75314	.74519	.74115	.73707	.69314	.64153	.57553	.45777
					1.01093	1.01659	1.02239	1.02535	1.02834	1.06066	1.09844	1.14456	1.20732
.775	.5669	.9028	.82237	0.50610	.75980	.75196	.74395	.73987	.73574	.69138	.63906	.57161	.44145
					1.01099	1.01668	1.02252	1.02550	1.02851	1.06105	1.09918	1.14589	1.21044
.774	.5684	.9023	.82371	0.50006	.75868	.75078	.74270	.73858	.73442	.68960	.63657	.56759	.40514
					1.01105	1.01678	1.02265	1.02564	1.02868	1.06145	1.09992	1.14724	1.21396
.773	.5700	.9019	.82504	0.49409	.75756	.74959	.74144	.73729	.73310	.68782	.63405	.56346	
					1.01111	1.01687	1.02278	1.02579	1.02884	1.06185	1.10068	1.14862	
.772	.5715	.9014	.82636	0.48819	.75644	.74841	.74019	.73600	.73177	.68603	.63150	.55920	
					1.01117	1.01697	1.02291	1.02594	1.02901	1.06225	1.10144	1.15003	
.771	.5730	.9009	.82768	0.48236	.75532	.74722	.73893	.73471	.73044	.68422	.62892	.55481	
					1.01123	1.01706	1.02304	1.02609	1.02918	1.06266	1.10220	1.15146	
.770	.5745	.9005	.82899	0.47660	.75420	.74603	.73767	.73341	.72910	.68241	.62630	.55028	
					1.01129	1.01715	1.02317	1.02624	1.02935	1.06306	1.10298	1.15292	
.769	.5760	.9000	.83030	0.47090	.75308	.74484	.73641	.73212	.72776	.68059	.62366	.54558	
					1.01136	1.01725	1.02330	1.02639	1.02952	1.06347	1.10376	1.15441	
.768	.5775	.8995	.83160	0.46527	.75195	.74365	.73515	.73081	.72642	.67875	.62098	.54070	
					1.01142	1.01734	1.02343	1.02654	1.02969	1.06388	1.10455	1.15594	
.767	.5790	.8990	.83289	0.45971	.75083	.74246	.73388	.72951	.72508	.67691	.61827	.53562	
					1.01148	1.01744	1.02356	1.02669	1.02986	1.06430	1.10535	1.15750	
.766	.5805	.8986	.83418	0.45421	.74970	.74127	.73262	.72821	.72373	.67506	.61552	.53031	
					1.01154	1.01754	1.02370	1.02684	1.03003	1.06471	1.10616	1.15910	
.765	.5820	.8981	.83546	0.44877	.74858	.74007	.73135	.72690	.72239	.67319	.61274	.52473	
					1.01160	1.01763	1.02383	1.02699	1.03020	1.06513	1.10698	1.16075	
.764	.5835	.8976	.83673	0.44340	.74745	.73887	.73007	.72558	.72103	.67131	.60991	.51884	
					1.01166	1.01773	1.02396	1.02714	1.03037	1.06555	1.10780	1.16244	
.763	.5850	.8972	.83800	0.43809	.74632	.73768	.72880	.72427	.71968	.66943	.60704	.51257	
					1.01173	1.01782	1.02409	1.02730	1.03054	1.06598	1.10864	1.16418	
.762	.5865	.8967	.83926	0.43284	.74519	.73648	.72752	.72295	.71832	.66753	.60413	.50586	
					1.01179	1.01792	1.02423	1.02745	1.03072	1.06641	1.10948	1.16598	
.761	.5880	.8962	.84052	0.42765	.74406	.73527	.72624	.72163	.71696	.66561	.60117	.49859	
					1.01185	1.01802	1.02436	1.02760	1.03089	1.06683	1.11034	1.16785	

TABLE V. Specific Heat Ratio = 1.67

p/p_t	MACH	T/T_t	GAMMA	K_{crit}	.01	.02	.04	.06	.08	.09	.1	.2	.3
								R_2 and TPR versus Loss Coefficient K					
.760	.5894	.8957	.84176	0.42252	.75581	.75157	.74293	.73407	.72496	.72031	.71559	.66369	.59816
					1.00292	1.00588	1.01191	1.01812	1.02450	1.02776	1.03107	1.06727	1.11120
.759	.5909	.8953	.84301	0.41744	.75478	.75051	.74180	.73286	.72368	.71899	.71422	.66175	.59510
					1.00294	1.00591	1.01198	1.01821	1.02463	1.02791	1.03124	1.06770	1.11208
.758	.5924	.8948	.84424	0.41243	.75375	.74944	.74067	.73166	.72239	.71766	.71285	.65980	.59198
					1.00295	1.00594	1.01204	1.01831	1.02477	1.02807	1.03142	1.06814	1.11297
.757	.5939	.8943	.84547	0.40747	.75272	.74838	.73953	.73045	.72110	.71633	.71147	.65783	.58880
					1.00297	1.00597	1.01210	1.01841	1.02490	1.02822	1.03159	1.06858	1.11387
.756	.5954	.8938	.84670	0.40257	.75168	.74731	.73840	.72924	.71981	.71499	.71009	.65585	.58557
					1.00298	1.00600	1.01217	1.01851	1.02504	1.02838	1.03177	1.06903	1.11478
.755	.5969	.8934	.84792	0.39773	.75065	.74625	.73726	.72803	.71852	.71365	.70871	.65386	.58226
					1.00300	1.00603	1.01223	1.01861	1.02518	1.02854	1.03195	1.06947	1.11571
.754	.5984	.8929	.84913	0.39294	.74962	.74518	.73613	.72681	.71722	.71231	.70732	.65185	.57889
					1.00301	1.00607	1.01229	1.01871	1.02531	1.02869	1.03213	1.06992	1.11665
.753	.5999	.8924	.85033	0.38821	.74859	.74411	.73499	.72560	.71592	.71097	.70593	.64982	.57543
					1.00303	1.00610	1.01236	1.01881	1.02545	1.02885	1.03231	1.07038	1.11760
.752	.6013	.8919	.85153	0.38352	.74755	.74305	.73385	.72438	.71462	.70962	.70454	.64778	.57190
					1.00304	1.00613	1.01242	1.01890	1.02559	1.02901	1.03249	1.07083	1.11857
.751	.6028	.8915	.85273	0.37889	.74652	.74198	.73271	.72316	.71332	.70827	.70314	.64572	.56828
					1.00306	1.00616	1.01248	1.01900	1.02573	1.02917	1.03267	1.07130	1.11955
.750	.6043	.8910	.85392	0.37432	.74549	.74091	.73157	.72194	.71201	.70692	.70174	.64364	.56456
					1.00307	1.00619	1.01255	1.01910	1.02587	1.02933	1.03285	1.07176	1.12055
.749	.6058	.8905	.85510	0.36979	.74445	.73984	.73042	.72072	.71070	.70556	.70033	.64155	.56074
					1.00309	1.00622	1.01261	1.01921	1.02601	1.02949	1.03303	1.07223	1.12157
.748	.6073	.8900	.85627	0.36532	.74342	.73877	.72928	.71949	.70938	.70420	.69892	.63944	.55680
					1.00310	1.00625	1.01268	1.01931	1.02615	1.02965	1.03321	1.07270	1.12260
.747	.6088	.8896	.85744	0.36089	.74238	.73770	.72813	.71827	.70807	.70283	.69750	.63731	.55273
					1.00312	1.00628	1.01274	1.01941	1.02629	1.02981	1.03340	1.07317	1.12366
.746	.6102	.8891	.85861	0.35652	.74135	.73663	.72699	.71704	.70675	.70147	.69608	.63516	.54853
					1.00313	1.00631	1.01281	1.01951	1.02643	1.02997	1.03358	1.07365	1.12473
.745	.6117	.8886	.85977	0.35219	.74031	.73556	.72584	.71581	.70543	.70009	.69466	.63299	.54418
					1.00315	1.00635	1.01287	1.01961	1.02657	1.03014	1.03377	1.07413	1.12583
.744	.6132	.8881	.86092	0.34791	.73928	.73449	.72469	.71457	.70410	.69872	.69323	.63080	.53965
					1.00316	1.00638	1.01294	1.01971	1.02671	1.03030	1.03395	1.07462	1.12695
.743	.6147	.8876	.86207	0.34368	.73824	.73342	.72354	.71334	.70277	.69734	.69180	.62858	.53493
					1.00318	1.00641	1.01300	1.01981	1.02685	1.03046	1.03414	1.07511	1.12809
.742	.6162	.8872	.86321	0.33950	.73721	.73235	.72239	.71210	.70144	.69595	.69036	.62635	.52997
					1.00319	1.00644	1.01307	1.01992	1.02700	1.03063	1.03432	1.07560	1.12927
.741	.6176	.8867	.86434	0.33536	.73617	.73127	.72124	.71086	.70010	.69457	.68892	.62409	.52476
					1.00321	1.00647	1.01313	1.02002	1.02714	1.03079	1.03451	1.07610	1.13048
.740	.6191	.8862	.86547	0.33127	.73514	.73020	.72008	.70962	.69877	.69318	.68747	.62181	.51924
					1.00323	1.00650	1.01320	1.02012	1.02728	1.03096	1.03470	1.07661	1.13171
.739	.6206	.8857	.86660	0.32723	.73410	.72912	.71893	.70837	.69742	.69178	.68602	.61950	.51336
					1.00324	1.00653	1.01327	1.02023	1.02743	1.03113	1.03489	1.07711	1.13299
.738	.6221	.8852	.86772	0.32323	.73306	.72805	.71777	.70713	.69608	.69038	.68456	.61716	.50703
					1.00326	1.00657	1.01333	1.02033	1.02757	1.03129	1.03508	1.07763	1.13430
.737	.6235	.8848	.86883	0.31927	.73203	.72697	.71661	.70588	.69473	.68898	.68310	.61480	.50013
					1.00327	1.00660	1.01340	1.02044	1.02772	1.03146	1.03527	1.07814	1.13566
.736	.6250	.8843	.86994	0.31536	.73099	.72590	.71546	.70463	.69338	.68757	.68163	.61241	.49249
					1.00329	1.00663	1.01347	1.02054	1.02786	1.03163	1.03546	1.07866	1.13708
.735	.6265	.8838	.87104	0.31149	.72995	.72482	.71429	.70338	.69202	.68616	.68016	.60998	.48379
					1.00330	1.00666	1.01353	1.02064	1.02801	1.03180	1.03566	1.07919	1.13855
.734	.6279	.8833	.87213	0.30766	.72891	.72374	.71313	.70212	.69066	.68474	.67868	.60753	.47340
					1.00332	1.00669	1.01360	1.02075	1.02816	1.03197	1.03585	1.07972	1.14012
.733	.6294	.8828	.87322	0.30387	.72788	.72267	.71197	.70086	.68929	.68332	.67720	.60504	.45987
					1.00334	1.00673	1.01367	1.02086	1.02831	1.03214	1.03605	1.08026	1.14178
.732	.6309	.8824	.87431	0.30013	.72684	.72159	.71080	.69960	.68793	.68189	.67571	.60252	.43217
					1.00335	1.00676	1.01373	1.02096	1.02845	1.03231	1.03624	1.08080	1.14363
.731	.6324	.8819	.87539	0.29643	.72580	.72051	.70964	.69834	.68656	.68046	.67421	.59996	
					1.00337	1.00679	1.01380	1.02107	1.02860	1.03248	1.03644	1.08135	

TABLE V. Specific Heat Ratio = 1.67

p/p_t	MACH	T/T_t	GAMMA	K_{crit}	.01	.02	.04	.06	.07	.08	.09	.1	.2
.730	.6338	.8814	.87646	0.29277	.72476	.71943	.70847	.69707	.69119	.68518	.67902	.67271	.59736
					1.00338	1.00682	1.01387	1.02117	1.02493	1.02875	1.03266	1.03663	1.08191
.729	.6353	.8809	.87753	0.28914	.72372	.71835	.70730	.69580	.68987	.68380	.67758	.67121	.59472
					1.00340	1.00686	1.01394	1.02128	1.02506	1.02891	1.03283	1.03683	1.08247
.728	.6368	.8804	.87859	0.28556	.72268	.71727	.70613	.69453	.68854	.68241	.67613	.66969	.59204
					1.00341	1.00689	1.01401	1.02139	1.02519	1.02906	1.03300	1.03703	1.08304
.727	.6382	.8799	.87965	0.28202	.72164	.71619	.70496	.69326	.68721	.68102	.67468	.66817	.58931
					1.00343	1.00692	1.01407	1.02150	1.02532	1.02921	1.03318	1.03723	1.08361
.726	.6397	.8794	.88070	0.27852	.72060	.71510	.70378	.69198	.68588	.67963	.67322	.66665	.58654
					1.00345	1.00695	1.01414	1.02160	1.02544	1.02936	1.03335	1.03743	1.08419
.725	.6412	.8790	.88174	0.27505	.71956	.71402	.70261	.69070	.68454	.67823	.67176	.66511	.58371
					1.00346	1.00699	1.01421	1.02171	1.02557	1.02951	1.03353	1.03764	1.08478
.724	.6426	.8785	.88278	0.27162	.71852	.71294	.70143	.68942	.68320	.67683	.67029	.66357	.58083
					1.00348	1.00702	1.01428	1.02182	1.02571	1.02967	1.03371	1.03784	1.08538
.723	.6441	.8780	.88382	0.26823	.71748	.71185	.70025	.68814	.68186	.67542	.66882	.66202	.57789
					1.00349	1.00705	1.01435	1.02193	1.02584	1.02982	1.03389	1.03804	1.08598
.722	.6455	.8775	.88485	0.26488	.71644	.71077	.69907	.68685	.68051	.67401	.66734	.66047	.57489
					1.00351	1.00709	1.01442	1.02204	1.02597	1.02997	1.03407	1.03825	1.08659
.721	.6470	.8770	.88587	0.26156	.71540	.70968	.69789	.68556	.67916	.67260	.66585	.65891	.57182
					1.00353	1.00712	1.01449	1.02215	1.02610	1.03013	1.03425	1.03845	1.08721
.720	.6485	.8765	.88689	0.25828	.71435	.70859	.69671	.68426	.67780	.67117	.66436	.65734	.56868
					1.00354	1.00715	1.01456	1.02226	1.02623	1.03028	1.03443	1.03866	1.08784
.719	.6499	.8760	.88791	0.25503	.71331	.70751	.69552	.68296	.67644	.66975	.66286	.65576	.56546
					1.00356	1.00719	1.01463	1.02237	1.02637	1.03044	1.03461	1.03887	1.08848
.718	.6514	.8755	.88892	0.25182	.71227	.70642	.69433	.68166	.67508	.66832	.66136	.65418	.56216
					1.00358	1.00722	1.01470	1.02248	1.02650	1.03060	1.03479	1.03908	1.08912
.717	.6529	.8751	.88992	0.24864	.71122	.70533	.69314	.68036	.67371	.66688	.65984	.65258	.55877
					1.00359	1.00725	1.01477	1.02259	1.02663	1.03076	1.03497	1.03929	1.08978
.716	.6543	.8746	.89092	0.24550	.71018	.70424	.69195	.67905	.67234	.66544	.65833	.65098	.55528
					1.00361	1.00729	1.01484	1.02271	1.02677	1.03091	1.03516	1.03950	1.09045
.715	.6558	.8741	.89191	0.24239	.70914	.70315	.69076	.67774	.67096	.66399	.65680	.64937	.55167
					1.00362	1.00732	1.01491	1.02282	1.02690	1.03107	1.03534	1.03971	1.09113
.714	.6572	.8736	.89290	0.23932	.70809	.70206	.68956	.67643	.66958	.66254	.65527	.64775	.54793
					1.00364	1.00735	1.01498	1.02293	1.02704	1.03123	1.03553	1.03993	1.09183
.713	.6587	.8731	.89388	0.23628	.70705	.70096	.68836	.67511	.66820	.66108	.65373	.64613	.54407
					1.00366	1.00739	1.01505	1.02304	1.02717	1.03139	1.03572	1.04014	1.09254
.712	.6602	.8726	.89486	0.23327	.70600	.69987	.68717	.67379	.66680	.65961	.65218	.64449	.54005
					1.00367	1.00742	1.01512	1.02316	1.02731	1.03156	1.03590	1.04036	1.09326
.711	.6616	.8721	.89583	0.23030	.70496	.69878	.68596	.67246	.66541	.65814	.65063	.64284	.53586
					1.00369	1.00746	1.01520	1.02327	1.02745	1.03172	1.03609	1.04057	1.09400
.710	.6631	.8716	.89679	0.22735	.70391	.69768	.68476	.67113	.66401	.65666	.64906	.64118	.53146
					1.00371	1.00749	1.01527	1.02339	1.02759	1.03188	1.03628	1.04079	1.09475
.709	.6645	.8711	.89775	0.22444	.70286	.69658	.68356	.66979	.66260	.65518	.64749	.63952	.52684
					1.00372	1.00752	1.01534	1.02351	1.02773	1.03205	1.03647	1.04101	1.09552
.708	.6660	.8706	.89871	0.22156	.70182	.69549	.68235	.66846	.66119	.65369	.64591	.63784	.52194
					1.00374	1.00756	1.01541	1.02362	1.02786	1.03221	1.03666	1.04123	1.09631
.707	.6675	.8701	.89966	0.21871	.70077	.69439	.68114	.66712	.65978	.65219	.64433	.63615	.51671
					1.00376	1.00759	1.01548	1.02374	1.02800	1.03238	1.03686	1.04146	1.09712
.706	.6689	.8696	.90061	0.21589	.69972	.69329	.67993	.66577	.65836	.65069	.64273	.63445	.51108
					1.00377	1.00763	1.01556	1.02385	1.02814	1.03254	1.03705	1.04168	1.09796
.705	.6704	.8691	.90155	0.21310	.69867	.69219	.67871	.66442	.65693	.64918	.64113	.63274	.50494
					1.00379	1.00766	1.01563	1.02397	1.02829	1.03271	1.03725	1.04191	1.09883
.704	.6718	.8687	.90248	0.21035	.69763	.69109	.67750	.66307	.65550	.64766	.63951	.63102	.49808
					1.00381	1.00770	1.01570	1.02409	1.02843	1.03288	1.03744	1.04213	1.09973
.703	.6733	.8682	.90341	0.20762	.69658	.68999	.67627	.66171	.65406	.64613	.63789	.62928	.49024
					1.00382	1.00773	1.01578	1.02421	1.02857	1.03304	1.03764	1.04236	1.10067
.702	.6748	.8677	.90434	0.20492	.69553	.68889	.67505	.66035	.65262	.64460	.63625	.62754	.48082
					1.00384	1.00776	1.01585	1.02432	1.02871	1.03321	1.03784	1.04259	1.10165
.701	.6762	.8672	.90526	0.20225	.69448	.68778	.67382	.65898	.65117	.64306	.63461	.62578	.46812
					1.00386	1.00780	1.01593	1.02444	1.02886	1.03338	1.03804	1.04282	1.10271

TABLE V. Specific Heat Ratio = 1.67

R_t and TPR versus Loss Coefficient K

p/p_t	MACH	T/T_t	GAMMA	K_{crit}	.01	.02	.04	.05	.06	.07	.08	.09	.1
.700	.6777	.8667	.90618	0.19961	.69343	.68668	.67260	.66523	.65761	.64971	.64151	.63296	.62400
					1.00388	1.00783	1.01600	1.02023	1.02456	1.02900	1.03356	1.03824	1.04305
.699	.6791	.8662	.90709	0.19700	.69238	.68557	.67137	.66393	.65623	.64825	.63995	.63129	.62221
					1.00389	1.00787	1.01608	1.02033	1.02468	1.02915	1.03373	1.03844	1.04329
.698	.6806	.8657	.90799	0.19441	.69132	.68447	.67014	.66263	.65485	.64678	.63839	.62961	.62041
					1.00391	1.00790	1.01615	1.02042	1.02480	1.02929	1.03390	1.03864	1.04352
.697	.6820	.8652	.90889	0.19185	.69027	.68336	.66890	.66132	.65347	.64531	.63681	.62793	.61859
					1.00393	1.00794	1.01623	1.02052	1.02492	1.02944	1.03408	1.03885	1.04376
.696	.6835	.8647	.90979	0.18932	.68922	.68225	.66767	.66001	.65207	.64383	.63523	.62623	.61676
					1.00394	1.00798	1.01630	1.02062	1.02504	1.02958	1.03425	1.03905	1.04400
.695	.6849	.8642	.91068	0.18682	.68817	.68114	.66643	.65870	.65068	.64234	.63364	.62452	.61491
					1.00396	1.00801	1.01638	1.02072	1.02517	1.02973	1.03443	1.03926	1.04424
.694	.6864	.8637	.91156	0.18435	.68711	.68003	.66518	.65738	.64928	.64084	.63203	.62279	.61305
					1.00398	1.00805	1.01645	1.02081	1.02529	1.02988	1.03460	1.03947	1.04448
.693	.6879	.8632	.91244	0.18190	.68606	.67892	.66394	.65606	.64787	.63934	.63042	.62105	.61116
					1.00399	1.00808	1.01653	1.02091	1.02541	1.03003	1.03478	1.03968	1.04473
.692	.6893	.8627	.91332	0.17948	.68501	.67780	.66269	.65473	.64646	.63783	.62880	.61930	.60926
					1.00401	1.00812	1.01661	1.02101	1.02553	1.03018	1.03496	1.03989	1.04497
.691	.6908	.8622	.91419	0.17708	.68395	.67669	.66144	.65340	.64504	.63631	.62717	.61754	.60734
					1.00403	1.00815	1.01668	1.02111	1.02566	1.03033	1.03514	1.04010	1.04522
.690	.6922	.8617	.91505	0.17471	.68290	.67557	.66018	.65206	.64361	.63478	.62552	.61576	.60540
					1.00405	1.00819	1.01676	1.02121	1.02578	1.03048	1.03532	1.04031	1.04547
.689	.6937	.8612	.91591	0.17236	.68184	.67446	.65893	.65072	.64218	.63325	.62387	.61397	.60344
					1.00406	1.00822	1.01684	1.02131	1.02591	1.03064	1.03550	1.04053	1.04572
.688	.6951	.8607	.91677	0.17004	.68078	.67334	.65767	.64938	.64074	.63170	.62220	.61216	.60146
					1.00408	1.00826	1.01691	1.02141	1.02603	1.03079	1.03569	1.04075	1.04598
.687	.6966	.8602	.91762	0.16775	.67973	.67222	.65640	.64803	.63930	.63015	.62052	.61033	.59946
					1.00410	1.00830	1.01699	1.02151	1.02616	1.03094	1.03587	1.04096	1.04623
.686	.6980	.8597	.91847	0.16548	.67867	.67110	.65513	.64668	.63785	.62859	.61883	.60849	.59743
					1.00412	1.00833	1.01707	1.02161	1.02629	1.03110	1.03606	1.04118	1.04649
.685	.6995	.8592	.91931	0.16323	.67761	.66998	.65386	.64532	.63639	.62702	.61713	.60663	.59538
					1.00413	1.00837	1.01715	1.02172	1.02641	1.03125	1.03625	1.04141	1.04675
.684	.7009	.8587	.92015	0.16101	.67655	.66885	.65259	.64396	.63493	.62544	.61541	.60475	.59331
					1.00415	1.00841	1.01723	1.02182	1.02654	1.03141	1.03643	1.04163	1.04702
.683	.7024	.8582	.92098	0.15882	.67549	.66773	.65131	.64259	.63346	.62385	.61368	.60285	.59120
					1.00417	1.00844	1.01731	1.02192	1.02667	1.03157	1.03662	1.04185	1.04728
.682	.7039	.8577	.92180	0.15664	.67443	.66660	.65003	.64122	.63198	.62225	.61194	.60093	.58907
					1.00419	1.00848	1.01739	1.02203	1.02680	1.03173	1.03681	1.04208	1.04755
.681	.7053	.8572	.92263	0.15449	.67337	.66547	.64875	.63984	.63049	.62064	.61018	.59900	.58689
					1.00420	1.00852	1.01747	1.02213	1.02693	1.03189	1.03701	1.04231	1.04783
.680	.7068	.8566	.92344	0.15237	.67231	.66434	.64746	.63845	.62900	.61901	.60841	.59704	.58470
					1.00422	1.00855	1.01755	1.02223	1.02706	1.03205	1.03720	1.04254	1.04810
.679	.7082	.8561	.92425	0.15026	.67125	.66321	.64616	.63706	.62749	.61738	.60662	.59505	.58247
					1.00424	1.00859	1.01763	1.02234	1.02719	1.03221	1.03739	1.04277	1.04838
.678	.7097	.8556	.92506	0.14818	.67018	.66208	.64487	.63567	.62598	.61574	.60481	.59305	.58021
					1.00426	1.00863	1.01771	1.02244	1.02733	1.03237	1.03759	1.04301	1.04866
.677	.7111	.8551	.92587	0.14612	.66912	.66095	.64357	.63427	.62447	.61408	.60299	.59101	.57791
					1.00428	1.00866	1.01779	1.02255	1.02746	1.03253	1.03779	1.04325	1.04894
.676	.7126	.8546	.92666	0.14409	.66806	.65981	.64226	.63286	.62294	.61241	.60114	.58895	.57557
					1.00429	1.00870	1.01787	1.02265	1.02759	1.03270	1.03799	1.04349	1.04922
.675	.7140	.8541	.92746	0.14207	.66699	.65868	.64095	.63144	.62140	.61073	.59928	.58686	.57319
					1.00431	1.00874	1.01795	1.02276	1.02773	1.03286	1.03819	1.04373	1.04951
.674	.7155	.8536	.92825	0.14008	.66593	.65754	.63964	.63002	.61986	.60903	.59740	.58475	.57076
					1.00433	1.00878	1.01803	1.02287	1.02786	1.03303	1.03839	1.04397	1.04980
.673	.7169	.8531	.92903	0.13811	.66486	.65640	.63832	.62860	.61830	.60732	.59550	.58260	.56828
					1.00435	1.00881	1.01811	1.02298	1.02800	1.03320	1.03859	1.04422	1.05010
.672	.7184	.8526	.92981	0.13616	.66379	.65526	.63700	.62716	.61673	.60560	.59358	.58042	.56575
					1.00437	1.00885	1.01820	1.02308	1.02814	1.03337	1.03880	1.04447	1.05040
.671	.7199	.8521	.93058	0.13423	.66273	.65411	.63567	.62572	.61516	.60386	.59162	.57821	.56317
					1.00438	1.00889	1.01828	1.02319	1.02827	1.03354	1.03901	1.04472	1.05070

TABLE V. Specific Heat Ratio = 1.67

p/pₜ	MACH	T/Tₜ	GAMMA	Kcrit	.01	.02	.04	.05	.06	.07	.08	.09	.1

R₂ and TPR versus Loss Coefficient K

p/p	MACH	T/T	GAMMA	K_crit	.01	.02	.04	.05	.06	.07	.08	.09	.1
.670	.7213	.8516	.93135	0.13232	.66166	.65297	.63434	.62427	.61357	.60210	.58965	.57596	.56052
					1.00440	1.00893	1.01836	1.02330	1.02841	1.03371	1.03922	1.04497	1.05101
.669	.7228	.8511	.93212	0.13044	.66059	.65182	.63300	.62282	.61197	.60033	.58766	.57367	.55781
					1.00442	1.00897	1.01845	1.02341	1.02855	1.03388	1.03943	1.04523	1.05132
.668	.7242	.8506	.93288	0.12857	.65952	.65067	.63166	.62135	.61036	.59854	.58564	.57134	.55502
					1.00444	1.00900	1.01853	1.02352	1.02869	1.03406	1.03965	1.04549	1.05163
.667	.7257	.8500	.93363	0.12673	.65845	.64952	.63031	.61988	.60874	.59674	.58359	.56897	.55216
					1.00446	1.00904	1.01861	1.02363	1.02883	1.03423	1.03986	1.04575	1.05195
.666	.7271	.8495	.93439	0.12490	.65738	.64837	.62896	.61840	.60711	.59490	.58152	.56654	.54921
					1.00448	1.00908	1.01870	1.02374	1.02897	1.03441	1.04008	1.04602	1.05228
.665	.7286	.8490	.93513	0.12310	.65630	.64722	.62760	.61692	.60547	.59305	.57941	.56407	.54617
					1.00449	1.00912	1.01878	1.02386	1.02912	1.03459	1.04030	1.04629	1.05261
.664	.7300	.8485	.93587	0.12131	.65523	.64606	.62624	.61542	.60381	.59119	.57727	.56154	.54301
					1.00451	1.00916	1.01887	1.02397	1.02926	1.03477	1.04052	1.04656	1.05295
.663	.7315	.8480	.93661	0.11954	.65416	.64490	.62487	.61391	.60213	.58930	.57510	.55895	.53974
					1.00453	1.00920	1.01895	1.02408	1.02941	1.03495	1.04074	1.04683	1.05329
.662	.7330	.8475	.93734	0.11780	.65308	.64374	.62349	.61240	.60045	.58739	.57289	.55630	.53633
					1.00455	1.00924	1.01904	1.02420	1.02955	1.03513	1.04097	1.04711	1.05364
.661	.7344	.8470	.93807	0.11607	.65201	.64258	.62211	.61087	.59873	.58546	.57064	.55358	.53277
					1.00457	1.00927	1.01913	1.02431	1.02970	1.03532	1.04120	1.04740	1.05400
.660	.7359	.8465	.93879	0.11436	.65093	.64141	.62072	.60934	.59701	.58350	.56835	.55077	.52902
					1.00459	1.00931	1.01921	1.02443	1.02985	1.03550	1.04143	1.04768	1.05436
.659	.7373	.8459	.93951	0.11267	.64985	.64025	.61933	.60780	.59528	.58151	.56601	.54789	.52505
					1.00461	1.00935	1.01930	1.02454	1.03000	1.03569	1.04166	1.04798	1.05473
.658	.7388	.8454	.94023	0.11100	.64877	.63908	.61793	.60624	.59353	.57950	.56363	.54490	.52083
					1.00462	1.00939	1.01939	1.02466	1.03015	1.03588	1.04190	1.04827	1.05511
.657	.7402	.8449	.94094	0.10935	.64769	.63791	.61652	.60467	.59176	.57746	.56119	.54181	.51629
					1.00464	1.00943	1.01948	1.02478	1.03030	1.03607	1.04214	1.04858	1.05551
.656	.7417	.8444	.94164	0.10772	.64661	.63673	.61510	.60310	.58997	.57539	.55870	.53860	.51136
					1.00466	1.00947	1.01957	1.02490	1.03046	1.03627	1.04238	1.04888	1.05591
.655	.7432	.8439	.94234	0.10610	.64553	.63556	.61368	.60149	.58816	.57328	.55615	.53526	.50583
					1.00468	1.00951	1.01966	1.02502	1.03061	1.03646	1.04263	1.04920	1.05633
.654	.7446	.8434	.94304	0.10450	.64445	.63438	.61225	.59989	.58633	.57114	.55353	.53176	.49955
					1.00470	1.00955	1.01975	1.02514	1.03076	1.03666	1.04288	1.04952	1.05677
.653	.7461	.8428	.94373	0.10292	.64337	.63320	.61081	.59828	.58448	.56897	.55084	.52807	.49204
					1.00472	1.00959	1.01984	1.02526	1.03092	1.03686	1.04313	1.04985	1.05722
.652	.7475	.8423	.94442	0.10136	.64228	.63201	.60937	.59665	.58261	.56675	.54807	.52417	.48197
					1.00474	1.00963	1.01993	1.02538	1.03108	1.03706	1.04339	1.05018	1.05771
.651	.7490	.8418	.94510	0.09981	.64120	.63083	.60791	.59500	.58072	.56449	.54521	.52001	
					1.00476	1.00967	1.02002	1.02550	1.03124	1.03726	1.04365	1.05052	
.650	.7504	.8413	.94578	0.09828	.64011	.62964	.60645	.59335	.57880	.56218	.54225	.51552	
					1.00478	1.00971	1.02011	1.02563	1.03140	1.03747	1.04392	1.05087	
.649	.7519	.8408	.94645	0.09677	.63902	.62844	.60498	.59167	.57685	.55983	.53918	.51058	
					1.00479	1.00975	1.02020	1.02575	1.03156	1.03768	1.04419	1.05124	
.648	.7534	.8402	.94712	0.09528	.63793	.62725	.60349	.58998	.57488	.55742	.53599	.50509	
					1.00481	1.00979	1.02030	1.02588	1.03172	1.03789	1.04446	1.05162	
.647	.7548	.8397	.94779	0.09380	.63684	.62605	.60200	.58828	.57287	.55495	.53265	.49877	
					1.00483	1.00983	1.02039	1.02600	1.03189	1.03810	1.04474	1.05201	
.646	.7563	.8392	.94845	0.09234	.63575	.62485	.60050	.58655	.57084	.55243	.52915	.49107	
					1.00485	1.00988	1.02048	1.02613	1.03206	1.03832	1.04503	1.05242	
.645	.7577	.8387	.94910	0.09090	.63466	.62365	.59899	.58481	.56877	.54983	.52544	.48017	
					1.00487	1.00992	1.02058	1.02626	1.03222	1.03854	1.04532	1.05287	
.644	.7592	.8382	.94975	0.08947	.63356	.62244	.59747	.58305	.56666	.54715	.52151		
					1.00489	1.00996	1.02067	1.02639	1.03239	1.03877	1.04562		
.643	.7607	.8376	.95040	0.08806	.63247	.62123	.59593	.58127	.56452	.54439	.51728		
					1.00491	1.01000	1.02077	1.02652	1.03257	1.03899	1.04593		
.642	.7621	.8371	.95104	0.08666	.63137	.62002	.59439	.57947	.56234	.54154	.51264		
					1.00493	1.01004	1.02086	1.02665	1.03274	1.03922	1.04625		
.641	.7636	.8366	.95168	0.08528	.63027	.61880	.59283	.57765	.56011	.53858	.50754		
					1.00495	1.01008	1.02096	1.02678	1.03292	1.03946	1.04658		

TABLE V. Specific Heat Ratio = 1.67

p/p_t	MACH	T/T_t	GAMMA	K_{crit}	.01	.02	.03	.04	.05	.055	.06	.065	.07
.640	.7651	.8361	.95231	0.08392	.62917	.61758	.60503	.59126	.57580	.56722	.55784	.54743	.53549
					1.00497	1.01013	1.01547	1.02106	1.02691	1.02996	1.03309	1.03633	1.03970
.639	.7665	.8355	.95294	0.08257	.62807	.61636	.60365	.58967	.57393	.56515	.55552	.54473	.53227
					1.00499	1.01017	1.01554	1.02116	1.02705	1.03011	1.03327	1.03655	1.03994
.638	.7680	.8350	.95357	0.08124	.62697	.61513	.60226	.58807	.57203	.56304	.55314	.54196	.52889
					1.00501	1.01021	1.01561	1.02126	1.02718	1.03027	1.03345	1.03676	1.04019
.637	.7694	.8345	.95419	0.07992	.62587	.61390	.60087	.58646	.57010	.56089	.55070	.53909	.52532
					1.00503	1.01025	1.01568	1.02136	1.02732	1.03043	1.03364	1.03697	1.04045
.636	.7709	.8340	.95480	0.07862	.62476	.61266	.59947	.58484	.56815	.55871	.54819	.53610	.52152
					1.00505	1.01029	1.01575	1.02146	1.02746	1.03059	1.03383	1.03719	1.04071
.635	.7724	.8334	.95542	0.07733	.62365	.61142	.59806	.58319	.56616	.55647	.54559	.53299	.51744
					1.00507	1.01034	1.01582	1.02156	1.02760	1.03075	1.03402	1.03742	1.04098
.634	.7738	.8329	.95602	0.07606	.62255	.61017	.59664	.58153	.56414	.55418	.54292	.52973	.51300
					1.00509	1.01039	1.01589	1.02166	1.02774	1.03092	1.03422	1.03765	1.04125
.633	.7753	.8324	.95663	0.07480	.62144	.60892	.59521	.57986	.56208	.55184	.54016	.52629	.50809
					1.00511	1.01043	1.01596	1.02176	1.02788	1.03108	1.03441	1.03788	1.04154
.632	.7768	.8319	.95722	0.07355	.62032	.60767	.59377	.57816	.55999	.54944	.53730	.52264	.50250
					1.00513	1.01047	1.01603	1.02186	1.02803	1.03125	1.03462	1.03812	1.04183
.631	.7782	.8313	.95782	0.07232	.61921	.60641	.59233	.57645	.55785	.54695	.53432	.51873	.49579
					1.00515	1.01052	1.01610	1.02197	1.02817	1.03143	1.03482	1.03837	1.04214
.630	.7797	.8308	.95841	0.07111	.61810	.60515	.59087	.57471	.55567	.54441	.53121	.51450	.48695
					1.00517	1.01056	1.01617	1.02207	1.02832	1.03161	1.03503	1.03862	1.04247
.629	.7812	.8303	.95899	0.06991	.61698	.60388	.58940	.57296	.55344	.54178	.52794	.50984	
					1.00519	1.01060	1.01624	1.02218	1.02847	1.03179	1.03524	1.03888	
.628	.7826	.8297	.95957	0.06872	.61586	.60261	.58793	.57118	.55115	.53907	.52448	.50460	
					1.00521	1.01065	1.01632	1.02229	1.02862	1.03197	1.03546	1.03915	
.627	.7841	.8292	.96015	0.06755	.61474	.60133	.58644	.56938	.54881	.53625	.52080	.49841	
					1.00523	1.01068	1.01639	1.02239	1.02878	1.03215	1.03568	1.03944	
.626	.7856	.8287	.96072	0.06639	.61360	.60005	.58494	.56755	.54640	.53331	.51684	.49059	
					1.00526	1.01073	1.01646	1.02250	1.02893	1.03234	1.03591	1.03973	
.625	.7870	.8281	.96129	0.06525	.61248	.59876	.58342	.56570	.54393	.53024	.51252	.47760	
					1.00528	1.01077	1.01654	1.02261	1.02909	1.03253	1.03614	1.04006	
.624	.7885	.8276	.96185	0.06412	.61135	.59747	.58190	.56382	.54138	.52701	.50773		
					1.00530	1.01082	1.01661	1.02272	1.02925	1.03272	1.03639		
.623	.7900	.8271	.96241	0.06300	.61022	.59617	.58036	.56190	.53874	.52359	.50225		
					1.00532	1.01086	1.01668	1.02284	1.02941	1.03292	1.03664		
.622	.7915	.8266	.96297	0.06189	.60909	.59487	.57881	.55996	.53600	.51995	.49556		
					1.00534	1.01091	1.01676	1.02295	1.02958	1.03313	1.03691		
.621	.7929	.8260	.96352	0.06080	.60796	.59355	.57724	.55798	.53315	.51602	.48651		
					1.00536	1.01096	1.01684	1.02306	1.02975	1.03334	1.03719		
.620	.7944	.8255	.96407	0.05973	.60682	.59224	.57566	.55596	.53017	.51172			
					1.00538	1.01100	1.01691	1.02318	1.02992	1.03355			
.619	.7959	.8249	.96461	0.05866	.60568	.59091	.57406	.55390	.52704	.50693			
					1.00540	1.01105	1.01699	1.02330	1.03009	1.03377			
.618	.7974	.8244	.96515	0.05761	.60454	.58958	.57245	.55180	.52374	.50141			
					1.00542	1.01109	1.01707	1.02341	1.03027	1.03400			
.617	.7988	.8239	.96568	0.05657	.60340	.58824	.57082	.54965	.52018	.49454			
					1.00545	1.01114	1.01715	1.02353	1.03046	1.03426			
.616	.8003	.8233	.96621	0.05554	.60225	.58689	.56916	.54745	.51638	.48469			
					1.00547	1.01119	1.01722	1.02366	1.03065	1.03452			
.615	.8018	.8228	.96673	0.05453	.60111	.58554	.56749	.54519	.51223				
					1.00549	1.01124	1.01730	1.02378	1.03085				
.614	.8033	.8223	.96725	0.05353	.59995	.58418	.56580	.54287	.50761				
					1.00551	1.01128	1.01738	1.02390	1.03105				
.613	.8047	.8217	.96777	0.05254	.59880	.58280	.56409	.54047	.50230				
					1.00553	1.01133	1.01746	1.02403	1.03126				
.612	.8062	.8212	.96828	0.05156	.59764	.58142	.56235	.53800	.49580				
					1.00555	1.01138	1.01755	1.02416	1.03148				
.611	.8077	.8207	.96879	0.05060	.59648	.58004	.56059	.53545	.48672				
					1.00557	1.01143	1.01763	1.02429	1.03171				

TABLE V. Specific Heat Ratio = 1.67

p/p_t	MACH	T/T_t	GAMMA	K_{crit} T	.006	.008	.01	.015	.02	.025	.03	.035	.04
.610	.8092	.8201	.96930	0.04964	.60140	.59840	.59533	.58729	.57864	.56923	.55881	.54692	.53277
					1.00332	1.00445	1.00559	1.00850	1.01147	1.01454	1.01771	1.02100	1.02442
.609	.8106	.8196	.96979	0.04870	.60030	.59727	.59416	.58601	.57723	.56765	.55700	.54475	.53000
					1.00334	1.00447	1.00561	1.00853	1.01152	1.01460	1.01779	1.02110	1.02456
.608	.8121	.8190	.97029	0.04777	.59921	.59614	.59300	.58473	.57582	.56605	.55512	.54252	.52709
					1.00335	1.00449	1.00564	1.00857	1.01157	1.01467	1.01788	1.02121	1.02470
.607	.8136	.8185	.97078	0.04686	.59811	.59501	.59182	.58344	.57439	.56444	.55324	.54023	.52402
					1.00336	1.00450	1.00566	1.00860	1.01162	1.01474	1.01797	1.02132	1.02484
.606	.8151	.8180	.97127	0.04595	.59702	.59388	.59065	.58215	.57294	.56280	.55132	.53786	.52076
					1.00337	1.00452	1.00568	1.00864	1.01167	1.01480	1.01805	1.02143	1.02498
.605	.8166	.8174	.97175	0.04506	.59592	.59274	.58947	.58085	.57149	.56114	.54936	.53542	.51727
					1.00339	1.00454	1.00570	1.00867	1.01172	1.01487	1.01814	1.02155	1.02513
.604	.8181	.8169	.97223	0.04417	.59482	.59160	.58829	.57954	.57002	.55946	.54736	.53288	.51348
					1.00340	1.00456	1.00572	1.00871	1.01177	1.01494	1.01823	1.02166	1.02528
.603	.8195	.8163	.97270	0.04330	.59371	.59045	.58710	.57823	.56854	.55773	.54532	.53024	.50930
					1.00341	1.00457	1.00575	1.00874	1.01182	1.01501	1.01832	1.02178	1.02544
.602	.8210	.8158	.97317	0.04244	.59261	.58931	.58591	.57691	.56705	.55600	.54322	.52747	.50449
					1.00343	1.00459	1.00577	1.00878	1.01187	1.01508	1.01841	1.02190	1.02561
.601	.8225	.8152	.97364	0.04159	.59150	.58816	.58471	.57558	.56554	.55424	.54107	.52457	.49880
					1.00344	1.00461	1.00579	1.00882	1.01192	1.01515	1.01850	1.02202	1.02578
.600	.8240	.8147	.97410	0.04076	.59039	.58701	.58352	.57424	.56402	.55245	.53885	.52149	.49133
					1.00345	1.00463	1.00581	1.00885	1.01197	1.01522	1.01860	1.02214	1.02596
.599	.8255	.8141	.97456	0.03993	.58928	.58585	.58229	.57289	.56248	.55063	.53657	.51820	
					1.00346	1.00464	1.00584	1.00889	1.01203	1.01529	1.01869	1.02227	
.598	.8270	.8136	.97501	0.03911	.58816	.58469	.58109	.57153	.56092	.54877	.53421	.51466	
					1.00348	1.00466	1.00587	1.00892	1.01208	1.01536	1.01879	1.02240	
.597	.8285	.8131	.97546	0.03831	.58705	.58353	.57987	.57017	.55932	.54688	.53176	.51077	
					1.00349	1.00468	1.00589	1.00896	1.01214	1.01544	1.01889	1.02254	
.596	.8299	.8125	.97591	0.03751	.58593	.58235	.57865	.56879	.55773	.54494	.52921	.50635	
					1.00350	1.00470	1.00591	1.00900	1.01219	1.01551	1.01899	1.02268	
.595	.8314	.8120	.97635	0.03673	.58481	.58118	.57743	.56740	.55611	.54296	.52654	.50123	
					1.00352	1.00472	1.00594	1.00904	1.01225	1.01559	1.01909	1.02283	
.594	.8329	.8114	.97678	0.03596	.58368	.58000	.57620	.56601	.55447	.54094	.52374	.49480	
					1.00353	1.00474	1.00596	1.00907	1.01230	1.01566	1.01919	1.02298	
.593	.8344	.8109	.97722	0.03519	.58256	.57882	.57496	.56460	.55280	.53886	.52078	.48440	
					1.00354	1.00476	1.00598	1.00911	1.01236	1.01574	1.01930	1.02315	
.592	.8359	.8103	.97765	0.03444	.58143	.57764	.57372	.56317	.55111	.53671	.51761		
					1.00356	1.00478	1.00601	1.00915	1.01241	1.01582	1.01941		
.591	.8374	.8098	.97807	0.03370	.58030	.57646	.57248	.56174	.54939	.53451	.51419		
					1.00357	1.00479	1.00603	1.00919	1.01247	1.01589	1.01952		
.590	.8389	.8092	.97849	0.03297	.57916	.57526	.57122	.56027	.54764	.53222	.51045		
					1.00358	1.00481	1.00605	1.00923	1.01252	1.01597	1.01963		
.589	.8404	.8087	.97891	0.03224	.57803	.57407	.56996	.55880	.54586	.52985	.50618		
					1.00360	1.00483	1.00608	1.00927	1.01258	1.01606	1.01976		
.588	.8419	.8081	.97932	0.03153	.57688	.57287	.56870	.55732	.54404	.52739	.50123		
					1.00361	1.00485	1.00610	1.00931	1.01264	1.01614	1.01988		
.587	.8434	.8076	.97973	0.03083	.57574	.57166	.56742	.55582	.54219	.52481	.49498		
					1.00362	1.00487	1.00613	1.00935	1.01270	1.01622	1.02001		
.586	.8449	.8070	.98013	0.03014	.57457	.57045	.56614	.55431	.54029	.52209	.48457		
					1.00364	1.00489	1.00615	1.00939	1.01276	1.01631	1.02015		
.585	.8464	.8065	.98053	0.02946	.57342	.56923	.56485	.55277	.53835	.51921			
					1.00366	1.00491	1.00617	1.00943	1.01282	1.01640			
.584	.8479	.8059	.98092	0.02878	.57226	.56801	.56355	.55122	.53635	.51613			
					1.00367	1.00493	1.00620	1.00947	1.01288	1.01649			
.583	.8494	.8054	.98131	0.02812	.57110	.56678	.56224	.54964	.53430	.51280			
					1.00368	1.00494	1.00622	1.00951	1.01294	1.01658			
.582	.8509	.8048	.98170	0.02747	.56994	.56555	.56093	.54805	.53218	.50907			
					1.00370	1.00496	1.00625	1.00955	1.01300	1.01668			
.581	.8524	.8042	.98208	0.02682	.56877	.56431	.55960	.54642	.52999	.50487			
					1.00371	1.00498	1.00627	1.00959	1.01307	1.01678			

TABLE V. Specific Heat Ratio = 1.67

p/p,	MACH	T/T,	GAMMA	K_crit	R: and TPR versus Loss Coefficient K								
					.001	.002	.003	.004	.005	.006	.01	.015	.02
.580	.8539	.8037	.98246	0.02619	.57804	.57604	.57400	.57192	.56978	.56760	.55828	.54477	.52772
					1.00061	1.00123	1.00184	1.00247	1.00309	1.00373	1.00630	1.00964	1.01313
.579	.8554	.8031	.98284	0.02556	.57700	.57500	.57292	.57081	.56864	.56642	.55693	.54310	.52536
					1.00062	1.00123	1.00185	1.00248	1.00311	1.00374	1.00632	1.00968	1.01320
.578	.8569	.8026	.98321	0.02495	.57600	.57394	.57184	.56969	.56749	.56524	.55558	.54139	.52288
					1.00061	1.00123	1.00186	1.00249	1.00312	1.00375	1.00635	1.00972	1.01327
.577	.8584	.8020	.98358	0.02434	.57496	.57288	.57076	.56858	.56634	.56405	.55421	.53964	.52027
					1.00062	1.00124	1.00186	1.00249	1.00313	1.00377	1.00637	1.00977	1.01333
.576	.8599	.8015	.98394	0.02374	.57394	.57183	.56967	.56746	.56519	.56286	.55283	.53786	.51751
					1.00062	1.00124	1.00187	1.00250	1.00314	1.00378	1.00640	1.00981	1.01340
.575	.8614	.8009	.98430	0.02315	.57291	.57077	.56858	.56634	.56403	.56166	.55143	.53604	.51454
					1.00062	1.00125	1.00188	1.00251	1.00315	1.00380	1.00642	1.00986	1.01348
.574	.8629	.8003	.98465	0.02257	.57188	.56971	.56749	.56521	.56287	.56046	.54999	.53417	.51127
					1.00062	1.00125	1.00188	1.00252	1.00316	1.00381	1.00645	1.00990	1.01356
.573	.8645	.7998	.98500	0.02200	.57085	.56865	.56640	.56408	.56170	.55925	.54857	.53225	.50768
					1.00063	1.00126	1.00189	1.00253	1.00318	1.00383	1.00648	1.00995	1.01363
.572	.8660	.7992	.98535	0.02144	.56982	.56759	.56530	.56295	.56053	.55803	.54712	.53028	.50357
					1.00063	1.00126	1.00190	1.00254	1.00319	1.00384	1.00651	1.00999	1.01371
.571	.8675	.7987	.98569	0.02089	.56879	.56653	.56421	.56182·	.55935	.55681	.54566	.52824	.49861
					1.00063	1.00127	1.00191	1.00255	1.00320	1.00385	1.00653	1.01004	1.01379
.570	.8690	.7981	.98603	0.02034	.56776	.56547	.56311	.56068	.55817	.55558	.54418	.52612	.49163
					1.00063	1.00127	1.00191	1.00256	1.00321	1.00387	1.00656	1.01009	1.01389
.569	.8705	.7975	.98636	0.01981	.56673	.56440	.56200	.55953	.55698	.55435	.54268	.52393	
					1.00063	1.00127	1.00192	1.00257	1.00322	1.00388	1.00659	1.01014	
.568	.8720	.7970	.98669	0.01928	.56570	.56333	.56090	.55839	.55579	.55310	.54115	.52163	
					1.00064	1.00128	1.00193	1.00258	1.00324	1.00390	1.00662	1.01019	
.567	.8735	.7964	.98702	0.01876	.56467	.56226	.55979	.55723	.55459	.55185	.53961	.51922	
					1.00064	1.00128	1.00193	1.00259	1.00325	1.00391	1.00665	1.01024	
.566	.8751	.7958	.98734	0.01825	.56363	.56119	.55868	.55608	.55339	.55059	.53803	.51667	
					1.00064	1.00129	1.00194	1.00260	1.00326	1.00393	1.00667	1.01029	
.565	.8766	.7953	.98765	0.01775	.56260	.56012	.55756	.55492	.55217	.54932	.53643	.51395	
					1.00064	1.00129	1.00195	1.00261	1.00327	1.00394	1.00670	1.01034	
.564	.8781	.7947	.98797	0.01725	.56156	.55905	.55644	.55375	.55095	.54804	.53480	.51094	
					1.00065	1.00130	1.00195	1.00262	1.00328	1.00396	1.00673	1.01040	
.563	.8796	.7942	.98828	0.01677	.56053	.55797	.55532	.55258	.54973	.54675	.53313	.50766	
					1.00065	1.00130	1.00196	1.00263	1.00330	1.00397	1.00676	1.01046	
.562	.8811	.7936	.98858	0.01629	.55949	.55689	.55420	.55140	.54849	.54545	.53143	.50395	
					1.00065	1.00131	1.00197	1.00263	1.00331	1.00399	1.00679	1.01052	
.561	.8827	.7930	.98888	0.01582	.55845	.55581	.55307	.55022	.54725	.54414	.52968	.49953	
					1.00065	1.00131	1.00197	1.00264	1.00332	1.00400	1.00682	1.01058	
.560	.8842	.7925	.98918	0.01536	.55741	.55472	.55193	.54903	.54599	.54281	.52788	.49351	
					1.00065	1.00131	1.00198	1.00265	1.00333	1.00402	1.00685	1.01064	
.559	.8857	.7919	.98947	0.01491	.55637	.55364	.55080	.54783	.54473	.54147	.52604		
					1.00066	1.00132	1.00199	1.00266	1.00335	1.00403	1.00688		
.558	.8873	.7913	.98976	0.01446	.55533	.55255	.54965	.54663	.54346	.54012	.52413		
					1.00066	1.00132	1.00200	1.00267	1.00336	1.00405	1.00691		
.557	.8888	.7907	.99005	0.01403	.55428	.55146	.54851	.54542	.54218	.53875	.52215		
					1.00066	1.00133	1.00200	1.00268	1.00337	1.00407	1.00694		
.556	.8903	.7902	.99033	0.01360	.55324	.55036	.54735	.54420	.54088	.53737	.52009		
					1.00066	1.00133	1.00201	1.00269	1.00338	1.00408	1.00698		
.555	.8918	.7896	.99061	0.01317	.55219	.54926	.54620	.54298	.53958	.53593	.51794		
					1.00066	1.00134	1.00202	1.00270	1.00340	1.00410	1.00701		
.554	.8934	.7890	.99088	0.01276	.55115	.54816	.54503	.54174	.53826	.53450	.51568		
					1.00067	1.00134	1.00202	1.00271	1.00341	1.00412	1.00704		
.553	.8949	.7885	.99115	0.01235	.55010	.54706	.54386	.54050	.53689	.53305	.51327		
					1.00067	1.00135	1.00203	1.00272	1.00343	1.00414	1.00708		
.552	.8964	.7879	.99141	0.01196	.54905	.54595	.54269	.53924	.53553	.53158	.51070		
					1.00067	1.00135	1.00204	1.00273	1.00344	1.00415	1.00711		
.551	.8980	.7873	.99167	0.01156	.54800	.54484	.54151	.53798	.53416	.53008	.50780		
					1.00067	1.00136	1.00205	1.00274	1.00346	1.00417	1.00715		

TABLE V. Specific Heat Ratio = 1.67

p/p_t	MACH	T/T_t	GAMMA	K_{crit}	.001	.002	.003	.004	.005	.006	.007	.008	.009
.550	.8995	.7867	.99193	0.01118	.54694	.54372	.54032	.53666	.53277	.52855	.52390	.51866	.51250
					1.00068	1.00136	1.00205	1.00276	1.00347	1.00419	1.00492	1.00566	1.00641
.549	.9011	.7862	.99218	0.01080	.54589	.54260	.53912	.53537	.53136	.52700	.52214	.51660	.50995
					1.00068	1.00136	1.00206	1.00277	1.00348	1.00421	1.00494	1.00569	1.00645
.548	.9026	.7856	.99243	0.01043	.54483	.54148	.53792	.53406	.52993	.52540	.52032	.51444	.50718
					1.00068	1.00137	1.00207	1.00278	1.00350	1.00422	1.00496	1.00571	1.00648
.547	.9041	.7850	.99268	0.01007	.54377	.54035	.53666	.53274	.52848	.52377	.51844	.51215	.50399
					1.00068	1.00137	1.00208	1.00279	1.00351	1.00424	1.00498	1.00574	1.00652
.546	.9057	.7844	.99292	0.00972	.54271	.53921	.53544	.53140	.52699	.52209	.51647	.50969	.50032
					1.00068	1.00138	1.00209	1.00280	1.00353	1.00426	1.00501	1.00577	1.00655
.545	.9072	.7839	.99315	0.00937	.54165	.53804	.53420	.53004	.52548	.52036	.51440	.50701	.49568
					1.00069	1.00139	1.00210	1.00281	1.00354	1.00428	1.00503	1.00579	1.00659
.544	.9088	.7833	.99339	0.00903	.54059	.53689	.53295	.52867	.52393	.51857	.51222	.50394	.48736
					1.00069	1.00139	1.00211	1.00282	1.00355	1.00430	1.00505	1.00583	1.00663
.543	.9103	.7827	.99362	0.00870	.53952	.53573	.53169	.52727	.52235	.51671	.50989	.50043	
					1.00069	1.00140	1.00211	1.00284	1.00357	1.00432	1.00508	1.00586	
.542	.9119	.7821	.99384	0.00837	.53845	.53457	.53041	.52585	.52072	.51476	.50737	.49603	
					1.00069	1.00140	1.00212	1.00285	1.00358	1.00433	1.00510	1.00589	
.541	.9134	.7816	.99406	0.00805	.53734	.53340	.52912	.52440	.51904	.51272	.50450	.48861	
					1.00070	1.00141	1.00213	1.00286	1.00360	1.00435	1.00513	1.00593	
.540	.9150	.7810	.99428	0.00774	.53626	.53222	.52782	.52292	.51731	.51056	.50128		
					1.00070	1.00142	1.00214	1.00287	1.00362	1.00437	1.00516		
.539	.9165	.7804	.99449	0.00744	.53517	.53104	.52650	.52141	.51551	.50824	.49736		
					1.00071	1.00142	1.00215	1.00288	1.00363	1.00439	1.00519		
.538	.9181	.7798	.99470	0.00714	.53409	.52984	.52516	.51986	.51364	.50573	.49162		
					1.00071	1.00143	1.00215	1.00289	1.00365	1.00441	1.00522		
.537	.9196	.7792	.99490	0.00685	.53300	.52864	.52379	.51827	.51167	.50278			
					1.00071	1.00143	1.00216	1.00291	1.00366	1.00444			
.536	.9212	.7786	.99510	0.00656	.53190	.52742	.52241	.51663	.50958	.49945			
					1.00071	1.00144	1.00217	1.00292	1.00368	1.00447			
.535	.9227	.7781	.99530	0.00628	.53081	.52619	.52100	.51494	.50735	.49524			
					1.00072	1.00144	1.00218	1.00293	1.00369	1.00449			
.534	.9243	.7775	.99549	0.00601	.52970	.52495	.51956	.51317	.50482	.48749			
					1.00072	1.00145	1.00219	1.00294	1.00372	1.00451			
.533	.9259	.7769	.99568	0.00575	.52859	.52369	.51808	.51133	.50206				
					1.00072	1.00145	1.00220	1.00295	1.00374				
.532	.9274	.7763	.99586	0.00549	.52748	.52242	.51657	.50939	.49883				
					1.00072	1.00146	1.00220	1.00297	1.00375				
.531	.9290	.7757	.99604	0.00524	.52636	.52113	.51501	.50733	.49473				
					1.00073	1.00146	1.00221	1.00298	1.00377				
.530	.9306	.7751	.99622	0.00499	.52523	.51982	.51341	.50499					
					1.00073	1.00147	1.00222	1.00300					
.529	.9321	.7746	.99639	0.00475	.52410	.51849	.51174	.50249					
					1.00073	1.00147	1.00223	1.00301					
.528	.9337	.7740	.99656	0.00452	.52296	.51714	.51001	.49964					
					1.00073	1.00148	1.00224	1.00303					
.527	.9353	.7734	.99673	0.00430	.52181	.51575	.50819	.49617					
					1.00074	1.00148	1.00225	1.00304					
.526	.9368	.7728	.99689	0.00408	.52065	.51434	.50616	.49084					
					1.00074	1.00149	1.00226	1.00306					
.525	.9384	.7722	.99704	0.00386	.51948	.51289	.50404						
					1.00074	1.00150	1.00227						
.524	.9400	.7716	.99719	0.00366	.51830	.51140	.50171						
					1.00074	1.00150	1.00228						
.523	.9416	.7710	.99734	0.00345	.51711	.50985	.49906						
					1.00075	1.00151	1.00229						
.522	.9431	.7704	.99749	0.00326	.51591	.50825	.49587						
					1.00075	1.00151	1.00230						
.521	.9447	.7698	.99763	0.00307	.51468	.50648	.49102						
					1.00075	1.00152	1.00232						

TABLE V. Specific Heat Ratio = 1.67

R_1 and TPR versus Loss Coefficient K

p/p_t	MACH	T/T_t	GAMMA	K_{crit}	.001	.002	.003	.004	.005	.006	.007	.008	.009
.520	.9463	.7692	.99776	0.00289	.51345 1.00075	.50468 1.00153							
.519	.9479	.7686	.99789	0.00271	.51219 1.00075	.50275 1.00154							
.518	.9495	.7681	.99802	0.00254	.51091 1.00076	.50063 1.00154							
.517	.9511	.7675	.99815	0.00238	.50960 1.00076	.49824 1.00155							
.516	.9526	.7669	.99827	0.00222	.50827 1.00076	.49541 1.00155							
.515	.9542	.7663	.99838	0.00206	.50679 1.00077	.49113 1.00156							
.514	.9558	.7657	.99849	0.00192	.50535 1.00077								
.513	.9574	.7651	.99860	0.00177	.50385 1.00078								
.512	.9590	.7645	.99871	0.00164	.50226 1.00078								
.511	.9606	.7639	.99881	0.00151	.50058 1.00078								
.510	.9622	.7633	.99890	0.00138	.49875 1.00078								
.509	.9638	.7627	.99899	0.00127	.49673 1.00079								
.508	.9654	.7621	.99908	0.00115	.49439 1.00079								
.507	.9670	.7615	.99917	0.00104	.49077 1.00079								

TABLE VI. Specific Heat Ratio = 1.10

p/p_t	MACH	T/T_t	GAMMA	K_{crit}	SHOCK P_y/P_x	P_{tx}/P_{ty}	.1	.2	.3	.4	.5	.6	.7
.001	4.1805	.5337	.00955	1.32088	18.26093	51.90274	.00085	.00133	.00151	.00170	.00190	.00211	.00234
							1.00062	4.81692	8.64330	14.06147	21.14291	29.80289	39.79097
.002	3.8971	.5684	.01726	1.25488	15.86330	29.74485	.00230	.00262	.00295	.00330	.00368	.00408	.00453
							2.13077	3.96967	6.66038	10.26787	14.77617	20.07358	25.98322
.003	3.7302	.5897	.02432	1.21196	14.52924	21.58533	.00343	.00389	.00437	.00488	.00542	.00602	.00667
							2.00330	3.55627	5.74321	8.58291	12.03482	15.99683	20.32581
.004	3.6109	.6053	.03099	1.17928	13.61210	17.23799	.00456	.00515	.00577	.00644	.00715	.00792	.00878
							1.91975	3.29706	5.18108	7.57272	10.42481	13.64221	17.10518
.005	3.5179	.6178	.03735	1.15253	12.91704	14.50249	.00568	.00641	.00717	.00798	.00886	.00982	.01088
							1.85858	3.11125	4.78839	6.87967	9.33486	12.06966	14.97734
.006	3.4414	.6281	.04349	1.12969	12.35951	12.60758	.00681	.00766	.00856	.00952	.01056	.01170	.01297
							1.81081	2.96798	4.49289	6.36559	8.53628	10.92822	13.44497
.007	3.3764	.6369	.04943	1.10965	11.89528	11.20976	.00793	.00890	.00994	.01105	.01225	.01358	.01506
							1.77196	2.85459	4.25936	5.96370	7.91888	10.05235	12.27772
.008	3.3198	.6447	.05521	1.09172	11.49837	10.13176	.00904	.01014	.01132	.01257	.01394	.01545	.01714
							1.73918	2.76041	4.06836	5.63869	7.42292	9.35385	11.35149
.009	3.2697	.6517	.06084	1.07543	11.15225	9.27246	.01016	.01138	.01269	.01409	.01562	.01731	.01922
							1.71110	2.68056	3.90829	5.36803	7.01317	8.78038	10.59609
.010	3.2246	.6579	.06636	1.06048	10.84575	8.56971	.01127	.01262	.01406	.01561	.01730	.01917	.02130
							1.68661	2.61157	3.77111	5.13829	6.66751	8.29912	9.96431
.011	3.1837	.6637	.07175	1.04663	10.57102	7.98314	.01239	.01386	.01543	.01712	.01898	.02103	.02337
							1.66492	2.55107	3.65189	4.93987	6.37066	7.88780	9.42697
.012	3.1462	.6689	.07705	1.03371	10.32227	7.48532	.01350	.01509	.01679	.01863	.02065	.02289	.02545
							1.64551	2.49738	3.54661	4.76614	6.11210	7.53111	8.96279
.013	3.1116	.6738	.08225	1.02157	10.09518	7.05693	.01461	.01632	.01816	.02014	.02232	.02475	.02753
							1.62796	2.44923	3.45325	4.61231	5.88424	7.21804	8.55673
.014	3.0794	.6784	.08737	1.01013	9.88639	6.68396	.01572	.01756	.01952	.02165	.02399	.02660	.02961
							1.61197	2.40570	3.36928	4.47479	5.68144	6.94060	8.19786
.015	3.0493	.6826	.09240	.99927	9.69328	6.35597	.01683	.01879	.02088	.02315	.02565	.02846	.03170
							1.59731	2.36605	3.29321	4.35085	5.49942	6.69237	7.87783
.016	3.0210	.6867	.09736	.98895	9.51372	6.06502	.01795	.02002	.02224	.02466	.02732	.03032	.03379
							1.58378	2.32970	3.22384	4.23837	5.33496	6.46837	7.59021
.017	2.9944	.6904	.10226	.97910	9.34601	5.80498	.01905	.02124	.02360	.02616	.02898	.03217	.03588
							1.57124	2.29620	3.16023	4.13591	5.18524	6.26543	7.32596
.018	2.9693	.6940	.10708	.96967	9.18874	5.57100	.02016	.02247	.02495	.02766	.03065	.03403	.03798
							1.55912	2.26517	3.10158	4.04139	5.04824	6.08024	7.09306
.019	2.9454	.6975	.11185	.96062	9.04072	5.35922	.02127	.02370	.02631	.02916	.03231	.03589	.04008
							1.54823	2.23631	3.04727	3.95445	4.92225	5.91040	6.87627
.020	2.9226	.7007	.11655	.95191	8.90096	5.16652	.02238	.02492	.02766	.03066	.03398	.03775	.04219
							1.53802	2.20936	2.99677	3.87384	4.80586	5.75389	6.67693
.021	2.9010	.7038	.12120	.94352	8.76863	4.99035	.02349	.02615	.02902	.03216	.03564	.03961	.04430
							1.52839	2.18411	2.94964	3.79887	4.69790	5.60907	6.49285
.022	2.8802	.7068	.12580	.93542	8.64300	4.82858	.02460	.02737	.03037	.03365	.03731	.04148	.04642
							1.51931	2.16037	2.90534	3.72889	4.59740	5.47457	6.32222
.023	2.8603	.7097	.13035	.92759	8.52346	4.67946	.02570	.02860	.03172	.03515	.03897	.04334	.04855
							1.51070	2.13800	2.86397	3.66318	4.50354	5.34921	6.16348
.024	2.8412	.7124	.13484	.92000	8.40945	4.54152	.02681	.02982	.03308	.03665	.04064	.04521	.05068
							1.50253	2.11712	2.82499	3.60172	4.41561	5.23203	6.01534
.025	2.8229	.7151	.13930	.91263	8.30051	4.41349	.02792	.03104	.03443	.03815	.04231	.04709	.05283
							1.49476	2.09713	2.78818	3.54382	4.33301	5.12216	5.87667
.026	2.8052	.7176	.14370	.90548	8.19623	4.29430	.02902	.03226	.03578	.03964	.04397	.04896	.05497
							1.48735	2.07809	2.75333	3.48916	4.25521	5.01888	5.74653
.027	2.7881	.7201	.14807	.89852	8.09623	4.18305	.03013	.03349	.03713	.04114	.04564	.05084	.05713
							1.48028	2.05998	2.72028	3.43743	4.18182	4.92156	5.62408
.028	2.7716	.7225	.15239	.89175	8.00019	4.07892	.03123	.03471	.03848	.04264	.04731	.05272	.05930
							1.47357	2.04273	2.68886	3.38839	4.11231	4.82965	5.50861
.029	2.7557	.7248	.15668	.88515	7.90782	3.98124	.03234	.03593	.03983	.04413	.04898	.05460	.06148
							1.46710	2.02625	2.65895	3.34180	4.04644	4.74267	5.39947

TABLE VI. Specific Heat Ratio = 1.10

p/p_t	MACH	T/T_t	GAMMA	K_crit	P_y/P_x	P_ty/P_ty	.1	.2	.3	.4	.5	.6	.7
.030	2.7402	.7270	.16092	.87871	7.81887	3.88940	.03345	.03715	.04118	.04563	.05065	.05649	.06366
							1.46088	2.01050	2.63041	3.29747	3.98388	4.66019	5.29608
.031	2.7253	.7292	.16513	.87243	7.73309	3.80288	.03455	.03837	.04253	.04713	.05233	.05838	.06586
							1.45491	1.99542	2.60315	3.25521	3.92436	4.58184	5.19802
.032	2.7107	.7313	.16930	.86628	7.65028	3.72121	.03566	.03959	.04388	.04863	.05400	.06028	.06806
							1.44917	1.98095	2.57708	3.21487	3.86764	4.50729	5.10483
.033	2.6966	.7334	.17344	.86028	7.57024	3.64397	.03676	.04081	.04523	.05013	.05568	.06218	.07028
							1.44364	1.96706	2.55209	3.17630	3.81351	4.43624	5.01611
.034	2.6828	.7354	.17755	.85440	7.49281	3.57081	.03787	.04203	.04658	.05163	.05735	.06408	.07251
							1.43830	1.95370	2.52812	3.13937	3.76164	4.36842	4.93152
.035	2.6695	.7373	.18162	.84864	7.41782	3.50139	.03897	.04325	.04793	.05313	.05903	.06599	.07475
							1.43315	1.94085	2.50510	3.10397	3.71216	4.30370	4.85081
.036	2.6564	.7392	.18566	.84300	7.34514	3.43543	.04007	.04447	.04928	.05463	.06071	.06790	.07701
							1.42818	1.92846	2.48296	3.06998	3.66473	4.24158	4.77368
.037	2.6438	.7410	.18967	.83747	7.27463	3.37266	.04118	.04569	.05063	.05613	.06240	.06982	.07928
							1.42336	1.91650	2.46164	3.03733	3.61921	4.18211	4.69973
.038	2.6314	.7428	.19365	.83205	7.20616	3.31286	.04228	.04691	.05198	.05763	.06408	.07175	.08156
							1.41870	1.90496	2.44110	3.00592	3.57548	4.12506	4.62886
.039	2.6193	.7446	.19759	.82672	7.13963	3.25580	.04339	.04813	.05333	.05913	.06577	.07367	.08386
							1.41419	1.89381	2.42129	2.97566	3.53343	4.07027	4.56084
.040	2.6075	.7463	.20151	.82149	7.07494	3.20130	.04449	.04935	.05468	.06064	.06746	.07561	.08617
							1.40981	1.88301	2.40216	2.94650	3.49294	4.01759	4.49549
.041	2.5959	.7480	.20541	.81635	7.01199	3.14917	.04560	.05057	.05603	.06214	.06915	.07755	.08850
							1.40556	1.87256	2.38367	2.91836	3.45393	3.96689	4.43265
.042	2.5846	.7496	.20927	.81130	6.95069	3.09928	.04670	.05179	.05739	.06365	.07084	.07949	.09085
							1.40143	1.86244	2.36578	2.89108	3.41631	3.91804	4.37215
.043	2.5736	.7512	.21311	.80633	6.89096	3.05146	.04780	.05301	.05874	.06515	.07254	.08144	.09322
							1.39742	1.85262	2.34847	2.86483	3.37999	3.87093	4.31384
.044	2.5627	.7528	.21692	.80145	6.83273	3.00559	.04891	.05423	.06009	.06666	.07424	.08340	.09561
							1.39352	1.84309	2.33170	2.83944	3.34490	3.82538	4.25761
.045	2.5521	.7543	.22071	.79664	6.77593	2.96154	.05001	.05545	.06144	.06817	.07594	.08536	.09802
							1.38973	1.83384	2.31544	2.81486	3.31097	3.78149	4.20331
.046	2.5417	.7558	.22447	.79190	6.72049	2.91921	.05112	.05667	.06279	.06968	.07764	.08733	.10045
							1.38603	1.82485	2.29966	2.79105	3.27814	3.73905	4.15085
.047	2.5316	.7573	.22821	.78723	6.66634	2.87850	.05222	.05789	.06415	.07119	.07935	.08931	.10290
							1.38243	1.81610	2.28435	2.76796	3.24634	3.69799	4.10013
.048	2.5216	.7588	.23192	.78264	6.61344	2.83930	.05332	.05911	.06550	.07270	.08106	.09129	.10538
							1.37892	1.80759	2.26948	2.74556	3.21554	3.65824	4.05105
.049	2.5118	.7602	.23561	.77810	6.56173	2.80154	.05443	.06033	.06685	.07421	.08277	.09328	.10789
							1.37549	1.79931	2.25502	2.72381	3.18566	3.61971	4.00352
.050	2.5021	.7616	.23928	.77364	6.51116	2.76513	.05553	.06155	.06821	.07572	.08448	.09528	.11042
							1.37215	1.79124	2.24095	2.70269	3.15668	3.58237	3.95744
.051	2.4927	.7630	.24292	.76923	6.46167	2.73000	.05663	.06277	.06956	.07724	.08620	.09728	.11299
							1.36889	1.78338	2.22727	2.68215	3.12853	3.54613	3.91276
.052	2.4834	.7643	.24655	.76488	6.41324	2.69609	.05774	.06399	.07092	.07875	.08792	.09930	.11559
							1.36570	1.77572	2.21394	2.66219	3.10119	3.51100	3.86939
.053	2.4742	.7656	.25015	.76059	6.36581	2.66331	.05884	.06521	.07227	.08027	.08964	.10132	.11823
							1.36258	1.76824	2.20096	2.64276	3.07462	3.47683	3.82723
.054	2.4653	.7669	.25373	.75636	6.31935	2.63163	.05995	.06643	.07363	.08179	.09137	.10335	.12091
							1.35954	1.76094	2.18831	2.62384	3.04877	3.44363	3.78631
.055	2.4564	.7682	.25728	.75218	6.27381	2.60098	.06105	.06766	.07499	.08331	.09310	.10539	.12364
							1.35656	1.75381	2.17597	2.60542	3.02361	3.41134	3.74652
.056	2.4478	.7695	.26082	.74805	6.22917	2.57132	.06215	.06888	.07634	.08483	.09484	.10744	.12641
							1.35364	1.74684	2.16393	2.58747	2.99912	3.37993	3.70779
.057	2.4392	.7707	.26434	.74397	6.18540	2.54258	.06326	.07010	.07770	.08635	.09657	.10950	.12924
							1.35079	1.74004	2.15218	2.56996	2.97527	3.34935	3.67009
.058	2.4308	.7719	.26784	.73994	6.14245	2.51474	.06436	.07132	.07906	.08788	.09831	.11157	.13213
							1.34800	1.73338	2.14071	2.55289	2.95202	3.31956	3.63335
.059	2.4225	.7731	.27131	.73595	6.10030	2.48774	.06547	.07254	.08042	.08940	.10006	.11365	.13509
							1.34526	1.72687	2.12950	2.53622	2.92936	3.29054	3.59757

TABLE VI. Specific Heat Ratio = 1.10

					SHOCK		R₂ and TPR versus Loss Coefficient K						
p/p_t	MACH	T/T_t	GAMMA	K_{crit}	P_y/P_x	P_{ty}/P_{ty}	.1	.2	.3	.4	.5	.6	.7
.060	2.4143	.7743	.27477	.73202	6.05892	2.46155	.06657	.07376	.08178	.09093	.10181	.11574	.13814
							1.34258	1.72050	2.11855	2.51996	2.90717	3.26224	3.56264
.061	2.4063	.7755	.27821	.72812	6.01830	2.43613	.06767	.07499	.08314	.09246	.10356	.11784	.14128
							1.33995	1.71427	2.10784	2.50407	2.88561	3.23465	3.52855
.062	2.3984	.7766	.28163	.72427	5.97839	2.41145	.06878	.07621	.08450	.09399	.10531	.11995	.14453
							1.33737	1.70817	2.09737	2.48855	2.86455	3.20772	3.49525
.063	2.3905	.7778	.28504	.72046	5.93917	2.38747	.06988	.07743	.08586	.09552	.10708	.12207	.14792
							1.33484	1.70219	2.08712	2.47337	2.84399	3.18144	3.46271
.064	2.3828	.7789	.28842	.71670	5.90063	2.36417	.07099	.07865	.08722	.09706	.10884	.12421	.15149
							1.33236	1.69633	2.07709	2.45854	2.82389	3.15577	3.43089
.065	2.3752	.7800	.29179	.71297	5.86274	2.34151	.07209	.07988	.08859	.09859	.11061	.12636	.15529
							1.32992	1.69058	2.06727	2.44403	2.80425	3.13069	3.39973
.066	2.3677	.7811	.29514	.70928	5.82548	2.31946	.07319	.08110	.08995	.10013	.11238	.12852	.15944
							1.32753	1.68495	2.05766	2.42983	2.78505	3.10618	3.36922
.067	2.3603	.7821	.29847	.70563	5.78884	2.29801	.07430	.08232	.09131	.10167	.11416	.13070	.16415
							1.32518	1.67943	2.04823	2.41593	2.76626	3.08221	3.33927
.068	2.3530	.7832	.30178	.70201	5.75278	2.27712	.07540	.08355	.09268	.10321	.11594	.13289	.17017
							1.32287	1.67401	2.03909	2.40232	2.74788	3.05877	3.30982
.069	2.3458	.7842	.30508	.69843	5.71730	2.25679	.07651	.08477	.09405	.10476	.11773	.13510	
							1.32060	1.66870	2.03003	2.38899	2.72989	3.03583	
.070	2.3387	.7853	.30836	.69489	5.68237	2.23697	.07761	.08600	.09541	.10630	.11952	.13732	
							1.31838	1.66348	2.02115	2.37593	2.71227	3.01338	
.071	2.3317	.7863	.31162	.69138	5.64799	2.21765	.07872	.08722	.09678	.10785	.12131	.13956	
							1.31618	1.65835	2.01244	2.36313	2.69502	2.99140	
.072	2.3247	.7873	.31487	.68790	5.61413	2.19882	.07982	.08845	.09815	.10940	.12312	.14182	
							1.31403	1.65332	2.00389	2.35058	2.67811	2.96987	
.073	2.3179	.7883	.31810	.68445	5.58078	2.18046	.08093	.08967	.09952	.11095	.12492	.14410	
							1.31191	1.64837	1.99550	2.33828	2.66154	2.94877	
.074	2.3111	.7892	.32132	.68104	5.54792	2.16254	.08203	.09090	.10089	.11250	.12674	.14640	
							1.30983	1.64351	1.98727	2.32621	2.64530	2.92810	
.075	2.3044	.7902	.32452	.67765	5.51554	2.14505	.08314	.09212	.10226	.11406	.12855	.14871	
							1.30778	1.63874	1.97918	2.31437	2.62940	2.90778	
.076	2.2978	.7911	.32771	.67430	5.48364	2.12798	.08424	.09335	.10363	.11562	.13038	.15106	
							1.30576	1.63404	1.97124	2.30274	2.61378	2.88790	
.077	2.2912	.7921	.33088	.67098	5.45218	2.11131	.08535	.09458	.10501	.11718	.13221	.15342	
							1.30377	1.62942	1.96344	2.29133	2.59846	2.86839	
.078	2.2848	.7930	.33403	.66768	5.42117	2.09502	.08645	.09580	.10638	.11874	.13404	.15581	
							1.30181	1.62488	1.95578	2.28013	2.58342	2.84925	
.079	2.2784	.7939	.33717	.66441	5.39059	2.07910	.08756	.09703	.10776	.12030	.13588	.15822	
							1.29988	1.62042	1.94824	2.26912	2.56865	2.83046	
.080	2.2721	.7948	.34030	.66117	5.36043	2.06355	.08866	.09826	.10913	.12187	.13773	.16067	
							1.29798	1.61602	1.94083	2.25831	2.55415	2.81201	
.081	2.2658	.7957	.34341	.65796	5.33068	2.04834	.08977	.09949	.11051	.12344	.13958	.16314	
							1.29611	1.61169	1.93355	2.24769	2.53991	2.79389	
.082	2.2596	.7966	.34650	.65477	5.30132	2.03347	.09087	.10072	.11189	.12501	.14144	.16565	
							1.29427	1.60744	1.92639	2.23725	2.52592	2.77608	
.083	2.2535	.7975	.34958	.65161	5.27236	2.01892	.09198	.10195	.11327	.12659	.14331	.16819	
							1.29245	1.60324	1.91934	2.22699	2.51218	2.75858	
.084	2.2474	.7984	.35265	.64847	5.24377	2.00468	.09309	.10318	.11465	.12816	.14518	.17077	
							1.29066	1.59911	1.91241	2.21689	2.49866	2.74137	
.085	2.2414	.7992	.35571	.64536	5.21555	1.99075	.09419	.10441	.11603	.12974	.14707	.17339	
							1.28890	1.59505	1.90558	2.20697	2.48538	2.72444	
.086	2.2355	8001	.35875	.64227	5.18769	1.97711	.09530	.10564	.11741	.13132	.14895	.17605	
							1.28716	1.59104	1.89887	2.19720	2.47232	2.70779	
.087	2.2296	.8009	.36177	.63921	5.16018	1.96376	.09640	.10687	.11879	.13291	.15085	.17877	
							1.28544	1.58709	1.89225	2.18760	2.45948	2.69143	
.088	2.2238	.8018	.36479	.63616	5.13301	1.95068	.09751	.10810	.12018	.13450	.15275	.18155	
							1.28375	1.58320	1.88574	2.17814	2.44684	2.67531	
.089	2.2180	.8026	.36779	.63315	5.10618	1.93787	.09862	.10933	.12156	.13609	.15467	.18438	
							1.28207	1.57937	1.87933	2.16884	2.43441	2.65944	

TABLE VI. Specific Heat Ratio = 1.10

p/p_t	MACH	T/T_t	GAMMA	K_{crit}	SHOCK P_y/P_x	P_{ty}/P_{ty}	.05	.1	.2	.3	.4	.5	.6
.090	2.2123	.8034	.37077	.63015	5.07968	1.92531	.09474	.09972	.11056	.12295	.13768	.15659	.18728
							1.13762	1.28040	1.57559	1.87301	2.15968	2.42218	2.64380
.091	2.2066	.8042	.37375	.62717	5.05349	1.91301	.09579	.10083	.11180	.12434	.13928	.15852	.19027
							1.13686	1.27877	1.57186	1.86679	2.15066	2.41013	2.62840
.092	2.2010	.8050	.37671	.62422	5.02762	1.90095	.09684	.10194	.11303	.12573	.14088	.16045	.19334
							1.13612	1.27716	1.56818	1.86066	2.14177	2.39828	2.61321
.093	2.1955	.8058	.37965	.62129	5.00205	1.88912	.09790	.10304	.11426	.12712	.14248	.16240	.19653
							1.13539	1.27558	1.56456	1.85461	2.13302	2.38660	2.59824
.094	2.1900	.8066	.38259	.61838	4.97679	1.87753	.09895	.10415	.11550	.12851	.14408	.16436	.19984
							1.13466	1.27404	1.56098	1.84866	2.12440	2.37511	2.58348
.095	2.1845	.8074	.38551	.61548	4.95181	1.86615	.10000	.10526	.11673	.12990	.14569	.16632	.20332
							1.13395	1.27249	1.55746	1.84278	2.11591	2.36378	2.56890
.096	2.1791	.8081	.38842	.61261	4.92711	1.85499	.10105	.10637	.11797	.13130	.14730	.16830	.207C0
							1.13324	1.27097	1.55397	1.83699	2.10754	2.35262	2.55451
.097	2.1738	.8089	.39132	.60976	4.90270	1.84404	.10210	.10748	.11920	.13269	.14892	.17028	.21096
							1.13254	1.26946	1.55054	1.83128	2.09929	2.34162	2.54030
.098	2.1685	.8096	.39420	.60693	4.87856	1.83330	.10316	.10858	.12044	.13409	.15054	.17228	.21532
							1.13185	1.26797	1.54715	1.82565	2.09115	2.33078	2.52625
.099	2.1632	.8104	.39708	.60411	4.85469	1.82275	.10421	.10969	.12168	.13549	.15216	.17429	.22037
							1.13117	1.26650	1.54380	1.82009	2.08313	2.32010	2.51232
.100	2.1580	.8111	.39994	.60131	4.83107	1.81240	.10526	.11080	.12291	.13689	.15378	.17631	.22710
							1.13050	1.26505	1.54050	1.81461	2.07522	2.30956	2.49851
.101	2.1528	.8119	.40279	.59854	4.80772	1.80223	.10631	.11191	.12415	.13829	.15541	.17834	
							1.12983	1.26361	1.53724	1.80920	2.06742	2.29917	
.102	2.1477	.8126	.40562	.59578	4.78461	1.79224	.10736	.11302	.12539	.13969	.15704	.18038	
							1.12917	1.26219	1.53401	1.80386	2.05973	2.28893	
.103	2.1426	.8133	.40845	.59303	4.76176	1.78244	.10842	.11413	.12663	.14110	.15868	.18244	
							1.12852	1.26079	1.53083	1.79859	2.05214	2.27882	
.104	2.1376	.8140	.41126	.59031	4.73914	1.77280	.10947	.11524	.12787	.14250	.16032	.18451	
							1.12790	1.25940	1.52769	1.79339	2.04465	2.26885	
.105	2.1326	.8147	.41407	.58760	4.71676	1.76333	.11052	.11635	.12911	.14391	.16196	.18659	
							1.12726	1.25803	1.52458	1.78825	2.03726	2.25901	
.106	2.1276	.8154	.41686	.58491	4.69460	1.75403	.11157	.11745	.13035	.14532	.16361	.18869	
							1.12663	1.25667	1.52151	1.78318	2.02998	2.24930	
.107	2.1227	.8161	.41964	.58223	4.67268	1.74489	.11263	.11856	.13159	.14673	.16526	.19080	
							1.12601	1.25533	1.51848	1.77818	2.02278	2.23971	
.108	2.1178	.8168	.42241	.57957	4.65098	1.73590	.11368	.11967	.13283	.14814	.16692	.19293	
							1.12539	1.25400	1.51549	1.77323	2.01567	2.23024	
.109	2.1129	.8175	.42516	.57693	4.62950	1.72706	.11473	.12078	.13408	.14955	.16858	.19508	
							1.12478	1.25269	1.51253	1.76834	2.00865	2.22090	
.110	2.1081	.8182	.42791	.57430	4.60823	1.71837	.11578	.12189	.13532	.15097	.17024	.19724	
							1.12417	1.25139	1.50960	1.76352	2.00172	2.21167	
.111	2.1034	.8189	.43064	.57169	4.58717	1.70982	.11684	.12300	.13656	.15238	.17191	.19942	
							1.12357	1.25011	1.50671	1.75875	1.99492	2.20255	
.112	2.0986	.8195	.43337	.56909	4.56631	1.70142	.11789	.12412	.13781	.15380	.17358	.20163	
							1.12298	1.24884	1.50384	1.75404	1.98815	2.19355	
.113	2.0939	.8202	.43608	.56651	4.54566	1.69315	.11894	.12523	.13905	.15522	.17526	.20385	
							1.12240	1.24758	1.50102	1.74938	1.98147	2.18465	
.114	2.0893	.8209	.43878	.56394	4.52521	1.68501	.12000	.12634	.14030	.15664	.17694	.20609	
							1.12182	1.24634	1.49822	1.74478	1.97487	2.17586	
.115	2.0846	.8215	.44148	.56139	4.50495	1.67701	.12105	.12745	.14155	.15807	.17863	.20835	
							1.12124	1.24511	1.49545	1.74023	1.96834	2.16717	
.116	2.0800	.8221	.44416	.55885	4.48488	1.66913	.12210	.12856	.14279	.15949	.18032	.21064	
							1.12067	1.24389	1.49272	1.73573	1.96190	2.15858	
.117	2.0755	.8228	.44683	.55632	4.46501	1.66138	.12316	.12967	.14404	.16092	.18202	.21295	
							1.12011	1.24268	1.49001	1.73128	1.95553	2.15008	
.118	2.0709	.8234	.44949	.55381	4.44531	1.65374	.12421	.13078	.14529	.16235	.18372	.21529	
							1.11955	1.24149	1.48733	1.72689	1.94923	2.14169	
.119	2.0664	.8241	.45214	.55131	4.42580	1.64623	.12526	.13190	.14654	.16378	.18543	.21766	
							1.11900	1.24031	1.48468	1.72254	1.94300	2.13338	

TABLE VI. Specific Heat Ratio = 1.10

p/p_t	MACH	T/T_t	GAMMA	K_{crit}	SHOCK P_y/P_x	P_{ty}/P_{tx}	.01	.05	.1	.2	.3	.4	.5
.120	2.0619	.8247	.45478	.54883	4.40646	1.63883	.12124	.12632	.13301	.14779	.16521	.18714	.22006
							1.02347	1.11844	1.23914	1.48206	1.71824	1.93684	2.12516
.121	2.0575	.8253	.45741	.54636	4.38730	1.63155	.12225	.12737	.13412	.14904	.16664	.18886	.22249
							1.02336	1.11790	1.23798	1.47947	1.71399	1.93076	2.11704
.122	2.0531	.8259	.46003	.54390	4.36831	1.62438	.12326	.12842	.13523	.15029	.16808	.19058	.22495
							1.02326	1.11736	1.23683	1.47690	1.70978	1.92474	2.10899
.123	2.0487	.8265	.46264	.54145	4.34949	1.61731	.12427	.12948	.13635	.15155	.16952	.19231	.22746
							1.02316	1.11683	1.23569	1.47436	1.70562	1.91879	2.10105
.124	2.0444	.8271	.46524	.53902	4.33084	1.61035	.12528	.13053	.13746	.15280	.17096	.19404	.23000
							1.02306	1.11630	1.23457	1.47184	1.70150	1.91290	2.09318
.125	2.0400	.8278	.46783	.53660	4.31235	1.60350	.12629	.13159	.13857	.15405	.17240	.19578	.23259
							1.02296	1.11578	1.23345	1.46935	1.69743	1.90707	2.08538
.126	2.0358	.8284	.47041	.53419	4.29402	1.59674	.12730	.13264	.13969	.15531	.17384	.19753	.23522
							1.02287	1.11526	1.23235	1.46689	1.69340	1.90131	2.07766
.127	2.0315	.8289	.47298	.53180	4.27585	1.59008	.12831	.13369	.14080	.15656	.17529	.19928	.23791
							1.02277	1.11475	1.23125	1.46445	1.68941	1.89561	2.07001
.128	2.0273	.8295	.47554	.52941	4.25783	1.58352	.12932	.13475	.14191	.15782	.17674	.20104	.24066
							1.02267	1.11424	1.23017	1.46203	1.68546	1.88997	2.06244
.129	2.0230	.8301	.47809	.52704	4.23997	1.57706	.13033	.13580	.14303	.15908	.17819	.20281	.24347
							1.02258	1.11374	1.22910	1.45964	1.68155	1.88438	2.05494
.130	2.0189	.8307	.48063	.52468	4.22226	1.57069	.13134	.13686	.14414	.16033	.17964	.20458	.24636
							1.02248	1.11324	1.22803	1.45727	1.67768	1.87886	2.04750
.131	2.0147	.8313	.48316	.52233	4.20470	1.56441	.13235	.13791	.14526	.16159	.18110	.20636	.24933
							1.02239	1.11274	1.22698	1.45493	1.67385	1.87339	2.04013
.132	2.0106	.8319	.48569	.52000	4.18728	1.55822	.13336	.13897	.14637	.16285	.18255	.20815	.25240
							1.02230	1.11225	1.22593	1.45260	1.67006	1.86797	2.03283
.133	2.0065	.8324	.48820	.51767	4.17000	1.55211	.13437	.14002	.14749	.16411	.18401	.20994	.25559
							1.02221	1.11176	1.22490	1.45030	1.66631	1.86261	2.02558
.134	2.0024	.8330	.49070	.51536	4.15287	1.54609	.13538	.14108	.14860	.16537	.18547	.21174	.25891
							1.02211	1.11128	1.22387	1.44802	1.66259	1.85730	2.01839
.135	1.9983	.8336	.49320	.51305	4.13587	1.54016	.13639	.14213	.14972	.16663	.18694	.21355	.26239
							1.02202	1.11080	1.22285	1.44576	1.65891	1.85205	2.01125
.136	1.9943	.8341	.49568	.51076	4.11901	1.53430	.13741	.14319	.15084	.16790	.18840	.21537	.26610
							1.02193	1.11033	1.22184	1.44352	1.65526	1.84684	2.00417
.137	1.9903	.8347	.49816	.50848	4.10229	1.52853	.13842	.14424	.15195	.16916	.18987	.21720	.27009
							1.02185	1.10986	1.22085	1.44130	1.65165	1.84169	1.99713
.138	1.9863	.8352	.50062	.50621	4.08570	1.52283	.13943	.14530	.15307	.17043	.19134	.21903	.27447
							1.02176	1.10939	1.21985	1.43910	1.64807	1.83659	1.99013
.139	1.9824	.8358	.50308	.50395	4.06924	1.51722	.14044	.14635	.15419	.17169	.19282	.22087	.27951
							1.02167	1.10893	1.21887	1.43693	1.64453	1.83153	1.98315
.140	1.9784	.8363	.50553	.50169	4.05291	1.51167	.14145	.14741	.15530	.17296	.19429	.22272	.28592
							1.02158	1.10847	1.21790	1.43477	1.64102	1.82652	1.97620
.141	1.9745	.8369	.50797	.49945	4.03670	1.50621	.14246	.14847	.15642	.17423	.19577	.22458	
							1.02150	1.10801	1.21693	1.43263	1.63754	1.82156	
.142	1.9706	.8374	.51040	.49722	4.02062	1.50081	.14347	.14952	.15754	.17549	.19725	.22645	
							1.02141	1.10756	1.21597	1.43051	1.63409	1.81664	
.143	1.9667	.8379	.51282	.49500	4.00466	1.49549	.14448	.15058	.15866	.17676	.19874	.22833	
							1.02132	1.10711	1.21502	1.42840	1.63067	1.81176	
.144	1.9629	.8385	.51523	.49279	3.98883	1.49023	.14549	.15163	.15978	.17803	.20022	.23022	
							1.02124	1.10667	1.21408	1.42632	1.62729	1.80693	
.145	1.9591	.8390	.51764	.49059	3.97311	1.48505	.14650	.15269	.16090	.17930	.20171	.23212	
							1.02116	1.10623	1.21314	1.42426	1.62393	1.80215	
.146	1.9553	.8395	.52003	.48840	3.95751	1.47993	.14751	.15375	.16202	.18058	.20321	.23404	
							1.02107	1.10579	1.21222	1.42221	1.62063	1.79740	
.147	1.9515	.8400	.52242	.48622	3.94203	1.47488	.14852	.15480	.16313	.18185	.20471	.23596	
							1.02099	1.10536	1.21130	1.42018	1.61733	1.79270	
.148	1.9477	.8406	.52480	.48405	3.92666	1.46989	.14953	.15586	.16425	.18312	.20620	.23789	
							1.02091	1.10493	1.21038	1.41816	1.61407	1.78804	
.149	1.9440	.8411	.52717	.48188	3.91141	1.46496	.15055	.15692	.16537	.18440	.20771	.23984	
							1.02083	1.10450	1.20948	1.41617	1.61083	1.78341	

TABLE VI. Specific Heat Ratio = 1.10

p/p_t	MACH	T/T_t	GAMMA	K_crit	SHOCK P_y/P_x	P_ty/P_ty	.01	.05	.1	.15	.2	.3	.4
.150	1.9403	.8416	.52953	.47973	3.89626	1.46010	.15156	.15797	.16649	.17568	.18568	.20921	.24180
							1.02075	1.10407	1.20858	1.31231	1.41419	1.60762	1.77883
.151	1.9366	.8421	.53188	.47758	3.88123	1.45530	.15257	.15903	.16762	.17687	.18695	.21072	.24377
							1.02067	1.10367	1.20769	1.31091	1.41223	1.60443	1.77428
.152	1.9329	.8426	.53423	.47545	3.86631	1.45056	.15358	.16009	.16874	.17806	.18823	.21223	.24576
							1.02059	1.10326	1.20680	1.30951	1.41028	1.60128	1.76978
.153	1.9292	.8431	.53656	.47332	3.85149	1.44588	.15459	.16115	.16986	.17926	.18951	.21374	.24775
							1.02051	1.10284	1.20593	1.30813	1.40835	1.59815	1.76531
.154	1.9256	.8436	.53889	.47120	3.83677	1.44126	.15560	.16220	.17098	.18045	.19079	.21526	.24977
							1.02043	1.10243	1.20505	1.30675	1.40643	1.59504	1.76087
.155	1.9219	.8441	.54121	.46909	3.82216	1.43669	.15661	.16326	.17210	.18165	.19207	.21678	.25180
							1.02035	1.10202	1.20419	1.30539	1.40453	1.59197	1.75647
.156	1.9183	.8446	.54352	.46699	3.80765	1.43218	.15762	.16432	.17322	.18284	.19336	.21830	.25384
							1.02027	1.10162	1.20333	1.30404	1.40264	1.58891	1.75211
.157	1.9148	.8451	.54582	.46490	3.79325	1.42773	.15863	.16538	.17435	.18404	.19464	.21983	.25590
							1.02020	1.10121	1.20248	1.30270	1.40077	1.58589	1.74778
.158	1.9112	.8456	.54812	.46281	3.77894	1.42333	.15964	.16643	.17547	.18524	.19593	.22136	.25798
							1.02012	1.10081	1.20164	1.30136	1.39892	1.58288	1.74348
.159	1.9076	.8461	.55040	.46074	3.76473	1.41898	.16066	.16749	.17659	.18644	.19721	.22290	.26008
							1.02004	1.10042	1.20080	1.30004	1.39708	1.57990	1.73922
.160	1.9041	.8465	.55268	.45867	3.75062	1.41469	.16167	.16855	.17772	.18764	.19850	.22443	.26219
							1.01997	1.10002	1.19996	1.29873	1.39525	1.57695	1.73498
.161	1.9006	.8470	.55495	.45661	3.73661	1.41044	.16268	.16961	.17884	.18884	.19979	.22598	.26433
							1.01989	1.09963	1.19914	1.29743	1.39344	1.57401	1.73078
.162	1.8971	.8475	.55721	.45456	3.72269	1.40625	.16369	.17067	.17997	.19004	.20108	.22752	.26649
							1.01982	1.09924	1.19832	1.29614	1.39160	1.57110	1.72661
.163	1.8936	.8480	.55947	.45252	3.70886	1.40211	.16470	.17173	.18109	.19124	.20237	.22907	.26867
							1.01975	1.09886	1.19750	1.29485	1.38981	1.56822	1.72247
.164	1.8901	.8484	.56171	.45048	3.69513	1.39801	.16571	.17279	.18222	.19244	.20366	.23062	.27087
							1.01967	1.09848	1.19670	1.29358	1.38804	1.56535	1.71836
.165	1.8867	.8489	.56395	.44845	3.68148	1.39397	.16672	.17384	.18334	.19364	.20496	.23218	.27309
							1.01960	1.09809	1.19589	1.29232	1.38628	1.56251	1.71428
.166	1.8833	.8494	.56618	.44643	3.66793	1.38997	.16773	.17490	.18447	.19484	.20625	.23374	.27535
							1.01953	1.09772	1.19509	1.29106	1.38454	1.55969	1.71023
.167	1.8798	.8498	.56840	.44442	3.65446	1.38601	.16875	.17596	.18559	.19605	.20755	.23531	.27763
							1.01946	1.09734	1.19430	1.28982	1.38281	1.55689	1.70620
.168	1.8764	.8503	.57062	.44242	3.64109	1.38211	.16976	.17702	.18672	.19725	.20885	.23688	.27994
							1.01938	1.09697	1.19352	1.28858	1.38109	1.55411	1.70222
.169	1.8731	.8508	.57282	.44042	3.62779	1.37825	.17077	.17808	.18785	.19846	.21015	.23845	.28228
							1.01931	1.09660	1.19274	1.28736	1.37938	1.55135	1.69825
.170	1.8697	.8512	.57502	.43843	3.61459	1.37443	.17178	.17914	.18898	.19967	.21145	.24003	.28466
							1.01924	1.09623	1.19196	1.28614	1.37768	1.54861	1.69430
.171	1.8663	.8517	.57722	.43645	3.60147	1.37065	.17279	.18020	.19010	.20087	.21275	.24161	.28707
							1.01917	1.09587	1.19119	1.28493	1.37600	1.54590	1.69038
.172	1.8630	.8521	.57940	.43447	3.58843	1.36692	.17380	.18126	.19123	.20208	.21405	.24320	.28952
							1.01910	1.09550	1.19043	1.28372	1.37433	1.54320	1.68648
.173	1.8597	.8526	.58158	.43251	3.57548	1.36323	.17481	.18232	.19236	.20329	.21536	.24479	.29202
							1.01903	1.09510	1.18967	1.28253	1.37267	1.54052	1.68261
.174	1.8564	.8530	.58374	.43055	3.56260	1.35959	.17583	.18338	.19349	.20450	.21666	.24639	.29456
							1.01896	1.09474	1.18891	1.28135	1.37103	1.53786	1.67876
.175	1.8531	.8535	.58591	.42859	3.54981	1.35598	.17684	.18444	.19462	.20571	.21797	.24799	.29715
							1.01890	1.09438	1.18816	1.28017	1.36939	1.53522	1.67494
.176	1.8498	.8539	.58806	.42665	3.53710	1.35242	.17785	.18550	.19575	.20692	.21928	.24960	.29980
							1.01883	1.09403	1.18742	1.27900	1.36777	1.53259	1.67113
.177	1.8465	.8543	.59020	.42471	3.52446	1.34889	.17886	.18656	.19688	.20813	.22059	.25121	.30251
							1.01876	1.09368	1.18668	1.27784	1.36615	1.52999	1.66735
.178	1.8433	.8548	.59234	.42278	3.51190	1.34540	.17987	.18763	.19801	.20934	.22190	.25283	.30528
							1.01869	1.09333	1.18595	1.27669	1.36455	1.52740	1.66359
.179	1.8401	.8552	.59447	.42085	3.49942	1.34195	.18088	.18869	.19914	.21056	.22321	.25445	.30814
							1.01863	1.09299	1.18522	1.27554	1.36296	1.52484	1.65984

110

TABLE VI. Specific Heat Ratio = 1.10

p/p_t	MACH	T/T_t	GAMMA	K_{crit}	SHOCK P_x/P_y	P_{tx}/P_{ty}	.01	.05	.1	.15	.2	.3	.4
.180	1.8368	.8557	.59660	.41893	3.48702	1.33854	.18189	.18975	.20027	.21177	.22453	.25608	.31108
							1.01856	1.09264	1.18449	1.27440	1.36138	1.52228	1.65612
.181	1.8336	.8561	.59871	.41702	3.47469	1.33517	.18291	.19081	.20141	.21299	.22585	.25771	.31412
							1.01850	1.09230	1.18377	1.27327	1.35981	1.51975	1.65241
.182	1.8304	.8565	.60082	.41512	3.46243	1.33183	.18392	.19187	.20254	.21420	.22716	.25935	.31728
							1.01843	1.09196	1.18305	1.27215	1.35825	1.51723	1.64871
.183	1.8273	.8569	.60293	.41322	3.45025	1.32853	.18493	.19293	.20367	.21542	.22848	.26100	.32058
							1.01836	1.09162	1.18234	1.27103	1.35670	1.51473	1.64503
.184	1.8241	.8574	.60502	.41133	3.43814	1.32526	.18594	.19400	.20481	.21664	.22981	.26265	.32404
							1.01830	1.09129	1.18164	1.26993	1.35516	1.51225	1.64136
.185	1.8209	.8578	.60711	.40944	3.42610	1.32203	.18695	.19506	.20594	.21786	.23113	.26431	.32774
							1.01824	1.09095	1.18093	1.26882	1.35364	1.50978	1.63772
.186	1.8178	.8582	.60919	.40756	3.41413	1.31884	.18797	.19612	.20707	.21908	.23245	.26598	.33171
							1.01817	1.09062	1.18024	1.26773	1.35212	1.50733	1.63408
.187	1.8147	.8586	.61126	.40569	3.40224	1.31567	.18898	.19718	.20821	.22030	.23378	.26765	.33607
							1.01811	1.09029	1.17954	1.26664	1.35061	1.50490	1.63044
.188	1.8116	.8590	.61333	.40382	3.39041	1.31255	.18999	.19825	.20934	.22152	.23511	.26932	.34103
							1.01804	1.08996	1.17885	1.26556	1.34911	1.50247	1.62680
.189	1.8085	.8595	.61538	.40196	3.37865	1.30945	.19100	.19931	.21048	.22274	.23644	.27101	.34715
							1.01797	1.08964	1.17817	1.26449	1.34762	1.50007	1.62317
.190	1.8054	.8599	.61744	.40011	3.36696	1.30639	.19201	.20037	.21162	.22396	.23777	.27270	.35795
							1.01792	1.08932	1.17749	1.26342	1.34614	1.49768	1.61950
.191	1.8023	.8603	.61948	.39826	3.35533	1.30336	.19302	.20144	.21275	.22519	.23910	.27440	
							1.01786	1.08899	1.17681	1.26236	1.34467	1.49531	
.192	1.7992	.8607	.62152	.39642	3.34377	1.30036	.19404	.20250	.21389	.22641	.24044	.27611	
							1.01780	1.08867	1.17614	1.26131	1.34321	1.49295	
.193	1.7962	.8611	.62355	.39459	3.33228	1.29739	.19505	.20356	.21503	.22764	.24178	.27782	
							1.01773	1.08836	1.17547	1.26026	1.34176	1.49060	
.194	1.7931	.8615	.62557	.39276	3.32085	1.29446	.19606	.20463	.21617	.22887	.24312	.27954	
							1.01767	1.08804	1.17481	1.25922	1.34032	1.48827	
.195	1.7901	.8619	.62759	.39094	3.30948	1.29155	.19707	.20569	.21731	.23010	.24446	.28127	
							1.01761	1.08773	1.17415	1.25819	1.33888	1.48595	
.196	1.7871	.8623	.62960	.38912	3.29818	1.28868	.19809	.20676	.21845	.23133	.24580	.28301	
							1.01755	1.08741	1.17349	1.25716	1.33746	1.48365	
.197	1.7841	.8627	.63160	.38731	3.28694	1.28583	.19910	.20782	.21959	.23256	.24714	.28476	
							1.01749	1.08710	1.17284	1.25614	1.33604	1.48136	
.198	1.7811	.8631	.63359	.38551	3.27577	1.28301	.20011	.20889	.22073	.23379	.24849	.28651	
							1.01743	1.08679	1.17219	1.25512	1.33463	1.47908	
.199	1.7781	.8635	.63558	.38371	3.26465	1.28023	.20112	.20995	.22187	.23502	.24984	.28828	
							1.01737	1.08649	1.17154	1.25411	1.33323	1.47681	
.200	1.7752	.8639	.63756	.38191	3.25360	1.27747	.20213	.21102	.22301	.23625	.25119	.29005	
							1.01731	1.08618	1.17090	1.25311	1.33184	1.47456	
.201	1.7722	.8643	.63954	.38013	3.24260	1.27473	.20315	.21208	.22415	.23749	.25254	.29183	
							1.01726	1.08588	1.17027	1.25211	1.33046	1.47233	
.202	1.7692	.8647	.64151	.37834	3.23167	1.27203	.20416	.21315	.22529	.23873	.25390	.29363	
							1.01720	1.08557	1.16963	1.25112	1.32908	1.47010	
.203	1.7663	.8651	.64347	.37657	3.22079	1.26935	.20517	.21421	.22644	.23996	.25526	.29543	
							1.01714	1.08527	1.16900	1.25013	1.32772	1.46789	
.204	1.7634	.8654	.64542	.37480	3.20997	1.26670	.20618	.21528	.22758	.24120	.25662	.29725	
							1.01708	1.08498	1.16838	1.24915	1.32636	1.46569	
.205	1.7605	.8658	.64737	.37303	3.19921	1.26408	.20720	.21635	.22872	.24244	.25798	.29907	
							1.01702	1.08468	1.16775	1.24818	1.32501	1.46350	
.206	1.7576	.8662	.64931	.37127	3.18851	1.26148	.20821	.21741	.22987	.24368	.25934	.30091	
							1.01697	1.08438	1.16713	1.24721	1.32366	1.46132	
.207	1.7547	.8666	.65125	.36952	3.17786	1.25891	.20922	.21848	.23102	.24492	.26071	.30276	
							1.01691	1.08409	1.16652	1.24625	1.32233	1.45915	
.208	1.7518	.8670	.65317	.36777	3.16727	1.25637	.21023	.21955	.23216	.24617	.26208	.30462	
							1.01685	1.08380	1.16590	1.24529	1.32100	1.45700	
.209	1.7489	.8674	.65510	.36603	3.15674	1.25385	.21125	.22061	.23331	.24741	.26345	.30650	
							1.01680	1.08351	1.16529	1.24434	1.31968	1.45486	

111

TABLE VI. Specific Heat Ratio = 1.10

p/P_t	MACH	T/T_t	GAMMA	K_{crit}	SHOCK P_y/P_x	P_{ty}/P_{tv}	.01	.05	.1	.15	.2	.25	.3
.210	1.7461	.8677	.65701	.36429	3.14626	1.25135	.21226	.22168	.23446	.24866	.26482	.28396	.30838
							1.01674	1.08322	1.16469	1.24339	1.31837	1.38859	1.45273
.211	1.7432	.8681	.65892	.36256	3.13583	1.24888	.21327	.22275	.23560	.24990	.26620	.28552	.31029
							1.01668	1.08293	1.16409	1.24245	1.31706	1.38690	1.45060
.212	1.7404	.8685	.66082	.36083	3.12546	1.24643	.21428	.22382	.23675	.25115	.26758	.28709	.31220
							1.01663	1.08264	1.16349	1.24151	1.31576	1.38521	1.44849
.213	1.7375	.8688	.66271	.35911	3.11514	1.24401	.21530	.22488	.23790	.25240	.26896	.28866	.31413
							1.01657	1.08236	1.16289	1.24058	1.31447	1.38353	1.44639
.214	1.7347	.8692	.66460	.35739	3.10488	1.24161	.21631	.22595	.23905	.25365	.27034	.29024	.31608
							1.01652	1.08208	1.16230	1.23965	1.31319	1.38187	1.44430
.215	1.7319	.8696	.66648	.35568	3.09466	1.23924	.21732	.22702	.24020	.25490	.27173	.29183	.31804
							1.01646	1.08179	1.16171	1.23873	1.31191	1.38021	1.44222
.216	1.7291	.8700	.66836	.35398	3.08450	1.23688	.21834	.22809	.24135	.25616	.27312	.29342	.32002
							1.01641	1.08151	1.16112	1.23781	1.31064	1.37856	1.44015
.217	1.7263	.8703	.67023	.35227	3.07439	1.23455	.21935	.22916	.24250	.25741	.27451	.29502	.32202
							1.01636	1.08124	1.16054	1.23690	1.30938	1.37691	1.43809
.218	1.7235	.8707	.67209	.35058	3.06433	1.23224	.22036	.23023	.24366	.25867	.27591	.29662	.32403
							1.01630	1.08096	1.15996	1.23599	1.30812	1.37528	1.43604
.219	1.7207	.8710	.67395	.34889	3.05432	1.22996	.22137	.23130	.24481	.25993	.27730	.29823	.32607
							1.01625	1.08068	1.15938	1.23509	1.30687	1.37366	1.43400
.220	1.7180	.8714	.67580	.34720	3.04436	1.22770	.22239	.23237	.24596	.26118	.27870	.29985	.32812
							1.01620	1.08041	1.15881	1.23419	1.30563	1.37204	1.43197
.221	1.7152	.8718	.67764	.34552	3.03445	1.22545	.22340	.23344	.24712	.26244	.28011	.30147	.33020
							1.01614	1.08014	1.15823	1.23330	1.30439	1.37043	1.42994
.222	1.7125	.8721	.67948	.34385	3.02458	1.22323	.22441	.23451	.24827	.26371	.28151	.30310	.33230
							1.01609	1.07986	1.15767	1.23241	1.30316	1.36883	1.42793
.223	1.7097	.8725	.68131	.34218	3.01477	1.22104	.22543	.23558	.24943	.26497	.28292	.30473	.33443
							1.01604	1.07959	1.15710	1.23153	1.30194	1.36724	1.42592
.224	1.7070	.8728	.68313	.34051	3.00500	1.21886	.22644	.23665	.25059	.26623	.28433	.30638	.33657
							1.01599	1.07933	1.15654	1.23065	1.30072	1.36565	1.42393
.225	1.7043	.8732	.68495	.33885	2.99528	1.21670	.22745	.23772	.25174	.26750	.28575	.30803	.33875
							1.01593	1.07906	1.15598	1.22978	1.29951	1.36407	1.42193
.226	1.7016	.8735	.68676	.33719	2.98561	1.21456	.22847	.23879	.25290	.26877	.28717	.30968	.34096
							1.01588	1.07879	1.15542	1.22891	1.29830	1.36251	1.41995
.227	1.6989	.8739	.68857	.33554	2.97599	1.21245	.22948	.23987	.25406	.27004	.28859	.31135	.34319
							1.01583	1.07853	1.15487	1.22804	1.29710	1.36094	1.41798
.228	1.6962	.8742	.69037	.33390	2.96641	1.21035	.23050	.24094	.25522	.27131	.29001	.31302	.34546
							1.01579	1.07826	1.15432	1.22718	1.29591	1.35939	1.41601
.229	1.6935	.8746	.69216	.33225	2.95688	1.20827	.23151	.24201	.25638	.27258	.29144	.31470	.34776
							1.01574	1.07800	1.15377	1.22632	1.29472	1.35784	1.41405
.230	1.6908	.8749	.69394	.33062	2.94739	1.20621	.23252	.24308	.25754	.27386	.29287	.31639	.35010
							1.01569	1.07774	1.15322	1.22547	1.29354	1.35630	1.41210
.231	1.6882	.8753	.69572	.32898	2.93795	1.20417	.23354	.24416	.25870	.27513	.29431	.31809	.35248
							1.01564	1.07748	1.15268	1.22462	1.29236	1.35477	1.41015
.232	1.6855	.8756	.69750	.32736	2.92855	1.20215	.23455	.24523	.25987	.27641	.29575	.31980	.35491
							1.01559	1.07722	1.15214	1.22378	1.29119	1.35322	1.40821
.233	1.6828	.8760	.69927	.32573	2.91919	1.20015	.23556	.24630	.26103	.27769	.29719	.32151	.35738
							1.01554	1.07696	1.15160	1.22294	1.29003	1.35170	1.40627
.234	1.6802	.8763	.70103	.32412	2.90988	1.19817	.23658	.24738	.26220	.27897	.29864	.32324	.35991
							1.01550	1.07671	1.15107	1.22210	1.28887	1.35019	1.40434
.235	1.6776	.8766	.70278	.32250	2.90061	1.19621	.23759	.24845	.26336	.28025	.30009	.32498	.36249
							1.01545	1.07645	1.15053	1.22127	1.28771	1.34869	1.40242
.236	1.6749	.8770	.70453	.32089	2.89139	1.19426	.23861	.24953	.26453	.28154	.30154	.32672	.36513
							1.01540	1.07620	1.15000	1.22044	1.28657	1.34720	1.40050
.237	1.6723	.8773	.70628	.31929	2.88221	1.19234	.23962	.25060	.26569	.28282	.30300	.32848	.36785
							1.01535	1.07595	1.14948	1.21962	1.28542	1.34570	1.39859
.238	1.6697	.8777	.70802	.31769	2.87307	1.19042	.24063	.25168	.26686	.28411	.30446	.33025	.37064
							1.01530	1.07570	1.14895	1.21880	1.28428	1.34422	1.39667
.239	1.6671	.8780	.70975	.31609	2.86397	1.18853	.24165	.25275	.26803	.28540	.30593	.33202	.37354
							1.01525	1.07545	1.14843	1.21798	1.28315	1.34274	1.39478

TABLE VI. Specific Heat Ratio = 1.10

p/p_t	MACH	T/T_t	GAMMA	K_{crit}	SHOCK p_y/p_x	p_{ty}/p_{tx}	.01	.05	.1	.15	.2	.25	.3
.240	1.6645	.8783	.71147	.31450	2.85491	1.18666	.24266	.25383	.26920	.28669	.30740	.33381	.37653
							1.01520	1.07520	1.14791	1.21717	1.28203	1.34126	1.39287
.241	1.6619	.8787	.71319	.31292	2.84590	1.18480	.24368	.25490	.27037	.28799	.30887	.33561	.37964
							1.01516	1.07495	1.14739	1.21636	1.28090	1.33980	1.39097
.242	1.6593	.8790	.71490	.31134	2.83692	1.18296	.24469	.25598	.27154	.28928	.31035	.33743	.38290
							1.01511	1.07470	1.14688	1.21556	1.27979	1.33834	1.38908
.243	1.6568	.8793	.71661	.30976	2.82799	1.18114	.24570	.25706	.27271	.29058	.31183	.33925	.38633
							1.01506	1.07446	1.14636	1.21476	1.27868	1.33688	1.38718
.244	1.6542	.8796	.71831	.30819	2.81909	1.17933	.24672	.25813	.27388	.29188	.31332	.34109	.38997
							1.01501	1.07421	1.14585	1.21396	1.27757	1.33543	1.38528
.245	1.6517	.8800	.72001	.30662	2.81024	1.17754	.24773	.25921	.27506	.29318	.31481	.34295	.39390
							1.01497	1.07397	1.14534	1.21317	1.27647	1.33399	1.38338
.246	1.6491	.8803	.72170	.30505	2.80142	1.17576	.24875	.26029	.27623	.29449	.31631	.34482	.39822
							1.01492	1.07373	1.14484	1.21238	1.27537	1.33255	1.38148
.247	1.6466	.8806	.72338	.30349	2.79265	1.17400	.24976	.26137	.27741	.29579	.31781	.34670	.40315
							1.01487	1.07349	1.14434	1.21159	1.27428	1.33111	1.37958
.248	1.6440	.8809	.72506	.30194	2.78391	1.17226	.25078	.26245	.27859	.29710	.31932	.34860	.40912
							1.01483	1.07325	1.14383	1.21081	1.27319	1.32969	1.37766
.249	1.6415	.8813	.72673	.30039	2.77521	1.17053	.25179	.26353	.27976	.29841	.32083	.35052	.41824
							1.01478	1.07301	1.14334	1.21003	1.27211	1.32826	1.37573
.250	1.6390	.8816	.72840	.29884	2.76655	1.16882	.25280	.26460	.28094	.29973	.32235	.35245	
							1.01474	1.07277	1.14284	1.20925	1.27103	1.32685	
.251	1.6365	.8819	.73005	.29730	2.75793	1.16713	.25382	.26568	.28212	.30104	.32387	.35440	
							1.01469	1.07253	1.14234	1.20848	1.26995	1.32543	
.252	1.6340	.8822	.73171	.29576	2.74934	1.16545	.25483	.26676	.28330	.30236	.32540	.35637	
							1.01464	1.07230	1.14185	1.20771	1.26888	1.32403	
.253	1.6315	.8825	.73336	.29422	2.74079	1.16378	.25585	.26784	.28448	.30368	.32693	.35836	
							1.01460	1.07206	1.14136	1.20695	1.26782	1.32262	
.254	1.6290	.8829	.73500	.29269	2.73228	1.16213	.25686	.26892	.28567	.30500	.32847	.36037	
							1.01455	1.07183	1.14087	1.20619	1.26676	1.32122	
.255	1.6265	.8832	.73664	.29117	2.72380	1.16049	.25788	.27001	.28685	.30632	.33002	.36240	
							1.01451	1.07160	1.14039	1.20543	1.26570	1.31983	
.256	1.6240	.8835	.73827	.28964	2.71536	1.15887	.25889	.27109	.28804	.30765	.33157	.36446	
							1.01446	1.07137	1.13991	1.20467	1.26465	1.31844	
.257	1.6215	.8838	.73989	.28813	2.70696	1.15727	.25991	.27217	.28922	.30898	.33313	.36654	
							1.01442	1.07113	1.13942	1.20392	1.26360	1.31705	
.258	1.6191	.8841	.74151	.28661	2.69859	1.15567	.26092	.27325	.29041	.31031	.33469	.36864	
							1.01438	1.07090	1.13894	1.20317	1.26256	1.31567	
.259	1.6166	.8844	.74312	.28510	2.69026	1.15409	.26194	.27433	.29160	.31165	.33626	.37077	
							1.01433	1.07068	1.13847	1.20242	1.26152	1.31429	
.260	1.6142	.8847	.74473	.28360	2.68196	1.15253	.26295	.27542	.29279	.31298	.33784	.37294	
							1.01429	1.07045	1.13799	1.20168	1.26048	1.31292	
.261	1.6117	.8850	.74633	.28210	2.67370	1.15098	.26397	.27650	.29398	.31432	.33942	.37513	
							1.01424	1.07022	1.13752	1.20094	1.25945	1.31155	
.262	1.6093	.8854	.74793	.28060	2.66547	1.14944	.26498	.27758	.29517	.31567	.34102	.37736	
							1.01420	1.07000	1.13705	1.20020	1.25842	1.31018	
.263	1.6068	.8857	.74952	.27910	2.65727	1.14792	.26600	.27867	.29636	.31701	.34261	.37962	
							1.01416	1.06977	1.13658	1.19947	1.25740	1.30882	
.264	1.6044	.8860	.75110	.27762	2.64911	1.14641	.26702	.27975	.29755	.31836	.34422	.38192	
							1.01411	1.06955	1.13611	1.19874	1.25638	1.30746	
.265	1.6020	.8863	.75268	.27613	2.64099	1.14491	.26803	.28084	.29875	.31971	.34583	.38426	
							1.01407	1.06932	1.13565	1.19801	1.25536	1.30610	
.266	1.5996	.8866	.75425	.27465	2.63289	1.14343	.26905	.28192	.29995	.32106	.34745	.38665	
							1.01403	1.06910	1.13518	1.19729	1.25435	1.30474	
.267	1.5972	.8869	.75582	.27317	2.62483	1.14196	.27006	.28301	.30114	.32242	.34909	.38909	
							1.01398	1.06888	1.13472	1.19656	1.25335	1.30339	
.268	1.5948	.8872	.75738	.27170	2.61681	1.14050	.27108	.28410	.30234	.32378	.35073	.39158	
							1.01394	1.06866	1.13426	1.19584	1.25234	1.30204	
.269	1.5924	.8875	.75894	.27023	2.60881	1.13906	.27209	.28518	.30354	.32514	.35238	.39413	
							1.01390	1.06844	1.13380	1.19513	1.25134	1.30069	

TABLE VI. Specific Heat Ratio = 1.10

p/p_t	MACH	T/T_t	GAMMA	K_crit	SHOCK p_y/p_x	p_tx/p_ty	.01	.05	.07	.1	.15	.2	.25
.270	1.5900	.8878	.76049	.26876	2.60085	1.13763	.27311	.28627	.29335	.30474	.32651	.35403	.39674
							1.01386	1.06822	1.09470	1.13335	1.19441	1.25034	1.29934
.271	1.5876	.8881	.76204	.26730	2.59292	1.13621	.27413	.28736	.29448	.30595	.32788	.35570	.39943
							1.01382	1.06801	1.09439	1.13289	1.19370	1.24935	1.29800
.272	1.5852	.8884	.76357	.26584	2.58502	1.13480	.27514	.28844	.29561	.30715	.32925	.35737	.40221
							1.01377	1.06779	1.09408	1.13244	1.19300	1.24836	1.29666
.273	1.5829	.8887	.76511	.26439	2.57715	1.13341	.27616	.28953	.29674	.30836	.33063	.35906	.40508
							1.01373	1.06757	1.09377	1.13199	1.19229	1.24737	1.29532
.274	1.5805	.8890	.76664	.26294	2.56932	1.13203	.27717	.29062	.29787	.30956	.33201	.36075	.40806
							1.01369	1.06736	1.09347	1.13154	1.19159	1.24639	1.29398
.275	1.5781	.8893	.76816	.26149	2.56151	1.13066	.27819	.29171	.29900	.31077	.33339	.36246	.41116
							1.01365	1.06715	1.09316	1.13110	1.19089	1.24540	1.29263
.276	1.5758	.8896	.76968	.26005	2.55374	1.12930	.27921	.29280	.30013	.31198	.33478	.36417	.41442
							1.01361	1.06693	1.09286	1.13065	1.19019	1.24443	1.29129
.277	1.5734	.8898	.77119	.25861	2.54600	1.12796	.28022	.29389	.30127	.31319	.33617	.36590	.41786
							1.01357	1.06672	1.09255	1.13021	1.18950	1.24345	1.28995
.278	1.5711	.8901	.77269	.25718	2.53828	1.12662	.28124	.29498	.30240	.31440	.33756	.36764	.42153
							1.01353	1.06651	1.09225	1.12976	1.18880	1.24248	1.28860
.279	1.5688	.8904	.77419	.25575	2.53060	1.12530	.28226	.29607	.30354	.31562	.33896	.36939	.42551
							1.01349	1.06630	1.09195	1.12933	1.18811	1.24151	1.28725
.280	1.5664	.8907	.77569	.25432	2.52295	1.12399	.28327	.29717	.30467	.31683	.34036	.37116	.42991
							1.01344	1.06609	1.09165	1.12889	1.18743	1.24054	1.28589
.281	1.5641	.8910	.77718	.25290	2.51533	1.12269	.28429	.29826	.30581	.31805	.34176	.37293	.43503
							1.01340	1.06588	1.09136	1.12845	1.18674	1.23958	1.28453
.282	1.5618	.8913	.77866	.25148	2.50773	1.12140	.28531	.29935	.30695	.31927	.34317	.37472	.44137
							1.01336	1.06567	1.09106	1.12802	1.18606	1.23862	1.28316
.283	1.5595	.8916	.78014	.25006	2.50017	1.12012	.28632	.30045	.30809	.32049	.34459	.37653	.45320
							1.01332	1.06546	1.09076	1.12758	1.18538	1.23766	1.28177
.284	1.5572	.8919	.78161	.24865	2.49264	1.11886	.28734	.30154	.30923	.32171	.34600	.37835	
							1.01328	1.06526	1.09047	1.12715	1.18470	1.23670	
.285	1.5549	.8922	.78308	.24724	2.48513	1.11760	.28836	.30263	.31037	.32293	.34743	.38018	
							1.01324	1.06505	1.09018	1.12672	1.18403	1.23575	
.286	1.5526	.8924	.78454	.24583	2.47765	1.11636	.28937	.30373	.31151	.32416	.34885	.38204	
							1.01320	1.06485	1.08988	1.12629	1.18335	1.23480	
.287	1.5503	.8927	.78600	.24443	2.47021	1.11513	.29039	.30482	.31265	.32538	.35028	.38391	
							1.01316	1.06464	1.08959	1.12587	1.18268	1.23385	
.288	1.5480	.8930	.78745	.24304	2.46278	1.11390	.29141	.30592	.31380	.32661	.35172	.38579	
							1.01313	1.06444	1.08930	1.12544	1.18201	1.23290	
.289	1.5457	.8933	.78890	.24164	2.45539	1.11269	.29243	.30702	.31494	.32784	.35316	.38770	
							1.01309	1.06424	1.08901	1.12502	1.18135	1.23196	
.290	1.5434	.8936	.79034	.24025	2.44803	1.11149	.29344	.30812	.31609	.32907	.35460	.38962	
							1.01305	1.06403	1.08872	1.12460	1.18068	1.23102	
.291	1.5412	.8938	.79177	.23887	2.44069	1.11030	.29446	.30921	.31723	.33031	.35605	.39157	
							1.01301	1.06383	1.08844	1.12417	1.18002	1.23008	
.292	1.5389	.8941	.79320	.23748	2.43338	1.10912	.29548	.31031	.31838	.33154	.35751	.39353	
							1.01297	1.06363	1.08815	1.12376	1.17936	1.22914	
.293	1.5366	.8944	.79462	.23610	2.42610	1.10795	.29650	.31141	.31953	.33278	.35897	.39552	
							1.01293	1.06343	1.08787	1.12334	1.17871	1.22820	
.294	1.5344	.8947	.79604	.23473	2.41885	1.10679	.29751	.31251	.32068	.33402	.36043	.39754	
							1.01289	1.06323	1.08758	1.12292	1.17805	1.22727	
.295	1.5321	.8950	.79746	.23336	2.41162	1.10564	.29853	.31361	.32183	.33526	.36190	.39958	
							1.01285	1.06304	1.08730	1.12251	1.17740	1.22634	
.296	1.5299	.8952	.79886	.23199	2.40442	1.10450	.29955	.31471	.32298	.33650	.36338	.40164	
							1.01282	1.06284	1.08702	1.12209	1.17675	1.22541	
.297	1.5277	.8955	.80027	.23062	2.39724	1.10337	.30057	.31581	.32413	.33774	.36486	.40374	
							1.01278	1.06264	1.08674	1.12168	1.17610	1.22448	
.298	1.5254	.8958	.80166	.22926	2.39009	1.10225	.30159	.31691	.32529	.33899	.36635	.40586	
							1.01274	1.06244	1.08647	1.12127	1.17545	1.22355	
.299	1.5232	.8961	.80306	.22790	2.38297	1.10114	.30260	.31802	.32644	.34024	.36784	.40802	
							1.01270	1.06225	1.08619	1.12086	1.17481	1.22263	

TABLE VI. Specific Heat Ratio = 1.10

p/p_t	MACH	T/T_t	GAMMA	K_{crit}	SHOCK P_y/P_x	P_{ty}/P_{tx}	.01	.05	.07	.09	.1	.15	.2
.300	1.5210	.8963	.80444	.22655	2.37587	1.10004	.30362	.31912	.32760	.33668	.34149	.36934	.41022
							1.01267	1.06205	1.08591	1.10912	1.12046	1.17416	1.22170
.301	1.5187	.8966	.80582	.22520	2.36880	1.09895	.30464	.32023	.32875	.33790	.34274	.37085	.41245
							1.01263	1.06186	1.08564	1.10875	1.12005	1.17352	1.22078
.302	1.5165	.8969	.80720	.22385	2.36176	1.09787	.30566	.32133	.32991	.33912	.34400	.37237	.41472
							1.01259	1.06167	1.08536	1.10839	1.11964	1.17289	1.21985
.303	1.5143	.8971	.80857	.22251	2.35474	1.09680	.30668	.32244	.33107	.34034	.34525	.37389	.417C4
							1.01255	1.06148	1.08509	1.10803	1.11924	1.17225	1.21893
.304	1.5121	.8974	.80994	.22117	2.34774	1.09573	.30770	.32354	.33223	.34156	.34651	.37542	.41941
							1.01252	1.06128	1.08481	1.10768	1.11884	1.17162	1.218C1
.305	1.5099	.8977	.81130	.21983	2.34077	1.09468	.30872	.32465	.33339	.34278	.34778	.37695	.42183
							1.01248	1.06109	1.08454	1.10732	1.11844	1.17098	1.21709
.306	1.5077	.8979	.81265	.21850	2.33383	1.09363	.30973	.32576	.33455	.34401	.34904	.37849	.42430
							1.01244	1.06090	1.08427	1.10696	1.11804	1.17035	1.21617
.307	1.5055	.8982	.81400	.21717	2.32691	1.09260	.31075	.32686	.33571	.34524	.35030	.38004	.42685
							1.01241	1.06071	1.08400	1.10661	1.11764	1.16972	1.21525
.308	1.5033	.8985	.81534	.21585	2.32001	1.09157	.31177	.32797	.33688	.34647	.35157	.38160	.42948
							1.01237	1.06052	1.08373	1.10625	1.11724	1.16909	1.21433
.309	1.5012	.8987	.81668	.21452	2.31314	1.09055	.31279	.32908	.33804	.34770	.35284	.38316	.43218
							1.01233	1.06033	1.08346	1.10590	1.11685	1.16847	1.21341
.310	1.4990	.8990	.81802	.21321	2.30630	1.08955	.31381	.33019	.33921	.34894	.35412	.38474	.43497
							1.01230	1.06015	1.08319	1.10555	1.11646	1.16784	1.21249
.311	1.4968	.8993	.81935	.21189	2.29947	1.08855	.31483	.33130	.34038	.35017	.35539	.38632	.43787
							1.01226	1.05996	1.08292	1.10520	1.11606	1.16722	1.21157
.312	1.4946	.8995	.82067	.21058	2.29267	1.08755	.31585	.33242	.34155	.35141	.35667	.38791	.44090
							1.01222	1.05977	1.08266	1.10485	1.11567	1.16660	1.21065
.313	1.4925	.8998	.82199	.20927	2.28590	1.08657	.31687	.33353	.34272	.35265	.35795	.38951	.44407
							1.01219	1.05959	1.08239	1.10450	1.11528	1.16598	1.20972
.314	1.4903	.9000	.82330	.20797	2.27915	1.08560	.31789	.33464	.34389	.35389	.35924	.39112	.44743
							1.01215	1.05940	1.08213	1.10416	1.11489	1.16536	1.20880
.315	1.4881	.9003	.82461	.20667	2.27242	1.08463	.31891	.33576	.34506	.35514	.36052	.39273	.45102
							1.01212	1.05922	1.08186	1.10381	1.11450	1.16474	1.20786
.316	1.4860	.9006	.82591	.20537	2.26572	1.08368	.31993	.33687	.34624	.35638	.36181	.39436	.45490
							1.01208	1.05903	1.08160	1.10347	1.11412	1.16412	1.20693
.317	1.4838	.9008	.82721	.20407	2.25903	1.08273	.32095	.33799	.34741	.35763	.36310	.39600	.45922
							1.01205	1.05885	1.08134	1.10312	1.11373	1.16351	1.20599
.318	1.4817	.9011	.82850	.20278	2.25237	1.08179	.32197	.33910	.34859	.35888	.36440	.39765	.46416
							1.01201	1.05867	1.08108	1.10278	1.11335	1.16290	1.20505
.319	1.4796	.9013	.82979	.20149	2.24574	1.08085	.32299	.34022	.34977	.36014	.36570	.39931	.47026
							1.01198	1.05848	1.08082	1.10244	1.11296	1.16228	1.20409
.320	1.4774	.9016	.83107	.20021	2.23913	1.07993	.32401	.34134	.35095	.36139	.36700	.40098	.48029
							1.01194	1.05830	1.08056	1.10210	1.11258	1.16167	1.20312
.321	1.4753	.9019	.83235	.19893	2.23254	1.07901	.32503	.34246	.35213	.36265	.36830	.40267	
							1.01191	1.05812	1.08030	1.10176	1.11220	1.16106	
.322	1.4732	.9021	.83362	.19765	2.22597	1.07811	.32605	.34358	.35332	.36391	.36961	.40437	
							1.01187	1.05794	1.08004	1.10142	1.11182	1.16046	
.323	1.4711	.9024	.83488	.19638	2.21942	1.07721	.32707	.34470	.35450	.36518	.37092	.40608	
							1.01184	1.05776	1.07979	1.10109	1.11144	1.15985	
.324	1.4689	.9026	.83615	.19511	2.21290	1.07631	.32810	.34582	.35569	.36644	.37223	.40780	
							1.01180	1.05758	1.07953	1.10075	1.11106	1.15924	
.325	1.4668	.9029	.83740	.19384	2.20639	1.07543	.32912	.34694	.35688	.36771	.37355	.40954	
							1.01177	1.05740	1.07927	1.10042	1.11069	1.15864	
.326	1.4647	.9031	.83865	.19258	2.19991	1.07455	.33014	.34807	.35806	.36898	.37487	.41129	
							1.01173	1.05722	1.07902	1.10008	1.11031	1.15804	
.327	1.4626	.9034	.83990	.19132	2.19345	1.07368	.33116	.34919	.35926	.37026	.37619	.41306	
							1.01170	1.05705	1.07877	1.09975	1.10994	1.15743	
.328	1.4605	.9036	.84114	.19006	2.18702	1.07282	.33218	.35032	.36045	.37153	.37752	.41485	
							1.01166	1.05687	1.07851	1.09942	1.10957	1.15683	
.329	1.4584	.9039	.84238	.18881	2.18060	1.07197	.33320	.35144	.36164	.37281	.37885	.41665	
							1.01163	1.05669	1.07826	1.09908	1.10919	1.15623	

Header note: R_1 and TPR versus Loss Coefficient K

TABLE VI. Specific Heat Ratio = 1.10

p/p_t	MACH	T/T_t	GAMMA	K_{crit}	SHOCK P_y/P_x	P_{tx}/P_{ty}	.01	.03	.05	.07	.09	.1	.15
.330	1.4563	.9041	.84361	.18756	2.17421	1.07112	.33422	.34308	.35257	.36284	.37409	.38018	.41847
							1.01160	1.03437	1.05652	1.07801	1.09875	1.10882	1.15563
.331	1.4542	.9044	.84483	.18631	2.16783	1.07029	.33525	.34415	.35370	.36404	.37538	.38152	.42031
							1.01156	1.03426	1.05634	1.07776	1.09843	1.10845	1.15503
.332	1.4521	.9046	.84605	.18507	2.16148	1.06945	.33627	.34522	.35483	.36524	.37667	.38286	.42216
							1.01153	1.03416	1.05617	1.07751	1.09810	1.10808	1.15444
.333	1.4500	.9049	.84727	.18383	2.15515	1.06863	.33729	.34630	.35596	.36644	.37796	.38421	.42404
							1.01150	1.03406	1.05599	1.07726	1.09777	1.10771	1.15384
.334	1.4480	.9051	.84848	.18259	2.14884	1.06781	.33831	.34737	.35709	.36764	.37925	.38556	.42594
							1.01146	1.03395	1.05582	1.07701	1.09744	1.10735	1.15324
.335	1.4459	.9054	.84968	.18136	2.14255	1.06701	.33934	.34844	.35823	.36885	.38055	.38691	.42787
							1.01143	1.03385	1.05565	1.07676	1.09712	1.10698	1.15265
.336	1.4438	.9056	.85089	.18013	2.13628	1.06620	.34036	.34952	.35936	.37006	.38185	.38827	.42982
							1.01140	1.03375	1.05547	1.07652	1.09679	1.10661	1.15205
.337	1.4418	.9059	.85208	.17891	2.13003	1.06541	.34138	.35059	.36049	.37126	.38315	.38963	.43179
							1.01136	1.03365	1.05530	1.07627	1.09647	1.10625	1.15145
.338	1.4397	.9061	.85327	.17768	2.12380	1.06462	.34240	.35166	.36163	.37248	.38446	.39100	.43379
							1.01133	1.03355	1.05513	1.07603	1.09615	1.10589	1.15086
.339	1.4376	.9063	.85446	.17646	2.11759	1.06384	.34343	.35274	.36277	.37369	.38577	.39237	.43582
							1.01130	1.03345	1.05496	1.07578	1.09582	1.10552	1.15027
.340	1.4356	.9066	.85564	.17525	2.11140	1.06307	.34445	.35382	.36391	.37490	.38709	.39375	.43789
							1.01126	1.03335	1.05479	1.07554	1.09550	1.10516	1.14967
.341	1.4335	.9068	.85681	.17403	2.10523	1.06230	.34547	.35489	.36504	.37612	.38840	.39513	.43999
							1.01123	1.03325	1.05462	1.07529	1.09518	1.10480	1.14908
.342	1.4315	.9071	.85798	.17283	2.09908	1.06154	.34650	.35597	.36619	.37734	.38973	.39651	.44212
							1.01120	1.03315	1.05445	1.07505	1.09486	1.10444	1.14848
.343	1.4294	.9073	.85915	.17162	2.09295	1.06079	.34752	.35705	.36733	.37856	.39105	.39790	.44429
							1.01117	1.03305	1.05428	1.07481	1.09454	1.10408	1.14789
.344	1.4274	.9075	.86031	.17042	2.08684	1.06004	.34855	.35812	.36847	.37979	.39238	.39930	.44651
							1.01113	1.03295	1.05411	1.07457	1.09422	1.10372	1.14729
.345	1.4253	.9078	.86146	.16922	2.08075	1.05930	.34957	.35920	.36961	.38101	.39371	.40070	.44877
							1.01110	1.03285	1.05394	1.07431	1.09391	1.10337	1.14670
.346	1.4233	.9080	.86261	.16802	2.07468	1.05857	.35059	.36028	.37076	.38224	.39505	.40211	.45108
							1.01107	1.03275	1.05377	1.07409	1.09359	1.10301	1.14610
.347	1.4213	.9083	.86376	.16683	2.06862	1.05784	.35162	.36136	.37191	.38347	.39639	.40352	.45345
							1.01104	1.03265	1.05361	1.07383	1.09328	1.10265	1.14551
.348	1.4193	.9085	.86490	.16564	2.06259	1.05712	.35264	.36244	.37306	.38470	.39774	.40494	.45590
							1.01100	1.03256	1.05344	1.07359	1.09296	1.10230	1.14492
.349	1.4172	.9087	.86603	.16446	2.05657	1.05641	.35367	.36352	.37420	.38594	.39909	.40636	.45840
							1.01099	1.03246	1.05327	1.07335	1.09265	1.10194	1.14432
.350	1.4152	.9090	.86716	.16327	2.05058	1.05570	.35469	.36461	.37536	.38718	.40044	.40779	.46099
							1.01096	1.03236	1.05311	1.07311	1.09233	1.10159	1.14373
.351	1.4132	.9092	.86829	.16209	2.04460	1.05500	.35572	.36569	.37651	.38842	.40180	.40923	.46366
							1.01091	1.03226	1.05294	1.07288	1.09202	1.10124	1.14313
.352	1.4112	.9094	.86941	.16092	2.03864	1.05431	.35674	.36677	.37766	.38966	.40317	.41067	.46643
							1.01090	1.03217	1.05278	1.07264	1.09171	1.10088	1.14253
.353	1.4092	.9097	.87052	.15975	2.03270	1.05362	.35777	.36786	.37882	.39091	.40453	.41212	.46932
							1.01087	1.03207	1.05261	1.07241	1.09140	1.10053	1.14193
.354	1.4072	.9099	.87163	.15858	2.02678	1.05294	.35879	.36894	.37997	.39216	.40591	.41358	.47236
							1.01084	1.03197	1.05245	1.07217	1.09109	1.10018	1.14133
.355	1.4052	.9101	.87274	.15741	2.02087	1.05227	.35982	.37003	.38113	.39341	.40729	.41504	.47556
							1.01080	1.03188	1.05229	1.07194	1.09078	1.09983	1.14072
.356	1.4032	.9104	.87384	.15625	2.01499	1.05160	.36085	.37111	.38229	.39466	.40867	.41651	.47898
							1.01077	1.03178	1.05212	1.07170	1.09047	1.09948	1.14011
.357	1.4012	.9106	.87494	.15509	2.00912	1.05093	.36187	.37220	.38345	.39592	.41006	.41799	.48267
							1.01074	1.03169	1.05196	1.07147	1.09016	1.09913	1.13950
.358	1.3992	.9108	.87603	.15394	2.00327	1.05028	.36290	.37329	.38462	.39718	.41146	.41948	.48676
							1.01071	1.03159	1.05180	1.07124	1.08985	1.09879	1.13888
.359	1.3972	.9111	.87711	.15278	1.99743	1.04963	.36393	.37437	.38578	.39844	.41286	.42097	.49139
							1.01068	1.03150	1.05164	1.07101	1.08954	1.09844	1.13826

TABLE VI. Specific Heat Ratio = 1.10

p/p_t	MACH	T/T_t	GAMMA	K_crit	SHOCK p_y/p_x	p_tx/p_ty	.01	.03	.05	.07	.09	.1	.15
.360	1.3952	.9113	.87819	.15163	1.99162	1.04898	.36495	.37546	.38695	.39971	.41426	.42249	.49698
							1.01065	1.03140	1.05148	1.07077	1.08924	1.09810	1.13763
.361	1.3932	.9115	.87927	.15049	1.98582	1.04834	.36598	.37655	.38811	.40098	.41568	.42400	.50503
							1.01062	1.03131	1.05132	1.07054	1.08893	1.09775	1.13699
.362	1.3912	.9118	.88034	.14935	1.98004	1.04771	.36701	.37764	.38928	.40225	.41710	.42552	
							1.01059	1.03121	1.05116	1.07031	1.08862	1.09741	
.363	1.3892	.9120	.88141	.14821	1.97428	1.04709	.36804	.37873	.39045	.40353	.41852	.42705	
							1.01056	1.03112	1.05100	1.07008	1.08832	1.09706	
.364	1.3873	.9122	.88247	.14707	1.96854	1.04646	.36906	.37983	.39163	.40481	.41995	.42859	
							1.01052	1.03103	1.05084	1.06985	1.08801	1.09672	
.365	1.3853	.9124	.88352	.14594	1.96281	1.04585	.37009	.38092	.39280	.40609	.42139	.43014	
							1.01049	1.03093	1.05068	1.06962	1.08771	1.09637	
.366	1.3833	.9127	.88458	.14481	1.95710	1.04524	.37112	.38201	.39398	.40738	.42284	.43170	
							1.01046	1.03084	1.05052	1.06940	1.08741	1.09603	
.367	1.3814	.9129	.88562	.14369	1.95140	1.04464	.37215	.38311	.39516	.40867	.42429	.43327	
							1.01043	1.03075	1.05036	1.06917	1.08710	1.09569	
.368	1.3794	.9131	.88666	.14257	1.94573	1.04404	.37318	.38420	.39634	.40996	.42575	.43485	
							1.01040	1.03066	1.05020	1.06894	1.08680	1.09534	
.369	1.3774	.9134	.88770	.14145	1.94007	1.04345	.37421	.38530	.39752	.41126	.42722	.43644	
							1.01037	1.03056	1.05004	1.06871	1.08650	1.09500	
.370	1.3755	.9136	.88873	.14033	1.93442	1.04286	.37523	.38640	.39870	.41256	.42870	.43805	
							1.01034	1.03047	1.04989	1.06849	1.08620	1.09466	
.371	1.3735	.9138	.88976	.13922	1.92880	1.04228	.37626	.38750	.39989	.41386	.43018	.43967	
							1.01031	1.03038	1.04973	1.06826	1.08590	1.09432	
.372	1.3716	.9140	.89078	.13811	1.92319	1.04171	.37729	.38859	.40108	.41517	.43167	.44130	
							1.01028	1.03029	1.04957	1.06804	1.08560	1.09398	
.373	1.3696	.9142	.89180	.13701	1.91759	1.04114	.37832	.38969	.40227	.41649	.43318	.44294	
							1.01025	1.03020	1.04942	1.06781	1.08530	1.09364	
.374	1.3677	.9145	.89281	.13591	1.91201	1.04057	.37935	.39080	.40346	.41781	.43469	.44460	
							1.01022	1.03011	1.04926	1.06759	1.08500	1.09329	
.375	1.3657	.9147	.89382	.13481	1.90645	1.04001	.38038	.39190	.40465	.41913	.43621	.44628	
							1.01019	1.03002	1.04911	1.06736	1.08470	1.09295	
.376	1.3638	.9149	.89482	.13372	1.90091	1.03946	.38141	.39300	.40585	.42046	.43774	.44797	
							1.01016	1.02993	1.04895	1.06714	1.08440	1.09261	
.377	1.3619	.9151	.89582	.13262	1.89538	1.03891	.38245	.39410	.40705	.42179	.43928	.44967	
							1.01013	1.02984	1.04880	1.06692	1.08410	1.09227	
.378	1.3599	.9154	.89681	.13154	1.88987	1.03837	.38348	.39521	.40825	.42313	.44083	.45140	
							1.01011	1.02975	1.04864	1.06669	1.08380	1.09193	
.379	1.3580	.9156	.89780	.13045	1.88437	1.03783	.38451	.39632	.40946	.42447	.44240	.45314	
							1.01008	1.02966	1.04849	1.06647	1.08350	1.09159	
.380	1.3561	.9158	.89878	.12937	1.87889	1.03730	.38554	.39742	.41066	.42581	.44397	.45490	
							1.01005	1.02957	1.04834	1.06625	1.08321	1.09126	
.381	1.3541	.9160	.89976	.12829	1.87342	1.03677	.38657	.39853	.41187	.42717	.44556	.45669	
							1.01002	1.02948	1.04818	1.06603	1.08291	1.09092	
.382	1.3522	.9162	.90074	.12722	1.86797	1.03625	.38760	.39964	.41308	.42852	.44716	.45849	
							1.00999	1.02939	1.04803	1.06581	1.08261	1.09058	
.383	1.3503	.9165	.90170	.12615	1.86254	1.03573	.38864	.40075	.41430	.42989	.44877	.46032	
							1.00996	1.02930	1.04788	1.06559	1.08231	1.09024	
.384	1.3484	.9167	.90267	.12508	1.85712	1.03522	.38967	.40186	.41551	.43126	.45040	.46217	
							1.00993	1.02921	1.04773	1.06537	1.08202	1.08990	
.385	1.3465	.9169	.90363	.12402	1.85172	1.03472	.39070	.40298	.41673	.43263	.45204	.46405	
							1.00990	1.02912	1.04757	1.06515	1.08172	1.08956	
.386	1.3446	.9171	.90458	.12296	1.84633	1.03421	.39174	.40409	.41795	.43401	.45369	.46596	
							1.00987	1.02903	1.04742	1.06493	1.08142	1.08922	
.387	1.3427	.9173	.90553	.12190	1.84095	1.03372	.39277	.40521	.41918	.43540	.45537	.46789	
							1.00984	1.02895	1.04727	1.06471	1.08113	1.08888	
.388	1.3408	.9175	.90648	.12085	1.83560	1.03323	.39380	.40632	.42041	.43679	.45706	.46986	
							1.00982	1.02886	1.04712	1.06449	1.08083	1.08854	
.389	1.3388	.9177	.90742	.11980	1.83025	1.03274	.39484	.40744	.42164	.43819	.45876	.47186	
							1.00979	1.02877	1.04697	1.06427	1.08054	1.08820	

117

TABLE VI. Specific Heat Ratio = 1.10

p/p_t	MACH	T/T_t	GAMMA	K_{crit}	SHOCK P_y/P_x	P_{ty}/P_{ty}	.01	.03	.05	.07	.08	.09	.1
.390	1.3369	.9180	.90835	.11875	1.82493	1.03226	.39587	.40856	.42287	.43960	.44934	.46049	.47390
							1.00976	1.02868	1.04682	1.06405	1.07230	1.08024	1.08786
.391	1.3350	.9182	.90928	.11771	1.81961	1.03178	.39691	.40968	.42411	.44102	.45089	.46224	.47598
							1.00973	1.02860	1.04667	1.06383	1.07204	1.07995	1.08751
.392	1.3332	.9184	.91021	.11667	1.81431	1.03131	.39794	.41080	.42535	.44244	.45245	.46400	.47811
							1.00970	1.02851	1.04652	1.06361	1.07179	1.07965	1.08717
.393	1.3313	.9186	.91113	.11564	1.80903	1.03084	.39898	.41192	.42659	.44387	.45402	.46579	.48028
							1.00967	1.02842	1.04637	1.06340	1.07153	1.07936	1.08683
.394	1.3294	.9188	.91204	.11460	1.80376	1.03038	.40002	.41305	.42784	.44530	.45561	.46760	.48250
							1.00965	1.02834	1.04622	1.06318	1.07128	1.07906	1.08649
.395	1.3275	.9190	.91296	.11357	1.79851	1.02992	.40105	.41417	.42909	.44675	.45720	.46944	.48479
							1.00962	1.02825	1.04607	1.06296	1.07102	1.07876	1.08614
.396	1.3256	.9192	.91386	.11255	1.79327	1.02947	.40209	.41530	.43035	.44820	.45882	.47130	.48714
							1.00959	1.02816	1.04593	1.06275	1.07077	1.07847	1.08580
.397	1.3237	.9194	.91476	.11153	1.78804	1.02902	.40313	.41643	.43160	.44967	.46045	.47319	.48956
							1.00956	1.02808	1.04578	1.06253	1.07051	1.07817	1.08545
.398	1.3218	.9197	.91566	.11051	1.78283	1.02858	.40416	.41756	.43287	.45114	.46209	.47511	.49207
							1.00953	1.02799	1.04563	1.06231	1.07026	1.07787	1.08510
.399	1.3200	.9199	.91655	.10949	1.77763	1.02814	.40520	.41869	.43413	.45262	.46375	.47707	.49467
							1.00951	1.02791	1.04548	1.06210	1.07000	1.07758	1.08475
.400	1.3181	.9201	.91744	.10848	1.77245	1.02771	.40624	.41983	.43540	.45412	.46543	.47906	.49738
							1.00948	1.02782	1.04533	1.06188	1.06975	1.07728	1.08440
.401	1.3162	.9203	.91832	.10748	1.76728	1.02728	.40728	.42096	.43667	.45562	.46713	.48108	.50023
							1.00945	1.02774	1.04519	1.06166	1.06950	1.07698	1.08405
.402	1.3143	.9205	.91920	.10647	1.76213	1.02685	.40832	.42210	.43795	.45713	.46884	.48315	.50326
							1.00942	1.02765	1.04504	1.06145	1.06924	1.07668	1.08370
.403	1.3125	.9207	.92007	.10547	1.75699	1.02643	.40936	.42324	.43923	.45866	.47058	.48527	.50664
							1.00939	1.02757	1.04489	1.06123	1.06899	1.07638	1.08335
.404	1.3106	.9209	.92094	.10447	1.75186	1.02601	.41040	.42438	.44052	.46019	.47233	.48743	.50995
							1.00937	1.02748	1.04475	1.06102	1.06873	1.07608	1.08299
.405	1.3087	.9211	.92181	.10348	1.74675	1.02560	.41144	.42552	.44181	.46174	.47411	.48966	.51378
							1.00934	1.02740	1.04460	1.06080	1.06848	1.07578	1.08262
.406	1.3069	.9213	.92266	.10249	1.74165	1.02519	.41248	.42666	.44311	.46330	.47592	.49194	.51811
							1.00931	1.02731	1.04446	1.06059	1.06822	1.07548	1.08225
.407	1.3050	.9215	.92352	.10150	1.73656	1.02479	.41352	.42781	.44441	.46488	.47775	.49429	.52335
							1.00929	1.02723	1.04431	1.06037	1.06797	1.07518	1.08188
.408	1.3032	.9217	.92437	.10052	1.73149	1.02439	.41456	.42896	.44571	.46647	.47961	.49672	.53061
							1.00926	1.02715	1.04416	1.06016	1.06771	1.07487	1.08150
.409	1.3013	.9219	.92521	.09954	1.72643	1.02400	.41560	.43011	.44702	.46807	.48150	.49924	
							1.00923	1.02706	1.04402	1.05994	1.06746	1.07457	
.410	1.2995	.9221	.92605	.09857	1.72139	1.02361	.41664	.43126	.44834	.46969	.48342	.50186	
							1.00920	1.02698	1.04387	1.05973	1.06720	1.07426	
.411	1.2976	.9223	.92689	.09760	1.71636	1.02322	.41769	.43241	.44966	.47133	.48538	.50460	
							1.00918	1.02690	1.04373	1.05951	1.06694	1.07395	
.412	1.2958	.9226	.92772	.09663	1.71134	1.02284	.41873	.43357	.45099	.47298	.48737	.50752	
							1.00915	1.02681	1.04358	1.05930	1.06669	1.07365	
.413	1.2939	.9228	.92854	.09566	1.70634	1.02246	.41978	.43472	.45232	.47465	.48941	.51059	
							1.00912	1.02673	1.04344	1.05908	1.06643	1.07334	
.414	1.2921	.9230	.92936	.09470	1.70134	1.02209	.42082	.43588	.45366	.47634	.49149	.51390	
							1.00910	1.02665	1.04330	1.05886	1.06617	1.07302	
.415	1.2902	.9232	.93018	.09374	1.69637	1.02172	.42186	.43705	.45501	.47805	.49362	.51750	
							1.00907	1.02656	1.04315	1.05865	1.06591	1.07271	
.416	1.2884	.9234	.93099	.09279	1.69140	1.02136	.42291	.43821	.45636	.47978	.49581	.52153	
							1.00904	1.02648	1.04301	1.05843	1.06565	1.07238	
.417	1.2866	.9236	.93180	.09184	1.68645	1.02100	.42396	.43938	.45772	.48154	.49806	.52622	
							1.00902	1.02640	1.04286	1.05822	1.06539	1.07206	
.418	1.2847	.9238	.93260	.09090	1.68151	1.02064	.42500	.44055	.45909	.48332	.50037	.53231	
							1.00899	1.02632	1.04272	1.05800	1.06513	1.07173	
.419	1.2829	.9240	.93340	.08995	1.67658	1.02029	.42605	.44172	.46046	.48514	.50277		
							1.00896	1.02623	1.04258	1.05779	1.06487		

TABLE VI. Specific Heat Ratio = 1.10

					SHOCK		R₂ and TPR versus Loss Coefficient K						
p/p_t	MACH	T/T_t	GAMMA	K_{crit}	P_y/P_x	P_{tx}/P_{ty}	.01	.03	.04	.05	.06	.07	.08
.420	1.2811	.9242	.93419	.08901	1.67167	1.01994	.42710	.44289	.45186	.46184	.47327	.48698	.50525
							1.00894	1.02615	1.03442	1.04243	1.05016	1.05757	1.06460
.421	1.2792	.9244	.93498	.08808	1.66677	1.01959	.42815	.44407	.45313	.46323	.47483	.48885	.50784
							1.00891	1.02607	1.03431	1.04229	1.04999	1.05736	1.06434
.422	1.2774	.9246	.93576	.08715	1.66188	1.01925	.42919	.44525	.45440	.46463	.47641	.49075	.51057
							1.00888	1.02599	1.03420	1.04215	1.04981	1.05714	1.06408
.423	1.2756	.9248	.93654	.08622	1.65700	1.01892	.43024	.44643	.45567	.46603	.47801	.49268	.51343
							1.00886	1.02591	1.03409	1.04200	1.04963	1.05692	1.06381
.424	1.2738	.9250	.93731	.08530	1.65214	1.01858	.43129	.44762	.45695	.46744	.47962	.49466	.51648
							1.00883	1.02583	1.03398	1.04186	1.04945	1.05670	1.06354
.425	1.2720	.9252	.93808	.08438	1.64729	1.01826	.43234	.44880	.45824	.46887	.48125	.49668	.51975
							1.00881	1.02575	1.03387	1.04172	1.04928	1.05649	1.06326
.426	1.2701	.9254	.93884	.08346	1.64245	1.01793	.43340	.44999	.45953	.47030	.48290	.49874	.52334
							1.00878	1.02566	1.03375	1.04158	1.04910	1.05627	1.06299
.427	1.2683	.9256	.93960	.08255	1.63763	1.01761	.43445	.45119	.46083	.47174	.48457	.50086	.52737
							1.00875	1.02558	1.03364	1.04143	1.04892	1.05605	1.06271
.428	1.2665	.9258	.94036	.08164	1.63281	1.01729	.43550	.45239	.46213	.47319	.48627	.50303	.53209
							1.00873	1.02550	1.03353	1.04129	1.04874	1.05582	1.06242
.429	1.2647	.9259	.94111	.08074	1.62801	1.01698	.43655	.45359	.46344	.47466	.48798	.50527	.53838
							1.00870	1.02542	1.03342	1.04115	1.04856	1.05560	1.06214
.430	1.2629	.9261	.94185	.07983	1.62322	1.01667	.43761	.45479	.46476	.47613	.48972	.50758	
							1.00868	1.02534	1.03331	1.04100	1.04838	1.05538	
.431	1.2611	.9263	.94259	.07894	1.61845	1.01636	.43866	.45600	.46608	.47762	.49148	.50997	
							1.00865	1.02526	1.03320	1.04086	1.04820	1.05516	
.432	1.2593	.9265	.94333	.07804	1.61368	1.01606	.43972	.45721	.46741	.47912	.49327	.51246	
							1.00863	1.02518	1.03309	1.04072	1.04802	1.05493	
.433	1.2575	.9267	.94406	.07715	1.60893	1.01576	.44077	.45842	.46874	.48063	.49509	.51508	
							1.00860	1.02510	1.03298	1.04058	1.04784	1.05471	
.434	1.2557	.9269	.94478	.07627	1.60419	1.01547	.44183	.45964	.47008	.48216	.49694	.51782	
							1.00857	1.02502	1.03287	1.04043	1.04766	1.05448	
.435	1.2539	.9271	.94551	.07539	1.59946	1.01518	.44289	.46086	.47143	.48370	.49883	.52072	
							1.00855	1.02494	1.03276	1.04029	1.04748	1.05425	
.436	1.2521	.9273	.94622	.07451	1.59474	1.01489	.44395	.46209	.47279	.48525	.50075	.52383	
							1.00852	1.02486	1.03265	1.04015	1.04730	1.05402	
.437	1.2503	.9275	.94693	.07364	1.59004	1.01460	.44500	.46331	.47415	.48682	.50271	.52721	
							1.00850	1.02478	1.03254	1.04001	1.04712	1.05379	
.438	1.2485	.9277	.94764	.07277	1.58534	1.01432	.44606	.46455	.47552	.48841	.50472	.53096	
							1.00847	1.02470	1.03243	1.03986	1.04694	1.05355	
.439	1.2467	.9279	.94834	.07190	1.58066	1.01405	.44712	.46579	.47690	.49002	.50678	.53527	
							1.00845	1.02462	1.03232	1.03972	1.04676	1.05331	
.440	1.2449	.9281	.94904	.07104	1.57599	1.01377	.44819	.46703	.47829	.49164	.50890	.54068	
							1.00842	1.02454	1.03221	1.03958	1.04657	1.05306	
.441	1.2431	.9283	.94974	.07018	1.57133	1.01350	.44925	.46828	.47969	.49329	.51108	.54932	
							1.00840	1.02446	1.03210	1.03943	1.04639	1.05281	
.442	1.2413	.9285	.95042	.06933	1.56669	1.01324	.45031	.46953	.48110	.49495	.51332		
							1.00837	1.02438	1.03199	1.03929	1.04620		
.443	1.2396	.9287	.95111	.06848	1.56205	1.01297	.45138	.47079	.48252	.49665	.51565		
							1.00835	1.02431	1.03189	1.03915	1.04602		
.444	1.2378	.9288	.95179	.06763	1.55743	1.01272	.45244	.47205	.48395	.49836	.51807		
							1.00832	1.02423	1.03178	1.03900	1.04583		
.445	1.2360	.9290	.95246	.06679	1.55281	1.01246	.45351	.47332	.48539	.50010	.52062		
							1.00830	1.02415	1.03167	1.03886	1.04564		
.446	1.2342	.9292	.95313	.06595	1.54821	1.01221	.45458	.47459	.48684	.50187	.52328		
							1.00827	1.02407	1.03156	1.03871	1.04546		
.447	1.2324	.9294	.95380	.06512	1.54362	1.01196	.45564	.47587	.48830	.50367	.52611		
							1.00825	1.02399	1.03145	1.03857	1.04526		
.448	1.2307	.9296	.95446	.06429	1.53904	1.01171	.45671	.47715	.48978	.50551	.52913		
							1.00822	1.02391	1.03134	1.03842	1.04507		
.449	1.2289	.9298	.95511	.06346	1.53447	1.01147	.45778	.47845	.49127	.50738	.53241		
							1.00820	1.02383	1.03123	1.03828	1.04488		

119

TABLE VI. Specific Heat Ratio = 1.10

p/p_t	MACH	T/T_t	GAMMA	K_{crit}	P_y/P_x	P_{ty}/P_{ty}	.005	.01	.02	.03	.04	.05	.06
					SHOCK				**R₂ and TPR versus Loss Coefficient K**				
.450	1.2271	.9300	.95576	.06264	1.52992	1.01123	.45433	.45886	.46866	.47975	.49278	.50929	.53604
							1.00412	1.00818	1.01609	1.02376	1.03112	1.03813	1.04468
.451	1.2254	.9302	.95641	.06182	1.52537	1.01099	.45536	.45994	.46983	.48105	.49430	.51125	.54021
							1.00410	1.00815	1.01604	1.02368	1.03101	1.03798	1.04447
.452	1.2236	.9304	.95705	.06101	1.52083	1.01076	.45640	.46101	.47100	.48236	.49583	.51325	.54544
							1.00409	1.00813	1.01599	1.02360	1.03090	1.03784	1.04427
.453	1.2218	.9305	.95769	.06020	1.51631	1.01053	.45743	.46209	.47218	.48368	.49739	.51531	.55359
							1.00408	1.00810	1.01594	1.02352	1.03079	1.03769	1.04406
.454	1.2201	.9307	.95832	.05940	1.51180	1.01030	.45847	.46316	.47336	.48501	.49896	.51743	
							1.00407	1.00808	1.01589	1.02344	1.03068	1.03754	
.455	1.2183	.9309	.95895	.05860	1.50730	1.01008	.45951	.46424	.47454	.48634	.50055	.51962	
							1.00405	1.00806	1.01584	1.02336	1.03057	1.03739	
.456	1.2165	.9311	.95957	.05780	1.50280	1.00986	.46054	.46532	.47572	.48768	.50216	.52188	
							1.00404	1.00803	1.01579	1.02328	1.03046	1.03724	
.457	1.2148	.9313	.96019	.05701	1.49832	1.00964	.46158	.46640	.47691	.48904	.50379	.52427	
							1.00403	1.00801	1.01574	1.02321	1.03035	1.03709	
.458	1.2130	.9315	.96080	.05622	1.49385	1.00943	.46262	.46748	.47811	.49040	.50545	.52674	
							1.00402	1.00798	1.01569	1.02313	1.03024	1.03694	
.459	1.2113	.9317	.96141	.05544	1.48939	1.00922	.46366	.46856	.47930	.49177	.50713	.52934	
							1.00401	1.00796	1.01564	1.02305	1.03013	1.03679	
.460	1.2095	.9318	.96201	.05466	1.48494	1.00901	.46470	.46965	.48051	.49315	.50884	.53210	
							1.00400	1.00793	1.01559	1.02297	1.03001	1.03663	
.461	1.2078	.9320	.96261	.05388	1.48051	1.00881	.46574	.47073	.48171	.49454	.51058	.53506	
							1.00398	1.00791	1.01554	1.02289	1.02990	1.03647	
.462	1.2060	.9322	.96321	.05311	1.47608	1.00861	.46678	.47182	.48292	.49594	.51235	.53828	
							1.00397	1.00789	1.01549	1.02281	1.02979	1.03631	
.463	1.2043	.9324	.96380	.05234	1.47166	1.00841	.46782	.47291	.48414	.49735	.51416	.54186	
							1.00396	1.00786	1.01544	1.02274	1.02968	1.03615	
.464	1.2025	.9326	.96438	.05158	1.46725	1.00821	.46886	.47400	.48536	.49878	.51600	.54606	
							1.00395	1.00784	1.01539	1.02266	1.02956	1.03599	
.465	1.2008	.9328	.96496	.05082	1.46286	1.00802	.46990	.47509	.48658	.50021	.51789	.55132	
							1.00394	1.00781	1.01534	1.02258	1.02945	1.03582	
.466	1.1990	.9329	.96554	.05007	1.45847	1.00783	.47095	.47619	.48781	.50166	.51982	.56053	
							1.00392	1.00779	1.01529	1.02250	1.02934	1.03564	
.467	1.1973	.9331	.96611	.04932	1.45409	1.00764	.47199	.47728	.48904	.50313	.52181		
							1.00391	1.00777	1.01524	1.02242	1.02922		
.468	1.1955	.9333	.96668	.04858	1.44973	1.00746	.47303	.47838	.49028	.50461	.52385		
							1.00390	1.00774	1.01519	1.02234	1.02911		
.469	1.1938	.9335	.96724	.04784	1.44537	1.00728	.47408	.47948	.49153	.50611	.52596		
							1.00389	1.00772	1.01514	1.02226	1.02899		
.470	1.1920	.9337	.96780	.04710	1.44103	1.00710	.47513	.48058	.49278	.50762	.52817		
							1.00388	1.00769	1.01509	1.02218	1.02888		
.471	1.1903	.9338	.96835	.04637	1.43669	1.00693	.47617	.48169	.49404	.50916	.53045		
							1.00387	1.00767	1.01505	1.02210	1.02876		
.472	1.1886	.9340	.96890	.04565	1.43237	1.00675	.47722	.48279	.49531	.51071	.53283		
							1.00385	1.00765	1.01500	1.02203	1.02864		
.473	1.1868	.9342	.96944	.04492	1.42805	1.00658	.47827	.48390	.49658	.51228	.53534		
							1.00384	1.00762	1.01495	1.02195	1.02853		
.474	1.1851	.9344	.96998	.04421	1.42374	1.00642	.47932	.48501	.49786	.51388	.53801		
							1.00383	1.00760	1.01490	1.02187	1.02840		
.475	1.1834	.9346	.97051	.04349	1.41945	1.00625	.48037	.48612	.49915	.51550	.54087		
							1.00382	1.00758	1.01485	1.02179	1.02828		
.476	1.1816	.9347	.97104	.04279	1.41516	1.00609	.48142	.48724	.50044	.51715	.54399		
							1.00381	1.00755	1.01480	1.02171	1.02816		
.477	1.1799	.9349	.97156	.04208	1.41089	1.00593	.48248	.48835	.50174	.51883	.54747		
							1.00380	1.00753	1.01475	1.02162	1.02803		
.478	1.1782	.9351	.97208	.04138	1.40662	1.00578	.48353	.48947	.50306	.52053	.55158		
							1.00378	1.00751	1.01470	1.02154	1.02791		
.479	1.1765	.9353	.97260	.04069	1.40236	1.00562	.48459	.49059	.50438	.52228	.55678		
							1.00377	1.00748	1.01465	1.02146	1.02777		

TABLE VI. Specific Heat Ratio = 1.10

p/p_t	MACH	T/T_t	GAMMA	K_crit	SHOCK p_y/p_x	p_tx/p_ty	R₁ and TPR versus Loss Coefficient K .001	.002	.003	.005	.01	.015	.02
.480	1.1747	.9355	.97311	.04000	1.39812	1.00547	.48110	.48221	.48334	.48564	.49172	.49835	.50570
							1.00076	1.00151	1.00227	1.00376	1.00746	1.01106	1.01460
.481	1.1730	.9356	.97361	.03931	1.39388	1.00532	.48211	.48323	.48437	.48670	.49285	.49957	.50704
							1.00076	1.00151	1.00226	1.00375	1.00744	1.01103	1.01455
.482	1.1713	.9358	.97411	.03863	1.38965	1.00518	.48312	.48425	.48541	.48776	.49398	.50079	.50839
							1.00075	1.00150	1.00225	1.00374	1.00741	1.01099	1.01449
.483	1.1696	.9360	.97461	.03796	1.38543	1.00504	.48413	.48528	.48644	.48882	.49511	.50202	.50975
							1.00075	1.00150	1.00225	1.00373	1.00739	1.01096	1.01445
.484	1.1679	.9362	.97510	.03729	1.38123	1.00490	.48514	.48630	.48748	.48988	.49625	.50326	.51113
							1.00075	1.00150	1.00224	1.00372	1.00737	1.01092	1.01439
.485	1.1661	.9363	.97558	.03662	1.37703	1.00476	.48615	.48732	.48851	.49094	.49739	.50450	.51253
							1.00075	1.00149	1.00223	1.00370	1.00734	1.01088	1.01435
.486	1.1644	.9365	.97607	.03596	1.37284	1.00462	.48717	.48835	.48955	.49200	.49853	.50575	.51393
							1.00074	1.00149	1.00223	1.00369	1.00732	1.01085	1.01430
.487	1.1627	.9367	.97654	.03531	1.36866	1.00449	.48818	.48937	.49059	.49307	.49968	.50701	.51535
							1.00074	1.00148	1.00222	1.00368	1.00730	1.01081	1.01425
.488	1.1610	.9369	.97702	.03466	1.36449	1.00436	.48919	.49040	.49162	.49413	.50083	.50827	.51678
							1.00074	1.00148	1.00221	1.00367	1.00727	1.01077	1.01420
.489	1.1593	.9370	.97748	.03401	1.36032	1.00423	.49020	.49142	.49266	.49520	.50199	.50954	.51823
							1.00074	1.00147	1.00221	1.00366	1.00725	1.01074	1.01415
.490	1.1576	.9372	.97795	.03337	1.35617	1.00411	.49121	.49245	.49370	.49627	.50315	.51082	.51970
							1.00074	1.00147	1.00220	1.00365	1.00722	1.01070	1.01410
.491	1.1559	.9374	.97840	.03273	1.35203	1.00398	.49223	.49347	.49474	.49734	.50431	.51212	.52118
							1.00073	1.00146	1.00219	1.00362	1.00720	1.01067	1.01405
.492	1.1542	.9376	.97886	.03210	1.34789	1.00386	.49324	.49450	.49578	.49841	.50548	.51341	.52269
							1.00073	1.00146	1.00219	1.00362	1.00718	1.01063	1.01400
.493	1.1525	.9377	.97931	.03147	1.34377	1.00375	.49425	.49553	.49683	.49949	.50665	.51474	.52422
							1.00073	1.00146	1.00218	1.00361	1.00716	1.01060	1.01395
.494	1.1507	.9379	.97975	.03085	1.33965	1.00363	.49527	.49656	.49787	.50056	.50782	.51606	.52577
							1.00073	1.00145	1.00217	1.00360	1.00713	1.01056	1.01390
.495	1.1490	.9381	.98019	.03024	1.33555	1.00352	.49628	.49759	.49891	.50164	.50901	.51740	.52735
							1.00073	1.00145	1.00217	1.00359	1.00711	1.01053	1.01385
.496	1.1473	.9382	.98062	.02963	1.33145	1.00341	.49730	.49862	.49996	.50272	.51019	.51874	.52896
							1.00072	1.00144	1.00216	1.00358	1.00708	1.01049	1.01380
.497	1.1456	.9384	.98105	.02902	1.32736	1.00330	.49831	.49965	.50101	.50380	.51139	.52010	.53060
							1.00072	1.00144	1.00215	1.00357	1.00706	1.01046	1.01375
.498	1.1439	.9386	.98148	.02842	1.32328	1.00319	.49933	.50068	.50205	.50488	.51259	.52147	.53227
							1.00072	1.00143	1.00215	1.00356	1.00704	1.01042	1.01370
.499	1.1422	.9388	.98190	.02782	1.31921	1.00309	.50034	.50171	.50310	.50597	.51379	.52286	.53398
							1.00072	1.00143	1.00214	1.00354	1.00701	1.01038	1.01364
.500	1.1405	.9389	.98231	.02723	1.31515	1.00298	.50136	.50274	.50415	.50706	.51500	.52426	.53574
							1.00071	1.00143	1.00213	1.00353	1.00699	1.01035	1.01359
.501	1.1388	.9391	.98272	.02665	1.31110	1.00288	.50237	.50377	.50521	.50815	.51624	.52568	.53755
							1.00071	1.00142	1.00213	1.00352	1.00697	1.01031	1.01354
.502	1.1371	.9393	.98313	.02607	1.30705	1.00279	.50339	.50481	.50626	.50925	.51746	.52712	.53941
							1.00071	1.00142	1.00212	1.00352	1.00695	1.01027	1.01349
.503	1.1354	.9394	.98353	.02549	1.30302	1.00269	.50441	.50584	.50731	.51035	.51870	.52858	.54133
							1.00071	1.00141	1.00211	1.00351	1.00693	1.01024	1.01343
.504	1.1338	.9396	.98393	.02492	1.29899	1.00260	.50542	.50688	.50837	.51145	.51994	.53006	.54333
							1.00071	1.00141	1.00211	1.00350	1.00691	1.01020	1.01338
.505	1.1321	.9398	.98432	.02436	1.29497	1.00250	.50644	.50792	.50942	.51255	.52119	.53156	.54542
							1.00070	1.00140	1.00210	1.00348	1.00688	1.01016	1.01332
.506	1.1304	.9399	.98471	.02380	1.29096	1.00242	.50746	.50895	.51048	.51366	.52245	.53309	.54761
							1.00070	1.00140	1.00209	1.00347	1.00686	1.01013	1.01327
.507	1.1287	.9401	.98509	.02325	1.28696	1.00233	.50848	.50999	.51154	.51476	.52372	.53465	.54993
							1.00070	1.00140	1.00209	1.00346	1.00684	1.01009	1.01321
.508	1.1270	.9403	.98546	.02270	1.28297	1.00224	.50950	.51103	.51261	.51588	.52500	.53624	.55241
							1.00070	1.00139	1.00208	1.00345	1.00681	1.01005	1.01315
.509	1.1253	.9405	.98584	.02216	1.27899	1.00216	.51052	.51207	.51367	.51699	.52629	.53786	.55516
							1.00070	1.00139	1.00208	1.00344	1.00679	1.01001	1.01310

TABLE VI. Specific Heat Ratio = 1.10

p/p_t	MACH	T/T_t	GAMMA	K_{crit}	SHOCK P_y/P_x	P_{ty}/P_{ty}	.001	.002	.003	.005	.01	.015	.02
.510	1.1236	.9406	.98621	.02162	1.27501	1.00208	.51154	.51312	.51473	.51811	.52760	.53952	.55818
							1.00069	1.00138	1.00207	1.00343	1.00677	1.00998	1.01304
.511	1.1219	.9408	.98657	.02109	1.27105	1.00200	.51256	.51416	.51580	.51923	.52891	.54123	.56168
							1.00069	1.00138	1.00206	1.00342	1.00674	1.00994	1.01298
.512	1.1202	.9410	.98693	.02057	1.26709	1.00192	.51358	.51520	.51687	.52036	.53024	.54298	.56603
							1.00069	1.00137	1.00206	1.00341	1.00672	1.00990	1.01292
.513	1.1186	.9411	.98728	.02005	1.26314	1.00185	.51460	.51625	.51794	.52149	.53159	.54479	.57399
							1.00069	1.00137	1.00205	1.00340	1.00670	1.00986	1.01286
.514	1.1169	.9413	.98763	.01953	1.25920	1.00177	.51563	.51730	.51902	.52262	.53295	.54667	
							1.00069	1.00137	1.00204	1.00339	1.00667	1.00982	
.515	1.1152	.9415	.98797	.01902	1.25527	1.00170	.51665	.51834	.52009	.52376	.53433	.54862	
							1.00068	1.00136	1.00204	1.00337	1.00665	1.00978	
.516	1.1135	.9416	.98831	.01852	1.25134	1.00163	.51767	.51939	.52117	.52491	.53573	.55067	
							1.00068	1.00136	1.00203	1.00336	1.00663	1.00974	
.517	1.1118	.9418	.98865	.01803	1.24742	1.00157	.51870	.52045	.52225	.52605	.53714	.55283	
							1.00068	1.00135	1.00202	1.00335	1.00660	1.00970	
.518	1.1102	.9420	.98898	.01753	1.24352	1.00150	.51972	.52150	.52334	.52721	.53858	.55512	
							1.00068	1.00135	1.00202	1.00334	1.00658	1.00966	
.519	1.1085	.9421	.98930	.01705	1.23962	1.00144	.52075	.52255	.52442	.52837	.54005	.55766	
							1.00067	1.00134	1.00201	1.00333	1.00655	1.00962	
.520	1.1068	.9423	.98962	.01657	1.23573	1.00137	.52178	.52361	.52551	.52954	.54154	.56042	
							1.00067	1.00134	1.00200	1.00332	1.00653	1.00958	
.521	1.1051	.9424	.98994	.01610	1.23184	1.00131	.52280	.52467	.52660	.53071	.54307	.56357	
							1.00067	1.00134	1.00200	1.00331	1.00651	1.00953	
.522	1.1035	.9426	.99025	.01563	1.22797	1.00125	.52383	.52573	.52770	.53189	.54463	.56739	
							1.00067	1.00133	1.00199	1.00330	1.00648	1.00949	
.523	1.1018	.9428	.99055	.01517	1.22410	1.00120	.52486	.52679	.52879	.53308	.54622	.57314	
							1.00067	1.00133	1.00198	1.00329	1.00646	1.00944	
.524	1.1001	.9429	.99086	.01471	1.22024	1.00114	.52589	.52786	.52990	.53428	.54786		
							1.00066	1.00132	1.00198	1.00328	1.00643		
.525	1.0984	.9431	.99115	.01426	1.21639	1.00109	.52693	.52893	.53101	.53549	.54956		
							1.00066	1.00132	1.00197	1.00327	1.00641		
.526	1.0968	.9433	.99144	.01382	1.21255	1.00104	.52796	.53000	.53212	.53670	.55131		
							1.00066	1.00131	1.00197	1.00325	1.00638		
.527	1.0951	.9434	.99173	.01338	1.20871	1.00099	.52900	.53107	.53324	.53793	.55313		
							1.00066	1.00131	1.00196	1.00324	1.00636		
.528	1.0934	.9436	.99201	.01295	1.20489	1.00094	.53003	.53214	.53436	.53916	.55503		
							1.00066	1.00131	1.00195	1.00323	1.00633		
.529	1.0918	.9438	.99229	.01252	1.20107	1.00089	.53106	.53322	.53549	.54041	.55703		
							1.00065	1.00130	1.00195	1.00322	1.00630		
.530	1.0901	.9439	.99256	.01210	1.19725	1.00084	.53210	.53430	.53662	.54168	.55922		
							1.00065	1.00130	1.00194	1.00321	1.00628		
.531	1.0884	.9441	.99283	.01169	1.19345	1.00080	.53314	.53539	.53775	.54295	.56155		
							1.00065	1.00129	1.00193	1.00320	1.00625		
.532	1.0868	.9442	.99309	.01128	1.18966	1.00076	.53418	.53648	.53890	.54424	.56412		
							1.00065	1.00129	1.00193	1.00319	1.00623		
.533	1.0851	.9444	.99335	.01088	1.18587	1.00072	.53522	.53757	.54005	.54555	.56705		
							1.00064	1.00128	1.00192	1.00318	1.00620		
.534	1.0834	.9446	.99360	.01049	1.18209	1.00068	.53627	.53866	.54121	.54688	.57071		
							1.00064	1.00128	1.00191	1.00316	1.00617		
.535	1.0818	.9447	.99385	.01010	1.17832	1.00064	.53731	.53976	.54237	.54823	.57619		
							1.00064	1.00128	1.00191	1.00315	1.00613		
.536	1.0801	.9449	.99409	.00972	1.17455	1.00060	.53836	.54087	.54355	.54960			
							1.00064	1.00127	1.00190	1.00314			
.537	1.0784	.9450	.99433	.00935	1.17079	1.00057	.53941	.54198	.54473	.55100			
							1.00064	1.00127	1.00189	1.00313			
.538	1.0768	.9452	.99456	.00898	1.16704	1.00053	.54046	.54309	.54592	.55243			
							1.00063	1.00126	1.00189	1.00312			
.539	1.0751	.9454	.99479	.00862	1.16330	1.00050	.54152	.54422	.54713	.55389			
							1.00063	1.00126	1.00188	1.00310			

TABLE VI. Specific Heat Ratio = 1.10

p/p_t	MACH	T/T_t	GAMMA	K_{crit}	P_y/P_x (SHOCK)	P_{ty}/P_{ty} (SHOCK)	.001	.002	.003	.004	.005	.01	.015
.540	1.0735	.9455	.99502	.00827	1.15957	1.00047	.54258	.54534	.54834	.55169	.55543		
							1.00063	1.00125	1.00187	1.00249	1.00310		
.541	1.0718	.9457	.99524	.00792	1.15584	1.00044	.54363	.54648	.54961	.55306	.55699		
							1.00063	1.00125	1.00187	1.00248	1.00309		
.542	1.0701	.9458	.99545	.00758	1.15212	1.00041	.54470	.54762	.55086	.55445	.55860		
							1.00062	1.00124	1.00186	1.00247	1.00307		
.543	1.0685	.9460	.99566	.00724	1.14841	1.00038	.54576	.54877	.55213	.55588	.56028		
							1.00062	1.00124	1.00186	1.00246	1.00306		
.544	1.0668	.9462	.99586	.00691	1.14471	1.00036	.54683	.54996	.55342	.55735	.56204		
							1.00062	1.00124	1.00185	1.00245	1.00305		
.545	1.0652	.9463	.99606	.00659	1.14101	1.00033	.54790	.55114	.55473	.55886	.56391		
							1.00062	1.00124	1.00184	1.00244	1.00304		
.546	1.0635	.9465	.99626	.00628	1.13732	1.00031	.54898	.55233	.55606	.56043	.56591		
							1.00061	1.00123	1.00184	1.00244	1.00302		
.547	1.0619	.9466	.99645	.00597	1.13364	1.00029	.55009	.55353	.55743	.56206	.56810		
							1.00062	1.00123	1.00183	1.00243	1.00301		
.548	1.0602	.9468	.99663	.00567	1.12996	1.00026	.55119	.55474	.55883	.56378	.57069		
							1.00061	1.00122	1.00182	1.00242	1.00300		
.549	1.0586	.9469	.99681	.00538	1.12629	1.00024	.55228	.55597	.56026	.56559	.57374		
							1.00061	1.00122	1.00182	1.00240	1.00298		
.550	1.0569	.9471	.99699	.00510	1.12263	1.00022	.55338	.55722	.56175	.56755	.57834		
							1.00061	1.00121	1.00181	1.00239	1.00297		
.551	1.0553	.9473	.99716	.00482	1.11898	1.00021	.55449	.55849	.56329	.56978			
							1.00061	1.00121	1.00180	1.00239			
.552	1.0536	.9474	.99732	.00455	1.11534	1.00019	.55561	.55979	.56490	.57227			
							1.00061	1.00121	1.00180	1.00238			
.553	1.0520	.9476	.99748	.00428	1.11170	1.00017	.55674	.56111	.56659	.57536			
							1.00060	1.00120	1.00179	1.00236			
.554	1.0503	.9477	.99764	.00403	1.10807	1.00016	.55787	.56247	.56840	.58065			
							1.00060	1.00120	1.00178	1.00235			
.555	1.0487	.9479	.99779	.00378	1.10444	1.00014	.55902	.56387	.57047				
							1.00060	1.00119	1.00178				
.556	1.0470	.9480	.99794	.00354	1.10082	1.00013	.56018	.56531	.57274				
							1.00060	1.00119	1.00177				
.557	1.0454	.9482	.99808	.00330	1.09721	1.00012	.56136	.56682	.57547				
							1.00060	1.00118	1.00176				
.558	1.0437	.9483	.99821	.00307	1.09361	1.00010	.56255	.56840	.57958				
							1.00059	1.00118	1.00175				
.559	1.0421	.9485	.99834	.00285	1.09001	1.00009	.56376	.57016					
							1.00059	1.00118					
.560	1.0404	.9487	.99847	.00264	1.08643	1.00008	.56499	.57202					
							1.00059	1.00117					
.561	1.0388	.9488	.99859	.00244	1.08284	1.00007	.56625	.57412					
							1.00059	1.00116					
.562	1.0371	.9490	.99871	.00224	1.07927	1.00006	.56754	.57662					
							1.00058	1.00116					
.563	1.0355	.9491	.99882	.00205	1.07570	1.00006	.56888	.58045					
							1.00058	1.00115					
.564	1.0339	.9493	.99892	.00187	1.07214	1.00005	.57034						
							1.00058						
.565	1.0322	.9494	.99902	.00170	1.06859	1.00004	.57182						
							1.00058						
.566	1.0306	.9496	.99912	.00154	1.06504	1.00004	.57341						
							1.00058						
.567	1.0289	.9497	.99921	.00138	1.06150	1.00003	.57515						
							1.00057						
.568	1.0273	.9499	.99930	.00123	1.05796	1.00003	.57713						
							1.00057						
.569	1.0257	.9500	.99938	.00109	1.05444	1.00002	.57985						
							1.00057						

TABLE VII. Specific Heat Ratio = 1.20

p/p_t	MACH	T/T_t	GAMMA	K_{crit}	SHOCK P_y/P_x	P_{ty}/P_{ty}	.05	.1	.2	.3	.4	.5	.6
.001	4.6500	.3162	.01397	1.05482	23.49757	39.56309	.00117	.00134	.00169	.00207	.00246	.00288	.00335
							1.80350	2.95777	6.46493	11.66247	18.43020	26.44930	35.26899
.002	4.2629	.3550	.02417	1.00870	19.73384	23.45076	.00230	.00260	.00323	.00389	.00459	.00535	.00619
							1.64931	2.52479	5.00189	8.43937	12.71233	17.59388	22.80665
.003	4.0413	.3798	.03323	.97790	17.72555	17.35098	.00341	.00383	.00471	.00563	.00662	.00770	.00889
							1.57125	2.31458	4.33551	7.03778	10.30242	13.94909	17.77317
.004	3.8857	.3984	.04159	.95408	16.38074	14.04624	.00452	.00506	.00616	.00734	.00860	.00997	.01151
							1.52061	2.18200	3.93133	6.20788	8.90361	11.86635	14.92850
.005	3.7660	.4135	.04946	.93435	15.38114	11.94155	.00563	.00627	.00760	.00901	.01053	.01220	.01408
							1.48381	2.08748	3.65019	5.64408	7.96543	10.48478	13.05975
.006	3.6688	.4263	.05695	.91735	14.59155	10.46979	.00673	.00747	.00901	.01066	.01244	.01440	.01661
							1.45533	2.01527	3.44058	5.22765	7.28175	9.48690	11.71911
.007	3.5866	.4374	.06412	.90231	13.94244	9.37569	.00782	.00867	.01042	.01230	.01433	.01657	.01911
							1.43211	1.95727	3.27537	4.90365	6.75536	8.72432	10.70088
.008	3.5158	.4472	.07104	.88877	13.39347	8.52639	.00892	.00986	.01182	.01392	.01619	.01871	.02159
							1.41273	1.90930	3.14058	4.64240	6.33417	8.11806	9.89558
.009	3.4534	.4561	.07774	.87641	12.91928	7.84553	.01001	.01104	.01321	.01552	.01804	.02084	.02406
							1.39622	1.86863	3.02771	4.42610	5.98738	7.62199	9.23928
.010	3.3977	.4642	.08424	.86500	12.50292	7.28594	.01110	.01223	.01459	.01712	.01988	.02296	.02650
							1.38175	1.83350	2.93127	4.24276	5.69568	7.20640	8.69183
.011	3.3474	.4716	.09057	.85438	12.13253	6.81677	.01219	.01341	.01596	.01871	.02171	.02506	.02894
							1.36903	1.80270	2.84754	4.08475	5.44558	6.85196	8.22662
.012	3.3014	.4785	.09674	.84444	11.79948	6.41699	.01328	.01459	.01733	.02029	.02352	.02715	.03137
							1.35767	1.77536	2.77389	3.94682	5.22811	6.54517	7.82527
.013	3.2592	.4849	.10278	.83507	11.49735	6.07172	.01436	.01576	.01870	.02186	.02533	.02923	.03379
							1.34744	1.75085	2.70838	3.82484	5.03699	6.27644	7.47465
.014	3.2202	.4909	.10868	.82619	11.22118	5.77009	.01545	.01694	.02006	.02343	.02713	.03131	.03621
							1.33813	1.72867	2.64969	3.71598	4.86698	6.03837	7.16507
.015	3.1838	.4966	.11447	.81776	10.96713	5.50402	.01653	.01811	.02142	.02499	.02893	.03337	.03862
							1.32962	1.70848	2.59652	3.61780	4.71491	5.82572	6.88923
.016	3.1498	.5020	.12015	.80971	10.73211	5.26733	.01761	.01928	.02277	.02655	.03071	.03544	.04104
							1.32178	1.68996	2.54809	3.52908	4.57756	5.63429	6.64158
.017	3.1178	.5071	.12573	.80201	10.51363	5.05521	.01870	.02044	.02412	.02810	.03250	.03749	.04345
							1.31453	1.67289	2.50372	3.44815	4.45270	5.46077	6.41756
.018	3.0877	.5119	.13121	.79462	10.30966	4.86387	.01978	.02161	.02547	.02965	.03427	.03955	.04586
							1.30779	1.65708	2.46309	3.37391	4.33857	5.30264	6.21378
.019	3.0592	.5166	.13660	.78751	10.11849	4.69028	.02086	.02277	.02681	.03119	.03605	.04160	.04827
							1.30149	1.64210	2.42522	3.30549	4.23371	5.15772	6.02740
.020	3.0321	.5210	.14191	.78066	9.93872	4.53198	.02194	.02394	.02815	.03273	.03782	.04365	.05069
							1.29559	1.62839	2.39004	3.24214	4.13693	5.02414	5.85611
.021	3.0064	.5253	.14714	.77404	9.76915	4.38695	.02302	.02510	.02949	.03427	.03958	.04569	.05311
							1.29003	1.61554	2.35723	3.18326	4.04724	4.90085	5.69801
.022	2.9818	.5293	.15230	.76764	9.60874	4.25352	.02410	.02626	.03083	.03580	.04135	.04773	.05553
							1.28482	1.60345	2.32653	3.12835	3.96382	4.78637	5.55150
.023	2.9584	.5333	.15738	.76143	9.45662	4.13029	.02517	.02742	.03216	.03733	.04311	.04978	.05796
							1.27988	1.59206	2.29771	3.07697	3.88596	4.67974	5.41526
.024	2.9359	.5371	.16240	.75541	9.31203	4.01609	.02625	.02858	.03349	.03886	.04487	.05182	.06040
							1.27520	1.58128	2.27059	3.02874	3.81307	4.58010	5.28816
.025	2.9143	.5407	.16735	.74956	9.17430	3.90992	.02733	.02974	.03483	.04039	.04662	.05386	.06284
							1.27075	1.57107	2.24498	2.98337	3.74464	4.48673	5.16923
.026	2.8935	.5443	.17225	.74386	9.04286	3.81093	.02841	.03089	.03616	.04192	.04838	.05590	.06529
							1.26651	1.56137	2.22076	2.94056	3.68024	4.39900	5.05762
.027	2.8736	.5477	.17708	.73832	8.91718	3.71837	.02948	.03205	.03748	.04344	.05013	.05794	.06774
							1.26248	1.55215	2.19780	2.90008	3.61936	4.31637	4.95269
.028	2.8543	.5511	.18185	.73291	8.79682	3.63162	.03056	.03320	.03881	.04496	.05188	.05999	.07021
							1.25862	1.54335	2.17599	2.86173	3.56193	4.23836	4.85374
.029	2.8357	.5543	.18658	.72753	8.68138	3.55012	.03163	.03436	.04013	.04648	.05363	.06203	.07269
							1.25493	1.53495	2.15523	2.82532	3.50751	4.16457	4.76023

TABLE VII. Specific Heat Ratio = 1.20

p/p_t	MACH	T/T_t	GAMMA	K_{crit}	SHOCK P_y/P_x	P_{tx}/P_{ty}	.05	.1	.2	.3	.4	.5	.6
.030	2.8177	.5574	.19124	.72248	8.57049	3.47339	.03271	.03551	.04146	.04800	.05538	.06407	.07517
							1.25139	1.52691	2.13543	2.79069	3.45585	4.09462	4.67169
.031	2.8003	.5605	.19586	.71744	8.46383	3.40101	.03378	.03666	.04278	.04951	.05713	.06612	.07767
							1.24800	1.51921	2.11653	2.75770	3.40671	4.02820	4.58770
.032	2.7835	.5635	.20043	.71251	8.36111	3.33259	.03486	.03782	.04410	.05103	.05888	.06817	.08019
							1.24473	1.51183	2.09845	2.72621	3.35991	3.96502	4.50788
.033	2.7671	.5664	.20495	.70768	8.26207	3.26781	.03593	.03897	.04542	.05254	.06063	.07022	.08271
							1.24159	1.50473	2.08113	2.69612	3.31525	3.90482	4.43190
.034	2.7512	.5692	.20943	.70295	8.16647	3.20638	.03700	.04012	.04674	.05406	.06237	.07227	.08525
							1.23857	1.49791	2.06453	2.66733	3.27258	3.84738	4.35946
.035	2.7358	.5719	.21386	.69831	8.07409	3.14802	.03808	.04127	.04806	.05557	.06412	.07433	.08781
							1.23565	1.49134	2.04858	2.63973	3.23176	3.79249	4.29030
.036	2.7208	.5746	.21826	.69376	7.98475	3.09251	.03915	.04242	.04938	.05708	.06587	.07639	.09039
							1.23284	1.48500	2.03325	2.61325	3.19265	3.73997	4.22417
.037	2.7062	.5773	.22260	.68929	7.89825	3.03963	.04022	.04357	.05070	.05859	.06761	.07845	.09298
							1.23011	1.47889	2.01849	2.58781	3.15514	3.68957	4.16085
.038	2.6919	.5798	.22691	.68491	7.81444	2.98920	.04130	.04472	.05201	.06010	.06936	.08052	.09559
							1.22748	1.47299	2.00427	2.56335	3.11911	3.64132	4.10017
.039	2.6781	.5823	.23118	.68060	7.73316	2.94103	.04237	.04586	.05333	.06161	.07111	.08259	.09823
							1.22493	1.46728	1.99056	2.53980	3.08448	3.59498	4.04192
.040	2.6645	.5848	.23542	.67636	7.65428	2.89497	.04344	.04701	.05464	.06312	.07286	.08466	.10089
							1.22246	1.46176	1.97732	2.51710	3.05116	3.55044	3.98594
.041	2.6513	.5872	.23961	.67219	7.57767	2.85089	.04451	.04816	.05596	.06463	.07460	.08674	.10357
							1.22007	1.45641	1.96453	2.49521	3.01905	3.50757	3.93214
.042	2.6384	.5896	.24377	.66809	7.50321	2.80865	.04558	.04931	.05727	.06614	.07635	.08882	.10628
							1.21774	1.45123	1.95216	2.47407	2.98810	3.46627	3.88031
.043	2.6258	.5919	.24790	.66405	7.43078	2.76813	.04665	.05045	.05858	.06765	.07810	.09091	.10903
							1.21548	1.44620	1.94018	2.45365	2.95823	3.42646	3.83037
.044	2.6135	.5942	.25199	.66008	7.36030	2.72923	.04773	.05160	.05989	.06915	.07985	.09301	.11180
							1.21329	1.44132	1.92859	2.43390	2.92937	3.38808	3.78218
.045	2.6014	.5964	.25604	.65616	7.29166	2.69184	.04880	.05275	.06121	.07066	.08161	.09510	.11461
							1.21116	1.43658	1.91734	2.41479	2.90148	3.35098	3.73566
.046	2.5896	.5986	.26007	.65231	7.22478	2.65588	.04987	.05389	.06252	.07217	.08336	.09721	.11746
							1.20908	1.43197	1.90644	2.39627	2.87450	3.31511	3.69069
.047	2.5780	.6007	.26406	.64850	7.15957	2.62127	.05094	.05504	.06383	.07368	.08512	.09932	.12035
							1.20706	1.42749	1.89585	2.37841	2.84838	3.28041	3.64715
.048	2.5667	.6028	.26802	.64475	7.09596	2.58792	.05201	.05618	.06514	.07518	.08687	.10144	.12329
							1.20509	1.42313	1.88557	2.36100	2.82307	3.24683	3.60506
.049	2.5556	.6049	.27196	.64105	7.03388	2.55577	.05308	.05733	.06645	.07669	.08863	.10356	.12629
							1.20317	1.41889	1.87558	2.34410	2.79853	3.21429	3.56428
.050	2.5447	.6070	.27586	.63740	6.97326	2.52475	.05415	.05847	.06776	.07820	.09039	.10570	.12934
							1.20130	1.41475	1.86586	2.32769	2.77473	3.18275	3.52476
.051	2.5340	.6090	.27973	.63380	6.91404	2.49479	.05522	.05962	.06907	.07970	.09215	.10784	.13246
							1.19947	1.41072	1.85641	2.31174	2.75162	3.15215	3.48640
.052	2.5235	.6109	.28358	.63024	6.85616	2.46585	.05629	.06076	.07038	.08121	.09391	.10998	.13567
							1.19769	1.40679	1.84720	2.29624	2.72918	3.12245	3.44915
.053	2.5132	.6129	.28739	.62672	6.79956	2.43787	.05736	.06191	.07169	.08272	.09568	.11214	.13896
							1.19595	1.40295	1.83823	2.28115	2.70736	3.09361	3.41294
.054	2.5031	.6148	.29118	.62325	6.74419	2.41080	.05843	.06305	.07300	.08423	.09744	.11430	.14236
							1.19425	1.39920	1.82949	2.26647	2.68615	3.06557	3.37773
.055	2.4932	.6167	.29495	.61982	6.69001	2.38460	.05950	.06419	.07431	.08574	.09921	.11648	.14589
							1.19258	1.39555	1.82097	2.25218	2.66551	3.03832	3.34346
.056	2.4834	.6185	.29868	.61643	6.63697	2.35922	.06057	.06534	.07562	.08725	.10098	.11866	.14959
							1.19096	1.39197	1.81266	2.23825	2.64542	3.01180	3.31008
.057	2.4738	.6204	.30239	.61308	6.58501	2.33462	.06164	.06648	.07693	.08876	.10276	.12085	.15351
							1.18936	1.38848	1.80455	2.22467	2.62586	2.98599	3.27753
.058	2.4643	.6222	.30608	.60977	6.53412	2.31077	.06271	.06762	.07824	.09027	.10453	.12305	.15773
							1.18781	1.38507	1.79663	2.21142	2.60679	2.96085	3.24579
.059	2.4550	.6239	.30974	.60650	6.48423	2.28763	.06378	.06877	.07954	.09178	.10631	.12527	.16241
							1.18628	1.38172	1.78889	2.19850	2.58821	2.93635	3.21476

TABLE VII. Specific Heat Ratio = 1.20

p/p₁	MACH	T/T₁	GAMMA	K_{crit}	SHOCK P_x/P_x	P_{tx}/P_{ty}	.05	.1	.2	.3	.4	.5	.6
.060	2.4459	.6257	.31338	.60326	6.43532	2.26517	.06485	.06991	.08085	.09329	.10809	.12749	.16797
							1.18479	1.37845	1.78133	2.18589	2.57008	2.91247	3.18438
.061	2.4369	.6274	.31699	.60005	6.38736	2.24336	.06591	.07105	.08216	.09480	.10988	.12973	.17777
							1.18332	1.37525	1.77394	2.17358	2.55239	2.88918	3.15455
.062	2.4280	.6291	.32058	.59688	6.34030	2.22217	.06698	.07219	.08347	.09631	.11166	.13198	
							1.18189	1.37212	1.76671	2.16154	2.53513	2.86646	
.063	2.4193	.6308	.32415	.59374	6.29412	2.20157	.06805	.07334	.08478	.09782	.11345	.13424	
							1.18048	1.36905	1.75964	2.14979	2.51827	2.84428	
.064	2.4107	.6325	.32769	.59063	6.24879	2.18154	.06912	.07448	.08609	.09934	.11524	.13652	
							1.17910	1.36604	1.75271	2.13829	2.50179	2.82262	
.065	2.4022	.6341	.33121	.58756	6.20427	2.16205	.07019	.07562	.08740	.10085	.11704	.13881	
							1.17775	1.36309	1.74594	2.12705	2.48570	2.80146	
.066	2.3938	.6357	.33471	.58451	6.16055	2.14309	.07126	.07676	.08871	.10237	.11884	.14112	
							1.17642	1.36019	1.73930	2.11605	2.46996	2.78077	
.067	2.3856	.6373	.33819	.58149	6.11760	2.12462	.07233	.07791	.09002	.10389	.12064	.14344	
							1.17512	1.35736	1.73280	2.10528	2.45457	2.76055	
.068	2.3775	.6389	.34165	.57850	6.07538	2.10664	.07340	.07905	.09132	.10540	.12245	.14578	
							1.17384	1.35461	1.72643	2.09474	2.43951	2.74077	
.069	2.3695	.6404	.34508	.57554	6.03388	2.08911	.07446	.08019	.09263	.10692	.12426	.14813	
							1.17258	1.35188	1.72019	2.08442	2.42477	2.72141	
.070	2.3616	.6420	.34850	.57261	5.99308	2.07203	.07553	.08133	.09394	.10844	.12607	.15051	
							1.17134	1.34920	1.71406	2.07431	2.41034	2.70247	
.071	2.3538	.6435	.35189	.56970	5.95296	2.05538	.07660	.08248	.09525	.10996	.12789	.15290	
							1.17013	1.34656	1.70806	2.06440	2.39624	2.68392	
.072	2.3461	.6450	.35527	.56682	5.91349	2.03913	.07767	.08362	.09656	.11148	.12971	.15532	
							1.16894	1.34398	1.70217	2.05469	2.38240	2.66574	
.073	2.3385	.6465	.35862	.56397	5.87465	2.02328	.07874	.08476	.09787	.11300	.13154	.15775	
							1.16776	1.34143	1.69638	2.04517	2.36884	2.64794	
.074	2.3310	.6479	.36196	.56114	5.83643	2.00780	.07981	.08590	.09918	.11453	.13337	.16022	
							1.16661	1.33894	1.69071	2.03582	2.35554	2.63049	
.075	2.3236	.6494	.36527	.55833	5.79880	1.99270	.08087	.08704	.10049	.11605	.13520	.16270	
							1.16547	1.33648	1.68513	2.02666	2.34250	2.61337	
.076	2.3162	.6508	.36857	.55555	5.76176	1.97794	.08194	.08818	.10181	.11758	.13704	.16521	
							1.16436	1.33407	1.67966	2.01767	2.32972	2.59659	
.077	2.3090	.6523	.37185	.55279	5.72528	1.96353	.08301	.08933	.10312	.11911	.13888	.16776	
							1.16326	1.33169	1.67428	2.00884	2.31722	2.58012	
.078	2.3019	.6537	.37511	.55005	5.68935	1.94944	.08408	.09047	.10443	.12063	.14073	.17033	
							1.16221	1.32936	1.66900	2.00017	2.30491	2.56396	
.079	2.2948	.6550	.37835	.54734	5.65395	1.93567	.08515	.09161	.10574	.12216	.14258	.17293	
							1.16115	1.32706	1.66380	1.99166	2.29282	2.54809	
.080	2.2878	.6564	.38158	.54464	5.61908	1.92221	.08622	.09275	.10705	.12370	.14444	.17558	
							1.16010	1.32480	1.65870	1.98330	2.28095	2.53250	
.081	2.2809	.6578	.38478	.54197	5.58470	1.90904	.08728	.09389	.10836	.12523	.14631	.17825	
							1.15907	1.32257	1.65368	1.97509	2.26930	2.51719	
.082	2.2741	.6591	.38797	.53932	5.55082	1.89615	.08835	.09504	.10968	.12676	.14817	.18098	
							1.15805	1.32038	1.64874	1.96701	2.25785	2.50216	
.083	2.2674	.6605	.39114	.53668	5.51742	1.88355	.08942	.09618	.11099	.12830	.15005	.18375	
							1.15705	1.31823	1.64389	1.95908	2.24661	2.48738	
.084	2.2607	.6618	.39430	.53407	5.48448	1.87121	.09049	.09732	.11230	.12984	.15193	.18657	
							1.15606	1.31610	1.63911	1.95127	2.23556	2.47284	
.085	2.2541	.6631	.39744	.53148	5.45200	1.85913	.09156	.09846	.11362	.13138	.15382	.18945	
							1.15509	1.31401	1.63440	1.94360	2.22469	2.45854	
.086	2.2476	.6644	.40056	.52891	5.41996	1.84730	.09263	.09961	.11493	.13292	.15571	.19239	
							1.15413	1.31195	1.62978	1.93605	2.21401	2.44446	
.087	2.2411	.6657	.40366	.52635	5.38835	1.83571	.09369	.10075	.11625	.13446	.15761	.19540	
							1.15319	1.30992	1.62522	1.92863	2.20350	2.43061	
.088	2.2347	.6669	.40675	.52382	5.35717	1.82436	.09476	.10189	.11756	.13600	.15951	.19850	
							1.15226	1.30792	1.62073	1.92132	2.19317	2.41698	
.089	2.2284	.6682	.40982	.52130	5.32639	1.81323	.09583	.10303	.11888	.13755	.16143	.20168	
							1.15134	1.30594	1.61631	1.91413	2.18301	2.40355	

TABLE VII. Specific Heat Ratio = 1.20

p/p_t	MACH	T/T_t	GAMMA	K_{crit}	SHOCK P_y/P_x	P_{tx}/P_{ty}	.01	.05	.1	.2	.3	.4	.5
.090	2.2222	.6694	.41288	.51880	5.29602	1.80233	.09135	.09690	.10418	.12020	.13910	.16334	.20498
							1.02975	1.15041	1.30400	1.61196	1.90705	2.17300	2.39031
.091	2.2160	.6707	.41592	.51631	5.26603	1.79164	.09237	.09797	.10532	.12151	.14065	.16527	.20841
							1.02958	1.14952	1.30208	1.60767	1.90008	2.16316	2.37727
.092	2.2098	.6719	.41895	.51385	5.23643	1.78116	.09338	.09903	.10646	.12283	.14220	.16720	.21200
							1.02941	1.14864	1.30019	1.60345	1.89322	2.15346	2.36440
.093	2.2038	.6731	.42196	.51140	5.20720	1.77088	.09439	.10010	.10760	.12415	.14375	.16915	.21579
							1.02925	1.14777	1.29832	1.59928	1.88646	2.14392	2.35170
.094	2.1978	.6743	.42495	.50896	5.17834	1.76079	.09541	.10117	.10875	.12547	.14531	.17110	.21987
							1.02909	1.14691	1.29648	1.59518	1.87980	2.13452	2.33918
.095	2.1918	.6755	.42793	.50654	5.14983	1.75090	.09642	.10224	.10989	.12679	.14687	.17305	.22432
							1.02893	1.14606	1.29467	1.59113	1.87324	2.12526	2.32680
.096	2.1859	.6767	.43090	.50414	5.12167	1.74119	.09743	.10331	.11103	.12811	.14843	.17502	.22941
							1.02877	1.14522	1.29288	1.58715	1.86677	2.11614	2.31455
.097	2.1801	.6778	.43385	.50176	5.09385	1.73166	.09845	.10438	.11218	.12943	.14999	.17699	.23582
							1.02862	1.14439	1.29111	1.58321	1.86040	2.10716	2.30242
.098	2.1743	.6790	.43678	.49938	5.06636	1.72231	.09946	.10544	.11332	.13075	.15155	.17898	
							1.02846	1.14358	1.28936	1.57933	1.85412	2.09830	
.099	2.1685	.6802	.43970	.49703	5.03920	1.71312	.10047	.10651	.11446	.13207	.15312	.18097	
							1.02831	1.14277	1.28764	1.57550	1.84793	2.08957	
.100	2.1629	.6813	.44261	.49469	5.01236	1.70410	.10149	.10758	.11561	.13339	.15469	.18298	
							1.02816	1.14197	1.28594	1.57173	1.84182	2.08096	
.101	2.1572	.6824	.44550	.49236	4.98582	1.69525	.10250	.10865	.11675	.13472	.15626	.18499	
							1.02801	1.14118	1.28426	1.56800	1.83580	2.07248	
.102	2.1517	.6835	.44838	.49005	4.95960	1.68654	.10351	.10972	.11789	.13604	.15783	.18701	
							1.02786	1.14041	1.28260	1.56432	1.82986	2.06411	
.103	2.1461	.6847	.45125	.48775	4.93367	1.67800	.10453	.11079	.11904	.13737	.15941	.18905	
							1.02772	1.13964	1.28096	1.56070	1.82400	2.05585	
.104	2.1406	.6858	.45410	.48546	4.90803	1.66959	.10554	.11186	.12018	.13869	.16099	.19109	
							1.02758	1.13888	1.27934	1.55711	1.81822	2.04771	
.105	2.1352	.6869	.45694	.48319	4.88268	1.66134	.10655	.11292	.12133	.14002	.16257	.19315	
							1.02743	1.13813	1.27774	1.55358	1.81251	2.03967	
.106	2.1298	.6879	.45976	.48093	4.85761	1.65322	.10756	.11399	.12247	.14134	.16415	.19522	
							1.02729	1.13739	1.27616	1.55009	1.80688	2.03174	
.107	2.1245	.6890	.46257	.47869	4.83281	1.64525	.10858	.11506	.12362	.14267	.16574	.19730	
							1.02715	1.13665	1.27460	1.54664	1.80133	2.02391	
.108	2.1192	.6901	.46537	.47645	4.80828	1.63740	.10959	.11613	.12476	.14400	.16733	.19940	
							1.02702	1.13593	1.27306	1.54323	1.79587	2.01619	
.109	2.1139	.6911	.46816	.47424	4.78402	1.62969	.11060	.11720	.12591	.14533	.16892	.20151	
							1.02688	1.13521	1.27153	1.53987	1.79045	2.00856	
.110	2.1087	.6922	.47093	.47203	4.76001	1.62210	.11162	.11827	.12705	.14666	.17052	.20363	
							1.02675	1.13450	1.27003	1.53654	1.78510	2.00103	
.111	2.1035	.6932	.47369	.46983	4.73626	1.61464	.11263	.11934	.12820	.14799	.17212	.20577	
							1.02661	1.13380	1.26854	1.53326	1.77982	1.99359	
.112	2.0984	.6943	.47643	.46765	4.71275	1.60730	.11364	.12041	.12934	.14932	.17372	.20792	
							1.02648	1.13310	1.26706	1.53002	1.77461	1.98624	
.113	2.0933	.6953	.47917	.46548	4.68949	1.60007	.11466	.12148	.13049	.15066	.17532	.21009	
							1.02635	1.13242	1.26560	1.52681	1.76945	1.97898	
.114	2.0883	.6963	.48189	.46332	4.66647	1.59297	.11567	.12254	.13164	.15199	.17693	.21228	
							1.02622	1.13174	1.26416	1.52364	1.76437	1.97181	
.115	2.0833	.6973	.48460	.46117	4.64368	1.58597	.11668	.12361	.13278	.15332	.17854	.21448	
							1.02610	1.13107	1.26274	1.52051	1.75934	1.96472	
.116	2.0783	.6984	.48729	.45904	4.62112	1.57909	.11770	.12468	.13393	.15466	.18016	.21671	
							1.02597	1.13040	1.26133	1.51742	1.75437	1.95771	
.117	2.0734	.6994	.48998	.45691	4.59879	1.57231	.11871	.12575	.13508	.15600	.18177	.21895	
							1.02584	1.12974	1.25994	1.51436	1.74946	1.95079	
.118	2.0685	.7003	.49265	.45480	4.57668	1.56563	.11972	.12682	.13622	.15733	.18340	.22121	
							1.02572	1.12909	1.25856	1.51133	1.74460	1.94394	
.119	2.0636	.7013	.49531	.45270	4.55479	1.55906	.12074	.12789	.13737	.15867	.18502	.22350	
							1.02560	1.12845	1.25719	1.50834	1.73980	1.93717	

TABLE VII. Specific Heat Ratio = 1.20

p/p_t	MACH	T/T_t	GAMMA	K_crit	SHOCK P_y/P_x	SHOCK P_tx/P_ty	.01	.05	.1	.15	.2	.3	.4
.120	2.0588	.7023	.49796	.45061	4.53311	1.55259	.12175	.12896	.13852	.14880	.16001	.18665	.22581
							1.02548	1.12784	1.25584	1.38225	1.50539	1.73506	1.93048
.121	2.0540	.7033	.50059	.44852	4.51164	1.54622	.12276	.13003	.13967	.15004	.16135	.18828	.22814
							1.02536	1.12721	1.25450	1.38015	1.50246	1.73037	1.92386
.122	2.0493	.7042	.50322	.44645	4.49038	1.53995	.12378	.13110	.14081	.15127	.16269	.18992	.23050
							1.02524	1.12658	1.25318	1.37806	1.49957	1.72573	1.91731
.123	2.0446	.7052	.50583	.44439	4.46931	1.53376	.12479	.13217	.14196	.15251	.16404	.19156	.23289
							1.02512	1.12596	1.25187	1.37600	1.49671	1.72115	1.91082
.124	2.0399	.7062	.50843	.44234	4.44845	1.52767	.12581	.13324	.14311	.15375	.16538	.19320	.23531
							1.02500	1.12535	1.25058	1.37397	1.49388	1.71661	1.90441
.125	2.0352	.7071	.51102	.44030	4.42778	1.52167	.12682	.13431	.14426	.15499	.16672	.19485	.23776
							1.02489	1.12474	1.24930	1.37195	1.49103	1.71213	1.89808
.126	2.0306	.7080	.51360	.43827	4.40731	1.51576	.12783	.13538	.14541	.15623	.16807	.19650	.24025
							1.02477	1.12414	1.24803	1.36995	1.48826	1.70769	1.89180
.127	2.0260	.7090	.51617	.43625	4.38702	1.50993	.12885	.13646	.14656	.15747	.16942	.19815	.24277
							1.02466	1.12355	1.24677	1.36797	1.48552	1.70330	1.88558
.128	2.0215	.7099	.51872	.43424	4.36692	1.50418	.12986	.13753	.14771	.15871	.17077	.19981	.24533
							1.02455	1.12296	1.24552	1.36602	1.48280	1.69896	1.87942
.129	2.0170	.7108	.52127	.43224	4.34700	1.49852	.13087	.13860	.14886	.15995	.17211	.20148	.24793
							1.02443	1.12238	1.24429	1.36408	1.48012	1.69466	1.87333
.130	2.0125	.7117	.52380	.43025	4.32727	1.49294	.13189	.13967	.15001	.16119	.17347	.20315	.25058
							1.02432	1.12180	1.24307	1.36216	1.47746	1.69040	1.86729
.131	2.0080	.7127	.52633	.42827	4.30770	1.48744	.13290	.14074	.15116	.16243	.17482	.20482	.25328
							1.02421	1.12123	1.24186	1.36027	1.47483	1.68619	1.86130
.132	2.0036	.7136	.52884	.42630	4.28831	1.48201	.13391	.14181	.15231	.16368	.17617	.20650	.25604
							1.02411	1.12066	1.24066	1.35839	1.47222	1.68203	1.85537
.133	1.9992	.7145	.53134	.42433	4.26910	1.47666	.13493	.14288	.15346	.16492	.17752	.20819	.25886
							1.02400	1.12010	1.23948	1.35652	1.46964	1.67790	1.84950
.134	1.9948	.7153	.53383	.42238	4.25004	1.47139	.13594	.14395	.15461	.16617	.17888	.20987	.26175
							1.02389	1.11954	1.23830	1.35468	1.46708	1.67382	1.84367
.135	1.9905	.7162	.53631	.42043	4.23116	1.46619	.13695	.14502	.15576	.16741	.18024	.21157	.26472
							1.02379	1.11899	1.23714	1.35285	1.46455	1.66978	1.83790
.136	1.9861	.7171	.53878	.41850	4.21244	1.46105	.13797	.14609	.15692	.16866	.18160	.21327	.26778
							1.02368	1.11844	1.23599	1.35104	1.46205	1.66577	1.83217
.137	1.9818	.7180	.54124	.41657	4.19387	1.45599	.13898	.14717	.15807	.16990	.18296	.21497	.27094
							1.02358	1.11790	1.23484	1.34925	1.45957	1.66181	1.82649
.138	1.9776	.7189	.54369	.41465	4.17547	1.45100	.13999	.14824	.15922	.17115	.18432	.21668	.27422
							1.02347	1.11736	1.23371	1.34748	1.45711	1.65788	1.82085
.139	1.9734	.7197	.54613	.41274	4.15722	1.44607	.14101	.14931	.16037	.17240	.18568	.21840	.27765
							1.02337	1.11683	1.23259	1.34572	1.45467	1.65400	1.81525
.140	1.9691	.7206	.54856	.41083	4.13912	1.44121	.14202	.15038	.16153	.17365	.18704	.22012	.28125
							1.02327	1.11630	1.23148	1.34398	1.45226	1.65015	1.80969
.141	1.9650	.7214	.55097	.40894	4.12117	1.43641	.14304	.15145	.16268	.17490	.18841	.22184	.28510
							1.02317	1.11577	1.23038	1.34225	1.44987	1.64633	1.80418
.142	1.9608	.7223	.55338	.40705	4.10337	1.43168	.14405	.15252	.16384	.17615	.18977	.22358	.28924
							1.02307	1.11525	1.22928	1.34054	1.44751	1.64255	1.79870
.143	1.9567	.7231	.55578	.40517	4.08572	1.42701	.14506	.15360	.16499	.17740	.19114	.22532	.29381
							1.02297	1.11474	1.22820	1.33884	1.44516	1.63881	1.79323
.144	1.9526	.7240	.55817	.40330	4.06820	1.42240	.14608	.15467	.16615	.17865	.19251	.22706	.29906
							1.02287	1.11423	1.22713	1.33716	1.44284	1.63510	1.78779
.145	1.9485	.7248	.56055	.40144	4.05083	1.41785	.14709	.15574	.16730	.17990	.19388	.22882	.30578
							1.02278	1.11372	1.22606	1.33550	1.44053	1.63142	1.78237
.146	1.9444	.7256	.56291	.39959	4.03360	1.41336	.14810	.15681	.16846	.18116	.19526	.23058	
							1.02268	1.11322	1.22501	1.33385	1.43825	1.62778	
.147	1.9404	.7265	.56527	.39774	4.01651	1.40893	.14912	.15789	.16961	.18241	.19663	.23234	
							1.02259	1.11268	1.22396	1.33221	1.43599	1.62417	
.148	1.9364	.7273	.56762	.39590	3.99955	1.40455	.15013	.15896	.17077	.18366	.19801	.23412	
							1.02249	1.11219	1.22292	1.33059	1.43374	1.62059	
.149	1.9324	.7281	.56996	.39407	3.98273	1.40023	.15115	.16003	.17193	.18492	.19939	.23590	
							1.02240	1.11170	1.22190	1.32898	1.43152	1.61704	

TABLE VII. Specific Heat Ratio = 1.20

p/p₁	MACH	T/T₁	GAMMA	K_crit	SHOCK P_y/P_x	SHOCK P_ty/P_ty	.01	.05	.1	.15	.2	.25	.3
.150	1.9284	.7289	.57229	.39225	3.96603	1.39596	.15216	.16111	.17308	.18618	.20077	.21750	.23768
							1.02230	1.11121	1.22088	1.32738	1.42932	1.52525	1.61352
.151	1.9245	.7297	.57461	.39043	3.94947	1.39175	.15317	.16218	.17424	.18743	.20215	.21905	.23948
							1.02221	1.11073	1.21987	1.32580	1.42713	1.52243	1.61003
.152	1.9206	.7305	.57692	.38862	3.93303	1.38758	.15419	.16325	.17540	.18869	.20353	.22059	.24128
							1.02212	1.11025	1.21886	1.32423	1.42497	1.51963	1.60657
.153	1.9167	.7313	.57922	.38682	3.91672	1.38348	.15520	.16433	.17656	.18995	.20491	.22215	.24310
							1.02203	1.10978	1.21787	1.32268	1.42282	1.51686	1.60315
.154	1.9128	.7321	.58151	.38502	3.90053	1.37942	.15622	.16540	.17772	.19121	.20630	.22370	.24492
							1.02194	1.10931	1.21688	1.32114	1.42069	1.51411	1.59975
.155	1.9090	.7329	.58379	.38324	3.88447	1.37541	.15723	.16648	.17888	.19247	.20769	.22526	.24675
							1.02185	1.10884	1.21590	1.31961	1.41858	1.51139	1.59638
.156	1.9051	.7337	.58607	.38146	3.86852	1.37145	.15824	.16755	.18004	.19374	.20908	.22683	.24859
							1.02176	1.10838	1.21493	1.31809	1.41648	1.50869	1.59303
.157	1.9013	.7345	.58833	.37968	3.85270	1.36754	.15926	.16862	.18120	.19500	.21047	.22840	.25043
							1.02167	1.10792	1.21397	1.31659	1.41441	1.50601	1.58972
.158	1.8975	.7353	.59058	.37791	3.83699	1.36367	.16027	.16970	.18236	.19626	.21187	.22997	.25229
							1.02158	1.10747	1.21301	1.31510	1.41234	1.50335	1.58643
.159	1.8937	.7360	.59283	.37615	3.82140	1.35985	.16129	.17077	.18352	.19753	.21326	.23154	.25416
							1.02149	1.10702	1.21206	1.31362	1.41030	1.50071	1.58316
.160	1.8900	.7368	.59506	.37440	3.80592	1.35608	.16230	.17185	.18468	.19880	.21466	.23312	.25604
							1.02141	1.10657	1.21112	1.31215	1.40827	1.49810	1.57992
.161	1.8863	.7376	.59729	.37266	3.79055	1.35236	.16331	.17292	.18584	.20006	.21606	.23471	.25793
							1.02132	1.10612	1.21019	1.31069	1.40626	1.49551	1.57671
.162	1.8826	.7383	.59951	.37092	3.77529	1.34867	.16433	.17400	.18701	.20133	.21746	.23630	.25983
							1.02124	1.10568	1.20926	1.30925	1.40427	1.49293	1.57352
.163	1.8789	.7391	.60172	.36918	3.76015	1.34503	.16534	.17507	.18817	.20260	.21887	.23789	.26174
							1.02115	1.10524	1.20834	1.30781	1.40229	1.49038	1.57037
.164	1.8752	.7398	.60392	.36746	3.74511	1.34144	.16636	.17615	.18933	.20387	.22027	.23949	.26367
							1.02107	1.10480	1.20743	1.30639	1.40033	1.48785	1.56724
.165	1.8715	.7406	.60611	.36573	3.73018	1.33788	.16737	.17723	.19050	.20514	.22168	.24109	.26560
							1.02098	1.10437	1.20653	1.30498	1.39838	1.48533	1.56412
.166	1.8679	.7413	.60829	.36402	3.71535	1.33437	.16839	.17830	.19166	.20641	.22309	.24270	.26755
							1.02090	1.10394	1.20563	1.30357	1.39645	1.48284	1.56103
.167	1.8643	.7421	.61047	.36231	3.70063	1.33090	.16940	.17938	.19283	.20769	.22451	.24431	.26952
							1.02082	1.10352	1.20474	1.30218	1.39453	1.48036	1.55796
.168	1.8607	.7428	.61263	.36061	3.68601	1.32746	.17041	.18045	.19399	.20896	.22592	.24593	.27149
							1.02074	1.10309	1.20385	1.30080	1.39262	1.47791	1.55491
.169	1.8571	.7436	.61479	.35892	3.67149	1.32407	.17143	.18153	.19516	.21024	.22734	.24755	.27348
							1.02065	1.10267	1.20297	1.29943	1.39074	1.47547	1.55189
.170	1.8535	.7443	.61694	.35723	3.65707	1.32072	.17244	.18261	.19633	.21151	.22876	.24918	.27549
							1.02057	1.10226	1.20210	1.29807	1.38886	1.47305	1.54888
.171	1.8500	.7450	.61908	.35555	3.64275	1.31740	.17346	.18368	.19749	.21279	.23018	.25081	.27751
							1.02049	1.10184	1.20123	1.29673	1.38700	1.47065	1.54590
.172	1.8465	.7457	.62121	.35387	3.62852	1.31412	.17447	.18476	.19866	.21407	.23161	.25245	.27955
							1.02041	1.10143	1.20037	1.29539	1.38515	1.46827	1.54294
.173	1.8430	.7465	.62333	.35220	3.61440	1.31088	.17549	.18584	.19983	.21535	.23303	.25409	.28160
							1.02034	1.10102	1.19952	1.29406	1.38332	1.46590	1.54000
.174	1.8395	.7472	.62545	.35053	3.60037	1.30768	.17650	.18692	.20100	.21663	.23446	.25574	.28368
							1.02026	1.10062	1.19867	1.29274	1.38150	1.46355	1.53707
.175	1.8360	.7479	.62755	.34887	3.58643	1.30451	.17752	.18799	.20217	.21792	.23590	.25739	.28577
							1.02018	1.10022	1.19783	1.29142	1.37969	1.46122	1.53417
.176	1.8325	.7486	.62965	.34722	3.57258	1.30138	.17853	.18907	.20334	.21920	.23733	.25905	.28788
							1.02010	1.09982	1.19699	1.29012	1.37790	1.45890	1.53129
.177	1.8291	.7493	.63174	.34557	3.55883	1.29828	.17954	.19015	.20451	.22049	.23877	.26072	.29001
							1.02002	1.09942	1.19616	1.28883	1.37612	1.45660	1.52842
.178	1.8257	.7500	.63382	.34393	3.54516	1.29521	.18056	.19123	.20568	.22177	.24021	.26239	.29217
							1.01997	1.09902	1.19534	1.28755	1.37435	1.45432	1.52558
.179	1.8223	.7507	.63589	.34230	3.53159	1.29218	.18158	.19231	.20685	.22306	.24165	.26407	.29434
							1.01989	1.09863	1.19452	1.28627	1.37259	1.45205	1.52275

TABLE VII. Specific Heat Ratio = 1.20

p/p_t	MACH	T/T_t	GAMMA	K_crit	SHOCK p_y/p_x	p_ty/p_ty	.01	.05	.1	.15	.2	.25	.3
.180	1.8189	.7514	.63796	.34067	3.51810	1.28919	.18259	.19339	.20803	.22435	.24310	.26576	.29654
							1.01982	1.09824	1.19371	1.28501	1.37085	1.44980	1.51994
.181	1.8155	.7521	.64002	.33904	3.50470	1.28622	.18361	.19447	.20920	.22564	.24455	.26745	.29877
							1.01974	1.09786	1.19290	1.28375	1.36912	1.44756	1.51715
.182	1.8121	.7528	.64206	.33742	3.49139	1.28329	.18462	.19554	.21037	.22693	.24600	.26915	.30103
							1.01967	1.09747	1.19210	1.28251	1.36740	1.44534	1.51437
.183	1.8088	.7535	.64410	.33581	3.47816	1.28039	.18564	.19662	.21155	.22823	.24745	.27086	.30331
							1.01960	1.09709	1.19130	1.28127	1.36569	1.44311	1.51161
.184	1.8054	.7542	.64614	.33420	3.46502	1.27752	.18665	.19770	.21272	.22952	.24891	.27258	.30562
							1.01952	1.09671	1.19051	1.28004	1.36399	1.44092	1.50887
.185	1.8021	.7549	.64816	.33260	3.45196	1.27468	.18767	.19878	.21390	.23082	.25037	.27430	.30797
							1.01945	1.09634	1.18973	1.27881	1.36231	1.43874	1.50614
.186	1.7988	.7555	.65018	.33100	3.43898	1.27188	.18868	.19986	.21508	.23212	.25184	.27603	.31035
							1.01938	1.09596	1.18895	1.27760	1.36063	1.43657	1.50342
.187	1.7955	.7562	.65219	.32941	3.42608	1.26910	.18970	.20094	.21625	.23342	.25330	.27777	.31278
							1.01931	1.09559	1.18817	1.27639	1.35897	1.43443	1.50073
.188	1.7922	.7569	.65419	.32782	3.41326	1.26635	.19071	.20203	.21743	.23472	.25478	.27952	.31524
							1.01923	1.09522	1.18740	1.27519	1.35732	1.43229	1.49804
.189	1.7890	.7575	.65618	.32624	3.40052	1.26363	.19173	.20311	.21861	.23602	.25625	.28127	.31775
							1.01916	1.09485	1.18664	1.27400	1.35568	1.43017	1.49537
.190	1.7857	.7582	.65817	.32467	3.38786	1.26094	.19274	.20419	.21979	.23732	.25773	.28303	.32030
							1.01909	1.09449	1.18588	1.27282	1.35405	1.42806	1.49271
.191	1.7825	.7589	.66014	.32310	3.37528	1.25828	.19376	.20527	.22097	.23863	.25921	.28481	.32291
							1.01902	1.09413	1.18512	1.27165	1.35243	1.42596	1.49007
.192	1.7793	.7595	.66211	.32153	3.36278	1.25565	.19477	.20635	.22215	.23994	.26069	.28659	.32558
							1.01895	1.09377	1.18437	1.27048	1.35082	1.42388	1.48744
.193	1.7761	.7602	.66408	.31997	3.35035	1.25304	.19579	.20743	.22333	.24125	.26218	.28839	.32831
							1.01889	1.09341	1.18362	1.26932	1.34922	1.42182	1.48482
.194	1.7729	.7609	.66603	.31841	3.33799	1.25046	.19680	.20851	.22451	.24256	.26367	.29019	.33111
							1.01882	1.09305	1.18288	1.26817	1.34763	1.41977	1.48221
.195	1.7697	.7615	.66798	.31686	3.32571	1.24791	.19782	.20960	.22570	.24387	.26517	.29201	.33400
							1.01875	1.09270	1.18215	1.26702	1.34605	1.41772	1.47961
.196	1.7666	.7622	.66992	.31532	3.31350	1.24538	.19883	.21068	.22688	.24518	.26667	.29383	.33699
							1.01868	1.09235	1.18141	1.26589	1.34448	1.41569	1.47703
.197	1.7634	.7628	.67185	.31378	3.30137	1.24289	.19985	.21176	.22806	.24650	.26817	.29567	.34008
							1.01861	1.09200	1.18069	1.26476	1.34293	1.41367	1.47446
.198	1.7603	.7634	.67377	.31224	3.28930	1.24041	.20086	.21285	.22925	.24782	.26968	.29751	.34329
							1.01855	1.09165	1.17996	1.26363	1.34138	1.41166	1.47189
.199	1.7571	.7641	.67569	.31071	3.27731	1.23796	.20188	.21393	.23044	.24913	.27119	.29937	.34665
							1.01848	1.09131	1.17925	1.26252	1.33984	1.40966	1.46933
.200	1.7540	.7647	.67760	.30918	3.26539	1.23554	.20290	.21501	.23162	.25046	.27271	.30124	.35019
							1.01841	1.09097	1.17853	1.26141	1.33831	1.40767	1.46677
.201	1.7509	.7654	.67950	.30766	3.25354	1.23314	.20391	.21610	.23281	.25178	.27423	.30313	.35396
							1.01835	1.09062	1.17782	1.26030	1.33678	1.40570	1.46423
.202	1.7478	.7660	.68140	.30615	3.24175	1.23077	.20493	.21718	.23400	.25310	.27575	.30502	.35803
							1.01828	1.09029	1.17712	1.25921	1.33527	1.40373	1.46168
.203	1.7448	.7666	.68328	.30463	3.23003	1.22842	.20594	.21827	.23519	.25443	.27728	.30693	.36250
							1.01822	1.08995	1.17642	1.25812	1.33377	1.40178	1.45913
.204	1.7417	.7673	.68516	.30313	3.21838	1.22609	.20696	.21935	.23638	.25576	.27881	.30886	.36765
							1.01815	1.08961	1.17572	1.25703	1.33227	1.39984	1.45660
.205	1.7386	.7679	.68704	.30163	3.20680	1.22379	.20797	.22044	.23757	.25709	.28035	.31080	.37397
							1.01809	1.08928	1.17503	1.25596	1.33079	1.39791	1.45404
.206	1.7356	.7685	.68890	.30013	3.19528	1.22151	.20899	.22152	.23876	.25842	.28190	.31276	.38493
							1.01802	1.08895	1.17434	1.25489	1.32931	1.39598	1.45147
.207	1.7326	.7691	.69076	.29864	3.18383	1.21925	.21001	.22261	.23995	.25976	.28344	.31473	
							1.01796	1.08862	1.17365	1.25382	1.32784	1.39407	
.208	1.7296	.7697	.69261	.29715	3.17244	1.21702	.21102	.22370	.24115	.26109	.28500	.31672	
							1.01790	1.08829	1.17297	1.25277	1.32638	1.39217	
.209	1.7266	.7704	.69445	.29566	3.16112	1.21481	.21204	.22478	.24234	.26243	.28655	.31873	
							1.01783	1.08797	1.17230	1.25171	1.32493	1.39028	

TABLE VII. Specific Heat Ratio = 1.20

p/p_t	MACH	T/T_t	GAMMA	K_crit	SHOCK p_y/p_x	p_ty/p_ty	.01	.05	.07	.1	.15	.2	.25
.210	1.7236	.7710	.69629	.29418	3.14986	1.21262	.21305	.22587	.23269	.24354	.26377	.28812	.32076
							1.01777	1.08765	1.12175	1.17162	1.25067	1.32349	1.38839
.211	1.7206	.7716	.69812	.29271	3.13866	1.21045	.21407	.22696	.23381	.24473	.26512	.28969	.32280
							1.01771	1.08732	1.12130	1.17096	1.24963	1.32205	1.38652
.212	1.7176	.7722	.69994	.29124	3.12752	1.20830	.21509	.22804	.23494	.24593	.26646	.29126	.32487
							1.01765	1.08701	1.12084	1.17029	1.24860	1.32062	1.38465
.213	1.7147	.7728	.70176	.28977	3.11645	1.20618	.21610	.22913	.23607	.24713	.26781	.29284	.32696
							1.01759	1.08669	1.12039	1.16963	1.24757	1.31920	1.38280
.214	1.7117	.7734	.70356	.28831	3.10543	1.20407	.21712	.23022	.23720	.24833	.26916	.29443	.32907
							1.01753	1.08637	1.11994	1.16897	1.24655	1.31779	1.38095
.215	1.7088	.7740	.70537	.28686	3.09447	1.20199	.21814	.23131	.23833	.24953	.27051	.29602	.33121
							1.01746	1.08606	1.11949	1.16832	1.24553	1.31639	1.37911
.216	1.7059	.7746	.70716	.28540	3.08358	1.19993	.21915	.23240	.23946	.25073	.27187	.29761	.33337
							1.01740	1.08574	1.11905	1.16767	1.24452	1.31499	1.37728
.217	1.7029	.7752	.70895	.28396	3.07274	1.19788	.22017	.23349	.24059	.25193	.27322	.29922	.33556
							1.01734	1.08543	1.11860	1.16702	1.24352	1.31360	1.37545
.218	1.7000	.7758	.71073	.28251	3.06196	1.19586	.22119	.23458	.24172	.25313	.27458	.30083	.33778
							1.01728	1.08512	1.11816	1.16638	1.24252	1.31222	1.37364
.219	1.6971	.7764	.71250	.28107	3.05124	1.19386	.22220	.23567	.24285	.25434	.27594	.30245	.34003
							1.01722	1.08482	1.11772	1.16574	1.24152	1.31084	1.37183
.220	1.6943	.7770	.71427	.27964	3.04057	1.19187	.22322	.23676	.24399	.25554	.27731	.30407	.34232
							1.01717	1.08451	1.11729	1.16510	1.24053	1.30947	1.37003
.221	1.6914	.7776	.71603	.27821	3.02996	1.18991	.22424	.23785	.24512	.25675	.27868	.30570	.34464
							1.01711	1.08421	1.11686	1.16447	1.23955	1.30811	1.36824
.222	1.6885	.7781	.71778	.27678	3.01941	1.18796	.22525	.23894	.24625	.25796	.28005	.30734	.34699
							1.01705	1.08390	1.11643	1.16384	1.23857	1.30676	1.36645
.223	1.6857	.7787	.71953	.27536	3.00891	1.18604	.22627	.24003	.24739	.25917	.28142	.30898	.34939
							1.01699	1.08360	1.11600	1.16322	1.23760	1.30541	1.36467
.224	1.6828	.7793	.72127	.27394	2.99847	1.18413	.22729	.24112	.24852	.26038	.28279	.31064	.35183
							1.01693	1.08330	1.11557	1.16259	1.23663	1.30407	1.36289
.225	1.6800	.7799	.72300	.27252	2.98808	1.18224	.22830	.24222	.24966	.26159	.28417	.31230	.35433
							1.01687	1.08301	1.11515	1.16198	1.23567	1.30273	1.36113
.226	1.6772	.7805	.72472	.27111	2.97775	1.18037	.22932	.24331	.25080	.26280	.28555	.31397	.35687
							1.01682	1.08271	1.11473	1.16136	1.23471	1.30142	1.35936
.227	1.6744	.7810	.72644	.26971	2.96747	1.17851	.23034	.24440	.25193	.26401	.28693	.31565	.35947
							1.01676	1.08241	1.11431	1.16075	1.23376	1.30010	1.35761
.228	1.6716	.7816	.72815	.26831	2.95724	1.17668	.23135	.24549	.25307	.26523	.28832	.31733	.36214
							1.01670	1.08212	1.11389	1.16014	1.23281	1.29878	1.35585
.229	1.6688	.7822	.72986	.26691	2.94706	1.17486	.23237	.24659	.25421	.26644	.28971	.31903	.36488
							1.01665	1.08183	1.11347	1.15953	1.23187	1.29747	1.35411
.230	1.6660	.7827	.73156	.26551	2.93694	1.17305	.23339	.24768	.25535	.26766	.29110	.32073	.36771
							1.01659	1.08154	1.11306	1.15893	1.23093	1.29617	1.35237
.231	1.6632	.7833	.73325	.26412	2.92686	1.17127	.23441	.24878	.25649	.26888	.29250	.32244	.37062
							1.01653	1.08125	1.11265	1.15833	1.23000	1.29488	1.35063
.232	1.6605	.7839	.73494	.26274	2.91684	1.16950	.23542	.24987	.25763	.27010	.29390	.32416	.37364
							1.01648	1.08096	1.11224	1.15773	1.22907	1.29359	1.34890
.233	1.6577	.7844	.73662	.26136	2.90687	1.16775	.23644	.25097	.25877	.27132	.29530	.32590	.37679
							1.01642	1.08068	1.11184	1.15714	1.22815	1.29230	1.34717
.234	1.6550	.7850	.73829	.25998	2.89694	1.16602	.23746	.25206	.25991	.27254	.29671	.32764	.38008
							1.01637	1.08039	1.11143	1.15655	1.22723	1.29102	1.34544
.235	1.6522	.7856	.73995	.25860	2.88707	1.16430	.23847	.25316	.26105	.27376	.29811	.32939	.38354
							1.01631	1.08011	1.11103	1.15596	1.22631	1.28975	1.34371
.236	1.6495	7861	.74161	.25723	2.87725	1.16260	.23949	.25426	.26219	.27498	.29953	.33116	.38723
							1.01626	1.07983	1.11063	1.15538	1.22540	1.28848	1.34198
.237	1.6468	.7867	.74327	.25587	2.86747	1.16091	.24051	.25535	.26334	.27621	.30094	.33293	.39120
							1.01621	1.07955	1.11023	1.15479	1.22450	1.28722	1.34025
.238	1.6441	.7872	.74491	.25451	2.85774	1.15924	.24153	.25645	.26448	.27744	.30236	.33472	.39555
							1.01615	1.07927	1.10983	1.15422	1.22359	1.28597	1.33852
.239	1.6414	.7878	.74655	.25315	2.84806	1.15759	.24255	.25755	.26563	.27867	.30378	.33652	.40053
							1.01610	1.07899	1.10944	1.15364	1.22270	1.28471	1.33679

TABLE VII. Specific Heat Ratio = 1.20

p/p,	MACH	T/T,	GAMMA	K_{crit}	P_y/P_x	P_{tx}/P_{ty}	.01	.05	.07	.1	.15	.2	.25
					SHOCK				R_2 and TPR versus Loss Coefficient K				
.240	1.6387	.7883	.74819	.25179	2.83843	1.15595	.24356	.25865	.26677	.27990	.30521	.33833	.40651
							1.01604	1.07872	1.10905	1.15307	1.22180	1.28347	1.33505
.241	1.6360	.7889	.74981	.25044	2.82884	1.15432	.24458	.25975	.26792	.28113	.30664	.34016	.41534
							1.01599	1.07844	1.10866	1.15250	1.22092	1.28223	1.33329
.242	1.6333	.7894	.75143	.24910	2.81930	1.15271	.24560	.26085	.26907	.28236	.30808	.34200	
							1.01594	1.07817	1.10827	1.15193	1.22003	1.28099	
.243	1.6306	.7900	.75305	.24775	2.80981	1.15112	.24662	.26195	.27022	.28359	.30951	.34385	
							1.01589	1.07790	1.10788	1.15136	1.21915	1.27976	
.244	1.6280	.7905	.75466	.24641	2.80036	1.14954	.24764	.26305	.27136	.28483	.31096	.34572	
							1.01583	1.07763	1.10750	1.15080	1.21827	1.27854	
.245	1.6253	.7910	.75626	.24508	2.79095	1.14798	.24865	.26415	.27251	.28607	.31240	.34760	
							1.01578	1.07736	1.10711	1.15024	1.21740	1.27731	
.246	1.6227	.7916	.75785	.24375	2.78159	1.14643	.24967	.26525	.27366	.28730	.31385	.34950	
							1.01573	1.07709	1.10673	1.14969	1.21653	1.27610	
.247	1.6201	.7921	.75944	.24242	2.77228	1.14489	.25069	.26635	.27482	.28855	.31531	.35142	
							1.01568	1.07682	1.10635	1.14913	1.21567	1.27489	
.248	1.6174	.7926	.76103	.24109	2.76301	1.14337	.25171	.26745	.27597	.28979	.31677	.35336	
							1.01563	1.07656	1.10597	1.14858	1.21481	1.27368	
.249	1.6148	.7932	.76260	.23977	2.75378	1.14186	.25273	.26856	.27712	.29103	.31823	.35531	
							1.01557	1.07630	1.10560	1.14803	1.21395	1.27247	
.250	1.6122	.7937	.76417	.23846	2.74460	1.14037	.25375	.26966	.27828	.29228	.31970	.35729	
							1.01552	1.07603	1.10522	1.14748	1.21310	1.27127	
.251	1.6096	.7942	.76574	.23714	2.73545	1.13889	.25476	.27076	.27943	.29352	.32117	.35928	
							1.01547	1.07577	1.10485	1.14694	1.21225	1.27008	
.252	1.6070	.7948	.76730	.23583	2.72635	1.13742	.25578	.27187	.28059	.29477	.32265	.36130	
							1.01542	1.07551	1.10448	1.14640	1.21140	1.26889	
.253	1.6044	.7953	.76885	.23453	2.71730	1.13597	.25680	.27297	.28174	.29602	.32413	.36333	
							1.01537	1.07525	1.10411	1.14586	1.21056	1.26770	
.254	1.6019	.7958	.77039	.23323	2.70828	1.13453	.25782	.27408	.28290	.29727	.32562	.36540	
							1.01532	1.07499	1.10375	1.14532	1.20972	1.26652	
.255	1.5993	.7963	.77193	.23193	2.69931	1.13310	.25884	.27518	.28406	.29853	.32711	.36749	
							1.01527	1.07474	1.10338	1.14479	1.20889	1.26534	
.256	1.5967	.7968	.77347	.23063	2.69037	1.13169	.25986	.27629	.28522	.29978	.32861	.36960	
							1.01522	1.07448	1.10302	1.14426	1.20805	1.26416	
.257	1.5942	.7974	.77499	.22934	2.68148	1.13029	.26088	.27740	.28638	.30104	.33012	.37175	
							1.01517	1.07423	1.10265	1.14373	1.20723	1.26299	
.258	1.5916	.7979	.77651	.22805	2.67263	1.12890	.26190	.27851	.28754	.30230	.33163	.37392	
							1.01512	1.07397	1.10229	1.14320	1.20640	1.26182	
.259	1.5891	.7984	.77803	.22677	2.66382	1.12753	.26292	.27962	.28871	.30356	.33314	.37613	
							1.01508	1.07372	1.10195	1.14268	1.20558	1.26065	
.260	1.5865	.7989	.77954	.22549	2.65504	1.12617	.26394	.28072	.28987	.30482	.33466	.37838	
							1.01503	1.07347	1.10159	1.14215	1.20476	1.25949	
.261	1.5840	.7994	.78104	.22421	2.64631	1.12482	.26496	.28183	.29103	.30609	.33619	.38066	
							1.01498	1.07322	1.10124	1.14163	1.20395	1.25833	
.262	1.5815	.7999	.78254	.22294	2.63761	1.12348	.26598	.28294	.29220	.30735	.33772	.38298	
							1.01493	1.07297	1.10088	1.14112	1.20313	1.25717	
.263	1.5790	.8004	.78403	.22167	2.62896	1.12215	.26700	.28406	.29337	.30862	.33927	.38535	
							1.01488	1.07272	1.10053	1.14060	1.20234	1.25601	
.264	1.5765	.8009	.78551	.22040	2.62034	1.12084	.26802	.28517	.29453	.30989	.34082	.38776	
							1.01483	1.07248	1.10018	1.14009	1.20153	1.25486	
.265	1.5740	.8014	.78699	.21914	2.61176	1.11954	.26904	.28628	.29570	.31116	.34237	.39023	
							1.01479	1.07223	1.09983	1.13958	1.20073	1.25370	
.266	1.5715	.8019	.78846	.21788	2.60322	1.11825	.27006	.28739	.29687	.31244	.34393	.39275	
							1.01474	1.07199	1.09948	1.13907	1.19993	1.25255	
.267	1.5690	.8025	.78993	.21662	2.59471	1.11697	.27108	.28850	.29804	.31372	.34550	.39534	
							1.01469	1.07174	1.09914	1.13856	1.19914	1.25140	
.268	1.5665	.8030	.79139	.21537	2.58625	1.11570	.27210	.28962	.29921	.31500	.34707	.39802	
							1.01465	1.07150	1.09879	1.13806	1.19834	1.25027	
.269	1.5641	.8034	.79285	.21412	2.57782	1.11445	.27312	.29073	.30039	.31628	.34866	.40076	
							1.01460	1.07126	1.09845	1.13755	1.19755	1.24912	

TABLE VII. Specific Heat Ratio = 1.20

p/p_t	MACH	T/T_t	GAMMA	K_crit	SHOCK P_y/P_x	P_tx/P_ty	.01	.05	.07	.09	.1	.15	.2
.270	1.5616	.8039	.79430	.21287	2.56942	1.11320	.27414	.29185	.30156	.31201	.31756	.35025	.40359
							1.01455	1.07102	1.09811	1.12431	1.13705	1.19677	1.24798
.271	1.5592	.8044	.79574	.21163	2.56106	1.11197	.27516	.29296	.30274	.31325	.31885	.35184	.40653
							1.01451	1.07078	1.09777	1.12386	1.13655	1.19598	1.24683
.272	1.5567	.8049	.79718	.21039	2.55274	1.11075	.27618	.29408	.30391	.31450	.32014	.35345	.40959
							1.01446	1.07054	1.09743	1.12342	1.13606	1.19520	1.24569
.273	1.5543	.8054	.79861	.20915	2.54445	1.10954	.27720	.29520	.30509	.31575	.32143	.35506	.41280
							1.01441	1.07030	1.09709	1.12298	1.13556	1.19442	1.24455
.274	1.5518	.8059	.80004	.20792	2.53620	1.10834	.27822	.29632	.30627	.31700	.32272	.35669	.41617
							1.01437	1.07007	1.09675	1.12254	1.13507	1.19365	1.24340
.275	1.5494	.8064	.80146	.20669	2.52799	1.10715	.27924	.29744	.30745	.31825	.32402	.35832	.41976
							1.01432	1.06983	1.09642	1.12210	1.13458	1.19287	1.24225
.276	1.5470	.8069	.80287	.20547	2.51981	1.10597	.28026	.29856	.30863	.31951	.32531	.35996	.42363
							1.01428	1.06960	1.09608	1.12167	1.13409	1.19210	1.24110
.277	1.5446	.8074	.80428	.20424	2.51166	1.10481	.28128	.29968	.30981	.32077	.32662	.36161	.42786
							1.01423	1.06937	1.09575	1.12123	1.13360	1.19133	1.23995
.278	1.5422	.8079	.80568	.20302	2.50355	1.10365	.28230	.30080	.31100	.32202	.32792	.36328	.43269
							1.01419	1.06913	1.09542	1.12080	1.13312	1.19057	1.23880
.279	1.5398	.8084	.80708	.20181	2.49547	1.10250	.28332	.30192	.31218	.32329	.32923	.36495	.43844
							1.01414	1.06890	1.09509	1.12037	1.13264	1.18980	1.23763
.280	1.5374	.8088	.80847	.20060	2.48742	1.10137	.28435	.30304	.31337	.32455	.33054	.36663	.44649
							1.01410	1.06867	1.09476	1.11994	1.13216	1.18904	1.23645
.281	1.5350	.8093	.80985	.19939	2.47941	1.10024	.28537	.30417	.31455	.32582	.33185	.36833	
							1.01405	1.06844	1.09444	1.11951	1.13168	1.18829	
.282	1.5326	.8098	.81123	.19818	2.47143	1.09913	.28639	.30529	.31574	.32708	.33316	.37003	
							1.01401	1.06821	1.09411	1.11909	1.13120	1.18753	
.283	1.5302	.8103	.81261	.19698	2.46349	1.09802	.28741	.30641	.31693	.32835	.33448	.37175	
							1.01396	1.06799	1.09379	1.11866	1.13073	1.18678	
.284	1.5278	.8107	.81397	.19578	2.45557	1.09693	.28843	.30754	.31813	.32963	.33580	.37348	
							1.01392	1.06776	1.09347	1.11824	1.13025	1.18602	
.285	1.5255	.8112	.81534	.19458	2.44769	1.09584	.28945	.30867	.31932	.33090	.33712	.37522	
							1.01388	1.06753	1.09314	1.11782	1.12978	1.18527	
.286	1.5231	.8117	.81669	.19339	2.43985	1.09477	.29048	.30979	.32051	.33218	.33845	.37698	
							1.01383	1.06731	1.09282	1.11740	1.12931	1.18453	
.287	1.5208	.8122	.81804	.19220	2.43203	1.09370	.29150	.31092	.32171	.33346	.33978	.37875	
							1.01379	1.06709	1.09251	1.11698	1.12884	1.18378	
.288	1.5184	.8126	.81939	.19102	2.42425	1.09264	.29252	.31205	.32291	.33474	.34111	.38053	
							1.01375	1.06686	1.09219	1.11657	1.12837	1.18304	
.289	1.5161	.8131	.82073	.18983	2.41649	1.09160	.29354	.31318	.32411	.33603	.34245	.38233	
							1.01370	1.06664	1.09187	1.11615	1.12791	1.18230	
.290	1.5137	.8136	.82206	.18865	2.40877	1.09056	.29457	.31431	.32531	.33731	.34379	.38415	
							1.01366	1.06642	1.09156	1.11574	1.12745	1.18156	
.291	1.5114	.8140	.82339	.18748	2.40108	1.08953	.29559	.31544	.32651	.33860	.34513	.38598	
							1.01362	1.06620	1.09124	1.11533	1.12698	1.18082	
.292	1.5091	.8145	.82472	.18630	2.39342	1.08851	.29661	.31658	.32771	.33990	.34648	.38783	
							1.01358	1.06598	1.09093	1.11492	1.12652	1.18008	
.293	1.5068	.8150	.82603	.18513	2.38579	1.08750	.29763	.31771	.32892	.34119	.34783	.38970	
							1.01353	1.06576	1.09062	1.11451	1.12607	1.17935	
.294	1.5044	.8154	.82734	.18397	2.37819	1.08650	.29866	.31885	.33013	.34249	.34918	.39159	
							1.01349	1.06554	1.09031	1.11410	1.12561	1.17862	
.295	1.5021	.8159	.82865	.18280	2.37062	1.08551	.29968	.31998	.33134	.34379	.35054	.39350	
							1.01345	1.06533	1.08998	1.11370	1.12515	1.17789	
.296	1.4998	.8164	.82995	.18164	2.36308	1.08453	.30070	.32112	.33255	.34510	.35190	.39543	
							1.01341	1.06511	1.08967	1.11329	1.12470	1.17716	
.297	1.4975	.8168	.83125	.18049	2.35557	1.08355	.30173	.32226	.33376	.34640	.35327	.39738	
							1.01337	1.06489	1.08936	1.11289	1.12425	1.17643	
.298	1.4952	.8173	.83254	.17933	2.34809	1.08259	.30275	.32339	.33497	.34771	.35464	.39936	
							1.01332	1.06468	1.08906	1.11249	1.12380	1.17570	
.299	1.4930	.8177	.83382	.17818	2.34064	1.08163	.30377	.32453	.33619	.34903	.35601	.40136	
							1.01328	1.06447	1.08875	1.11209	1.12335	1.17498	

TABLE VII. Specific Heat Ratio = 1.20

p/p_t	MACH	T/T_t	GAMMA	K_{crit}	SHOCK P_y/P_x	P_{ty}/P_{tx}	.01	.03	.05	.07	.09	.1	.15
.300	1.4907	.8182	.83510	.17703	2.33322	1.08068	.30480	.31487	.32567	.33741	.35034	.35739	.40339
							1.01324	1.03916	1.06425	1.08845	1.11169	1.12290	1.17425
.301	1.4884	.8186	.83637	.17589	2.32583	1.07974	.30582	.31594	.32682	.33863	.35166	.35877	.40545
							1.01320	1.03903	1.06404	1.08815	1.11130	1.12246	1.17353
.302	1.4861	.8191	.83764	.17475	2.31846	1.07881	.30685	.31702	.32796	.33985	.35299	.36016	.40753
							1.01316	1.03891	1.06383	1.08784	1.11090	1.12201	1.17281
.303	1.4839	.8195	.83890	.17361	2.31113	1.07789	.30787	.31810	.32910	.34107	.35431	.36155	.40965
							1.01312	1.03878	1.06362	1.08754	1.11051	1.12157	1.17209
.304	1.4816	.8200	.84016	.17248	2.30382	1.07698	.30890	.31918	.33025	.34230	.35564	.36294	.41181
							1.01308	1.03866	1.06341	1.08725	1.11011	1.12113	1.17137
.305	1.4794	.8204	.84141	.17134	2.29654	1.07607	.30992	.32026	.33139	.34353	.35698	.36434	.41400
							1.01304	1.03854	1.06320	1.08695	1.10972	1.12069	1.17065
.306	1.4771	.8209	.84265	.17022	2.28929	1.07517	.31094	.32134	.33254	.34476	.35831	.36575	.41623
							1.01300	1.03841	1.06299	1.08665	1.10933	1.12025	1.16993
.307	1.4749	.8213	.84389	.16909	2.28206	1.07429	.31197	.32242	.33369	.34599	.35965	.36716	.41851
							1.01296	1.03829	1.06278	1.08635	1.10894	1.11981	1.16921
.308	1.4726	.8218	.84513	.16797	2.27487	1.07340	.31299	.32350	.33484	.34722	.36100	.36857	.42083
							1.01292	1.03817	1.06258	1.08606	1.10855	1.11937	1.16849
.309	1.4704	.8222	.84636	.16685	2.26770	1.07253	.31402	.32459	.33599	.34846	.36235	.36999	.42321
							1.01288	1.03805	1.06237	1.08576	1.10817	1.11894	1.16777
.310	1.4682	.8227	.84758	.16573	2.26055	1.07167	.31505	.32567	.33714	.34970	.36370	.37142	.42566
							1.01284	1.03793	1.06217	1.08547	1.10778	1.11850	1.16706
.311	1.4659	.8231	.84880	.16462	2.25344	1.07081	.31607	.32675	.33829	.35094	.36505	.37285	.42816
							1.01280	1.03781	1.06196	1.08518	1.10740	1.11807	1.16635
.312	1.4637	.8236	.85002	.16351	2.24635	1.06996	.31710	.32784	.33945	.35218	.36641	.37428	.43073
							1.01276	1.03769	1.06176	1.08489	1.10701	1.11764	1.16563
.313	1.4615	.8240	.85122	.16241	2.23929	1.06912	.31812	.32892	.34060	.35343	.36778	.37573	.43339
							1.01272	1.03757	1.06155	1.08460	1.10663	1.11721	1.16491
.314	1.4593	.8244	.85243	.16130	2.23225	1.06828	.31915	.33001	.34176	.35467	.36915	.37717	.43613
							1.01268	1.03745	1.06135	1.08431	1.10625	1.11678	1.16419
.315	1.4571	.8249	.85362	.16020	2.22524	1.06746	.32017	.33109	.34292	.35592	.37052	.37863	.43898
							1.01264	1.03733	1.06115	1.08402	1.10587	1.11635	1.16348
.316	1.4549	.8253	.85482	.15911	2.21825	1.06664	.32120	.33218	.34408	.35718	.37190	.38009	.44195
							1.01260	1.03721	1.06095	1.08373	1.10549	1.11593	1.16275
.317	1.4527	.8257	.85600	.15801	2.21130	1.06583	.32223	.33327	.34524	.35843	.37328	.38156	.44507
							1.01256	1.03709	1.06075	1.08345	1.10512	1.11550	1.16203
.318	1.4505	.8262	.85718	.15692	2.20436	1.06502	.32325	.33435	.34640	.35969	.37467	.38303	.44836
							1.01253	1.03698	1.06055	1.08316	1.10474	1.11508	1.16131
.319	1.4483	.8266	.85836	.15583	2.19746	1.06423	.32428	.33544	.34757	.36095	.37606	.38451	.45187
							1.01249	1.03686	1.06035	1.08288	1.10437	1.11466	1.16058
.320	1.4462	.8270	.85953	.15475	2.19057	1.06344	.32531	.33653	.34873	.36221	.37746	.38600	.45566
							1.01247	1.03674	1.06015	1.08259	1.10399	1.11423	1.15985
.321	1.4440	.8275	.86070	.15367	2.18371	1.06266	.32634	.33762	.34990	.36348	.37886	.38749	.45986
							1.01243	1.03663	1.05995	1.08231	1.10362	1.11381	1.15912
.322	1.4418	.8279	.86186	.15259	2.17688	1.06188	.32736	.33871	.35107	.36475	.38027	.38899	.46462
							1.01239	1.03651	1.05976	1.08203	1.10325	1.11339	1.15838
.323	1.4396	.8283	.86301	.15151	2.17007	1.06111	.32839	.33980	.35224	.36602	.38168	.39051	.47037
							1.01236	1.03639	1.05956	1.08175	1.10288	1.11298	1.15763
.324	1.4375	.8288	.86416	.15044	2.16329	1.06035	.32942	.34089	.35341	.36729	.38310	.39203	.47870
							1.01232	1.03628	1.05936	1.08147	1.10251	1.11257	1.15688
.325	1.4353	.8292	.86530	.14937	2.15653	1.05960	.33045	.34199	.35458	.36857	.38453	.39356	
							1.01228	1.03616	1.05917	1.08119	1.10214	1.11215	
.326	1.4332	.8296	.86644	.14831	2.14980	1.05885	.33147	.34308	.35575	.36985	.38596	.39509	
							1.01224	1.03605	1.05897	1.08091	1.10177	1.11174	
.327	1.4310	.8300	.86758	.14724	2.14309	1.05811	.33250	.34417	.35693	.37113	.38740	.39664	
							1.01221	1.03594	1.05878	1.08063	1.10140	1.11132	
.328	1.4289	.8304	.86871	.14618	2.13640	1.05738	.33353	.34527	.35811	.37242	.38884	.39819	
							1.01217	1.03582	1.05859	1.08035	1.10104	1.11091	
.329	1.4267	.8309	.86983	.14513	2.12974	1.05666	.33456	.34636	.35929	.37371	.39029	.39976	
							1.01213	1.03571	1.05840	1.08008	1.10067	1.11050	

TABLE VII. Specific Heat Ratio = 1.20

					SHOCK		R₂ and TPR versus Loss Coefficient K						
p/p_t	MACH	T/T_t	GAMMA	K_{crit}	P_y/P_x	P_{tx}/P_{ty}	.01	.03	.05	.07	.08	.09	.1
.330	1.4246	.8313	.87095	.14407	2.12310	1.05594	.33559	.34746	.36047	.37500	.38303	.39174	.40133
							1.01209	1.03560	1.05820	1.07980	1.09020	1.10030	1.11008
.331	1.4225	.8317	.87206	.14302	2.11648	1.05523	.33662	.34856	.36165	.37630	.38440	.39320	.40291
							1.01206	1.03548	1.05801	1.07953	1.08989	1.09993	1.10967
.332	1.4204	.8321	.87317	.14198	2.10989	1.05452	.33765	.34966	.36283	.37760	.38578	.39468	.40451
							1.01202	1.03537	1.05782	1.07926	1.08957	1.09958	1.10926
.333	1.4182	.8325	.87427	.14093	2.10332	1.05382	.33868	.35075	.36402	.37890	.38716	.39616	.40611
							1.01198	1.03526	1.05763	1.07898	1.08925	1.09922	1.10885
.334	1.4161	.8330	.87537	.13989	2.09677	1.05313	.33971	.35185	.36521	.38021	.38854	.39764	.40773
							1.01195	1.03515	1.05744	1.07871	1.08894	1.09886	1.10845
.335	1.4140	.8334	.87646	.13885	2.09025	1.05244	.34074	.35295	.36640	.38152	.38993	.39913	.40935
							1.01191	1.03504	1.05725	1.07844	1.08862	1.09850	1.10804
.336	1.4119	.8338	.87755	.13782	2.08375	1.05176	.34177	.35406	.36759	.38284	.39133	.40063	.41100
							1.01187	1.03493	1.05706	1.07817	1.08831	1.09814	1.10763
.337	1.4098	.8342	.87863	.13679	2.07727	1.05109	.34280	.35516	.36878	.38416	.39273	.40214	.41265
							1.01184	1.03482	1.05688	1.07790	1.08799	1.09778	1.10722
.338	1.4077	.8346	.87971	.13576	2.07081	1.05042	.34383	.35626	.36998	.38548	.39414	.40366	.41432
							1.01180	1.03471	1.05669	1.07763	1.08768	1.09742	1.10682
.339	1.4056	.8350	.88078	.13473	2.06438	1.04976	.34486	.35736	.37118	.38681	.39555	.40519	.41600
							1.01177	1.03460	1.05650	1.07736	1.08737	1.09707	1.10641
.340	1.4035	.8354	.88185	.13371	2.05796	1.04911	.34589	.35847	.37238	.38814	.39698	.40672	.41769
							1.01173	1.03449	1.05632	1.07709	1.08706	1.09671	1.10601
.341	1.4014	.8358	.88291	.13269	2.05157	1.04846	.34692	.35958	.37358	.38947	.39840	.40827	.41940
							1.01170	1.03438	1.05613	1.07683	1.08675	1.09636	1.10561
.342	1.3993	.8363	.88396	.13167	2.04521	1.04782	.34795	.36068	.37478	.39081	.39984	.40983	.42113
							1.01166	1.03427	1.05595	1.07656	1.08645	1.09600	1.10520
.343	1.3972	.8367	.88502	.13066	2.03886	1.04719	.34899	.36179	.37599	.39216	.40128	.41139	.42287
							1.01162	1.03416	1.05576	1.07629	1.08614	1.09565	1.10480
.344	1.3952	.8371	.88606	.12965	2.03254	1.04656	.35002	.36290	.37720	.39351	.40273	.41297	.42464
							1.01159	1.03406	1.05558	1.07603	1.08583	1.09529	1.10440
.345	1.3931	.8375	.88710	.12864	2.02623	1.04593	.35105	.36401	.37841	.39486	.40419	.41456	.42642
							1.01155	1.03395	1.05539	1.07577	1.08552	1.09494	1.10400
.346	1.3910	.8379	.88814	.12764	2.01995	1.04532	.35208	.36512	.37962	.39622	.40565	.41615	.42822
							1.01152	1.03384	1.05521	1.07550	1.08522	1.09459	1.10360
.347	1.3890	.8383	.88917	.12664	2.01369	1.04470	.35312	.36623	.38084	.39759	.40712	.41777	.43004
							1.01148	1.03373	1.05503	1.07524	1.08491	1.09424	1.10319
.348	1.3869	.8387	.89020	.12564	2.00745	1.04410	.35415	.36734	.38206	.39896	.40859	.41939	.43188
							1.01145	1.03363	1.05485	1.07498	1.08460	1.09388	1.10279
.349	1.3848	.8391	.89122	.12465	2.00123	1.04350	.35518	.36846	.38328	.40033	.41008	.42103	.43375
							1.01141	1.03352	1.05467	1.07471	1.08430	1.09353	1.10239
.350	1.3828	.8395	.89223	.12365	1.99503	1.04291	.35622	.36957	.38450	.40171	.41157	.42268	.43564
							1.01138	1.03342	1.05448	1.07445	1.08399	1.09318	1.10199
.351	1.3807	.8399	.89324	.12267	1.98885	1.04232	.35725	.37069	.38573	.40310	.41308	.42434	.43756
							1.01135	1.03331	1.05430	1.07419	1.08369	1.09283	1.10159
.352	1.3787	.8403	.89425	.12168	1.98270	1.04174	.35829	.37181	.38695	.40449	.41459	.42602	.43950
							1.01131	1.03321	1.05412	1.07393	1.08339	1.09248	1.10119
.353	1.3767	.8407	.89525	.12070	1.97656	1.04116	.35932	.37292	.38819	.40589	.41611	.42772	.44148
							1.01128	1.03310	1.05394	1.07367	1.08309	1.09214	1.10079
.354	1.3746	.8411	.89625	.11972	1.97044	1.04059	.36036	.37404	.38942	.40729	.41764	.42943	.44349
							1.01124	1.03300	1.05377	1.07341	1.08278	1.09179	1.10039
.355	1.3726	.8415	.89724	.11874	1.96435	1.04002	.36139	.37516	.39066	.40870	.41918	.43116	.44553
							1.01121	1.03289	1.05359	1.07316	1.08248	1.09144	1.09999
.356	1.3706	.8419	.89822	.11777	1.95827	1.03946	.36243	.37629	.39190	.41012	.42073	.43291	.44761
							1.01117	1.03279	1.05341	1.07290	1.08218	1.09109	1.09959
.357	1.3685	.8423	.89920	.11680	1.95221	1.03891	.36346	.37741	.39314	.41154	.42229	.43467	.44973
							1.01114	1.03268	1.05323	1.07264	1.08188	1.09074	1.09919
.358	1.3665	.8426	.90018	.11583	1.94618	1.03836	.36450	.37853	.39438	.41298	.42387	.43646	.45190
							1.01111	1.03258	1.05305	1.07238	1.08158	1.09040	1.09879
.359	1.3645	.8430	.90115	.11487	1.94016	1.03782	.36553	.37966	.39563	.41441	.42545	.43827	.45411
							1.01107	1.03248	1.05288	1.07213	1.08128	1.09005	1.09839

TABLE VII. Specific Heat Ratio = 1.20

p/p_t	MACH	T/T_t	GAMMA	K_{crit}	SHOCK P_y/P_x	P_{tx}/P_{ty}	.01	.03	.05	.07	.08	.09	.1
.360	1.3625	.8434	.90212	.11391	1.93416	1.03728	.36657	.38079	.39688	.41586	.42705	.44010	.45638
							1.01104	1.03238	1.05270	1.07187	1.08098	1.08970	1.09799
.361	1.3605	.8438	.90308	.11295	1.92818	1.03675	.36761	.38191	.39814	.41732	.42866	.44195	.45870
							1.01101	1.03227	1.05253	1.07162	1.08068	1.08936	1.09759
.362	1.3584	.8442	.90403	.11200	1.92222	1.03622	.36865	.38304	.39940	.41878	.43029	.44383	.46109
							1.01097	1.03217	1.05235	1.07136	1.08038	1.08901	1.09719
.363	1.3564	.8446	.90498	.11105	1.91628	1.03570	.36968	.38418	.40066	.42025	.43193	.44574	.46355
							1.01094	1.03207	1.05218	1.07111	1.08008	1.08866	1.09678
.364	1.3544	.8450	.90593	.11010	1.91036	1.03518	.37072	.38531	.40193	.42173	.43358	.44768	.46610
							1.01091	1.03197	1.05200	1.07085	1.07979	1.08832	1.09638
.365	1.3524	.8454	.90687	.10916	1.90446	1.03467	.37176	.38644	.40319	.42322	.43525	.44965	.46874
							1.01087	1.03187	1.05183	1.07060	1.07947	1.08797	1.09597
.366	1.3504	.8458	.90781	.10821	1.89858	1.03417	.37280	.38758	.40447	.42471	.43693	.45165	.47148
							1.01084	1.03177	1.05165	1.07034	1.07918	1.08762	1.09556
.367	1.3485	.8461	.90874	.10728	1.89271	1.03367	.37384	.38871	.40574	.42622	.43864	.45369	.47436
							1.01081	1.03167	1.05148	1.07009	1.07888	1.08727	1.09515
.368	1.3465	.8465	.90966	.10634	1.88687	1.03317	.37488	.38985	.40702	.42774	.44036	.45576	.47742
							1.01078	1.03157	1.05131	1.06984	1.07858	1.08693	1.09475
.369	1.3445	.8469	.91058	.10541	1.88104	1.03268	.37592	.39099	.40831	.42927	.44210	.45788	.48065
							1.01074	1.03147	1.05114	1.06959	1.07828	1.08658	1.09434
.370	1.3425	.8473	.91150	.10448	1.87523	1.03219	.37696	.39213	.40960	.43080	.44386	.46005	.48412
							1.01071	1.03137	1.05096	1.06933	1.07799	1.08623	1.09392
.371	1.3405	.8477	.91241	.10355	1.86944	1.03171	.37800	.39328	.41089	.43235	.44564	.46227	.48791
							1.01068	1.03127	1.05079	1.06908	1.07769	1.08588	1.09350
.372	1.3385	.8481	.91332	.10263	1.86367	1.03124	.37904	.39442	.41219	.43392	.44744	.46454	.49216
							1.01065	1.03117	1.05062	1.06883	1.07739	1.08553	1.09308
.373	1.3366	.8484	.91422	.10171	1.85791	1.03077	.38008	.39557	.41349	.43549	.44927	.46688	.49712
							1.01061	1.03107	1.05045	1.06858	1.07710	1.08518	1.09265
.374	1.3346	.8488	.91512	.10080	1.85218	1.03030	.38112	.39672	.41480	.43708	.45112	.46929	.50364
							1.01058	1.03097	1.05028	1.06833	1.07680	1.08483	1.09222
.375	1.3326	.8492	.91601	.09988	1.84646	1.02984	.38217	.39787	.41611	.43868	.45300	.47177	
							1.01055	1.03087	1.05011	1.06808	1.07650	1.08448	
.376	1.3307	.8496	.91690	.09897	1.84076	1.02938	.38321	.39902	.41742	.44029	.45491	.47435	
							1.01052	1.03077	1.04994	1.06783	1.07621	1.08412	
.377	1.3287	.8499	.91778	.09807	1.83508	1.02893	.38425	.40017	.41874	.44192	.45685	.47703	
							1.01049	1.03067	1.04977	1.06758	1.07591	1.08377	
.378	1.3268	.8503	.91866	.09717	1.82941	1.02849	.38530	.40133	.42007	.44357	.45883	.47986	
							1.01045	1.03058	1.04960	1.06733	1.07561	1.08342	
.379	1.3248	.8507	.91953	.09627	1.82376	1.02805	.38634	.40249	.42140	.44523	.46084	.48281	
							1.01042	1.03048	1.04943	1.06708	1.07532	1.08306	
.380	1.3229	.8511	.92039	.09537	1.81813	1.02761	.38738	.40365	.42274	.44691	.46289	.48595	
							1.01039	1.03038	1.04926	1.06683	1.07502	1.08270	
.381	1.3209	.8514	.92126	.09448	1.81252	1.02718	.38843	.40481	.42408	.44861	.46498	.48932	
							1.01036	1.03028	1.04909	1.06658	1.07472	1.08234	
.382	1.3190	.8518	.92212	.09359	1.80692	1.02675	.38948	.40597	.42543	.45032	.46712	.49299	
							1.01033	1.03019	1.04893	1.06633	1.07442	1.08197	
.383	1.3170	.8522	.92297	.09270	1.80134	1.02633	.39052	.40714	.42678	.45206	.46932	.49707	
							1.01030	1.03009	1.04876	1.06608	1.07412	1.08160	
.384	1.3151	.8526	.92382	.09181	1.79578	1.02591	.39157	.40830	.42815	.45382	.47157	.50179	
							1.01026	1.02999	1.04859	1.06583	1.07382	1.08123	
.385	1.3132	.8529	.92466	.09093	1.79024	1.02549	.39261	.40948	.42951	.45560	.47388	.50783	
							1.01023	1.02990	1.04842	1.06557	1.07352	1.08085	
.386	1.3112	.8533	.92550	.09006	1.78471	1.02508	.39366	.41065	.43089	.45740	.47626	.51866	
							1.01020	1.02980	1.04826	1.06532	1.07322	1.08046	
.387	1.3093	.8537	.92633	.08918	1.77919	1.02468	.39471	.41182	.43227	.45925	.47873		
							1.01017	1.02971	1.04809	1.06508	1.07291		
.388	1.3074	.8540	.92716	.08831	1.77370	1.02428	.39576	.41300	.43365	.46111	.48129		
							1.01014	1.02961	1.04792	1.06483	1.07261		
.389	1.3055	.8544	.92798	.08744	1.76822	1.02388	.39681	.41418	.43505	.46300	.48395		
							1.01011	1.02952	1.04776	1.06458	1.07230		

TABLE VII. Specific Heat Ratio = 1.20

p/p_t	MACH	T/T_t	GAMMA	K_{crit}	P_y/P_x (SHOCK)	P_{ty}/P_{ty}	.01	.03	.04	.05	.06	.07	.08
.390	1.3035	.8548	.92880	.08658	1.76276	1.02349	.39786	.41536	.42532	.43645	.44929	.46492	.48678
							1.01008	1.02942	1.03867	1.04759	1.05616	1.06433	1.07200
.391	1.3016	.8551	.92962	.08572	1.75731	1.02310	.39891	.41655	.42660	.43787	.45089	.46688	.48973
							1.01005	1.02933	1.03854	1.04743	1.05596	1.06408	1.07169
.392	1.2997	.8555	.93043	.08486	1.75188	1.02272	.39996	.41773	.42789	.43928	.45252	.46888	.49288
							1.01002	1.02923	1.03841	1.04726	1.05575	1.06383	1.07138
.393	1.2978	.8559	.93123	.08401	1.74647	1.02234	.40101	.41892	.42918	.44071	.45416	.47091	.49627
							1.00999	1.02914	1.03828	1.04709	1.05555	1.06358	1.07107
.394	1.2959	.8562	.93203	.08316	1.74107	1.02197	.40206	.42012	.43047	.44215	.45581	.47300	.49999
							1.00996	1.02904	1.03815	1.04693	1.05534	1.06332	1.07075
.395	1.2940	.8566	.93283	.08231	1.73569	1.02160	.40311	.42131	.43177	.44360	.45749	.47513	.50417
							1.00993	1.02895	1.03802	1.04676	1.05514	1.06307	1.07043
.396	1.2921	.8569	.93362	.08147	1.73032	1.02123	.40417	.42251	.43308	.44505	.45919	.47731	.50916
							1.00990	1.02886	1.03789	1.04660	1.05493	1.06282	1.07011
.397	1.2902	.8573	.93440	.08063	1.72497	1.02087	.40522	.42371	.43439	.44652	.46090	.47956	.51574
							1.00987	1.02876	1.03777	1.04643	1.05472	1.06256	1.06978
.398	1.2883	.8577	.93518	.07979	1.71964	1.02051	.40627	.42492	.43570	.44799	.46264	.48187	
							1.00984	1.02867	1.03764	1.04627	1.05452	1.06231	
.399	1.2864	.8580	.93596	.07895	1.71432	1.02016	.40733	.42612	.43703	.44948	.46440	.48426	
							1.00981	1.02858	1.03751	1.04611	1.05431	1.06205	
.400	1.2845	.8584	.93673	.07812	1.70902	1.01981	.40838	.42733	.43835	.45098	.46619	.48674	
							1.00978	1.02848	1.03738	1.04594	1.05411	1.06179	
.401	1.2826	.8587	.93750	.07730	1.70373	1.01947	.40944	.42855	.43969	.45249	.46801	.48932	
							1.00975	1.02839	1.03726	1.04578	1.05390	1.06154	
.402	1.2807	.8591	.93826	.07647	1.69846	1.01912	.41050	.42977	.44103	.45401	.46985	.49205	
							1.00972	1.02830	1.03713	1.04561	1.05369	1.06128	
.403	1.2788	.8594	.93901	.07565	1.69320	1.01879	.41155	.43099	.44238	.45555	.47173	.49490	
							1.00969	1.02821	1.03700	1.04545	1.05349	1.06102	
.404	1.2770	.8598	.93977	.07483	1.68796	1.01845	.41261	.43221	.44373	.45710	.47364	.49792	
							1.00966	1.02811	1.03688	1.04529	1.05328	1.06076	
.405	1.2751	.8602	.94051	.07402	1.68273	1.01812	.41367	.43344	.44509	.45866	.47558	.50117	
							1.00963	1.02802	1.03675	1.04512	1.05307	1.06049	
.406	1.2732	.8605	.94126	.07321	1.67752	1.01780	.41473	.43467	.44646	.46025	.47757	.50472	
							1.00960	1.02793	1.03662	1.04496	1.05287	1.06023	
.407	1.2713	.8609	.94199	.07240	1.67232	1.01748	.41579	.43591	.44784	.46184	.47960	.50868	
							1.00957	1.02784	1.03650	1.04479	1.05266	1.05995	
.408	1.2695	.8612	.94273	.07160	1.66714	1.01716	.41685	.43715	.44922	.46345	.48168	.51332	
							1.00954	1.02775	1.03637	1.04463	1.05245	1.05968	
.409	1.2676	.8616	.94346	.07080	1.66197	1.01685	.41792	.43839	.45061	.46509	.48381	.51921	
							1.00951	1.02765	1.03625	1.04446	1.05224	1.05940	
.410	1.2657	.8619	.94418	.07000	1.65682	1.01654	.41898	.43964	.45201	.46674	.48600	.53139	
							1.00948	1.02756	1.03612	1.04430	1.05203	1.05911	
.411	1.2639	.8623	.94490	.06921	1.65168	1.01623	.42004	.44090	.45342	.46841	.48826		
							1.00945	1.02747	1.03600	1.04414	1.05182		
.412	1.2620	.8626	.94561	.06842	1.64656	1.01593	.42111	.44216	.45484	.47010	.49059		
							1.00942	1.02738	1.03587	1.04397	1.05160		
.413	1.2602	.8630	.94632	.06764	1.64145	1.01563	.42217	.44342	.45627	.47181	.49300		
							1.00939	1.02729	1.03574	1.04381	1.05139		
.414	1.2583	.8633	.94703	.06685	1.63636	1.01534	.42324	.44469	.45771	.47355	.49555		
							1.00937	1.02720	1.03562	1.04364	1.05118		
.415	1.2565	.8637	.94773	.06607	1.63128	1.01505	.42431	.44596	.45916	.47532	.49819		
							1.00934	1.02711	1.03549	1.04348	1.05097		
.416	1.2546	.8640	.94842	.06530	1.62621	1.01476	.42537	.44724	.46063	.47711	.50097		
							1.00931	1.02702	1.03537	1.04331	1.05075		
.417	1.2528	.8643	.94911	.06453	1.62116	1.01447	.42644	.44852	.46210	.47894	.50393		
							1.00928	1.02693	1.03524	1.04315	1.05053		
.418	1.2509	.8647	.94980	.06376	1.61612	1.01419	.42752	.44981	.46359	.48080	.50711		
							1.00926	1.02684	1.03512	1.04298	1.05031		
.419	1.2491	.8650	.95048	.06299	1.61110	1.01392	.42859	.45111	.46509	.48269	.51058		
							1.00923	1.02675	1.03499	1.04282	1.05009		

TABLE VII. Specific Heat Ratio = 1.20

p/p_t	MACH	T/T_t	GAMMA	K_crit	SHOCK p_y/p_x	p_tx/p_ty	.005	.01	.02	.03	.04	.05	.06
.420	1.2472	.8654	.95115	.06223	1.60609	1.01365	.42472	.42967	.44034	.45241	.46660	.48462	.51445
							1.00463	1.00920	1.01809	1.02666	1.03487	1.04265	1.04986
.421	1.2454	.8657	.95183	.06148	1.60110	1.01338	.42576	.43074	.44151	.45372	.46813	.48660	.51903
							1.00462	1.00917	1.01803	1.02657	1.03474	1.04248	1.04964
.422	1.2436	.8661	.95249	.06072	1.59611	1.01311	.42679	.43181	.44269	.45503	.46967	.48862	.52485
							1.00461	1.00915	1.01797	1.02648	1.03462	1.04231	1.04940
.423	1.2417	.8664	.95316	.05997	1.59115	1.01285	.42783	.43289	.44386	.45635	.47123	.49070	
							1.00459	1.00912	1.01791	1.02639	1.03449	1.04215	
.424	1.2399	.8668	.95381	.05923	1.58619	1.01259	.42886	.43397	.44504	.45768	.47280	.49283	
							1.00458	1.00909	1.01786	1.02630	1.03437	1.04198	
.425	1.2381	.8671	.95447	.05848	1.58125	1.01233	.42990	.43505	.44623	.45902	.47440	.49502	
							1.00456	1.00906	1.01780	1.02621	1.03424	1.04181	
.426	1.2362	.8674	.95511	.05774	1.57632	1.01208	.43094	.43612	.44741	.46036	.47601	.49730	
							1.00455	1.00903	1.01774	1.02612	1.03412	1.04164	
.427	1.2344	.8678	.95576	.05701	1.57141	1.01183	.43198	.43720	.44860	.46171	.47764	.49968	
							1.00454	1.00901	1.01769	1.02603	1.03399	1.04147	
.428	1.2326	.8681	.95640	.05628	1.56651	1.01159	.43301	.43829	.44979	.46307	.47930	.50214	
							1.00452	1.00898	1.01763	1.02594	1.03386	1.04130	
.429	1.2308	.8684	.95703	.05555	1.56162	1.01135	.43405	.43937	.45099	.46444	.48097	.50472	
							1.00451	1.00895	1.01757	1.02585	1.03374	1.04113	
.430	1.2290	.8688	.95766	.05482	1.55675	1.01111	.43509	.44045	.45219	.46582	.48268	.50744	
							1.00450	1.00892	1.01751	1.02576	1.03361	1.04095	
.431	1.2272	.8691	.95828	.05410	1.55189	1.01087	.43613	.44154	.45340	.46721	.48441	.51034	
							1.00448	1.00890	1.01746	1.02568	1.03349	1.04077	
.432	1.2253	.8695	.95890	.05339	1.54704	1.01064	.43717	.44263	.45461	.46860	.48617	.51346	
							1.00447	1.00887	1.01740	1.02559	1.03336	1.04060	
.433	1.2235	.8698	.95952	.05267	1.54221	1.01041	.43821	.44372	.45582	.47001	.48796	.51687	
							1.00446	1.00884	1.01734	1.02550	1.03323	1.04042	
.434	1.2217	.8701	.96013	.05196	1.53739	1.01019	.43925	.44481	.45704	.47143	.48978	.52071	
							1.00444	1.00881	1.01729	1.02541	1.03310	1.04023	
.435	1.2199	.8705	.96074	.05126	1.53258	1.00996	.44030	.44590	.45826	.47286	.49164	.52530	
							1.00443	1.00879	1.01723	1.02532	1.03298	1.04005	
.436	1.2181	.8708	.96134	.05056	1.52778	1.00975	.44134	.44699	.45948	.47430	.49354	.53129	
							1.00442	1.00876	1.01717	1.02523	1.03285	1.03986	
.437	1.2163	.8711	.96194	.04986	1.52300	1.00953	.44238	.44808	.46071	.47576	.49549		
							1.00440	1.00873	1.01712	1.02514	1.03272		
.438	1.2145	.8715	.96253	.04916	1.51823	1.00932	.44343	.44918	.46195	.47723	.49749		
							1.00439	1.00870	1.01706	1.02505	1.03259		
.439	1.2127	.8718	.96312	.04847	1.51347	1.00911	.44447	.45028	.46319	.47871	.49954		
							1.00438	1.00868	1.01700	1.02496	1.03246		
.440	1.2109	.8721	.96370	.04779	1.50873	1.00890	.44552	.45138	.46444	.48021	.50165		
							1.00436	1.00865	1.01695	1.02487	1.03233		
.441	1.2091	.8724	.96428	.04710	1.50400	1.00870	.44656	.45248	.46569	.48173	.50386		
							1.00435	1.00862	1.01689	1.02478	1.03220		
.442	1.2073	.8728	.96485	.04642	1.49928	1.00850	.44761	.45358	.46695	.48326	.50613		
							1.00434	1.00860	1.01684	1.02469	1.03207		
.443	1.2056	.8731	.96542	.04575	1.49457	1.00830	.44866	.45469	.46821	.48481	.50850		
							1.00432	1.00857	1.01678	1.02460	1.03194		
.444	1.2038	.8734	.96598	.04508	1.48988	1.00811	.44971	.45580	.46949	.48638	.51098		
							1.00431	1.00854	1.01672	1.02451	1.03181		
.445	1.2020	.8738	.96654	.04441	1.48519	1.00791	.45076	.45691	.47076	.48797	.51360		
							1.00430	1.00852	1.01667	1.02442	1.03167		
.446	1.2002	.8741	.96710	.04375	1.48052	1.00773	.45181	.45802	.47205	.48958	.51639		
							1.00428	1.00849	1.01661	1.02433	1.03154		
.447	1.1984	.8744	.96765	.04309	1.47587	1.00754	.45286	.45913	.47334	.49121	.51939		
							1.00427	1.00846	1.01656	1.02424	1.03140		
.448	1.1966	.8747	.96819	.04243	1.47122	1.00736	.45391	.46025	.47464	.49287	.52268		
							1.00426	1.00843	1.01650	1.02415	1.03126		
.449	1.1949	.8751	.96873	.04178	1.46659	1.00718	.45497	.46137	.47594	.49456	.52637		
							1.00424	1.00841	1.01644	1.02406	1.03111		

TABLE VII. Specific Heat Ratio = 1.20

p/p_t	MACH	T/T_t	GAMMA	K_{crit}	SHOCK		R₁ and TPR versus Loss Coefficient K						
					P_y/P_x	P_{ty}/P_{ty}	.001	.005	.01	.015	.02	.03	.04
.450	1.1931	.8754	.96927	.04114	1.46197	1.00700	.45117	.45602	.46249	.46951	.47725	.49628	.53081
							1.00085	1.00423	1.00838	1.01243	1.01638	1.02397	1.03097
.451	1.1913	.8757	.96980	.04049	1.45736	1.00683	.45218	.45708	.46361	.47072	.47857	.49803	.53661
							1.00085	1.00422	1.00835	1.01239	1.01633	1.02388	1.03082
.452	1.1895	.8760	.97033	.03985	1.45276	1.00666	.45319	.45813	.46473	.47193	.47990	.49982	
							1.00085	1.00421	1.00833	1.01234	1.01627	1.02379	
.453	1.1878	.8764	.97085	.03922	1.44817	1.00649	.45420	.45919	.46586	.47315	.48125	.50165	
							1.00085	1.00419	1.00830	1.01230	1.01621	1.02370	
.454	1.1860	.8767	.97137	.03859	1.44360	1.00632	.45522	.46025	.46699	.47437	.48260	.50352	
							1.00084	1.00418	1.00828	1.01226	1.01616	1.02360	
.455	1.1843	.8770	.97188	.03796	1.43904	1.00616	.45623	.46131	.46813	.47560	.48396	.50543	
							1.00084	1.00417	1.00825	1.01222	1.01610	1.02351	
.456	1.1825	.8773	.97239	.03734	1.43449	1.00600	.45724	.46237	.46926	.47684	.48533	.50741	
							1.00084	1.00415	1.00822	1.01218	1.01605	1.02342	
.457	1.1807	.8776	.97289	.03672	1.42995	1.00584	.45825	.46343	.47040	.47808	.48674	.50947	
							1.00084	1.00414	1.00820	1.01214	1.01600	1.02333	
.458	1.1790	.8780	.97339	.03611	1.42542	1.00569	.45926	.46450	.47155	.47933	.48813	.51158	
							1.00083	1.00413	1.00817	1.01210	1.01594	1.02324	
.459	1.1772	.8783	.97389	.03550	1.42090	1.00554	.46027	.46556	.47269	.48058	.48955	.51378	
							1.00083	1.00412	1.00814	1.01206	1.01588	1.02314	
.460	1.1755	.8786	.97438	.03489	1.41640	1.00539	.46129	.46663	.47384	.48184	.49097	.51606	
							1.00083	1.00410	1.00812	1.01202	1.01583	1.02305	
.461	1.1737	.8789	.97486	.03429	1.41191	1.00524	.46230	.46769	.47500	.48311	.49241	.51847	
							1.00083	1.00409	1.00809	1.01198	1.01577	1.02295	
.462	1.1720	.8792	.97534	.03369	1.40742	1.00510	.46331	.46876	.47615	.48439	.49387	.52101	
							1.00082	1.00408	1.00806	1.01194	1.01572	1.02285	
.463	1.1702	.8796	.97582	.03310	1.40295	1.00496	.46432	.46983	.47731	.48568	.49534	.52373	
							1.00082	1.00406	1.00804	1.01190	1.01566	1.02275	
.464	1.1685	.8799	.97629	.03251	1.39850	1.00482	.46534	.47090	.47848	.48697	.49684	.52668	
							1.00082	1.00405	1.00801	1.01186	1.01560	1.02265	
.465	1.1667	.8802	.97676	.03193	1.39405	1.00468	.46635	.47198	.47965	.48829	.49835	.52993	
							1.00082	1.00404	1.00799	1.01182	1.01555	1.02255	
.466	1.1650	.8805	.97722	.03135	1.38961	1.00455	.46736	.47305	.48082	.48960	.49988	.53372	
							1.00081	1.00403	1.00796	1.01178	1.01549	1.02245	
.467	1.1632	.8808	.97768	.03077	1.38519	1.00442	.46838	.47413	.48200	.49093	.50144	.53830	
							1.00081	1.00401	1.00793	1.01174	1.01543	1.02234	
.468	1.1615	.8811	.97813	.03020	1.38077	1.00429	.46939	.47521	.48318	.49226	.50302	.54520	
							1.00081	1.00400	1.00791	1.01170	1.01538	1.02223	
.469	1.1597	.8814	.97858	.02963	1.37637	1.00416	.47041	.47629	.48437	.49360	.50463		
							1.00081	1.00399	1.00788	1.01166	1.01532		
.470	1.1580	.8818	.97902	.02907	1.37197	1.00404	.47142	.47737	.48556	.49496	.50626		
							1.00080	1.00398	1.00785	1.01162	1.01526		
.471	1.1563	.8821	.97946	.02851	1.36759	1.00392	.47243	.47845	.48676	.49633	.50793		
							1.00080	1.00396	1.00783	1.01158	1.01521		
.472	1.1545	.8824	.97990	.02796	1.36322	1.00380	.47345	.47954	.48796	.49771	.50963		
							1.00080	1.00395	1.00780	1.01154	1.01515		
.473	1.1528	.8827	.98033	.02741	1.35886	1.00368	.47447	.48062	.48917	.49911	.51138		
							1.00080	1.00394	1.00778	1.01150	1.01509		
.474	1.1511	.8830	.98075	.02686	1.35451	1.00356	.47548	.48171	.49040	.50052	.51316		
							1.00079	1.00393	1.00776	1.01146	1.01503		
.475	1.1493	.8833	.98117	.02632	1.35017	1.00345	.47650	.48281	.49162	.50195	.51500		
							1.00079	1.00391	1.00773	1.01142	1.01497		
.476	1.1476	.8836	.98159	.02579	1.34585	1.00334	.47752	.48390	.49285	.50340	.51689		
							1.00079	1.00390	1.00771	1.01138	1.01491		
.477	1.1459	.8839	.98200	.02526	1.34153	1.00324	.47853	.48500	.49409	.50487	.51885		
							1.00079	1.00389	1.00768	1.01134	1.01485		
.478	1.1442	.8842	.98241	.02473	1.33722	1.00313	.47955	.48610	.49534	.50636	.52087		
							1.00078	1.00388	1.00766	1.01129	1.01479		
.479	1.1424	.8846	.98281	.02421	1.33292	1.00303	.48057	.48720	.49659	.50786	.52299		
							1.00078	1.00386	1.00763	1.01125	1.01473		

TABLE VII. Specific Heat Ratio = 1.20

					SHOCK		R₂ and TPR versus Loss Coefficient K						
p/P_t	MACH	T/T_t	GAMMA	K_{crit}	P_y/P_x	P_{ty}/P_{ty}	.001	.002	.003	.005	.01	.015	.02
.480	1.1407	.8849	.98321	.02369	1.32864	1.00293	.48159	.48321	.48487	.48831	.49785	.50940	.52520
							1.00078	1.00155	1.00233	1.00385	1.00760	1.01121	1.01467
.481	1.1390	.8852	.98360	.02318	1.32436	1.00283	.48260	.48425	.48593	.48942	.49912	.51096	.52754
							1.00078	1.00155	1.00232	1.00384	1.00758	1.01117	1.01461
.482	1.1373	.8855	.98399	.02267	1.32010	1.00273	.48362	.48529	.48699	.49053	.50041	.51255	.53003
							1.00077	1.00154	1.00231	1.00383	1.00755	1.01113	1.01454
.483	1.1356	.8858	.98437	.02217	1.31584	1.00264	.48464	.48633	.48805	.49164	.50170	.51417	.53278
							1.00077	1.00154	1.00230	1.00382	1.00753	1.01109	1.01449
.484	1.1339	.8861	.98475	.02167	1.31160	1.00254	.48566	.48737	.48912	.49276	.50300	.51583	.53578
							1.00077	1.00154	1.00230	1.00381	1.00750	1.01105	1.01442
.485	1.1321	.8864	.98512	.02118	1.30736	1.00245	.48668	.48841	.49018	.49388	.50431	.51753	.53919
							1.00077	1.00153	1.00229	1.00379	1.00747	1.01100	1.01436
.486	1.1304	.8867	.98549	.02069	1.30314	1.00237	.48771	.48946	.49125	.49501	.50564	.51927	.54331
							1.00077	1.00153	1.00228	1.00378	1.00745	1.01096	1.01428
.487	1.1287	.8870	.98586	.02020	1.29893	1.00228	.48873	.49050	.49232	.49614	.50698	.52106	.54939
							1.00076	1.00152	1.00227	1.00377	1.00742	1.01092	1.01422
.488	1.1270	.8873	.98622	.01973	1.29472	1.00220	.48975	.49155	.49340	.49727	.50833	.52292	
							1.00076	1.00152	1.00227	1.00376	1.00740	1.01087	
.489	1.1253	.8876	.98658	.01925	1.29053	1.00212	.49077	.49259	.49447	.49841	.50970	.52483	
							1.00076	1.00151	1.00226	1.00374	1.00737	1.01083	
.490	1.1236	.8879	.98693	.01878	1.28634	1.00204	.49180	.49364	.49555	.49955	.51109	.52683	
							1.00076	1.00151	1.00225	1.00373	1.00734	1.01079	
.491	1.1219	.8882	.98727	.01832	1.28217	1.00196	.49282	.49469	.49663	.50070	.51250	.52892	
							1.00075	1.00150	1.00225	1.00372	1.00732	1.01074	
.492	1.1202	.8885	.98762	.01786	1.27800	1.00188	.49385	.49575	.49771	.50185	.51392	.53112	
							1.00075	1.00150	1.00224	1.00371	1.00729	1.01070	
.493	1.1185	.8888	.98795	.01740	1.27385	1.00181	.49487	.49680	.49880	.50300	.51537	.53351	
							1.00075	1.00149	1.00223	1.00370	1.00726	1.01065	
.494	1.1168	.8891	.98829	.01695	1.26971	1.00174	.49590	.49786	.49988	.50417	.51684	.53606	
							1.00075	1.00149	1.00222	1.00368	1.00724	1.01061	
.495	1.1151	.8894	.98861	.01651	1.26557	1.00167	.49692	.49892	.50097	.50533	.51834	.53887	
							1.00074	1.00149	1.00222	1.00367	1.00721	1.01056	
.496	1.1134	.8897	.98894	.01607	1.26145	1.00160	.49795	.49995	.50206	.50651	.51986	.54205	
							1.00074	1.00147	1.00221	1.00366	1.00718	1.01051	
.497	1.1117	.8900	.98926	.01564	1.25733	1.00153	.49900	.50103	.50316	.50769	.52142	.54587	
							1.00075	1.00147	1.00220	1.00365	1.00716	1.01046	
.498	1.1100	.8903	.98957	.01521	1.25323	1.00147	.50000	.50209	.50426	.50888	.52301	.55140	
							1.00073	1.00147	1.00220	1.00363	1.00713	1.01041	
.499	1.1083	.8906	.98988	.01478	1.24913	1.00140	.50104	.50316	.50536	.51007	.52465		
							1.00073	1.00146	1.00219	1.00362	1.00710		
.500	1.1066	.8909	.99018	.01437	1.24504	1.00134	.50207	.50422	.50647	.51128	.52633		
							1.00073	1.00146	1.00218	1.00361	1.00708		
.501	1.1049	.8912	.99049	.01395	1.24097	1.00128	.50310	.50529	.50758	.51249	.52806		
							1.00073	1.00145	1.00217	1.00360	1.00705		
.502	1.1032	.8915	.99078	.01354	1.23690	1.00122	.50414	.50636	.50869	.51371	.52985		
							1.00073	1.00145	1.00217	1.00359	1.00702		
.503	1.1016	.8918	.99107	.01314	1.23284	1.00117	.50517	.50744	.50981	.51494	.53172		
							1.00072	1.00145	1.00216	1.00357	1.00699		
.504	1.0999	.8921	.99136	.01274	1.22879	1.00111	.50621	.50852	.51093	.51619	.53367		
							1.00072	1.00144	1.00215	1.00356	1.00696		
.505	1.0982	.8924	.99164	.01235	1.22475	1.00106	.50724	.50959	.51206	.51744	.53577		
							1.00072	1.00144	1.00215	1.00355	1.00694		
.506	1.0965	.8927	.99192	.01196	1.22072	1.00101	.50828	.51068	.51319	.51871	.53798		
							1.00072	1.00143	1.00214	1.00354	1.00691		
.507	1.0948	.8930	.99219	.01158	1.21670	1.00096	.50932	.51176	.51433	.51999	.54036		
							1.00072	1.00143	1.00213	1.00352	1.00688		
.508	1.0931	.8933	.99246	.01121	1.21269	1.00091	.51036	.51285	.51548	.52128	.54300		
							1.00071	1.00142	1.00212	1.00351	1.00685		
.509	1.0915	.8936	.99272	.01084	1.20869	1.00087	.51141	.51394	.51663	.52259	.54600		
							1.00071	1.00142	1.00212	1.00350	1.00681		

TABLE VII. Specific Heat Ratio = 1.20

p/p_t	MACH	T/T_t	GAMMA	K_{crit}	SHOCK p_y/p_x	p_{ty}/p_{tv}	.001	.002	.003	.004	.005	.01	.015
.510	1.0898	.8938	.99298	.01047	1.20470	1.00082	.51245	.51504	.51779	.52073	.52394	.54974	
							1.00071	1.00141	1.00211	1.00280	1.00349	1.00678	
.511	1.0881	.8941	.99323	.01011	1.20071	1.00078	.51350	.51614	.51895	.52197	.52529	.55523	
							1.00071	1.00141	1.00210	1.00279	1.00348	1.00675	
.512	1.0864	.8944	.99348	.00976	1.19674	1.00074	.51454	.51724	.52013	.52323	.52666		
							1.00070	1.00140	1.00209	1.00278	1.00346		
.513	1.0848	.8947	.99373	.00941	1.19277	1.00070	.51559	.51835	.52131	.52454	.52806		
							1.00070	1.00140	1.00209	1.00278	1.00345		
.514	1.0831	.8950	.99397	.00907	1.18881	1.00066	.51664	.51947	.52250	.52583	.52948		
							1.00070	1.00139	1.00208	1.00277	1.00344		
.515	1.0814	.8953	.99420	.00873	1.18487	1.00062	.51770	.52059	.52370	.52714	.53094		
							1.00070	1.00139	1.00207	1.00276	1.00343		
.516	1.0797	.8956	.99443	.00840	1.18093	1.00059	.51875	.52171	.52495	.52847	.53243		
							1.00069	1.00138	1.00207	1.00275	1.00341		
.517	1.0781	.8959	.99466	.00807	1.17700	1.00055	.51981	.52284	.52617	.52983	.53397		
							1.00069	1.00138	1.00206	1.00274	1.00340		
.518	1.0764	.8962	.99488	.00775	1.17308	1.00052	.52087	.52398	.52742	.53121	.53555		
							1.00069	1.00137	1.00206	1.00273	1.00339		
.519	1.0747	.8965	.99510	.00744	1.16916	1.00049	.52194	.52516	.52868	.53262	.53719		
							1.00068	1.00137	1.00205	1.00272	1.00337		
.520	1.0731	.8967	.99531	.00713	1.16526	1.00046	.52300	.52632	.52995	.53406	.53890		
							1.00068	1.00137	1.00204	1.00271	1.00336		
.521	1.0714	.8970	.99552	.00683	1.16137	1.00043	.52407	.52748	.53124	.53554	.54069		
							1.00068	1.00136	1.00203	1.00270	1.00335		
.522	1.0697	.8973	.99572	.00653	1.15748	1.00040	.52518	.52866	.53256	.53706	.54258		
							1.00068	1.00136	1.00203	1.00269	1.00333		
.523	1.0681	.8976	.99592	.00624	1.15360	1.00037	.52626	.52985	.53390	.53864	.54460		
							1.00068	1.00135	1.00202	1.00267	1.00332		
.524	1.0664	.8979	.99611	.00596	1.14973	1.00035	.52734	.53105	.53526	.54028	.54689		
							1.00068	1.00135	1.00201	1.00266	1.00331		
.525	1.0648	.8982	.99630	.00568	1.14587	1.00032	.52843	.53226	.53666	.54199	.54941		
							1.00068	1.00134	1.00200	1.00265	1.00329		
.526	1.0631	.8985	.99649	.00541	1.14202	1.00030	.52953	.53349	.53810	.54380	.55242		
							1.00067	1.00134	1.00200	1.00264	1.00328		
.527	1.0614	.8987	.99667	.00514	1.13818	1.00028	.53063	.53474	.53957	.54573	.55647		
							1.00067	1.00134	1.00199	1.00263	1.00325		
.528	1.0598	.8990	.99684	.00488	1.13435	1.00026	.53174	.53601	.54110	.54791			
							1.00067	1.00133	1.00198	1.00262			
.529	1.0581	.8993	.99701	.00463	1.13052	1.00023	.53285	.53729	.54269	.55030			
							1.00067	1.00133	1.00197	1.00261			
.530	1.0565	.8996	.99718	.00438	1.12670	1.00022	.53398	.53861	.54435	.55314			
							1.00066	1.00132	1.00196	1.00260			
.531	1.0548	.8999	.99734	.00414	1.12289	1.00020	.53511	.53995	.54611	.55689			
							1.00066	1.00132	1.00195	1.00258			
.532	1.0532	.9002	.99749	.00390	1.11909	1.00018	.53625	.54133	.54808				
							1.00066	1.00131	1.00195				
.533	1.0515	.9004	.99764	.00368	1.11530	1.00017	.53740	.54274	.55019				
							1.00066	1.00131	1.00194				
.534	1.0499	.9007	.99779	.00345	1.11151	1.00015	.53857	.54421	.55259				
							1.00066	1.00130	1.00193				
.535	1.0482	.9010	.99793	.00324	1.10774	1.00014	.53975	.54573	.55553				
							1.00065	1.00129	1.00192				
.536	1.0466	.9013	.99807	.00303	1.10397	1.00012	.54094	.54733	.56043				
							1.00065	1.00129	1.00191				
.537	1.0449	.9016	.99820	.00283	1.10021	1.00011	.54216	.54913					
							1.00065	1.00129					
.538	1.0433	.9018	.99833	.00263	1.09646	1.00010	.54339	.55101					
							1.00065	1.00128					
.539	1.0416	.9021	.99845	.00244	1.09272	1.00009	.54465	.55311					
							1.00064	1.00128					

TABLE VII. Specific Heat Ratio = 1.20

p/p_t	MACH	T/T_t	GAMMA	K_{crit}	P_y/P_x	P_{tx}/P_{ty}	.001	.002	.003	.004	.005	.01	.015
.540	1.0400	.9024	.99857	.00226	1.08898	1.00008	.54594 1.00064	.55558 1.00127					
.541	1.0383	.9027	.99869	.00208	1.08525	1.00007	.54727 1.00064	.55915 1.00126					
.542	1.0367	.9030	.99880	.00191	1.08154	1.00006	.54872 1.00064						
.543	1.0351	.9032	.99890	.00175	1.07782	1.00005	.55018 1.00064						
.544	1.0334	.9035	.99900	.00160	1.07412	1.00005	.55173 1.00063						
.545	1.0318	.9038	.99910	.00145	1.07043	1.00004	.55340 1.00063						
.546	1.0301	.9041	.99919	.00130	1.06674	1.00003	.55525 1.00063						
.547	1.0285	.9043	.99927	.00117	1.06306	1.00003	.55740 1.00062						
.548	1.0269	.9046	.99935	.00104	1.05939	1.00002	.56068 1.00062						

The header note over the last columns reads: **R_1 and TPR versus Loss Coefficient K**. The note "SHOCK" spans the P_y/P_x and P_{tx}/P_{ty} columns.

TABLE VIII. Specific Heat Ratio = 1.30

p/p_t	MACH	T/T_t	GAMMA	K_crit	SHOCK P_Y/P_X	P_ty/P_ty	.1	.2	.3	.4	.5	.6	.7
.001	5.1146	.2031	.01939	.86121	29.44085	30.93404	.00161	.00225	.00292	.00365	.00445	.00537	.00648
							3.62905	8.11987	14.16915	21.25170	28.78352	36.20906	43.04468
.002	4.6159	.2383	.03231	.82791	23.95563	18.92560	.00303	.00411	.00526	.00651	.00790	.00952	.01154
							2.95663	5.99102	9.86755	14.24661	18.77990	23.14988	27.07399
.003	4.3368	.2617	.04346	.80513	21.13098	14.25885	.00440	.00587	.00745	.00916	.01109	.01336	.01628
							2.64619	5.06504	8.05521	11.35985	14.72260	17.91379	20.72867
.004	4.1439	.2797	.05356	.78724	19.28124	11.69024	.00574	.00757	.00954	.01170	.01413	.01703	.02087
							2.45595	4.51533	7.00342	9.70811	12.42472	14.97127	17.18285
.005	3.9969	.2944	.06293	.77226	17.92869	10.03566	.00706	.00923	.01157	.01415	.01707	.02059	.02540
							2.32305	4.14126	6.29730	8.61169	10.91171	13.04460	14.87215
.006	3.8785	.3071	.07176	.75924	16.87412	8.86835	.00836	.01086	.01356	.01654	.01994	.02407	.02992
							2.22297	3.86487	5.78295	7.81944	9.82457	11.66740	13.22520
.007	3.7794	.3182	.08014	.74764	16.01647	7.99425	.00965	.01247	.01551	.01888	.02275	.02751	.03448
							2.14382	3.64974	5.38690	7.21351	8.99779	10.62397	11.98090
.008	3.6943	.3282	.08816	.73713	15.29781	7.31154	.01093	.01405	.01743	.02119	.02552	.03090	.03911
							2.07903	3.47617	5.06977	6.73156	8.34322	9.80067	11.00122
.009	3.6199	.3372	.09586	.72749	14.68203	6.76134	.01221	.01562	.01933	.02346	.02825	.03427	.04389
							2.02460	3.33201	4.80859	6.33674	7.80920	9.13104	10.20553
.010	3.5537	.3455	.10330	.71855	14.14519	6.30701	.01347	.01718	.02121	.02571	.03096	.03761	.04891
							1.97809	3.20977	4.58846	6.00604	7.36334	8.57353	9.54345
.011	3.4941	.3532	.11051	.71020	13.67069	5.92451	.01473	.01872	.02307	.02794	.03364	.04095	.05438
							1.93753	3.10436	4.40006	5.72384	6.98413	8.10043	8.98162
.012	3.4400	.3604	.11750	.70234	13.24651	5.59736	.01599	.02026	.02491	.03014	.03630	.04428	.06109
							1.90176	3.01220	4.23624	5.47946	6.65665	7.69283	8.49673
.013	3.3904	.3671	.12431	.69492	12.86377	5.31384	.01724	.02178	.02674	.03233	.03894	.04762	
							1.86988	2.93068	4.09207	5.26523	6.37050	7.33718	
.014	3.3447	.3734	.13094	.68787	12.51564	5.06540	.01848	.02329	.02856	.03450	.04157	.05096	
							1.84121	2.85808	3.96393	5.07547	6.11763	7.02348	
.015	3.3023	.3794	.13741	.68114	12.19686	4.84561	.01972	.02480	.03036	.03666	.04419	.05431	
							1.81523	2.79248	3.84905	4.90588	5.89218	6.74418	
.016	3.2627	.3851	.14374	.67471	11.90321	4.64957	.02096	.02630	.03216	.03881	.04680	.05768	
							1.79152	2.73300	3.74528	4.75317	5.68958	6.49366	
.017	3.2256	.3905	.14994	.66854	11.63133	4.47344	.02220	.02779	.03395	.04095	.04940	.06108	
							1.76977	2.67872	3.65095	4.61462	5.50629	6.26727	
.018	3.1908	.3957	.15601	.66261	11.37845	4.31420	.02343	.02928	.03572	.04308	.05200	.06450	
							1.74951	2.62892	3.56470	4.48840	5.33945	6.06143	
.019	3.1579	.4007	.16197	.65689	11.14230	4.16942	.02466	.03076	.03750	.04520	.05459	.06796	
							1.73093	2.58299	3.48543	4.37267	5.18678	5.87326	
.020	3.1267	.4054	.16781	.65136	10.92097	4.03712	.02588	.03224	.03926	.04732	.05717	.07145	
							1.71365	2.54046	3.41225	4.26608	5.04639	5.70038	
.021	3.0971	.4100	.17356	.64602	10.71286	3.91567	.02711	.03371	.04102	.04942	.05976	.07501	
							1.69750	2.50092	3.34442	4.16749	4.91683	5.54084	
.022	3.0690	.4145	.17920	.64083	10.51661	3.80374	.02833	.03517	.04277	.05152	.06235	.07862	
							1.68237	2.46402	3.28130	4.07595	4.79665	5.39302	
.023	3.0421	.4187	.18476	.63580	10.33103	3.70018	.02955	.03664	.04452	.05362	.06493	.08231	
							1.66815	2.42947	3.22238	3.99067	4.68477	5.25555	
.024	3.0164	.4229	.19023	.63091	10.15514	3.60405	.03076	.03810	.04626	.05571	.06752	.08609	
							1.65474	2.39705	3.16721	3.91096	4.58044	5.12733	
.025	2.9918	.4269	.19562	.62615	9.98804	3.51454	.03198	.03955	.04800	.05780	.07011	.08998	
							1.64208	2.36652	3.11540	3.83624	4.48277	5.00726	
.026	2.9682	.4307	.20092	.62151	9.82896	3.43096	.03319	.04101	.04973	.05988	.07271	.09401	
							1.63009	2.33771	3.06662	3.76602	4.39109	4.89453	
.027	2.9455	.4345	.20616	.61699	9.67725	3.35270	.03440	.04246	.05146	.06196	.07531	.09824	
							1.61870	2.31045	3.02059	3.69987	4.30481	4.78839	
.028	2.9237	.4382	.21132	.61257	9.53230	3.27924	.03561	.04390	.05318	.06404	.07792	.10273	
							1.60788	2.28462	2.97706	3.63740	4.22343	4.68817	
.029	2.9026	.4417	.21641	.60826	9.39359	3.21015	.03682	.04535	.05490	.06612	.08053	.10765	
							1.59756	2.26009	2.93580	3.57830	4.14650	4.59327	

TABLE VIII. Specific Heat Ratio = 1.30

P/P$_t$	MACH	T/T$_t$	GAMMA	K$_{crit}$	SHOCK P$_y$/P$_x$	P$_{tx}$/P$_{ty}$	R₂ and TPR versus Loss Coefficient K .05	.1	.2	.3	.4	.5	.6
.030	2.8823	.4452	.22144	.60403	9.26064	3.14501	.03394	.03803	.04679	.05662	.06819	.08315	.11336
							1.28298	1.58770	2.23675	2.89662	3.52226	4.07363	4.50319
.031	2.8626	.4486	.22640	.59990	9.13303	3.08349	.03504	.03924	.04823	.05834	.07026	.08578	
							1.27878	1.57829	2.21450	2.85936	3.46903	4.00449	
.032	2.8436	.4519	.23130	.59585	9.01040	3.02527	.03614	.04044	.04967	.06005	.07234	.08842	
							1.27474	1.56929	2.19326	2.82386	3.41838	3.93875	
.033	2.8252	.4551	.23614	.59188	8.89239	2.97009	.03724	.04164	.05110	.06177	.07441	.09108	
							1.27087	1.56065	2.17296	2.79007	3.37012	3.87616	
.034	2.8073	.4583	.24093	.58799	8.77871	2.91770	.03834	.04285	.05253	.06347	.07648	.09374	
							1.26714	1.55237	2.15352	2.75768	3.32406	3.81647	
.035	2.7900	.4613	.24566	.58417	8.66906	2.86788	.03943	.04405	.05397	.06518	.07855	.09642	
							1.26356	1.54441	2.13489	2.72669	3.28003	3.75946	
.036	2.7732	.4643	.25034	.58042	8.56321	2.82045	.04053	.04525	.05539	.06689	.08062	.09912	
							1.26010	1.53675	2.11701	2.69700	3.23790	3.70493	
.037	2.7568	.4673	.25497	.57673	8.46091	2.77521	.04163	.04645	.05682	.06859	.08269	.10183	
							1.25677	1.52937	2.09982	2.66851	3.19752	3.65272	
.038	2.7409	.4702	.25955	.57311	8.36197	2.73203	.04272	.04764	.05825	.07029	.08476	.10456	
							1.25356	1.52225	2.08329	2.64115	3.15878	3.60265	
.039	2.7254	.4730	.26408	.56954	8.26617	2.69075	.04382	.04884	.05967	.07199	.08684	.10731	
							1.25045	1.51539	2.06737	2.61485	3.12158	3.55459	
.040	2.7103	.4758	.26857	.56604	8.17336	2.65125	.04491	.05004	.06110	.07369	.08891	.11008	
							1.24744	1.50876	2.05202	2.58953	3.08581	3.50839	
.041	2.6956	.4785	.27301	.56258	8.08335	2.61340	.04601	.05123	.06252	.07539	.09099	.11287	
							1.24453	1.50234	2.03722	2.56513	3.05138	3.46395	
.042	2.6812	.4812	.27740	.55918	7.99601	2.57711	.04710	.05243	.06394	.07709	.09307	.11569	
							1.24170	1.49614	2.02292	2.54161	3.01825	3.42115	
.043	2.6672	.4838	.28175	.55583	7.91119	2.54227	.04820	.05362	.06536	.07879	.09516	.11854	
							1.23897	1.49013	2.00910	2.51890	2.98627	3.37989	
.044	2.6535	.4864	.28606	.55253	7.82876	2.50879	.04929	.05482	.06678	.08048	.09724	.12142	
							1.23632	1.48431	1.99573	2.49697	2.95540	3.34008	
.045	2.6401	.4889	.29034	.54928	7.74861	2.47659	.05038	.05601	.06819	.08218	.09933	.12433	
							1.23374	1.47866	1.98279	2.47577	2.92558	3.30163	
.046	2.6270	.4914	.29457	.54607	7.67062	2.44560	.05148	.05720	.06961	.08387	.10142	.12728	
							1.23123	1.47318	1.97025	2.45525	2.89675	3.26446	
.047	2.6141	.4938	.29876	.54290	7.59469	2.41575	.05257	.05839	.07103	.08557	.10351	.13027	
							1.22880	1.46786	1.95810	2.43538	2.86886	3.22850	
.048	2.6016	.4962	.30292	.53978	7.52072	2.38697	.05366	.05958	.07244	.08726	.10561	.13331	
							1.22643	1.46273	1.94631	2.41613	2.84186	3.19368	
.049	2.5893	.4986	.30703	.53669	7.44863	2.35920	.05475	.06078	.07385	.08896	.10771	.13640	
							1.22412	1.45771	1.93486	2.39747	2.81570	3.15998	
.050	2.5773	.5009	.31112	.53365	7.37832	2.33239	.05585	.06196	.07527	.09065	.10982	.13955	
							1.22188	1.45281	1.92374	2.37936	2.79033	3.12727	
.051	2.5655	.5032	.31517	.53064	7.30973	2.30648	.05694	.06315	.07668	.09235	.11193	.14276	
							1.21969	1.44805	1.91293	2.36177	2.76572	3.09552	
.052	2.5539	.5055	.31918	.52767	7.24277	2.28144	.05803	.06434	.07809	.09404	.11405	.14604	
							1.21756	1.44341	1.90242	2.34469	2.74184	3.06470	
.053	2.5426	.5077	.32316	.52473	7.17737	2.25721	.05912	.06553	.07950	.09574	.11617	.14941	
							1.21547	1.43889	1.89220	2.32809	2.71868	3.03473	
.054	2.5314	.5099	.32711	.52183	7.11348	2.23375	.06021	.06672	.08091	.09743	.11830	.15288	
							1.21344	1.43448	1.88224	2.31195	2.69612	3.00559	
.055	2.5205	.5121	.33103	.51896	7.05103	2.21104	.06130	.06791	.08232	.09913	.12043	.15646	
							1.21146	1.43018	1.87254	2.29624	2.67419	2.97723	
.056	2.5097	.5142	.33491	.51612	6.98996	2.18902	.06239	.06909	.08373	.10083	.12257	.16020	
							1.20952	1.42599	1.86309	2.28094	2.65284	2.94960	
.057	2.4992	.5163	.33877	.51331	6.93021	2.16767	.06348	.07028	.08514	.10252	.12471	.16411	
							1.20767	1.42189	1.85388	2.26604	2.63207	2.92267	
.058	2.4888	.5184	.34259	.51054	6.87175	2.14696	.06457	.07147	.08655	.10422	.12687	.16827	
							1.20582	1.41789	1.84489	2.25152	2.61183	2.89638	
.059	2.4786	.5204	.34639	.50779	6.81451	2.12686	.06566	.07265	.08796	.10592	.12903	.17278	
							1.20401	1.41398	1.83612	2.23737	2.59211	2.87075	

144

TABLE VIII. Specific Heat Ratio = 1.30

p/p_t	MACH	T/T_t	GAMMA	K_crit	SHOCK P_y/P_x	P_ty/P_tx	.01	.05	.1	.2	.3	.4	.5
.060	2.4686	.5224	.35016	.50507	6.75845	2.10734	.06133	.06675	.07384	.08937	.10762	.13119	.17783
							1.03982	1.20223	1.41016	1.82755	2.22356	2.57289	2.84566
.061	2.4587	.5244	.35389	.50238	6.70353	2.08837	.06234	.06784	.07502	.09077	.10932	.13337	.18394
							1.03950	1.20049	1.40642	1.81919	2.21009	2.55414	2.82109
.062	2.4490	.5264	.35760	.49971	6.64971	2.06993	.06336	.06893	.07621	.09218	.11102	.13555	
							1.03918	1.19880	1.40276	1.81102	2.19693	2.53585	
.063	2.4395	.5284	.36129	.49707	6.59694	2.05200	.06438	.07001	.07739	.09359	.11273	.13774	
							1.03887	1.19713	1.39918	1.80303	2.18409	2.51799	
.064	2.4301	.5303	.36494	.49446	6.54520	2.03455	.06540	.07110	.07858	.09500	.11443	.13994	
							1.03857	1.19551	1.39568	1.79522	2.17154	2.50054	
.065	2.4208	.5322	.36858	.49187	6.49444	2.01757	.06641	.07219	.07976	.09641	.11614	.14216	
							1.03827	1.19391	1.39225	1.78758	2.15927	2.48350	
.066	2.4117	.5341	.37218	.48931	6.44464	2.00104	.06743	.07328	.08095	.09781	.11784	.14438	
							1.03798	1.19235	1.38889	1.78010	2.14727	2.46685	
.067	2.4027	.5359	.37576	.48676	6.39575	1.98493	.06845	.07437	.08213	.09922	.11955	.14661	
							1.03769	1.19081	1.38559	1.77279	2.13554	2.45056	
.068	2.3939	.5377	.37931	.48425	6.34776	1.96924	.06946	.07545	.08331	.10063	.12126	.14885	
							1.03741	1.18931	1.38237	1.76562	2.12406	2.43463	
.069	2.3852	.5396	.38284	.48175	6.30062	1.95394	.07048	.07654	.08450	.10204	.12298	.15111	
							1.03714	1.18783	1.37920	1.75860	2.11283	2.41905	
.070	2.3766	.5414	.38635	.47928	6.25432	1.93903	.07150	.07763	.08568	.10344	.12469	.15337	
							1.03686	1.18638	1.37609	1.75173	2.10183	2.40379	
.071	2.3681	.5431	.38983	.47682	6.20883	1.92448	.07251	.07872	.08686	.10485	.12641	.15565	
							1.03660	1.18496	1.37305	1.74499	2.09106	2.38885	
.072	2.3597	.5449	.39329	.47439	6.16412	1.91028	.07353	.07981	.08804	.10626	.12813	.15794	
							1.03633	1.18356	1.37006	1.73838	2.08050	2.37422	
.073	2.3515	.5466	.39673	.47198	6.12016	1.89642	.07455	.08089	.08923	.10767	.12985	.16025	
							1.03608	1.18219	1.36712	1.73190	2.07016	2.35988	
.074	2.3433	.5483	.40014	.46959	6.07694	1.88288	.07557	.08198	.09041	.10908	.13157	.16257	
							1.03582	1.18084	1.36424	1.72554	2.06002	2.34583	
.075	2.3353	.5500	.40353	.46722	6.03444	1.86966	.07658	.08307	.09159	.11049	.13330	.16491	
							1.03557	1.17952	1.36141	1.71931	2.05011	2.33206	
.076	2.3273	.5517	.40690	.46486	5.99262	1.85675	.07760	.08415	.09277	.11189	.13502	.16727	
							1.03533	1.17821	1.35863	1.71319	2.04036	2.31854	
.077	2.3195	.5534	.41025	.46253	5.95148	1.84413	.07862	.08524	.09396	.11330	.13675	.16964	
							1.03509	1.17693	1.35590	1.70718	2.03079	2.30529	
.078	2.3118	.5550	.41357	.46021	5.91099	1.83179	.07963	.08633	.09514	.11471	.13848	.17203	
							1.03485	1.17567	1.35321	1.70128	2.02140	2.29228	
.079	2.3041	.5567	.41688	.45791	5.87113	1.81972	.08065	.08741	.09632	.11612	.14022	.17444	
							1.03462	1.17443	1.35057	1.69548	2.01219	2.27951	
.080	2.2966	.5583	.42016	.45563	5.83189	1.80792	.08167	.08850	.09750	.11753	.14196	.17687	
							1.03439	1.17322	1.34798	1.68979	2.00314	2.26698	
.081	2.2891	.5599	.42343	.45337	5.79325	1.79638	.08268	.08959	.09868	.11894	.14370	.17932	
							1.03416	1.17202	1.34542	1.68420	1.99425	2.25466	
.082	2.2818	.5615	.42667	.45112	5.75519	1.78508	.08370	.09067	.09987	.12036	.14544	.18180	
							1.03394	1.17084	1.34291	1.67870	1.98552	2.24257	
.083	2.2745	.5631	.42989	.44889	5.71770	1.77402	.08472	.09176	.10105	.12177	.14719	.18430	
							1.03372	1.16967	1.34044	1.67329	1.97695	2.23068	
.084	2.2673	.5646	.43310	.44667	5.68076	1.76319	.08573	.09285	.10223	.12318	.14894	.18682	
							1.03350	1.16853	1.33801	1.66798	1.96852	2.21899	
.085	2.2602	.5662	.43628	.44447	5.64436	1.75258	.08675	.09393	.10341	.12459	.15069	.18938	
							1.03329	1.16740	1.33562	1.66275	1.96023	2.20750	
.086	2.2532	.5677	.43945	.44229	5.60848	1.74219	.08777	.09502	.10459	.12601	.15245	.19197	
							1.03308	1.16629	1.33326	1.65761	1.95208	2.19620	
.087	2.2462	.5692	.44260	.44012	5.57311	1.73202	.08878	.09611	.10577	.12742	.15421	.19458	
							1.03287	1.16520	1.33094	1.65255	1.94407	2.18508	
.088	2.2393	.5707	.44572	.43796	5.53824	1.72204	.08980	.09719	.10696	.12884	.15597	.19724	
							1.03267	1.16412	1.32866	1.64757	1.93619	2.17416	
.089	2.2325	.5722	.44883	.43582	5.50385	1.71226	.09082	.09828	.10814	.13025	.15774	.19994	
							1.03246	1.16306	1.32641	1.64267	1.92844	2.16340	

TABLE VIII. Specific Heat Ratio = 1.30

p/p_t	MACH	T/T_t	GAMMA	K_{crit}	SHOCK P_y/P_x	P_ty/P_ty	.01	.05	.1	.15	.2	.3	.4
.090	2.2258	.5737	.45193	.43370	5.46993	1.70268	.09183	.09936	.10932	.12001	.13167	.15951	.20287
							1.03227	1.16201	1.32419	1.48364	1.63786	1.92081	2.15280
.091	2.2191	.5751	.45500	.43159	5.43648	1.69328	.09285	.10045	.11050	.12130	.13309	.16128	.20545
							1.03207	1.16098	1.32200	1.48021	1.63311	1.91330	2.14236
.092	2.2126	.5766	.45806	.42949	5.40347	1.68406	.09387	.10154	.11168	.12259	.13450	.16306	.20828
							1.03188	1.15996	1.31985	1.47683	1.62843	1.90590	2.13208
.093	2.2060	.5780	.46110	.42741	5.37090	1.67502	.09488	.10262	.11287	.12389	.13592	.16484	.21116
							1.03169	1.15899	1.31773	1.47350	1.62382	1.89863	2.12195
.094	2.1996	.5795	.46412	.42533	5.33876	1.66614	.09590	.10371	.11405	.12518	.13734	.16663	.21411
							1.03150	1.15800	1.31563	1.47022	1.61928	1.89146	2.11196
.095	2.1932	.5809	.46712	.42328	5.30704	1.65743	.09692	.10480	.11523	.12647	.13876	.16842	.21712
							1.03131	1.15702	1.31357	1.46698	1.61481	1.88440	2.10212
.096	2.1869	.5823	.47011	.42123	5.27573	1.64888	.09793	.10588	.11641	.12776	.14018	.17022	.22021
							1.03113	1.15606	1.31154	1.46379	1.61040	1.87744	2.09241
.097	2.1806	.5837	.47308	.41920	5.24482	1.64049	.09895	.10697	.11760	.12905	.14160	.17202	.22339
							1.03095	1.15511	1.30953	1.46065	1.60606	1.87059	2.08283
.098	2.1744	.5851	.47603	.41718	5.21429	1.63225	.09996	.10806	.11878	.13035	.14303	.17383	.22666
							1.03077	1.15417	1.30755	1.45754	1.60174	1.86384	2.07338
.099	2.1683	.5864	.47897	.41517	5.18415	1.62416	.10098	.10914	.11996	.13164	.14445	.17564	.23006
							1.03059	1.15325	1.30560	1.45449	1.59752	1.85718	2.06405
.100	2.1622	.5878	.48189	.41318	5.15438	1.61621	.10200	.11023	.12114	.13293	.14588	.17746	.23360
							1.03042	1.15233	1.30367	1.45147	1.59336	1.85062	2.05484
.101	2.1562	.5892	.48480	.41119	5.12497	1.60841	.10301	.11131	.12233	.13423	.14730	.17928	.23731
							1.03025	1.15143	1.30177	1.44849	1.58925	1.84415	2.04573
.102	2.1502	.5905	.48769	.40922	5.09592	1.60074	.10403	.11240	.12351	.13552	.14873	.18110	.24126
							1.03008	1.15054	1.29989	1.44556	1.58520	1.83777	2.03675
.103	2.1443	.5918	.49057	.40726	5.06722	1.59320	.10505	.11349	.12469	.13682	.15016	.18294	.24549
							1.02991	1.14966	1.29804	1.44266	1.58121	1.83148	2.02785
.104	2.1384	.5931	.49342	.40531	5.03886	1.58579	.10606	.11457	.12588	.13811	.15159	.18478	.25015
							1.02974	1.14879	1.29621	1.43980	1.57727	1.82527	2.01905
.105	2.1326	.5945	.49627	.40337	5.01083	1.57851	.10708	.11566	.12706	.13941	.15302	.18662	.25548
							1.02958	1.14793	1.29441	1.43698	1.57338	1.81915	2.01031
.106	2.1269	.5958	.49910	.40144	4.98313	1.57135	.10810	.11675	.12824	.14070	.15445	.18847	.26227
							1.02942	1.14709	1.29262	1.43419	1.56954	1.81311	2.00166
.107	2.1212	.5971	.50191	.39952	4.95575	1.56430	.10911	.11783	.12943	.14200	.15588	.19033	
							1.02925	1.14625	1.29086	1.43144	1.56575	1.80715	
.108	2.1155	.5983	.50471	.39762	4.92869	1.55738	.11013	.11892	.13061	.14330	.15731	.19219	
							1.02910	1.14542	1.28912	1.42872	1.56201	1.80126	
.109	2.1099	.5996	.50749	.39572	4.90193	1.55057	.11115	.12001	.13180	.14459	.15875	.19406	
							1.02894	1.14460	1.28740	1.42604	1.55832	1.79545	
.110	2.1044	.6009	.51026	.39384	4.87547	1.54387	.11216	.12109	.13298	.14589	.16019	.19594	
							1.02879	1.14380	1.28571	1.42339	1.55467	1.78972	
.111	2.0988	.6021	.51302	.39196	4.84930	1.53728	.11318	.12218	.13416	.14719	.16162	.19783	
							1.02863	1.14300	1.28403	1.42077	1.55107	1.78406	
.112	2.0934	.6034	.51576	.39009	4.82342	1.53080	.11420	.12327	.13535	.14849	.16306	.19972	
							1.02848	1.14221	1.28237	1.41819	1.54752	1.77846	
.113	2.0880	.6046	.51849	.38824	4.79783	1.52442	.11521	.12435	.13654	.14979	.16450	.20162	
							1.02833	1.14143	1.28073	1.41563	1.54400	1.77294	
.114	2.0826	.6058	.52120	.38639	4.77251	1.51814	.11623	.12544	.13772	.15109	.16595	.20353	
							1.02818	1.14066	1.27911	1.41311	1.54053	1.76748	
.115	2.0773	.6071	.52390	.38456	4.74746	1.51196	.11725	.12653	.13891	.15239	.16739	.20544	
							1.02804	1.13989	1.27751	1.41062	1.53710	1.76209	
.116	2.0720	.6083	.52658	.38273	4.72269	1.50587	.11826	.12761	.14009	.15370	.16884	.20737	
							1.02789	1.13914	1.27593	1.40815	1.53371	1.75677	
.117	2.0668	.6095	.52926	.38091	4.69817	1.49988	.11928	.12870	.14128	.15500	.17028	.20930	
							1.02775	1.13839	1.27436	1.40571	1.53037	1.75150	
.118	2.0616	.6107	.53191	.37910	4.67391	1.49398	.12029	.12979	.14247	.15630	.17173	.21124	
							1.02761	1.13765	1.27282	1.40330	1.52706	1.74630	
.119	2.0564	.6119	.53456	.37730	4.64990	1.48817	.12131	.13087	.14365	.15761	.17318	.21320	
							1.02747	1.13692	1.27129	1.40092	1.52378	1.74116	

TABLE VIII. Specific Heat Ratio = 1.30

p/p_t	MACH	T/T_t	GAMMA	K_{crit}	SHOCK P_y/P_x	P_{ty}/P_{ty}	.01	.05	.1	.15	.2	.25	.3
.120	2.0513	.6131	.53719	.37551	4.62614	1.48245	.12233	.13196	.14484	.15891	.17463	.19280	.21516
							1.02733	1.13620	1.26977	1.39857	1.52055	1.63378	1.73606
.121	2.0462	.6142	.53981	.37373	4.60262	1.47681	.12334	.13305	.14603	.16022	.17609	.19445	.21713
							1.02719	1.13548	1.26828	1.39624	1.51735	1.62969	1.73104
.122	2.0412	.6154	.54242	.37195	4.57934	1.47126	.12436	.13414	.14721	.16153	.17754	.19611	.21911
							1.02705	1.13478	1.26680	1.39394	1.51419	1.62564	1.72606
.123	2.0362	.6166	.54501	.37019	4.55629	1.46579	.12538	.13522	.14840	.16283	.17900	.19777	.22110
							1.02692	1.13408	1.26533	1.39166	1.51106	1.62163	1.72115
.124	2.0312	.6177	.54759	.36843	4.53347	1.46040	.12639	.13631	.14959	.16414	.18046	.19943	.22311
							1.02678	1.13338	1.26388	1.38941	1.50797	1.61767	1.71629
.125	2.0263	.6189	.55016	.36668	4.51088	1.45509	.12741	.13740	.15078	.16545	.18192	.20110	.22513
							1.02665	1.13270	1.26245	1.38718	1.50491	1.61376	1.71150
.126	2.0214	.6200	.55271	.36494	4.48850	1.44986	.12843	.13849	.15197	.16676	.18338	.20278	.22716
							1.02652	1.13202	1.26103	1.38497	1.50189	1.60988	1.70674
.127	2.0165	.6211	.55525	.36321	4.46635	1.44470	.12944	.13958	.15316	.16807	.18484	.20445	.22920
							1.02639	1.13131	1.25963	1.38279	1.49890	1.60605	1.70203
.128	2.0117	.6223	.55778	.36149	4.44441	1.43961	.13046	.14066	.15435	.16938	.18631	.20614	.23126
							1.02626	1.13064	1.25824	1.38063	1.49594	1.60226	1.69738
.129	2.0069	.6234	.56030	.35977	4.42268	1.43460	.13148	.14175	.15554	.17070	.18778	.20782	.23333
							1.02614	1.12998	1.25686	1.37850	1.49301	1.59851	1.69277
.130	2.0022	.6245	.56281	.35806	4.40116	1.42966	.13249	.14284	.15673	.17201	.18925	.20951	.23541
							1.02601	1.12933	1.25550	1.37638	1.49011	1.59480	1.68821
.131	1.9975	.6256	.56530	.35636	4.37983	1.42479	.13351	.14393	.15792	.17332	.19072	.21121	.23751
							1.02589	1.12869	1.25415	1.37429	1.48724	1.59112	1.68369
.132	1.9928	.6267	.56778	.35467	4.35871	1.41998	.13453	.14502	.15911	.17464	.19219	.21291	.23962
							1.02576	1.12805	1.25282	1.37222	1.48440	1.58749	1.67922
.133	1.9881	.6278	.57025	.35298	4.33779	1.41525	.13554	.14611	.16030	.17596	.19367	.21462	.24175
							1.02564	1.12741	1.25150	1.37017	1.48159	1.58389	1.67479
.134	1.9835	.6289	.57271	.35131	4.31705	1.41057	.13656	.14720	.16150	.17727	.19515	.21633	.24390
							1.02552	1.12679	1.25019	1.36814	1.47881	1.58033	1.67041
.135	1.9789	.6300	.57515	.34963	4.29651	1.40597	.13758	.14828	.16269	.17859	.19663	.21805	.24607
							1.02540	1.12617	1.24889	1.36613	1.47605	1.57680	1.66607
.136	1.9744	.6310	.57759	.34797	4.27615	1.40142	.13859	.14937	.16388	.17991	.19811	.21977	.24825
							1.02528	1.12555	1.24761	1.36414	1.47333	1.57331	1.66177
.137	1.9698	.6321	.58001	.34632	4.25598	1.39694	.13961	.15046	.16507	.18123	.19960	.22150	.25045
							1.02519	1.12494	1.24634	1.36217	1.47063	1.56985	1.65751
.138	1.9654	.6332	.58242	.34467	4.23599	1.39251	.14063	.15155	.16627	.18255	.20109	.22324	.25268
							1.02507	1.12434	1.24508	1.36022	1.46796	1.56643	1.65329
.139	1.9609	.6342	.58482	.34302	4.21617	1.38815	.14165	.15264	.16746	.18388	.20258	.22498	.25493
							1.02496	1.12374	1.24384	1.35829	1.46531	1.56304	1.64911
.140	1.9565	.6353	.58721	.34139	4.19653	1.38384	.14266	.15373	.16866	.18520	.20407	.22672	.25720
							1.02484	1.12315	1.24260	1.35637	1.46269	1.55968	1.64497
.141	1.9520	.6363	.58959	.33976	4.17706	1.37959	.14368	.15482	.16985	.18652	.20557	.22848	.25949
							1.02473	1.12256	1.24138	1.35448	1.46009	1.55635	1.64087
.142	1.9477	.6373	.59196	.33814	4.15776	1.37539	.14470	.15591	.17105	.18785	.20707	.23024	.26181
							1.02462	1.12198	1.24016	1.35260	1.45752	1.55306	1.63680
.143	1.9433	.6384	.59431	.33653	4.13863	1.37125	.14572	.15700	.17225	.18918	.20857	.23200	.26416
							1.02451	1.12140	1.23896	1.35074	1.45497	1.54980	1.63276
.144	1.9390	.6394	.59666	.33492	4.11966	1.36717	.14673	.15809	.17344	.19051	.21007	.23378	.26654
							1.02440	1.12083	1.23777	1.34890	1.45244	1.54654	1.62876
.145	1.9347	.6404	.59899	.33332	4.10085	1.36313	.14775	.15918	.17464	.19184	.21158	.23556	.26895
							1.02429	1.12026	1.23659	1.34707	1.44994	1.54333	1.62480
.146	1.9304	.6414	.60132	.33172	4.08220	1.35915	.14877	.16027	.17584	.19317	.21309	.23735	.27140
							1.02418	1.11970	1.23542	1.34526	1.44747	1.54016	1.62087
.147	1.9262	.6425	.60363	.33013	4.06371	1.35522	.14978	.16136	.17704	.19450	.21460	.23914	.27388
							1.02407	1.11914	1.23427	1.34347	1.44501	1.53701	1.61697
.148	1.9220	.6435	.60593	.32855	4.04537	1.35134	.15080	.16245	.17823	.19583	.21612	.24095	.27639
							1.02396	1.11859	1.23312	1.34169	1.44258	1.53389	1.61310
.149	1.9178	.6445	.60822	.32698	4.02718	1.34751	.15182	.16355	.17943	.19717	.21763	.24276	.27896
							1.02386	1.11804	1.23198	1.33993	1.44016	1.53080	1.60926

TABLE VIII. Specific Heat Ratio = 1.30

p/p_t	MACH	T/T_t	GAMMA	K_crit	SHOCK p_y/p_x	p_ty/p_ty	.01	.05	.1	.15	.2	.25	.3
.150	1.9136	.6455	.61050	.32541	4.00915	1.34373	.15284	.16464	.18063	.19850	.21916	.24458	.28156
							1.02375	1.11750	1.23085	1.33818	1.43777	1.52774	1.60545
.151	1.9095	.6464	.61277	.32384	3.99126	1.33999	.15385	.16573	.18184	.19984	.22068	.24641	.28422
							1.02365	1.11696	1.22973	1.33645	1.43541	1.52470	1.60167
.152	1.9054	.6474	.61503	.32229	3.97351	1.33630	.15487	.16682	.18304	.20118	.22221	.24824	.28693
							1.02355	1.11643	1.22862	1.33473	1.43306	1.52169	1.59792
.153	1.9013	.6484	.61728	.32074	3.95591	1.33266	.15589	.16791	.18424	.20252	.22374	.25009	.28970
							1.02344	1.11590	1.22752	1.33303	1.43073	1.51870	1.59419
.154	1.8972	.6494	.61952	.31919	3.93845	1.32906	.15690	.16900	.18544	.20386	.22528	.25195	.29254
							1.02334	1.11538	1.22643	1.33135	1.42842	1.51574	1.59049
.155	1.8932	.6504	.62175	.31765	3.92113	1.32550	.15792	.17010	.18664	.20521	.22681	.25382	.29545
							1.02324	1.11485	1.22535	1.32967	1.42613	1.51282	1.58682
.156	1.8891	.6513	.62397	.31612	3.90395	1.32199	.15894	.17119	.18785	.20655	.22836	.25570	.29845
							1.02314	1.11434	1.22428	1.32802	1.42386	1.50991	1.58317
.157	1.8852	.6523	.62618	.31459	3.88690	1.31852	.15996	.17228	.18905	.20790	.22990	.25758	.30156
							1.02304	1.11383	1.22322	1.32637	1.42161	1.50702	1.57955
.158	1.8812	.6532	.62838	.31307	3.86998	1.31509	.16097	.17337	.19026	.20924	.23145	.25948	.30477
							1.02294	1.11332	1.22216	1.32474	1.41938	1.50415	1.57595
.159	1.8772	.6542	.63057	.31156	3.85320	1.31171	.16199	.17447	.19146	.21059	.23300	.26139	.30811
							1.02285	1.11281	1.22112	1.32313	1.41717	1.50131	1.57237
.160	1.8733	.6551	.63275	.31005	3.83654	1.30836	.16301	.17556	.19267	.21194	.23456	.26331	.31161
							1.02275	1.11231	1.22008	1.32152	1.41497	1.49849	1.56881
.161	1.8694	.6561	.63492	.30854	3.82001	1.30506	.16403	.17666	.19388	.21330	.23612	.26524	.31530
							1.02265	1.11182	1.21905	1.31993	1.41280	1.49569	1.56526
.162	1.8655	.6570	.63708	.30704	3.80361	1.30179	.16504	.17775	.19508	.21465	.23768	.26719	.31924
							1.02256	1.11133	1.21803	1.31835	1.41064	1.49292	1.56173
.163	1.8616	.6580	.63923	.30555	3.78733	1.29856	.16606	.17884	.19629	.21601	.23925	.26915	.32351
							1.02247	1.11084	1.21702	1.31679	1.40849	1.49016	1.55821
.164	1.8578	.6589	.64138	.30406	3.77118	1.29537	.16708	.17994	.19750	.21737	.24083	.27112	.32824
							1.02237	1.11035	1.21601	1.31524	1.40637	1.48743	1.55470
.165	1.8540	.6598	.64351	.30258	3.75514	1.29222	.16810	.18103	.19871	.21872	.24240	.27311	.33378
							1.02228	1.10987	1.21501	1.31370	1.40426	1.48472	1.55122
.166	1.8502	.6607	.64563	.30110	3.73923	1.28911	.16911	.18213	.19992	.22009	.24398	.27511	.34087
							1.02219	1.10939	1.21403	1.31217	1.40217	1.48202	1.54771
.167	1.8464	.6616	.64774	.29963	3.72343	1.28603	.17013	.18322	.20113	.22145	.24557	.27712	
							1.02209	1.10892	1.21304	1.31065	1.40009	1.47935	
.168	1.8426	.6626	.64985	.29817	3.70775	1.28298	.17115	.18432	.20234	.22281	.24716	.27916	
							1.02200	1.10845	1.21207	1.30915	1.39803	1.47670	
.169	1.8389	.6635	.65194	.29670	3.69218	1.27997	.17217	.18541	.20356	.22418	.24876	.28121	
							1.02191	1.10798	1.21110	1.30766	1.39599	1.47406	
.170	1.8352	.6644	.65403	.29525	3.67673	1.27700	.17319	.18651	.20477	.22555	.25036	.28327	
							1.02182	1.10752	1.21014	1.30618	1.39396	1.47144	
.171	1.8315	.6653	.65610	.29380	3.66138	1.27406	.17420	.18760	.20598	.22692	.25196	.28536	
							1.02173	1.10706	1.20919	1.30471	1.39195	1.46885	
.172	1.8278	.6662	.65817	.29235	3.64615	1.27115	.17522	.18870	.20720	.22829	.25357	.28746	
							1.02165	1.10661	1.20825	1.30325	1.38995	1.46627	
.173	1.8241	.6671	.66023	.29091	3.63103	1.26827	.17624	.18980	.20842	.22967	.25519	.28959	
							1.02156	1.10615	1.20731	1.30180	1.38797	1.46370	
.174	1.8205	.6679	.66228	.28948	3.61601	1.26543	.17726	.19089	.20963	.23104	.25681	.29173	
							1.02147	1.10570	1.20638	1.30036	1.38600	1.46116	
.175	1.8169	.6688	.66432	.28805	3.60110	1.26262	.17828	.19199	.21085	.23242	.25844	.29390	
							1.02138	1.10526	1.20546	1.29894	1.38404	1.45863	
.176	1.8132	.6697	.66635	.28662	3.58629	1.25984	.17929	.19309	.21207	.23380	.26007	.29610	
							1.02130	1.10481	1.20454	1.29752	1.38210	1.45612	
.177	1.8097	.6706	.66837	.28520	3.57159	1.25710	.18031	.19419	.21329	.23519	.26171	.29831	
							1.02121	1.10437	1.20363	1.29612	1.38017	1.45362	
.178	1.8061	.6715	.67038	.28378	3.55699	1.25438	.18133	.19528	.21451	.23657	.26335	.30056	
							1.02113	1.10394	1.20273	1.29472	1.37826	1.45114	
.179	1.8025	.6723	.67239	.28237	3.54249	1.25169	.18235	.19638	.21573	.23796	.26500	.30283	
							1.02104	1.10350	1.20183	1.29333	1.37636	1.44867	

TABLE VIII. Specific Heat Ratio = 1.30

p/p_t	MACH	T/T_t	GAMMA	K_{crit}	SHOCK P_y/P_x	P_{tx}/P_{ty}	.01	.05	.07	.1	.15	.2	.25
.180	1.7990	.6732	.67438	.28097	3.52809	1.24903	.18337	.19748	.20499	.21695	.23935	.26666	.30513
							1.02096	1.10307	1.14294	1.20094	1.29196	1.37447	1.44622
.181	1.7955	.6741	.67637	.27957	3.51378	1.24641	.18438	.19858	.20613	.21818	.24074	.26832	.30747
							1.02088	1.10265	1.14233	1.20005	1.29059	1.37260	1.44379
.182	1.7920	.6749	.67835	.27817	3.49958	1.24381	.18540	.19968	.20728	.21940	.24214	.26999	.30984
							1.02080	1.10222	1.14173	1.19917	1.28924	1.37074	1.44137
.183	1.7885	.6758	.68032	.27678	3.48547	1.24123	.18642	.20078	.20843	.22062	.24353	.27166	.31224
							1.02071	1.10180	1.14113	1.19830	1.28789	1.36889	1.43896
.184	1.7850	.6766	.68228	.27539	3.47145	1.23869	.18744	.20188	.20957	.22185	.24493	.27335	.31469
							1.02063	1.10138	1.14053	1.19744	1.28655	1.36706	1.43657
.185	1.7816	.6775	.68423	.27401	3.45753	1.23617	.18846	.20298	.21072	.22308	.24633	.27503	.31717
							1.02055	1.10096	1.13994	1.19658	1.28523	1.36523	1.43418
.186	1.7781	.6783	.68618	.27263	3.44370	1.23369	.18948	.20408	.21187	.22430	.24774	.27673	.31971
							1.02047	1.10055	1.13936	1.19572	1.28391	1.36342	1.43182
.187	1.7747	.6791	.68811	.27125	3.42996	1.23122	.19050	.20518	.21301	.22553	.24915	.27844	.32229
							1.02039	1.10014	1.13877	1.19487	1.28260	1.36162	1.42946
.188	1.7713	.6800	.69004	.26989	3.41631	1.22879	.19151	.20628	.21416	.22676	.25056	.28015	.32494
							1.02031	1.09973	1.13819	1.19403	1.28130	1.35983	1.42711
.189	1.7679	.6808	.69196	.26852	3.40275	1.22638	.19253	.20738	.21531	.22799	.25197	.28188	.32764
							1.02023	1.09933	1.13762	1.19319	1.28000	1.35807	1.42478
.190	1.7645	.6816	.69387	.26716	3.38928	1.22399	.19355	.20848	.21646	.22923	.25338	.28361	.33041
							1.02016	1.09893	1.13705	1.19236	1.27872	1.35631	1.42246
.191	1.7612	.6825	.69578	.26581	3.37589	1.22164	.19457	.20959	.21761	.23046	.25480	.28534	.33326
							1.02008	1.09853	1.13648	1.19154	1.27745	1.35456	1.42016
.192	1.7578	.6833	.69767	.26445	3.36259	1.21930	.19559	.21069	.21876	.23169	.25622	.28709	.33620
							1.02000	1.09813	1.13592	1.19072	1.27618	1.35281	1.41785
.193	1.7545	.6841	.69956	.26311	3.34938	1.21699	.19661	.21179	.21991	.23293	.25765	.28885	.33923
							1.01992	1.09774	1.13536	1.18990	1.27492	1.35108	1.41556
.194	1.7512	.6849	.70144	.26176	3.33625	1.21471	.19763	.21290	.22106	.23417	.25907	.29061	.34238
							1.01985	1.09734	1.13480	1.18909	1.27367	1.34936	1.41328
.195	1.7479	.6857	.70331	.26043	3.32320	1.21245	.19865	.21400	.22222	.23540	.26050	.29238	.34566
							1.01977	1.09695	1.13425	1.18829	1.27243	1.34765	1.41100
.196	1.7446	.6866	.70517	.25909	3.31024	1.21021	.19966	.21510	.22337	.23664	.26193	.29417	.34909
							1.01970	1.09657	1.13370	1.18749	1.27119	1.34595	1.40874
.197	1.7413	.6874	.70702	.25776	3.29735	1.20799	.20068	.21621	.22452	.23788	.26337	.29596	.35273
							1.01962	1.09618	1.13315	1.18670	1.26997	1.34426	1.40647
.198	1.7381	.6882	.70887	.25644	3.28455	1.20580	.20170	.21731	.22568	.23912	.26481	.29777	.35660
							1.01955	1.09580	1.13261	1.18591	1.26875	1.34258	1.40421
.199	1.7348	.6890	.71071	.25511	3.27183	1.20364	.20272	.21842	.22683	.24036	.26625	.29959	.36080
							1.01948	1.09542	1.13207	1.18512	1.26754	1.34091	1.40196
.200	1.7316	.6898	.71254	.25380	3.25918	1.20149	.20374	.21952	.22799	.24161	.26770	.30141	.36549
							1.01940	1.09504	1.13154	1.18434	1.26633	1.33925	1.39971
.201	1.7284	.6906	.71436	.25248	3.24661	1.19937	.20476	.22063	.22914	.24285	.26915	.30325	.37089
							1.01933	1.09467	1.13101	1.18357	1.26514	1.33760	1.39745
.202	1.7252	.6913	.71617	.25118	3.23412	1.19726	.20578	.22174	.23030	.24410	.27060	.30511	.37772
							1.01926	1.09430	1.13048	1.18280	1.26395	1.33596	1.39518
.203	1.7220	.6921	.71798	.24987	3.22171	1.19518	.20680	.22284	.23146	.24535	.27206	.30697	
							1.01919	1.09393	1.12995	1.18204	1.26277	1.33433	
.204	1.7188	.6929	.71978	.24857	3.20936	1.19312	.20782	.22395	.23262	.24660	.27352	.30885	
							1.01911	1.09356	1.12943	1.18128	1.26159	1.33270	
.205	1.7157	.6937	.72157	.24727	3.19710	1.19108	.20884	.22506	.23378	.24785	.27498	.31074	
							1.01904	1.09319	1.12891	1.18052	1.26042	1.33109	
.206	1.7125	.6945	.72336	.24598	3.18491	1.18907	.20986	.22616	.23494	.24910	.27645	.31265	
							1.01897	1.09283	1.12839	1.17977	1.25926	1.32948	
.207	1.7094	.6953	.72513	.24469	3.17279	1.18707	.21088	.22727	.23610	.25035	.27792	.31457	
							1.01890	1.09247	1.12788	1.17903	1.25811	1.32788	
.208	1.7063	.6960	.72690	.24341	3.16074	1.18509	.21190	.22838	.23726	.25160	.27940	.31651	
							1.01883	1.09211	1.12737	1.17828	1.25696	1.32630	
.209	1.7032	.6968	.72866	.24212	3.14876	1.18314	.21292	.22949	.23842	.25286	.28087	.31846	
							1.01876	1.09175	1.12687	1.17755	1.25582	1.32471	

TABLE VIII. Specific Heat Ratio = 1.30

p/p_t	MACH	T/T_t	GAMMA	K_crit	SHOCK P_y/P_x	P_ty/P_ty	.01	.05	.07	.09	.1	.15	.2
.210	1.7001	.6976	.73042	.24085	3.13685	1.18120	.21394	.23060	.23958	.24911	.25412	.28236	.32043
							1.01869	1.09140	1.12636	1.16029	1.17681	1.25469	1.32314
.211	1.6970	.6983	.73216	.23957	3.12502	1.17928	.21496	.23171	.24075	.25034	.25537	.28385	.32242
							1.01862	1.09105	1.12586	1.15963	1.17609	1.25356	1.32158
.212	1.6939	.6991	.73390	.23830	3.11325	1.17738	.21598	.23282	.24191	.25156	.25663	.28534	.32443
							1.01856	1.09070	1.12536	1.15899	1.17536	1.25244	1.32002
.213	1.6909	.6999	.73563	.23704	3.10155	1.17550	.21700	.23393	.24308	.25279	.25790	.28684	.32646
							1.01849	1.09035	1.12487	1.15834	1.17464	1.25133	1.31847
.214	1.6878	.7006	.73736	.23578	3.08991	1.17364	.21802	.23504	.24424	.25402	.25916	.28834	.32851
							1.01842	1.09000	1.12438	1.15770	1.17393	1.25022	1.31693
.215	1.6848	.7014	.73907	.23452	3.07835	1.17180	.21904	.23616	.24541	.25525	.26042	.28984	.33058
							1.01835	1.08966	1.12389	1.15707	1.17322	1.24912	1.31539
.216	1.6818	.7021	.74078	.23327	3.06685	1.16997	.22006	.23727	.24657	.25648	.26169	.29135	.33267
							1.01829	1.08931	1.12340	1.15643	1.17251	1.24802	1.31387
.217	1.6788	.7029	.74248	.23202	3.05541	1.16817	.22108	.23838	.24774	.25771	.26296	.29287	.33479
							1.01822	1.08897	1.12292	1.15581	1.17181	1.24693	1.31234
.218	1.6758	.7036	.74418	.23077	3.04404	1.16638	.22210	.23950	.24891	.25894	.26423	.29439	.33693
							1.01816	1.08864	1.12244	1.15518	1.17111	1.24585	1.31083
.219	1.6728	.7044	.74586	.22953	3.03274	1.16461	.22312	.24061	.25008	.26018	.26550	.29591	.33910
							1.01809	1.08830	1.12196	1.15456	1.17041	1.24477	1.30932
.220	1.6698	.7051	.74754	.22829	3.02149	1.16286	.22414	.24172	.25125	.26142	.26677	.29744	.34130
							1.01802	1.08796	1.12149	1.15394	1.16972	1.24370	1.30782
.221	1.6668	.7058	.74922	.22705	3.01031	1.16112	.22516	.24284	.25242	.26265	.26805	.29898	.34353
							1.01796	1.08763	1.12101	1.15333	1.16903	1.24263	1.30633
.222	1.6639	.7066	.75088	.22582	2.99920	1.15940	.22618	.24395	.25360	.26389	.26932	.30052	.34580
							1.01790	1.08730	1.12054	1.15271	1.16835	1.24157	1.30484
.223	1.6609	.7073	.75254	.22459	2.98814	1.15770	.22720	.24507	.25477	.26513	.27060	.30207	.34809
							1.01783	1.08697	1.12008	1.15211	1.16767	1.24052	1.30335
.224	1.6580	.7080	.75419	.22337	2.97714	1.15601	.22822	.24619	.25595	.26637	.27188	.30362	.35043
							1.01777	1.08664	1.11963	1.15150	1.16699	1.23947	1.30187
.225	1.6551	.7088	.75584	.22215	2.96621	1.15434	.22924	.24730	.25713	.26762	.27316	.30518	.35281
							1.01770	1.08632	1.11917	1.15090	1.16632	1.23842	1.30040
.226	1.6522	.7095	.75747	.22093	2.95533	1.15269	.23026	.24842	.25830	.26886	.27444	.30675	.35523
							1.01764	1.08600	1.11871	1.15030	1.16565	1.23738	1.29893
.227	1.6493	.7102	.75910	.21971	2.94452	1.15105	.23129	.24954	.25948	.27011	.27573	.30832	.35770
							1.01758	1.08567	1.11825	1.14971	1.16499	1.23636	1.29747
.228	1.6464	.7109	.76073	.21850	2.93376	1.14943	.23231	.25066	.26066	.27135	.27702	.30990	.36022
							1.01752	1.08535	1.11780	1.14912	1.16433	1.23534	1.29601
.229	1.6435	.7117	.76234	.21730	2.92306	1.14783	.23333	.25178	.26184	.27260	.27831	.31149	.36280
							1.01745	1.08503	1.11735	1.14853	1.16367	1.23431	1.29456
.230	1.6407	.7124	.76395	.21609	2.91241	1.14624	.23435	.25290	.26302	.27386	.27960	.31308	.36546
							1.01739	1.08472	1.11690	1.14794	1.16301	1.23329	1.29312
.231	1.6378	.7131	.76556	.21489	2.90183	1.14466	.23537	.25402	.26420	.27511	.28089	.31468	.36818
							1.01733	1.08440	1.11645	1.14736	1.16236	1.23228	1.29167
.232	1.6350	.7138	.76715	.21370	2.89130	1.14310	.23639	.25514	.26538	.27636	.28219	.31628	.37097
							1.01727	1.08409	1.11601	1.14678	1.16171	1.23127	1.29023
.233	1.6321	.7145	.76874	.21251	2.88082	1.14156	.23741	.25626	.26656	.27762	.28349	.31789	.37386
							1.01721	1.08378	1.11556	1.14621	1.16107	1.23027	1.28879
.234	1.6293	.7152	.77032	.21132	2.87040	1.14003	.23844	.25739	.26775	.27888	.28479	.31951	.37685
							1.01715	1.08347	1.11512	1.14563	1.16043	1.22927	1.28736
.235	1.6265	.7159	.77190	.21013	2.86004	1.13852	.23946	.25851	.26893	.28013	.28609	.32114	.37996
							1.01709	1.08316	1.11469	1.14506	1.15979	1.22828	1.28592
.236	1.6237	.7166	.77346	.20895	2.84973	1.13701	.24048	.25963	.27012	.28140	.28739	.32278	.38321
							1.01703	1.08285	1.11425	1.14450	1.15916	1.22729	1.28449
.237	1.6209	.7173	.77503	.20777	2.83947	1.13553	.24150	.26076	.27131	.28266	.28870	.32442	.38663
							1.01697	1.08255	1.11382	1.14393	1.15852	1.22630	1.28306
.238	1.6181	.7180	.77658	.20659	2.82927	1.13406	.24252	.26188	.27250	.28392	.29001	.32608	.39026
							1.01691	1.08224	1.11339	1.14337	1.15790	1.22532	1.28163
.239	1.6153	.7187	.77813	.20542	2.81912	1.13260	.24355	.26301	.27369	.28519	.29132	.32774	.39416
							1.01685	1.08194	1.11296	1.14281	1.15727	1.22435	1.28020

TABLE VIII. Specific Heat Ratio = 1.30

					SHOCK		R1 and TPR versus Loss Coefficient K						
p/p_t	MACH	T/T_t	GAMMA	K_{crit}	P_y/P_x	P_{ty}/P_{ty}	.01	.05	.07	.09	.1	.15	.2
.240	1.6125	.7194	.77967	.20425	2.80902	1.13116	.24457	.26413	.27488	.28646	.29263	.32941	.39841
							1.01679	1.08164	1.11253	1.14226	1.15665	1.22337	1.27876
.241	1.6098	.7201	.78120	.20309	2.79897	1.12973	.24559	.26526	.27607	.28773	.29395	.33109	.40322
							1.01674	1.08134	1.11211	1.14170	1.15603	1.22241	1.27733
.242	1.6070	.7208	.78273	.20193	2.78898	1.12831	.24661	.26639	.27726	.28900	.29527	.33278	.40886
							1.01668	1.08105	1.11168	1.14115	1.15542	1.22144	1.27589
.243	1.6043	.7215	.78425	.20077	2.77903	1.12691	.24764	.26752	.27846	.29028	.29659	.33448	.41641
							1.01662	1.08075	1.11126	1.14061	1.15480	1.22048	1.27444
.244	1.6016	.7222	.78577	.19961	2.76914	1.12552	.24866	.26865	.27965	.29155	.29791	.33619	
							1.01656	1.08046	1.11085	1.14006	1.15419	1.21953	
.245	1.5988	.7228	.78728	.19846	2.75929	1.12414	.24968	.26978	.28085	.29283	.29924	.33791	
							1.01651	1.08016	1.11043	1.13952	1.15359	1.21858	
.246	1.5961	.7235	.78878	.19731	2.74950	1.12278	.25070	.27091	.28205	.29411	.30057	.33965	
							1.01645	1.07987	1.11002	1.13898	1.15298	1.21763	
.247	1.5934	.7242	.79027	.19617	2.73975	1.12143	.25173	.27204	.28325	.29539	.30190	.34139	
							1.01639	1.07958	1.10961	1.13844	1.15238	1.21669	
.248	1.5907	.7249	.79176	.19502	2.73005	1.12009	.25275	.27317	.28445	.29668	.30323	.34315	
							1.01634	1.07929	1.10920	1.13791	1.15178	1.21575	
.249	1.5880	.7255	.79324	.19388	2.72040	1.11876	.25377	.27430	.28565	.29797	.30457	.34492	
							1.01628	1.07900	1.10879	1.13738	1.15119	1.21481	
.250	1.5854	.7262	.79472	.19275	2.71080	1.11745	.25480	.27544	.28686	.29925	.30591	.34670	
							1.01623	1.07872	1.10838	1.13685	1.15060	1.21388	
.251	1.5827	.7269	.79619	.19162	2.70124	1.11615	.25582	.27657	.28806	.30055	.30725	.34850	
							1.01617	1.07843	1.10798	1.13632	1.15001	1.21295	
.252	1.5800	.7275	.79765	.19049	2.69173	1.11486	.25684	.27771	.28927	.30184	.30860	.35031	
							1.01611	1.07815	1.10756	1.13580	1.14942	1.21203	
.253	1.5774	.7282	.79911	.18936	2.68227	1.11358	.25787	.27884	.29048	.30314	.30995	.35213	
							1.01606	1.07787	1.10716	1.13527	1.14883	1.21111	
.254	1.5747	.7289	.80056	.18824	2.67285	1.11232	.25889	.27998	.29168	.30444	.31130	.35397	
							1.01601	1.07759	1.10676	1.13475	1.14825	1.21019	
.255	1.5721	.7295	.80200	.18712	2.66348	1.11107	.25991	.28112	.29290	.30574	.31266	.35582	
							1.01595	1.07731	1.10636	1.13424	1.14767	1.20927	
.256	1.5695	.7302	.80344	.18600	2.65416	1.10983	.26094	.28226	.29411	.30704	.31402	.35770	
							1.01590	1.07703	1.10597	1.13372	1.14710	1.20836	
.257	1.5669	.7309	.80487	.18489	2.64488	1.10860	.26196	.28340	.29532	.30835	.31538	.35959	
							1.01584	1.07676	1.10558	1.13321	1.14652	1.20745	
.258	1.5643	.7315	.80629	.18378	2.63564	1.10738	.26299	.28454	.29654	.30966	.31675	.36150	
							1.01579	1.07648	1.10519	1.13270	1.14595	1.20655	
.259	1.5617	.7322	.80771	.18267	2.62645	1.10618	.26401	.28568	.29775	.31097	.31811	.36342	
							1.01574	1.07621	1.10480	1.13219	1.14538	1.20564	
.260	1.5591	.7328	.80912	.18157	2.61730	1.10498	.26503	.28682	.29897	.31228	.31949	.36537	
							1.01568	1.07594	1.10441	1.13169	1.14482	1.20474	
.261	1.5565	.7335	.81053	.18047	2.60819	1.10380	.26606	.28796	.30019	.31360	.32086	.36734	
							1.01563	1.07566	1.10402	1.13118	1.14425	1.20385	
.262	1.5539	.7341	.81193	.17937	2.59913	1.10263	.26708	.28911	.30141	.31492	.32224	.36933	
							1.01558	1.07539	1.10364	1.13068	1.14369	1.20295	
.263	1.5513	.7348	.81332	.17828	2.59011	1.10146	.26811	.29025	.30264	.31624	.32363	.37135	
							1.01553	1.07513	1.10326	1.13018	1.14313	1.20206	
.264	1.5488	.7354	.81471	.17718	2.58113	1.10031	.26913	.29140	.30386	.31757	.32502	.37339	
							1.01547	1.07486	1.10288	1.12969	1.14257	1.20117	
.265	1.5462	.7360	.81609	.17610	2.57219	1.09917	.27016	.29254	.30509	.31890	.32641	.37546	
							1.01542	1.07459	1.10250	1.12919	1.14202	1.20028	
.266	1.5437	.7367	.81747	.17501	2.56329	1.09804	.27118	.29369	.30632	.32023	.32780	.37755	
							1.01537	1.07433	1.10212	1.12870	1.14147	1.19940	
.267	1.5411	.7373	.81884	.17393	2.55444	1.09693	.27221	.29484	.30755	.32156	.32920	.37968	
							1.01532	1.07406	1.10175	1.12821	1.14092	1.19852	
.268	1.5386	.7380	.82020	.17285	2.54563	1.09582	.27323	.29599	.30878	.32290	.33061	.38183	
							1.01527	1.07380	1.10138	1.12772	1.14037	1.19764	
.269	1.5361	.7386	.82156	.17177	2.53685	1.09472	.27426	.29714	.31002	.32424	.33202	.38403	
							1.01522	1.07354	1.10100	1.12724	1.13982	1.19676	

TABLE VIII. Specific Heat Ratio = 1.30

p/p_t	MACH	T/T_t	GAMMA	K_{crit}	SHOCK p_y/p_x	p_{ty}/p_{ty}	.01	.03	.05	.07	.09	.1	.15
.270	1.5336	.7392	.82291	.17070	2.52812	1.09363	.27529	.28638	.29829	.31125	.32559	.33343	.38626
							1.01517	1.04476	1.07328	1.10063	1.12675	1.13928	1.19588
.271	1.5310	.7399	.82425	.16963	2.51943	1.09255	.27631	.28746	.29945	.31249	.32694	.33485	.38852
							1.01512	1.04460	1.07302	1.10027	1.12627	1.13874	1.19501
.272	1.5285	.7405	.82559	.16857	2.51077	1.09149	.27734	.28855	.30060	.31373	.32829	.33627	.39084
							1.01507	1.04445	1.07276	1.09990	1.12579	1.13820	1.19413
.273	1.5261	.7411	.82692	.16750	2.50216	1.09043	.27836	.28963	.30175	.31497	.32964	.33770	.39319
							1.01502	1.04430	1.07250	1.09953	1.12531	1.13766	1.19326
.274	1.5236	.7417	.82825	.16644	2.49358	1.08938	.27939	.29072	.30291	.31621	.33100	.33913	.39560
							1.01497	1.04415	1.07225	1.09917	1.12483	1.13713	1.19239
.275	1.5211	.7424	.82957	.16538	2.48504	1.08834	.28042	.29180	.30407	.31746	.33236	.34057	.39809
							1.01492	1.04400	1.07199	1.09881	1.12436	1.13660	1.19153
.276	1.5186	.7430	.83088	.16433	2.47654	1.08732	.28144	.29289	.30523	.31871	.33373	.34201	.40062
							1.01487	1.04385	1.07174	1.09845	1.12389	1.13607	1.19066
.277	1.5161	.7436	.83219	.16328	2.46808	1.08630	.28247	.29397	.30639	.31996	.33510	.34346	.40322
							1.01482	1.04370	1.07149	1.09809	1.12342	1.13554	1.18979
.278	1.5137	.7442	.83349	.16223	2.45965	1.08529	.28350	.29506	.30755	.32121	.33648	.34491	.40590
							1.01477	1.04355	1.07124	1.09773	1.12295	1.13501	1.18893
.279	1.5112	.7448	.83479	.16118	2.45127	1.08429	.28452	.29615	.30871	.32247	.33785	.34637	.40867
							1.01472	1.04340	1.07099	1.09737	1.12248	1.13448	1.18806
.280	1.5088	.7455	.83608	.16014	2.44292	1.08330	.28555	.29724	.30987	.32372	.33924	.34783	.41153
							1.01467	1.04326	1.07074	1.09702	1.12201	1.13396	1.18719
.281	1.5063	.7461	.83737	.15910	2.43460	1.08232	.28658	.29833	.31104	.32498	.34062	.34930	.41452
							1.01462	1.04311	1.07049	1.09666	1.12155	1.13344	1.18633
.282	1.5039	.7467	.83865	.15806	2.42633	1.08135	.28760	.29942	.31220	.32624	.34201	.35078	.41763
							1.01458	1.04296	1.07024	1.09631	1.12109	1.13292	1.18546
.283	1.5015	.7473	.83992	.15703	2.41808	1.08038	.28863	.30051	.31337	.32751	.34341	.35226	.42092
							1.01453	1.04282	1.06999	1.09596	1.12063	1.13240	1.18459
.284	1.4991	.7479	.84119	.15600	2.40988	1.07943	.28966	.30160	.31454	.32877	.34481	.35375	.42440
							1.01448	1.04267	1.06975	1.09561	1.12017	1.13189	1.18372
.285	1.4967	.7485	.84245	.15497	2.40171	1.07848	.29069	.30269	.31571	.33004	.34621	.35524	.42813
							1.01443	1.04253	1.06951	1.09526	1.11971	1.13137	1.18285
.286	1.4942	.7491	.84370	.15395	2.39357	1.07755	.29172	.30378	.31688	.33131	.34762	.35675	.43222
							1.01439	1.04238	1.06926	1.09492	1.11926	1.13086	1.18198
.287	1.4918	.7497	.84495	.15293	2.38547	1.07662	.29274	.30488	.31805	.33259	.34904	.35826	.43678
							1.01434	1.04224	1.06902	1.09457	1.11881	1.13035	1.18110
.288	1.4895	.7503	.84620	.15191	2.37741	1.07570	.29377	.30597	.31922	.33387	.35046	.35978	.44210
							1.01429	1.04210	1.06878	1.09423	1.11835	1.12985	1.18021
.289	1.4871	.7509	.84744	.15089	2.36938	1.07479	.29480	.30706	.32040	.33515	.35188	.36131	.44898
							1.01425	1.04196	1.06854	1.09389	1.11790	1.12935	1.17932
.290	1.4847	.7515	.84867	.14988	2.36138	1.07389	.29583	.30816	.32157	.33643	.35331	.36284	
							1.01420	1.04182	1.06830	1.09355	1.11745	1.12884	
.291	1.4823	.7521	.84990	.14887	2.35342	1.07300	.29686	.30925	.32275	.33771	.35475	.36438	
							1.01415	1.04168	1.06806	1.09321	1.11701	1.12834	
.292	1.4799	.7527	.85112	.14786	2.34549	1.07211	.29789	.31035	.32393	.33900	.35619	.36593	
							1.01411	1.04154	1.06782	1.09287	1.11656	1.12784	
.293	1.4776	.7533	.85233	.14686	2.33759	1.07124	.29892	.31145	.32511	.34029	.35764	.36748	
							1.01406	1.04140	1.06759	1.09253	1.11612	1.12734	
.294	1.4752	.7539	.85354	.14586	2.32973	1.07037	.29995	.31254	.32629	.34158	.35909	.36905	
							1.01404	1.04126	1.06735	1.09219	1.11567	1.12684	
.295	1.4729	.7545	.85475	.14486	2.32190	1.06951	.30098	.31364	.32748	.34288	.36055	.37062	
							1.01399	1.04112	1.06712	1.09186	1.11523	1.12634	
.296	1.4705	.7551	.85595	.14386	2.31410	1.06866	.30201	.31474	.32866	.34418	.36201	.37221	
							1.01395	1.04098	1.06688	1.09152	1.11479	1.12585	
.297	1.4682	.7557	.85714	.14287	2.30634	1.06781	.30304	.31584	.32985	.34548	.36349	.37380	
							1.01390	1.04084	1.06665	1.09119	1.11436	1.12535	
.298	1.4659	.7562	.85833	.14188	2.29860	1.06698	.30407	.31694	.33104	.34679	.36496	.37540	
							1.01386	1.04071	1.06642	1.09086	1.11392	1.12486	
.299	1.4635	.7568	.85951	.14089	2.29090	1.06615	.30510	.31804	.33223	.34810	.36645	.37701	
							1.01381	1.04057	1.06619	1.09053	1.11348	1.12437	

TABLE VIII. Specific Heat Ratio $= 1.30$

p/p_t	MACH	T/T_t	GAMMA	K_{crit}	SHOCK p_y/p_x	p_{ty}/p_{ty}	.01	.03	.05	.07	.08	.09	.1
.300	1.4612	.7574	.86068	.13991	2.28323	1.06533	.30613	.31915	.33342	.34941	.35828	.36794	.37864
							1.01377	1.04044	1.06596	1.09020	1.10181	1.11305	1.12388
.301	1.4589	.7580	.86186	.13893	2.27559	1.06452	.30716	.32025	.33462	.35073	.35968	.36944	.38027
							1.01372	1.04030	1.06573	1.08987	1.10143	1.11262	1.12339
.302	1.4566	.7586	.86302	.13795	2.26799	1.06371	.30819	.32135	.33581	.35205	.36108	.37095	.38192
							1.01368	1.04017	1.06550	1.08955	1.10105	1.11219	1.12290
.303	1.4543	.7592	.86418	.13698	2.26041	1.06292	.30922	.32246	.33701	.35338	.36249	.37246	.38358
							1.01363	1.04003	1.06527	1.08922	1.10068	1.11175	1.12241
.304	1.4520	.7597	.86533	.13601	2.25287	1.06213	.31025	.32356	.33821	.35470	.36390	.37399	.38525
							1.01359	1.03990	1.06504	1.08890	1.10030	1.11133	1.12193
.305	1.4497	.7603	.86648	.13504	2.24535	1.06134	.31128	.32467	.33941	.35603	.36532	.37552	.38694
							1.01355	1.03976	1.06482	1.08857	1.09992	1.11090	1.12145
.306	1.4474	.7609	.86762	.13407	2.23787	1.06057	.31232	.32577	.34061	.35737	.36674	.37706	.38863
							1.01350	1.03963	1.06459	1.08825	1.09955	1.11047	1.12096
.307	1.4451	.7615	.86876	.13311	2.23041	1.05980	.31335	.32688	.34182	.35871	.36817	.37861	.39035
							1.01346	1.03950	1.06437	1.08793	1.09918	1.11005	1.12048
.308	1.4429	.7620	.86989	.13215	2.22299	1.05904	.31438	.32799	.34302	.36005	.36962	.38016	.39207
							1.01342	1.03937	1.06415	1.08761	1.09882	1.10962	1.12000
.309	1.4406	.7626	.87102	.13119	2.21559	1.05829	.31541	.32910	.34423	.36140	.37106	.38173	.39381
							1.01337	1.03924	1.06392	1.08729	1.09845	1.10920	1.11952
.310	1.4383	.7632	.87214	.13023	2.20823	1.05755	.31645	.33021	.34545	.36275	.37251	.38331	.39557
							1.01333	1.03911	1.06370	1.08697	1.09808	1.10878	1.11904
.311	1.4361	.7637	.87326	.12928	2.20089	1.05681	.31748	.33132	.34666	.36411	.37397	.38490	.39735
							1.01329	1.03898	1.06348	1.08665	1.09771	1.10836	1.11856
.312	1.4338	.7643	.87437	.12833	2.19358	1.05608	.31851	.33243	.34787	.36547	.37543	.38649	.39914
							1.01324	1.03885	1.06326	1.08634	1.09734	1.10794	1.11809
.313	1.4316	.7649	.87547	.12739	2.18630	1.05535	.31955	.33355	.34909	.36684	.37690	.38810	.40096
							1.01320	1.03872	1.06304	1.08602	1.09698	1.10752	1.11761
.314	1.4293	.7654	.87657	.12644	2.17905	1.05463	.32058	.33466	.35031	.36821	.37837	.38972	.40279
							1.01316	1.03859	1.06282	1.08571	1.09661	1.10710	1.11714
.315	1.4271	.7660	.87766	.12550	2.17183	1.05392	.32161	.33578	.35153	.36958	.37986	.39135	.40464
							1.01312	1.03846	1.06260	1.08540	1.09623	1.10668	1.11666
.316	1.4249	.7666	.87875	.12457	2.16464	1.05322	.32265	.33689	.35276	.37096	.38135	.39300	.40652
							1.01308	1.03833	1.06239	1.08508	1.09587	1.10627	1.11619
.317	1.4226	.7671	.87983	.12363	2.15747	1.05252	.32368	.33801	.35398	.37235	.38285	.39466	.40841
							1.01303	1.03821	1.06217	1.08477	1.09551	1.10585	1.11572
.318	1.4204	.7677	.88091	.12270	2.15034	1.05183	.32472	.33913	.35521	.37374	.38436	.39633	.41034
							1.01299	1.03808	1.06195	1.08446	1.09515	1.10544	1.11524
.319	1.4182	.7682	.88198	.12177	2.14322	1.05115	.32575	.34025	.35645	.37514	.38587	.39801	.41229
							1.01295	1.03795	1.06174	1.08415	1.09479	1.10502	1.11477
.320	1.4160	.7688	.88305	.12084	2.13614	1.05047	.32679	.34137	.35768	.37654	.38740	.39971	.41426
							1.01291	1.03783	1.06152	1.08384	1.09443	1.10461	1.11430
.321	1.4138	.7693	.88411	.11992	2.12909	1.04980	.32782	.34249	.35892	.37795	.38893	.40142	.41627
							1.01287	1.03770	1.06131	1.08354	1.09407	1.10420	1.11383
.322	1.4116	.7699	.88517	.11900	2.12206	1.04914	.32886	.34361	.36016	.37936	.39048	.40315	.41831
							1.01283	1.03758	1.06110	1.08323	1.09372	1.10379	1.11336
.323	1.4094	.7704	.88622	.11808	2.11506	1.04848	.32990	.34474	.36140	.38078	.39203	.40490	.42038
							1.01279	1.03745	1.06089	1.08292	1.09336	1.10338	1.11289
.324	1.4072	.7710	.88726	.11717	2.10808	1.04783	.33093	.34586	.36264	.38221	.39359	.40666	.42249
							1.01275	1.03733	1.06067	1.08262	1.09300	1.10297	1.11242
.325	1.4050	.7715	.88830	.11626	2.10113	1.04719	.33197	.34699	.36389	.38364	.39517	.40845	.42464
							1.01271	1.03721	1.06046	1.08231	1.09265	1.10256	1.11195
.326	1.4028	.7721	.88934	.11535	2.09421	1.04655	.33301	.34811	.36514	.38508	.39675	.41025	.42684
							1.01267	1.03708	1.06025	1.08201	1.09229	1.10215	1.11148
.327	1.4007	.7726	.89037	.11444	2.08731	1.04592	.33404	.34924	.36639	.38653	.39835	.41207	.42908
							1.01263	1.03696	1.06004	1.08171	1.09194	1.10174	1.11101
.328	1.3985	.7732	.89139	.11354	2.08044	1.04530	.33508	.35037	.36765	.38798	.39996	.41392	.43137
							1.01259	1.03684	1.05983	1.08140	1.09159	1.10134	1.11054
.329	1.3963	.7737	.89241	.11264	2.07360	1.04468	.33612	.35150	.36891	.38944	.40158	.41579	.43372
							1.01255	1.03672	1.05963	1.08110	1.09124	1.10093	1.11007

TABLE VIII. Specific Heat Ratio = 1.30

p/p_t	MACH	T/T_t	GAMMA	K_{crit}	P_y/P_x	P_{tx}/P_{ty}	.01	.03	.05	.07	.08	.09	.1
.330	1.3942	.7743	.89342	.11174	2.06678	1.04406	.33716	.35263	.37017	.39091	.40321	.41768	.43613
							1.01251	1.03660	1.05942	1.08080	1.09089	1.10052	1.10960
.331	1.3920	.7748	.89443	.11085	2.05998	1.04346	.33820	.35377	.37144	.39239	.40486	.41960	.43861
							1.01247	1.03647	1.05921	1.08050	1.09054	1.10012	1.10913
.332	1.3899	.7753	.89544	.10996	2.05321	1.04286	.33923	.35490	.37271	.39387	.40652	.42154	.44118
							1.01243	1.03635	1.05900	1.08020	1.09019	1.09971	1.10866
.333	1.3877	.7759	.89643	.10907	2.04647	1.04226	.34027	.35604	.37398	.39537	.40819	.42352	.44383
							1.01239	1.03623	1.05880	1.07990	1.08984	1.09930	1.10818
.334	1.3856	.7764	.89743	.10818	2.03975	1.04167	.34131	.35718	.37526	.39687	.40988	.42553	.44658
							1.01235	1.03611	1.05859	1.07961	1.08949	1.09890	1.10771
.335	1.3834	.7770	.89841	.10730	2.03306	1.04109	.34235	.35831	.37654	.39838	.41159	.42757	.44945
							1.01231	1.03599	1.05839	1.07931	1.08914	1.09849	1.10723
.336	1.3813	.7775	.89940	.10642	2.02639	1.04051	.34339	.35945	.37782	.39990	.41331	.42964	.45251
							1.01227	1.03588	1.05818	1.07901	1.08879	1.09809	1.10677
.337	1.3792	.7780	.90037	.10554	2.01974	1.03994	.34443	.36060	.37911	.40143	.41505	.43176	.45571
							1.01223	1.03576	1.05798	1.07872	1.08844	1.09768	1.10629
.338	1.3770	.7786	.90135	.10467	2.01312	1.03938	.34547	.36174	.38040	.40297	.41681	.43392	.45913
							1.01219	1.03564	1.05778	1.07842	1.08810	1.09728	1.10581
.339	1.3749	.7791	.90231	.10380	2.00653	1.03882	.34652	.36288	.38169	.40452	.41859	.43613	.46283
							1.01215	1.03552	1.05758	1.07813	1.08775	1.09687	1.10532
.340	1.3728	.7796	.90328	.10293	1.99995	1.03826	.34756	.36403	.38299	.40608	.42039	.43839	.46691
							1.01212	1.03540	1.05737	1.07783	1.08741	1.09647	1.10484
.341	1.3707	.7801	.90423	.10206	1.99340	1.03771	.34860	.36518	.38429	.40766	.42221	.44070	.47157
							1.01208	1.03529	1.05717	1.07754	1.08706	1.09606	1.10434
.342	1.3686	.7807	.90518	.10120	1.98688	1.03717	.34964	.36633	.38560	.40924	.42406	.44308	.47727
							1.01204	1.03517	1.05697	1.07724	1.08671	1.09565	1.10385
.343	1.3665	.7812	.90613	.10034	1.98038	1.03663	.35068	.36748	.38691	.41084	.42593	.44553	.48558
							1.01200	1.03505	1.05677	1.07695	1.08637	1.09525	1.10334
.344	1.3644	.7817	.90707	.09948	1.97390	1.03610	.35173	.36863	.38823	.41245	.42783	.44806	
							1.01196	1.03494	1.05657	1.07666	1.08602	1 09484	
.345	1.3623	.7822	.90801	.09863	1.96744	1.03557	.35277	.36979	.38955	.41407	.42975	.45067	
							1.01193	1.03482	1.05637	1.07637	1.08568	1.09443	
.346	1.3602	.7828	.90894	.09778	1.96101	1.03505	.35382	.37094	.39087	.41571	.43170	.45339	
							1.01189	1.03471	1.05617	1.07607	1.08533	1.09402	
.347	1.3581	.7833	.90986	.09693	1.95460	1.03454	.35486	.37210	.39220	.41736	.43369	.45626	
							1.01185	1.03459	1.05598	1.07578	1.08499	1.09361	
.348	1.3560	.7838	.91079	.09609	1.94821	1.03403	.35590	.37326	.39354	.41903	.43571	.45925	
							1.01181	1.03448	1.05578	1.07549	1.08464	1.09320	
.349	1.3539	.7843	.91170	.09524	1.94185	1.03352	.35695	.37442	.39488	.42072	.43777	.46242	
							1.01178	1.03436	1.05558	1.07520	1.08430	1.09278	
.350	1.3519	.7848	.91261	.09440	1.93550	1.03302	.35799	.37558	.39623	.42242	.43987	.46581	
							1.01174	1.03425	1.05538	1.07491	1.08395	1.09237	
.351	1.3498	.7854	.91352	.09357	1.92918	1.03253	.35904	.37675	.39758	.42414	.44202	.46949	
							1.01170	1.03413	1.05519	1.07462	1.08361	1.09195	
.352	1.3477	.7859	.91442	.09274	1.92289	1.03204	.36009	.37792	.39893	.42588	.44421	.47355	
							1.01167	1.03402	1.05499	1.07433	1.08326	1.09152	
.353	1.3457	.7864	.91531	.09190	1.91661	1.03156	.36113	.37909	.40030	.42763	.44645	.47820	
							1.01163	1.03391	1.05480	1.07404	1.08292	1.09109	
.354	1.3436	.7869	.91620	.09108	1.91036	1.03108	.36218	.38026	.40167	.42941	.44876	.48397	
							1.01159	1.03380	1.05460	1.07375	1.08257	1.09066	
.355	1.3416	.7874	.91709	.09025	1.90412	1.03060	.36323	.38143	.40304	.43122	.45113	.49269	
							1.01156	1.03368	1.05441	1.07346	1.08222	1.09022	
.356	1.3395	.7879	.91797	.08943	1.89791	1.03014	.36428	.38261	.40442	.43306	.45357		
							1.01152	1.03357	1.05421	1.07318	1.08187		
.357	1.3375	.7884	.91885	.08861	1.89172	1.02967	.36533	.38378	.40581	.43491	.45610		
							1.01148	1.03346	1.05402	1.07289	1.08152		
.358	1.3354	.7890	.91972	.08780	1.88556	1.02921	.36638	.38496	.40720	.43679	.45872		
							1.01145	1.03335	1.05383	1.07260	1.08117		
.359	1.3334	.7895	.92058	.08698	1.87941	1.02876	.36743	.38615	.40860	.43870	.46148		
							1.01141	1.03324	1.05363	1.07232	1.08083		

TABLE VIII. Specific Heat Ratio = 1.30

p/p_t	MACH	T/T_t	GAMMA	K_crit	SHOCK P_y/P_x	SHOCK P_tx/P_ty	.01	.03	.04	.05	.06	.07	.08
.360	1.3314	.7900	.92144	.08617	1.87328	1.02831	.36848	.38733	.39805	.41002	.42381	.44064	.46435
							1.01138	1.03313	1.04348	1.05344	1.06298	1.07203	1.08047
.361	1.3293	.7905	.92230	.08537	1.86718	1.02787	.36953	.38852	.39933	.41143	.42543	.44262	.46738
							1.01134	1.03302	1.04333	1.05325	1.06274	1.07174	1.08012
.362	1.3273	.7910	.92315	.08456	1.86109	1.02743	.37058	.38971	.40062	.41286	.42706	.44463	.47061
							1.01130	1.03291	1.04318	1.05306	1.06250	1.07145	1.07976
.363	1.3253	.7915	.92400	.08376	1.85503	1.02699	.37163	.39090	.40191	.41429	.42870	.44668	.47408
							1.01127	1.03280	1.04303	1.05287	1.06227	1.07116	1.07940
.364	1.3233	.7920	.92484	.08296	1.84899	1.02656	.37268	.39209	.40321	.41573	.43037	.44877	.47789
							1.01123	1.03269	1.04288	1.05268	1.06203	1.07087	1.07904
.365	1.3212	.7925	.92567	.08217	1.84297	1.02614	.37373	.39329	.40451	.41718	.43205	.45091	.48218
							1.01120	1.03258	1.04273	1.05249	1.06179	1.07058	1.07868
.366	1.3192	.7930	.92650	.08138	1.83696	1.02572	.37479	.39449	.40582	.41864	.43375	.45311	.48730
							1.01116	1.03247	1.04259	1.05230	1.06156	1.07029	1.07831
.367	1.3172	.7935	.92733	.08059	1.83098	1.02530	.37584	.39570	.40713	.42011	.43547	.45536	.49403
							1.01113	1.03236	1.04244	1.05211	1.06132	1.07000	1.07793
.368	1.3152	.7940	.92815	.07980	1.82502	1.02489	.37690	.39690	.40845	.42158	.43721	.45767	
							1.01109	1.03225	1.04229	1.05192	1.06108	1.06970	
.369	1.3132	.7945	.92897	.07902	1.81908	1.02449	.37795	.39811	.40977	.42307	.43897	.46005	
							1.01106	1.03215	1.04214	1.05173	1.06085	1.06941	
.370	1.3112	.7950	.92978	.07824	1.81315	1.02408	.37901	.39932	.41110	.42457	.44076	.46252	
							1.01102	1.03204	1.04200	1.05154	1.06061	1.06912	
.371	1.3092	.7955	.93058	.07746	1.80725	1.02369	.38006	.40054	.41244	.42608	.44257	.46508	
							1.01099	1.03193	1.04185	1.05135	1.06037	1.06882	
.372	1.3073	.7960	.93138	.07669	1.80137	1.02329	.38112	.40175	.41378	.42760	.44441	.46777	
							1.01095	1.03182	1.04170	1.05116	1.06014	1.06853	
.373	1.3053	.7965	.93218	.07592	1.79550	1.02291	.38218	.40298	.41512	.42914	.44627	.47057	
							1.01092	1.03172	1.04156	1.05097	1.05990	1.06824	
.374	1.3033	.7970	.93297	.07515	1.78966	1.02252	.38324	.40420	.41648	.43068	.44817	.47353	
							1.01088	1.03161	1.04141	1.05078	1.05966	1.06794	
.375	1.3013	.7974	.93376	.07439	1.78383	1.02214	.38430	.40543	.41783	.43224	.45010	.47667	
							1.01085	1.03150	1.04126	1.05059	1.05943	1.06764	
.376	1.2993	.7979	.93454	.07363	1.77803	1.02177	.38536	.40666	.41920	.43382	.45207	.48005	
							1.01082	1.03140	1.04112	1.05041	1.05919	1.06733	
.377	1.2974	.7984	.93532	.07287	1.77224	1.02140	.38642	.40789	.42057	.43540	.45407	.48376	
							1.01078	1.03129	1.04097	1.05022	1.05895	1.06703	
.378	1.2954	.7989	.93609	.07211	1.76647	1.02103	.38748	.40913	.42195	.43701	.45612	.48792	
							1.01075	1.03119	1.04083	1.05003	1.05871	1.06672	
.379	1.2934	.7994	.93686	.07136	1.76072	1.02067	.38854	.41038	.42334	.43863	.45822	.49287	
							1.01071	1.03108	1.04068	1.04984	1.05848	1.06641	
.380	1.2915	.7999	.93762	.07061	1.75499	1.02031	.38960	.41162	.42474	.44027	.46036	.49931	
							1.01068	1.03097	1.04054	1.04965	1.05824	1.06609	
.381	1.2895	.8004	.93838	.06986	1.74928	1.01996	.39067	.41287	.42614	.44192	.46256		
							1.01065	1.03087	1.04039	1.04947	1.05800		
.382	1.2876	.8009	.93913	.06912	1.74359	1.01961	.39173	.41413	.42756	.44359	.46483		
							1.01061	1.03076	1.04025	1.04928	1.05776		
.383	1.2856	.8013	.93988	.06838	1.73791	1.01926	.39280	.41539	.42898	.44529	.46716		
							1.01058	1.03066	1.04011	1.04909	1.05751		
.384	1.2837	.8018	.94062	.06764	1.73225	1.01892	.39386	.41665	.43041	.44700	.46958		
							1.01055	1.03056	1.03996	1.04890	1.05727		
.385	1.2817	.8023	.94136	.06691	1.72661	1.01858	.39493	.41792	.43185	.44874	.47212		
							1.01051	1.03045	1.03982	1.04871	1.05704		
.386	1.2798	.8028	.94209	.06618	1.72099	1.01825	.39600	.41919	.43330	.45051	.47474		
							1.01048	1.03035	1.03968	1.04853	1.05679		
.387	1.2778	.8033	.94282	.06545	1.71539	1.01792	.39706	.42047	.43476	.45230	.47749		
							1.01045	1.03024	1.03953	1.04834	1.05655		
.388	1.2759	.8037	.94355	.06473	1.70980	1.01760	.39813	.42176	.43623	.45411	.48040		
							1.01041	1.03014	1.03939	1.04815	1.05630		
.389	1.2740	.8042	.94427	.06401	1.70424	1.01727	.39921	.42304	.43771	.45596	.48350		
							1.01039	1.03004	1.03925	1.04796	1.05605		

TABLE VIII. Specific Heat Ratio = 1.30

p/p_t	MACH	T/T_t	GAMMA	K_crit	SHOCK P_y/P_x	P_ty/P_ty	.005	.01	.02	.03	.04	.05	.06
.390	1.2720	.8047	.94498	.06329	1.69868	1.01696	.39503	.40029	.41160	.42434	.43921	.45784	.48685
							1.00522	1.01036	1.02034	1.02994	1.03910	1.04777	1.05580
.391	1.2701	.8052	.94569	.06257	1.69315	1.01664	.39606	.40136	.41277	.42564	.44072	.45976	.49053
							1.00520	1.01032	1.02027	1.02983	1.03896	1.04758	1.05555
.392	1.2682	.8056	.94639	.06186	1.68764	1.01633	.39710	.40243	.41394	.42695	.44224	.46171	.49468
							1.00519	1.01029	1.02020	1.02973	1.03882	1.04740	1.05529
.393	1.2663	.8061	.94709	.06115	1.68214	1.01603	.39813	.40351	.41511	.42826	.44377	.46371	.49971
							1.00517	1.01026	1.02014	1.02963	1.03868	1.04721	1.05503
.394	1.2644	.8066	.94779	.06045	1.67666	1.01573	.39917	.40458	.41629	.42958	.44532	.46575	.50645
							1.00515	1.01023	1.02007	1.02952	1.03853	1.04702	1.05476
.395	1.2624	.8071	.94848	.05975	1.67119	1.01543	.40020	.40566	.41747	.43090	.44689	.46784	
							1.00514	1.01019	1.02000	1.02942	1.03839	1.04682	
.396	1.2605	.8075	.94917	.05905	1.66575	1.01514	.40124	.40673	.41865	.43223	.44847	.46998	
							1.00512	1.01016	1.01994	1.02932	1.03825	1.04663	
.397	1.2586	.8080	.94985	.05835	1.66032	1.01485	.40228	.40781	.41983	.43357	.45006	.47219	
							1.00511	1.01013	1.01987	1.02922	1.03811	1.04644	
.398	1.2567	.8085	.95052	.05766	1.65490	1.01456	.40331	.40889	.42102	.43491	.45168	.47447	
							1.00509	1.01010	1.01981	1.02912	1.03796	1.04625	
.399	1.2548	.8089	.95119	.05697	1.64951	1.01428	.40435	.40997	.42221	.43627	.45331	.47686	
							1.00508	1.01007	1.01974	1.02901	1.03782	1.04606	
.400	1.2529	.8094	.95186	.05629	1.64413	1.01400	.40539	.41106	.42340	.43763	.45497	.47932	
							1.00506	1.01003	1.01968	1.02891	1.03768	1.04587	
.401	1.2510	.8099	.95252	.05560	1.63876	1.01372	.40643	.41214	.42460	.43900	.45664	.48189	
							1.00504	1.01000	1.01961	1.02881	1.03754	1.04567	
.402	1.2491	.8103	.95318	.05492	1.63342	1.01345	.40747	.41322	.42580	.44037	.45834	.48459	
							1.00503	1.00997	1.01955	1.02871	1.03739	1.04547	
.403	1.2472	.8108	.95383	.05425	1.62809	1.01318	.40851	.41431	.42701	.44176	.46007	.48745	
							1.00501	1.00994	1.01948	1.02861	1.03725	1.04528	
.404	1.2454	.8113	.95448	.05358	1.62277	1.01292	.40955	.41540	.42821	.44315	.46182	.49051	
							1.00500	1.00991	1.01942	1.02851	1.03711	1.04508	
.405	1.2435	.8117	.95512	.05291	1.61747	1.01266	.41059	.41648	.42943	.44456	.46360	.49382	
							1.00498	1.00988	1.01935	1.02841	1.03696	1.04488	
.406	1.2416	.8122	.95576	.05224	1.61219	1.01240	.41163	.41757	.43064	.44597	.46541	.49747	
							1.00497	1.00985	1.01929	1.02830	1.03682	1.04467	
.407	1.2397	.8127	.95640	.05158	1.60693	1.01214	.41267	.41866	.43186	.44740	.46725	.50169	
							1.00495	1.00981	1.01922	1.02820	1.03667	1.04447	
.408	1.2378	.8131	.95703	.05092	1.60167	1.01189	.41372	.41976	.43309	.44884	.46913	.50681	
							1.00494	1.00978	1.01916	1.02810	1.03653	1.04426	
.409	1.2360	.8136	.95765	.05026	1.59644	1.01165	.41476	.42085	.43431	.45028	.47105	.51432	
							1.00492	1.00975	1.01909	1.02800	1.03639	1.04404	
.410	1.2341	.8140	.95827	.04961	1.59122	1.01140	.41580	.42195	.43555	.45174	.47302		
							1.00491	1.00972	1.01903	1.02790	1.03624		
.411	1.2322	.8145	.95888	.04896	1.58602	1.01116	.41685	.42304	.43679	.45322	.47503		
							1.00489	1.00969	1.01897	1.02780	1.03609		
.412	1.2304	.8149	.95950	.04832	1.58083	1.01092	.41789	.42414	.43803	.45470	.47709		
							1.00487	1.00966	1.01890	1.02770	1.03595		
.413	1.2285	.8154	.96010	.04767	1.57566	1.01069	.41894	.42524	.43928	.45621	.47922		
							1.00486	1.00963	1.01884	1.02760	1.03580		
.414	1.2267	.8159	.96070	.04703	1.57050	1.01046	.41999	.42634	.44053	.45772	.48144		
							1.00484	1.00960	1.01877	1.02750	1.03566		
.415	1.2248	.8163	.96130	.04640	1.56536	1.01023	.42103	.42745	.44179	.45925	.48371		
							1.00483	1.00957	1.01871	1.02740	1.03551		
.416	1.2229	.8168	.96189	.04577	1.56024	1.01001	.42208	.42855	.44305	.46080	.48608		
							1.00481	1.00954	1.01865	1.02730	1.03536		
.417	1.2211	.8172	.96248	.04514	1.55512	1.00979	.42313	.42966	.44432	.46237	.48855		
							1.00480	1.00951	1.01858	1.02719	1.03521		
.418	1.2192	.8177	.96306	.04451	1.55003	1.00957	.42418	.43077	.44560	.46396	.49114		
							1.00478	1.00948	1.01852	1.02709	1.03506		
.419	1.2174	.8181	.96364	.04389	1.54495	1.00936	.42523	.43188	.44688	.46556	.49389		
							1.00477	1.00945	1.01846	1.02699	1.03491		

TABLE VIII. Specific Heat Ratio = 1.30

p/p_t	MACH	T/T_t	GAMMA	K_crit	SHOCK P_y/P_x	P_ty/P_ty	.001	.005	.01	.015	.02	.03	.04
.420	1.2156	.8186	.96421	.04327	1.53988	1.00915	.42123	.42628	.43299	.44024	.44816	.46719	.49683
							1.00096	1.00476	1.00942	1.01395	1.01839	1.02689	1.03476
.421	1.2137	.8190	.96478	.04266	1.53483	1.00894	.42224	.42733	.43411	.44143	.44946	.46885	.50001
							1.00096	1.00474	1.00938	1.01391	1.01833	1.02679	1.03460
.422	1.2119	.8195	.96534	.04205	1.52979	1.00873	.42325	.42839	.43523	.44263	.45076	.47052	.50352
							1.00095	1.00473	1.00935	1.01386	1.01826	1.02669	1.03444
.423	1.2101	.8199	.96590	.04144	1.52477	1.00853	.42426	.42944	.43635	.44383	.45207	.47223	.50759
							1.00095	1.00471	1.00932	1.01382	1.01820	1.02658	1.03429
.424	1.2082	.8204	.96646	.04084	1.51977	1.00833	.42527	.43050	.43747	.44504	.45339	.47396	.51251
							1.00095	1.00470	1.00929	1.01377	1.01814	1.02648	1.03412
.425	1.2064	.8208	.96701	.04024	1.51477	1.00814	.42628	.43155	.43859	.44625	.45472	.47573	.51978
							1.00094	1.00468	1.00926	1.01372	1.01807	1.02638	1.03396
.426	1.2046	.8213	.96755	.03964	1.50979	1.00795	.42729	.43261	.43972	.44747	.45605	.47753	
							1.00094	1.00467	1.00923	1.01368	1.01801	1.02628	
.427	1.2027	.8217	.96809	.03905	1.50483	1.00776	.42830	.43367	.44085	.44869	.45740	.47936	
							1.00094	1.00465	1.00920	1.01363	1.01795	1.02617	
.428	1.2009	.8221	.96863	.03846	1.49988	1.00757	.42931	.43473	.44198	.44991	.45875	.48124	
							1.00094	1.00464	1.00917	1.01359	1.01788	1.02607	
.429	1.1991	.8226	.96916	.03787	1.49495	1.00739	.43032	.43579	.44311	.45114	.46012	.48317	
							1.00093	1.00462	1.00914	1.01354	1.01782	1.02597	
.430	1.1973	.8230	.96969	.03729	1.49002	1.00721	.43133	.43685	.44425	.45238	.46151	.48514	
							1.00093	1.00461	1.00912	1.01349	1.01777	1.02586	
.431	1.1955	.8235	.97021	.03671	1.48512	1.00703	.43234	.43791	.44539	.45362	.46290	.48721	
							1.00093	1.00459	1.00909	1.01345	1.01770	1.02576	
.432	1.1937	.8239	.97073	.03613	1.48022	1.00685	.43335	.43897	.44653	.45487	.46429	.48932	
							1.00092	1.00458	1.00906	1.01340	1.01764	1.02566	
.433	1.1918	.8244	.97124	.03556	1.47535	1.00668	.43437	.44004	.44768	.45612	.46570	.49150	
							1.00092	1.00457	1.00903	1.01336	1.01758	1.02555	
.434	1.1900	.8248	.97175	.03499	1.47048	1.00651	.43538	.44110	.44883	.45739	.46713	.49378	
							1.00092	1.00455	1.00900	1.01331	1.01751	1.02545	
.435	1.1882	.8252	.97225	.03443	1.46563	1.00635	.43639	.44217	.44998	.45865	.46857	.49615	
							1.00092	1.00454	1.00897	1.01326	1.01745	1.02534	
.436	1.1864	.8257	.97275	.03387	1.46079	1.00618	.43740	.44324	.45114	.45993	.47002	.49865	
							1.00091	1.00452	1.00894	1.01322	1.01739	1.02523	
.437	1.1846	.8261	.97325	.03331	1.45597	1.00602	.43842	.44431	.45230	.46121	.47149	.50130	
							1.00091	1.00451	1.00891	1.01317	1.01733	1.02512	
.438	1.1828	.8265	.97374	.03276	1.45116	1.00586	.43943	.44538	.45346	.46250	.47297	.50414	
							1.00091	1.00449	1.00888	1.01313	1.01726	1.02501	
.439	1.1810	.8270	.97422	.03221	1.44636	1.00571	.44044	.44645	.45463	.46382	.47447	.50724	
							1.00091	1.00448	1.00885	1.01309	1.01720	1.02490	
.440	1.1792	.8274	.97470	.03167	1.44157	1.00556	.44146	.44752	.45580	.46512	.47599	.51067	
							1.00090	1.00447	1.00882	1.01305	1.01714	1.02478	
.441	1.1775	.8278	.97518	.03113	1.43680	1.00541	.44247	.44860	.45697	.46644	.47753	.51475	
							1.00090	1.00445	1.00879	1.01300	1.01707	1.02467	
.442	1.1757	.8283	.97565	.03059	1.43205	1.00526	.44348	.44968	.45815	.46777	.47910	.51979	
							1.00090	1.00444	1.00876	1.01296	1.01701	1.02455	
.443	1.1739	.8287	.97612	.03005	1.42730	1.00511	.44450	.45076	.45933	.46910	.48068	.52852	
							1.00089	1.00442	1.00873	1.01291	1.01695	1.02442	
.444	1.1721	.8291	.97658	.02952	1.42257	1.00497	.44551	.45184	.46052	.47045	.48230		
							1.00089	1.00441	1.00870	1.01286	1.01688		
.445	1.1703	.8296	.97704	.02900	1.41786	1.00483	.44653	.45292	.46171	.47181	.48394		
							1.00089	1.00440	1.00867	1.01282	1.01682		
.446	1.1685	.8300	.97749	.02848	1.41315	1.00470	.44754	.45400	.46291	.47318	.48561		
							1.00089	1.00438	1.00864	1.01277	1.01675		
.447	1.1668	.8304	.97794	.02796	1.40846	1.00456	.44856	.45509	.46411	.47456	.48731		
							1.00088	1.00437	1.00861	1.01273	1.01669		
.448	1.1650	.8309	.97839	.02744	1.40378	1.00443	.44957	.45617	.46532	.47596	.48905		
							1.00088	1.00435	1.00859	1.01268	1.01663		
.449	1.1632	.8313	.97883	.02693	1.39911	1.00430	.45059	.45726	.46656	.47737	.49083		
							1.00088	1.00434	1.00856	1.01264	1.01656		

TABLE VIII. Specific Heat Ratio = 1.30

p/p_t	MACH	T/T_t	GAMMA	K_crit	SHOCK P_y/P_x	P_ix/P_iy	.001	.002	.003	.005	.01	.015	.02

R₂ and TPR versus Loss Coefficient K

p/p_t	MACH	T/T_t	GAMMA	K_crit	P_y/P_x	P_ix/P_iy	.001	.002	.003	.005	.01	.015	.02
.450	1.1614	.8317	.97926	.02643	1.39446	1.00417	.45160	.45324	.45491	.45836	.46778	.47880	.49266
							1.00088	1.00175	1.00261	1.00433	1.00854	1.01259	1.01650
.451	1.1597	.8321	.97969	.02593	1.38982	1.00405	.45262	.45427	.45596	.45946	.46901	.48024	.49453
							1.00087	1.00174	1.00260	1.00431	1.00851	1.01255	1.01643
.452	1.1579	.8326	.98012	.02543	1.38519	1.00393	.45364	.45531	.45702	.46055	.47024	.48170	.49647
							1.00087	1.00173	1.00259	1.00430	1.00848	1.01250	1.01636
.453	1.1561	.8330	.98054	.02493	1.38058	1.00381	.45466	.45635	.45807	.46165	.47149	.48318	.49847
							1.00087	1.00173	1.00259	1.00429	1.00845	1.01246	1.01630
.454	1.1544	.8334	.98096	.02444	1.37598	1.00369	.45567	.45738	.45913	.46275	.47274	.48468	.50054
							1.00086	1.00172	1.00258	1.00427	1.00842	1.01241	1.01623
.455	1.1526	.8338	.98137	.02396	1.37139	1.00357	.45669	.45842	.46019	.46386	.47399	.48620	.50270
							1.00086	1.00172	1.00257	1.00426	1.00839	1.01237	1.01616
.456	1.1509	.8343	.98178	.02347	1.36681	1.00346	.45771	.45946	.46125	.46496	.47526	.48774	.50496
							1.00086	1.00171	1.00256	1.00425	1.00836	1.01232	1.01610
.457	1.1491	.8347	.98218	.02300	1.36224	1.00335	.45873	.46050	.46231	.46607	.47653	.48931	.50735
							1.00086	1.00171	1.00255	1.00423	1.00833	1.01227	1.01603
.458	1.1474	.8351	.98258	.02252	1.35769	1.00324	.45975	.46154	.46337	.46718	.47782	.49091	.50990
							1.00085	1.00170	1.00255	1.00422	1.00831	1.01223	1.01595
.459	1.1456	.8355	.98297	.02205	1.35315	1.00314	.46077	.46258	.46443	.46830	.47911	.49254	.51272
							1.00085	1.00170	1.00254	1.00420	1.00828	1.01218	1.01589
.460	1.1439	.8359	.98336	.02159	1.34862	1.00304	.46179	.46362	.46550	.46942	.48042	.49421	.51578
							1.00085	1.00169	1.00253	1.00419	1.00825	1.01213	1.01582
.461	1.1421	.8364	.98375	.02113	1.34410	1.00293	.46281	.46466	.46657	.47054	.48173	.49591	.51926
							1.00085	1.00169	1.00252	1.00418	1.00822	1.01209	1.01574
.462	1.1404	.8368	.98413	.02067	1.33960	1.00284	.46383	.46571	.46764	.47166	.48306	.49765	.52344
							1.00084	1.00168	1.00251	1.00416	1.00819	1.01204	1.01567
.463	1.1386	.8372	.98451	.02022	1.33511	1.00274	.46485	.46675	.46871	.47279	.48440	.49945	.52951
							1.00084	1.00168	1.00251	1.00415	1.00816	1.01199	1.01559
.464	1.1369	.8376	.98488	.01977	1.33062	1.00264	.46587	.46780	.46978	.47393	.48575	.50129	
							1.00084	1.00167	1.00250	1.00414	1.00813	1.01194	
.465	1.1351	.8380	.98525	.01932	1.32616	1.00255	.46690	.46885	.47085	.47506	.48712	.50320	
							1.00084	1.00167	1.00249	1.00412	1.00810	1.01190	
.466	1.1334	.8384	.98561	.01888	1.32170	1.00246	.46792	.46989	.47193	.47620	.48850	.50518	
							1.00083	1.00166	1.00248	1.00411	1.00808	1.01185	
.467	1.1317	.8389	.98597	.01845	1.31725	1.00237	.46894	.47094	.47301	.47735	.48990	.50724	
							1.00083	1.00165	1.00247	1.00410	1.00805	1.01180	
.468	1.1299	.8393	.98632	.01802	1.31282	1.00229	.46997	.47200	.47409	.47850	.49132	.50940	
							1.00083	1.00165	1.00247	1.00408	1.00802	1.01175	
.469	1.1282	.8397	.98667	.01759	1.30840	1.00220	.47100	.47305	.47517	.47965	.49276	.51168	
							1.00083	1.00164	1.00246	1.00407	1.00799	1.01170	
.470	1.1265	.8401	.98701	.01717	1.30399	1.00212	.47200	.47410	.47626	.48081	.49422	.51418	
							1.00081	1.00164	1.00245	1.00406	1.00796	1.01165	
.471	1.1247	.8405	.98735	.01675	1.29959	1.00204	.47305	.47516	.47735	.48198	.49570	.51685	
							1.00082	1.00163	1.00244	1.00404	1.00793	1.01160	
.472	1.1230	.8409	.98769	.01633	1.29520	1.00196	.47407	.47622	.47844	.48315	.49721	.51979	
							1.00082	1.00163	1.00243	1.00403	1.00790	1.01155	
.473	1.1213	.8413	.98802	.01592	1.29083	1.00189	.47510	.47728	.47953	.48432	.49874	.52316	
							1.00081	1.00162	1.00243	1.00401	1.00787	1.01149	
.474	1.1196	.8417	.98835	.01552	1.28646	1.00181	.47613	.47834	.48063	.48551	.50031	.52737	
							1.00081	1.00162	1.00242	1.00400	1.00784	1.01144	
.475	1.1178	.8422	.98867	.01512	1.28211	1.00174	.47716	.47940	.48173	.48670	.50191	.53363	
							1.00081	1.00161	1.00241	1.00399	1.00781	1.01138	
.476	1.1161	.8426	.98899	.01472	1.27777	1.00167	.47819	.48046	.48283	.48789	.50355		
							1.00081	1.00161	1.00240	1.00397	1.00778		
.477	1.1144	.8430	.98930	.01433	1.27343	1.00160	.47922	.48153	.48394	.48910	.50524		
							1.00080	1.00160	1.00239	1.00396	1.00775		
.478	1.1127	.8434	.98961	.01394	1.26912	1.00154	.48025	.48260	.48505	.49031	.50697		
							1.00080	1.00160	1.00239	1.00395	1.00772		
.479	1.1110	.8438	.98991	.01356	1.26481	1.00147	.48129	.48367	.48616	.49153	.50876		
							1.00080	1.00159	1.00238	1.00393	1.00769		

TABLE VIII. Specific Heat Ratio = 1.30

p/p_t	MACH	T/T_t	GAMMA	K_{crit}	SHOCK p_y/p_x	p_{ty}/p_{tx}	.001	.002	.003	.004	.005	.01	.015
.480	1.1093	.8442	.99021	.01318	1.26051	1.00141	.48232	.48474	.48728	.48994	.49275	.51062	
							1.00080	1.00159	1.00237	1.00315	1.00392	1.00766	
.481	1.1075	.8446	.99051	.01281	1.25622	1.00135	.48336	.48582	.48840	.49112	.49399	.51255	
							1.00079	1.00158	1.00236	1.00314	1.00390	1.00762	
.482	1.1058	.8450	.99080	.01244	1.25195	1.00129	.48439	.48690	.48953	.49230	.49524	.51463	
							1.00079	1.00158	1.00235	1.00313	1.00389	1.00760	
.483	1.1041	.8454	.99108	.01208	1.24768	1.00123	.48543	.48798	.49066	.49349	.49650	.51679	
							1.00079	1.00157	1.00235	1.00311	1.00388	1.00757	
.484	1.1024	.8458	.99136	.01172	1.24343	1.00117	.48647	.48906	.49180	.49469	.49777	.51911	
							1.00078	1.00156	1.00234	1.00310	1.00386	1.00753	
.485	1.1007	.8462	.99164	.01137	1.23919	1.00112	.48751	.49015	.49294	.49589	.49906	.52163	
							1.00078	1.00156	1.00233	1.00309	1.00385	1.00750	
.486	1.0990	.8466	.99191	.01102	1.23495	1.00107	.48855	.49124	.49408	.49711	.50039	.52444	
							1.00078	1.00155	1.00232	1.00308	1.00384	1.00747	
.487	1.0973	.8470	.99218	.01067	1.23073	1.00101	.48959	.49234	.49524	.49834	.50171	.52768	
							1.00078	1.00155	1.00231	1.00307	1.00383	1.00743	
.488	1.0956	.8474	.99244	.01033	1.22652	1.00097	.49064	.49343	.49640	.49957	.50305	.53193	
							1.00077	1.00154	1.00231	1.00306	1.00381	1.00740	
.489	1.0939	.8478	.99270	.01000	1.22232	1.00092	.49169	.49453	.49757	.50086	.50441	.54084	
							1.00077	1.00154	1.00230	1.00306	1.00380	1.00735	
.490	1.0922	.8482	.99296	.00967	1.21813	1.00087	.49273	.49564	.49874	.50212	.50579		
							1.00077	1.00153	1.00229	1.00304	1.00379		
.491	1.0905	.8486	.99321	.00934	1.21395	1.00083	.49378	.49675	.49992	.50340	.50719		
							1.00077	1.00153	1.00228	1.00303	1.00377		
.492	1.0888	.8490	.99345	.00902	1.20978	1.00078	.49483	.49786	.50115	.50470	.50862		
							1.00076	1.00152	1.00228	1.00302	1.00376		
.493	1.0872	.8494	.99369	.00871	1.20563	1.00074	.49589	.49898	.50236	.50601	.51008		
							1.00076	1.00152	1.00227	1.00301	1.00374		
.494	1.0855	.8498	.99393	.00840	1.20148	1.00070	.49694	.50011	.50357	.50734	.51158		
							1.00076	1.00151	1.00226	1.00300	1.00373		
.495	1.0838	.8502	.99416	.00809	1.19734	1.00066	.49800	.50124	.50480	.50870	.51311		
							1.00075	1.00150	1.00226	1.00299	1.00372		
.496	1.0821	.8506	.99439	.00779	1.19321	1.00063	.49906	.50241	.50604	.51007	.51468		
							1.00075	1.00151	1.00225	1.00298	1.00370		
.497	1.0804	.8510	.99461	.00750	1.18910	1.00059	.50012	.50356	.50730	.51148	.51631		
							1.00075	1.00150	1.00224	1.00297	1.00369		
.498	1.0787	.8514	.99483	.00721	1.18499	1.00055	.50119	.50472	.50857	.51291	.51800		
							1.00075	1.00150	1.00223	1.00296	1.00367		
.499	1.0770	.8518	.99504	.00692	1.18089	1.00052	.50229	.50588	.50985	.51438	.51976		
							1.00075	1.00149	1.00222	1.00295	1.00366		
.500	1.0754	.8522	.99525	.00664	1.17681	1.00049	.50337	.50705	.51116	.51588	.52162		
							1.00075	1.00149	1.00222	1.00294	1.00364		
.501	1.0737	.8526	.99546	.00637	1.17273	1.00046	.50445	.50824	.51249	.51744	.52358		
							1.00075	1.00148	1.00221	1.00292	1.00363		
.502	1.0720	.8530	.99566	.00610	1.16866	1.00043	.50553	.50943	.51384	.51904	.52569		
							1.00074	1.00148	1.00220	1.00291	1.00361		
.503	1.0703	.8534	.99585	.00584	1.16461	1.00040	.50662	.51064	.51522	.52071	.52811		
							1.00074	1.00147	1.00219	1.00290	1.00360		
.504	1.0687	.8537	.99605	.00558	1.16056	1.00037	.50771	.51186	.51663	.52246	.53079		
							1.00074	1.00147	1.00218	1.00289	1.00358		
.505	1.0670	.8541	.99623	.00533	1.15652	1.00035	.50881	.51309	.51808	.52430	.53404		
							1.00074	1.00146	1.00218	1.00288	1.00356		
.506	1.0653	.8545	.99642	.00508	1.15250	1.00032	.50991	.51434	.51957	.52627	.53899		
							1.00073	1.00146	1.00217	1.00286	1.00355		
.507	1.0636	.8549	.99659	.00484	1.14848	1.00030	.51102	.51561	.52111	.52851			
							1.00073	1.00145	1.00216	1.00286			
.508	1.0620	.8553	.99677	.00460	1.14447	1.00028	.51214	.51690	.52270	.53095			
							1.00073	1.00145	1.00215	1.00284			
.509	1.0603	.8557	.99694	.00437	1.14047	1.00026	.51326	.51822	.52437	.53383			
							1.00073	1.00144	1.00214	1.00283			

TABLE VIII. Specific Heat Ratio = 1.30

p/p_t	MACH	T/T_t	GAMMA	K_{crit}	SHOCK P_y/P_x	P_{tx}/P_{ty}	.001	.002	R₂ and TPR versus Loss Coefficient K .003	.004	.005	.01	.015
.510	1.0586	.8561	.99710	.00414	1.13649	1.00024	.51439	.51956	.52613	.53778			
							1.00072	1.00143	1.00213	1.00282			
.511	1.0570	.8565	.99726	.00392	1.13251	1.00022	.51553	.52093	.52810				
							1.00072	1.00143	1.00213				
.512	1.0553	.8569	.99741	.00371	1.12854	1.00020	.51668	.52234	.53018				
							1.00072	1.00142	1.00212				
.513	1.0537	.8572	.99757	.00350	1.12458	1.00018	.51785	.52380	.53252				
							1.00071	1.00142	1.00211				
.514	1.0520	.8576	.99771	.00329	1.12063	1.00017	.51902	.52531	.53529				
							1.00071	1.00141	1.00210				
.515	1.0503	.8580	.99785	.00310	1.11669	1.00015	.52021	.52688	.53927				
							1.00071	1.00140	1.00209				
.516	1.0487	.8584	.99799	.00291	1.11276	1.00014	.52142	.52863					
							1.00071	1.00140					
.517	1.0470	.8588	.99812	.00272	1.10884	1.00013	.52264	.53044					
							1.00070	1.00140					
.518	1.0454	.8592	.99825	.00254	1.10492	1.00011	.52389	.53242					
							1.00070	1.00139					
.519	1.0437	.8595	.99837	.00236	1.10102	1.00010	.52516	.53466					
							1.00070	1.00138					
.520	1.0421	.8599	.99849	.00219	1.09713	1.00009	.52646	.53734					
							1.00069	1.00137					
.521	1.0404	.8603	.99861	.00203	1.09324	1.00008	.52780	.54168					
							1.00069	1.00137					
.522	1.0388	.8607	.99872	.00188	1.08937	1.00007	.52928						
							1.00069						
.523	1.0371	.8611	.99882	.00172	1.08550	1.00006	.53076						
							1.00069						
.524	1.0355	.8615	.99893	.00158	1.08164	1.00005	.53232						
							1.00069						
.525	1.0338	.8618	.99902	.00144	1.07779	1.00005	.53400						
							1.00068						
.526	1.0322	.8622	.99911	.00131	1.07395	1.00004	.53585						
							1.00068						
.527	1.0305	.8626	.99920	.00118	1.07012	1.00004	.53797						
							1.00067						
.528	1.0289	.8630	.99928	.00106	1.06630	1.00003	.54107						
							1.00067						

TABLE IX. Specific Heat Ratio = 1.40

p/p₁	MACH	T/T₁	GAMMA	K_crit	SHOCK P_y/P_x	P_tx/P_ty	.1	.2	.3	.4	.5	.6	.7
									R₂ and TPR versus Loss Coefficient K				
.001	5.5664	.1389	.02580	.71633	35.98166	24.77842	.00199	.00304	.00417	.00541	.00686	.00867	.01177
							4.32049	9.48939	15.70828	22.23413	28.46984	33.95104	38.23490
.002	4.9517	.1694	.04158	.69157	28.43904	15.60818	.00362	.00533	.00719	.00926	.01170	.01488	
							3.38771	6.80308	10.74919	14.78529	18.56326	21.80936	
.003	4.6141	.1902	.05485	.67424	24.67180	11.95703	.00515	.00744	.00993	.01273	.01606	.02056	
							2.97150	5.65682	8.68228	11.72535	14.53347	16.90489	
.004	4.3836	.2065	.06668	.66045	22.25156	9.91781	.00663	.00945	.01252	.01599	.02017	.02599	
							2.72121	4.98647	7.49168	9.97915	12.24882	14.13696	
.005	4.2095	.2201	.07753	.64877	20.50659	8.59037	.00808	.01138	.01500	.01911	.02410	.03131	
							2.54886	4.53443	6.69754	8.82293	10.74383	12.31968	
.006	4.0702	.2318	.08764	.63855	19.16116	7.64613	.00951	.01327	.01740	.02212	.02791	.03660	
							2.42076	4.20315	6.12119	7.98902	9.66254	11.01731	
.007	3.9544	.2423	.09718	.62938	18.07704	6.93432	.01091	.01511	.01974	.02505	.03164	.04195	
							2.32030	3.94707	5.67902	7.35257	8.84019	10.02880	
.008	3.8555	.2517	.10624	.62103	17.17576	6.37521	.01230	.01692	.02203	.02792	.03530	.04744	
							2.23872	3.74146	5.32653	6.84720	8.18938	9.24721	
.009	3.7693	.2603	.11489	.61333	16.40882	5.92241	.01367	.01870	.02428	.03074	.03890	.05322	
							2.17067	3.57165	5.03710	6.43388	7.65841	8.61001	
.010	3.6930	.2683	.12321	.60616	15.74430	5.54689	.01503	.02046	.02650	.03351	.04247	.05963	
							2.11272	3.42831	4.79406	6.08790	7.21515	8.07791	
.011	3.6245	.2757	.13122	.59944	15.16016	5.22954	.01638	.02220	.02868	.03625	.04601		
							2.06255	3.30534	4.58630	5.79324	6.83822		
.012	3.5626	.2826	.13896	.59309	14.64059	4.95716	.01772	.02392	.03084	.03896	.04952		
							2.01854	3.19804	4.40608	5.53836	6.51282		
.013	3.5060	.2892	.14646	.58708	14.17391	4.72038	.01906	.02563	.03298	.04164	.05303		
							1.97948	3.10346	4.24785	5.31516	6.22837		
.014	3.4539	.2953	.15375	.58135	13.75124	4.51229	.02039	.02732	.03510	.04429	.05652		
							1.94450	3.01927	4.10750	5.11766	5.97700		
.015	3.4058	.3012	.16084	.57587	13.36572	4.32772	.02171	.02900	.03720	.04693	.06001		
							1.91292	2.94367	3.98190	4.94133	5.75298		
.016	3.3609	.3068	.16776	.57062	13.01189	4.16268	.02302	.03067	.03928	.04954	.06351		
							1.88421	2.87530	3.86866	4.78268	5.55167		
.017	3.3190	.3122	.17450	.56557	12.68542	4.01407	.02433	.03233	.04135	.05215	.06701		
							1.85781	2.81305	3.76588	4.63897	5.36953		
.018	3.2797	.3173	.18109	.56071	12.38274	3.87942	.02564	.03397	.04340	.05473	.07053		
							1.83368	2.75607	3.67206	4.50802	5.20373		
.019	3.2427	.3223	.18754	.55601	12.10095	3.75676	.02694	.03561	.04545	.05731	.07407		
							1.81138	2.70364	3.58595	4.38806	5.05199		
.020	3.2077	.3270	.19386	.55146	11.83761	3.64446	.02823	.03724	.04748	.05988	.07763		
							1.79068	2.65518	3.50657	4.27766	4.91244		
.021	3.1745	.3316	.20004	.54705	11.59068	3.54119	.02952	.03887	.04950	.06244	.08123		
							1.77140	2.61021	3.43309	4.17568	4.78354		
.022	3.1430	.3361	.20612	.54277	11.35842	3.44585	.03081	.04048	.05151	.06499	.08487		
							1.75338	2.56833	3.36480	4.08099	4.66400		
.023	3.1130	.3403	.21208	.53861	11.13936	3.35751	.03210	.04210	.05352	.06753	.08855		
							1.73648	2.52920	3.30121	3.99283	4.55275		
.024	3.0844	.3445	.21793	.53456	10.93221	3.27538	.03338	.04370	.05552	.07007	.09230		
							1.72059	2.49252	3.24164	3.91049	4.44892		
.025	3.0570	.3486	.22369	.53061	10.73587	3.19879	.03466	.04530	.05751	.07261	.09612		
							1.70561	2.45804	3.18577	3.83334	4.35163		
.026	3.0307	.3525	.22935	.52675	10.54937	3.12718	.03593	.04689	.05949	.07514	.10002		
							1.69145	2.42556	3.13322	3.76088	4.26026		
.027	3.0055	.3563	.23492	.52299	10.37188	3.06004	.03721	.04848	.06147	.07768	.10403		
							1.67804	2.39488	3.08369	3.69265	4.17420		
.028	2.9813	.3600	.24040	.51931	10.20264	2.99694	.03848	.05007	.06344	.08021	.10817		
							1.66531	2.36584	3.03688	3.62825	4.09295		
.029	2.9579	.3637	.24580	.51571	10.04101	2.93752	.03975	.05164	.06541	.08274	.11248		
							1.65321	2.33830	2.99257	3.56733	4.01605		

TABLE IX. Specific Heat Ratio = 1.40

					SHOCK		R₁ and TPR versus Loss Coefficient K						
p/p_t	MACH	T/T_t	GAMMA	K_{crit}	P_y/P_x	P_{tx}/P_{ty}	.01	.05	.1	.2	.3	.4	.5
.030	2.9355	.3672	.25113	.51218	9.88638	2.88144	.03106	.03539	.04101	.05322	.06737	.08527	.11703
							1.06083	1.31284	1.64162	2.31214	2.95053	3.50961	3.94311
.031	2.9137	.3706	.25638	.50873	9.73824	2.82841	.03209	.03651	.04228	.05479	.06933	.08781	.12193
							1.05992	1.30782	1.63061	2.28723	2.91058	3.45480	3.87378
.032	2.8928	.3740	.26155	.50534	9.59613	2.77818	.03311	.03764	.04354	.05636	.07128	.09034	.12739
							1.05905	1.30302	1.62016	2.26348	2.87255	3.40267	3.80769
.033	2.8725	.3773	.26666	.50202	9.45961	2.73051	.03414	.03877	.04480	.05793	.07323	.09288	.13411
							1.05821	1.29841	1.61010	2.24081	2.83628	3.35301	3.74452
.034	2.8528	.3806	.27170	.49876	9.32831	2.68522	.03516	.03989	.04606	.05949	.07518	.09543	
							1.05741	1.29399	1.60046	2.21912	2.80166	3.30563	
.035	2.8338	.3837	.27667	.49555	9.20188	2.64210	.03618	.04102	.04732	.06105	.07713	.09798	
							1.05663	1.28975	1.59121	2.19836	2.76854	3.26035	
.036	2.8153	.3868	.28158	.49240	9.08002	2.60101	.03721	.04214	.04857	.06260	.07907	.10054	
							1.05588	1.28566	1.58232	2.17846	2.73684	3.21709	
.037	2.7973	.3899	.28644	.48930	8.96243	2.56180	.03823	.04326	.04982	.06416	.08101	.10310	
							1.05516	1.28173	1.57377	2.15935	2.70645	3.17558	
.038	2.7799	.3928	.29123	.48625	8.84885	2.52433	.03926	.04438	.05108	.06571	.08295	.10567	
							1.05446	1.27794	1.56554	2.14099	2.67728	3.13578	
.039	2.7629	.3958	.29597	.48325	8.73906	2.48848	.04028	.04550	.05233	.06726	.08488	.10825	
							1.05379	1.27428	1.55761	2.12332	2.64925	3.09755	
.040	2.7463	.3986	.30065	.48029	8.63283	2.45414	.04130	.04662	.05358	.06880	.08682	.11084	
							1.05314	1.27074	1.54995	2.10631	2.62228	3.06080	
.041	2.7302	.4015	.30528	.47738	8.52995	2.42123	.04233	.04774	.05482	.07035	.08875	.11344	
							1.05251	1.26732	1.54256	2.08992	2.59632	3.02544	
.042	2.7146	.4042	.30986	.47451	8.43026	2.38963	.04335	.04886	.05607	.07189	.09069	.11605	
							1.05189	1.26401	1.53542	2.07410	2.57130	2.99137	
.043	2.6992	.4070	.31439	.47168	8.33357	2.35928	.04437	.04998	.05732	.07343	.09262	.11867	
							1.05130	1.26081	1.52851	2.05882	2.54717	2.95853	
.044	2.6843	.4097	.31887	.46889	8.23973	2.33010	.04540	.05110	.05856	.07497	.09455	.12131	
							1.05072	1.25771	1.52182	2.04406	2.52387	2.92683	
.045	2.6697	.4123	.32331	.46613	8.14859	2.30201	.04642	.05222	.05981	.07651	.09649	.12396	
							1.05017	1.25470	1.51534	2.02978	2.50135	2.89622	
.046	2.6555	.4149	.32770	.46342	8.06002	2.27496	.04744	.05333	.06105	.07804	.09842	.12663	
							1.04962	1.25177	1.50906	2.01595	2.47957	2.86662	
.047	2.6415	.4174	.33204	.46073	7.97389	2.24889	.04846	.05445	.06229	.07958	.10035	.12931	
							1.04909	1.24894	1.50297	2.00256	2.45850	2.83798	
.048	2.6279	.4200	.33635	.45809	7.89009	2.22373	.04949	.05556	.06353	.08111	.10229	.13202	
							1.04858	1.24618	1.49706	1.98958	2.43813	2.81026	
.049	2.6145	.4224	.34061	.45547	7.80850	2.19945	.05051	.05668	.06477	.08264	.10422	.13474	
							1.04808	1.24350	1.49131	1.97699	2.41836	2.78340	
.050	2.6015	.4249	.34482	.45288	7.72902	2.17598	.05153	.05779	.06601	.08417	.10615	.13749	
							1.04759	1.24089	1.48573	1.96477	2.39918	2.75735	
.051	2.5887	.4273	.34900	.45033	7.65157	2.15330	.05255	.05891	.06725	.08570	.10809	.14026	
							1.04712	1.23835	1.48030	1.95291	2.38056	2.73208	
.052	2.5762	.4297	.35314	.44780	7.57604	2.13136	.05357	.06002	.06848	.08723	.11002	.14306	
							1.04666	1.23588	1.47501	1.94137	2.36249	2.70754	
.053	2.5639	.4320	.35724	.44531	7.50235	2.11012	.05460	.06113	.06972	.08876	.11196	.14588	
							1.04620	1.23347	1.46987	1.93016	2.34493	2.68370	
.054	2.5518	.4343	.36131	.44284	7.43043	2.08955	.05562	.06225	.07096	.09029	.11390	.14874	
							1.04576	1.23112	1.46485	1.91925	2.32786	2.66052	
.055	2.5400	.4366	.36533	.44039	7.36021	2.06962	.05664	.06336	.07219	.09181	.11584	.15162	
							1.04533	1.22883	1.45997	1.90863	2.31126	2.63798	
.056	2.5284	.4389	.36932	.43798	7.29160	2.05029	.05766	.06447	.07342	.09334	.11778	.15455	
							1.04491	1.22659	1.45521	1.89828	2.29510	2.61606	
.057	2.5170	.4411	.37328	.43558	7.22456	2.03154	.05868	.06558	.07466	.09486	.11973	.15752	
							1.04450	1.22441	1.45056	1.88821	2.27937	2.59469	
.058	2.5058	.4433	.37720	.43322	7.15901	2.01334	.05970	.06669	.07589	.09639	.12168	.16053	
							1.04410	1.22228	1.44603	1.87838	2.26404	2.57388	
.059	2.4948	.4455	.38109	.43087	7.09489	1.99566	.06073	.06780	.07712	.09791	.12363	.16359	
							1.04371	1.22019	1.44160	1.86880	2.24910	2.55358	

162

TABLE IX. Specific Heat Ratio $= 1.40$

p/p_t	MACH	T/T_t	GAMMA	K_{crit}	SHOCK p_y/p_x	p_{ty}/p_{tx}	.01	.05	.1	.15	.2	.3	.4
.060	2.4840	.4476	.38495	.42855	7.03216	1.97849	.06175	.06892	.07836	.08846	.09943	.12558	.16670
							1.04332	1.21816	1.43728	1.65242	1.85947	2.23454	2.53378
.061	2.4734	.4497	.38877	.42625	6.97076	1.96180	.06277	.07003	.07959	.08983	.10096	.12753	.16988
							1.04295	1.21616	1.43305	1.64580	1.85034	2.22033	2.51446
.062	2.4630	.4518	.39256	.42398	6.91064	1.94556	.06379	.07114	.08082	.09119	.10248	.12949	.17312
							1.04258	1.21421	1.42892	1.63933	1.84143	2.20647	2.49559
.063	2.4527	.4539	.39633	.42172	6.85175	1.92977	.06481	.07225	.08205	.09256	.10400	.13145	.17645
							1.04222	1.21231	1.42489	1.63300	1.83273	2.19293	2.47715
.064	2.4426	.4559	.40006	.41949	6.79406	1.91440	.06583	.07335	.08328	.09392	.10553	.13342	.17986
							1.04187	1.21044	1.42094	1.62682	1.82423	2.17971	2.45913
.065	2.4327	.4580	.40376	.41727	6.73751	1.89943	.06685	.07446	.08451	.09529	.10705	.13539	.18338
							1.04152	1.20861	1.41707	1.62077	1.81591	2.16679	2.44150
.066	2.4229	.4600	.40743	.41508	6.68206	1.88485	.06787	.07557	.08574	.09665	.10857	.13736	.187C3
							1.04118	1.20682	1.41329	1.61486	1.80778	2.15416	2.42425
.067	2.4132	.4619	.41108	.41290	6.62769	1.87064	.06890	.07668	.08697	.09802	.11009	.13934	.19083
							1.04085	1.20507	1.40958	1.60907	1.79982	2.14181	2.40736
.068	2.4038	.4639	.41469	.41075	6.57435	1.85679	.06992	.07779	.08819	.09938	.11162	.14132	.19482
							1.04052	1.20335	1.40595	1.60340	1.79204	2.12973	2.39080
.069	2.3944	.4658	.41828	.40861	6.52201	1.84329	.07094	.07890	.08942	.10075	.11314	.14330	.19908
							1.04020	1.20166	1.40239	1.59784	1.78442	2.11791	2.37459
.070	2.3852	.4678	.42184	.40649	6.47064	1.83011	.07196	.08001	.09065	.10211	.11466	.14529	.20369
							1.03989	1.20005	1.39891	1.59240	1.77695	2.10634	2.35867
.071	2.3761	.4697	.42538	.40438	6.42020	1.81726	.07298	.08112	.09188	.10347	.11619	.14729	.20882
							1.03958	1.19843	1.39549	1.58707	1.76964	2.09501	2.34303
.072	2.3672	.4715	.42889	.40230	6.37067	1.80471	.07400	.08222	.09311	.10484	.11771	.14929	.21497
							1.03929	1.19684	1.39214	1.58185	1.76248	2.08392	2.32765
.073	2.3583	.4734	.43237	.40023	6.32201	1.79245	.07502	.08333	.09433	.10620	.11923	.15129	.22470
							1.03898	1.19528	1.38885	1.57673	1.75545	2.07304	2.31246
.074	2.3496	.4753	.43583	.39817	6.27421	1.78048	.07604	.08444	.09556	.10756	.12076	.15331	
							1.03869	1.19374	1.38562	1.57170	1.74857	2.06239	
.075	2.3410	.4771	.43926	.39614	6.22722	1.76879	.07706	.08555	.09679	.10893	.12228	.15532	
							1.03840	1.19224	1.38246	1.56677	1.74182	2.05194	
.076	2.3326	.4789	.44267	.39411	6.18104	1.75736	.07808	.08665	.09801	.11029	.12381	.15735	
							1.03812	1.19076	1.37935	1.56193	1.73519	2.04170	
.077	2.3242	.4807	.44605	.39211	6.13563	1.74619	.07910	.08776	.09924	.11165	.12533	.15938	
							1.03784	1.18930	1.37629	1.55718	1.72865	2.03165	
.078	2.3160	.4825	.44941	.39012	6.09097	1.73526	.08012	.08887	.10046	.11301	.12686	.16141	
							1.03757	1.18788	1.37329	1.55252	1.72227	2.02179	
.079	2.3078	.4842	.45275	.38814	6.04704	1.72457	.08114	.08997	.10169	.11438	.12839	.16346	
							1.03730	1.18647	1.37035	1.54794	1.71601	2.01211	
.080	2.2998	.4860	.45606	.38618	6.00383	1.71412	.08216	.09108	.10292	.11574	.12992	.16551	
							1.03704	1.18509	1.36745	1.54344	1.70986	2.00261	
.081	2.2918	.4877	.45935	.38423	5.96130	1.70388	.08318	.09219	.10414	.11710	.13145	.16757	
							1.03678	1.18373	1.36461	1.53902	1.70382	1.99328	
.082	2.2840	.4894	.46262	.38229	5.91944	1.69387	.08420	.09329	.10537	.11847	.13298	.16963	
							1.03652	1.18240	1.36181	1.53468	1.69789	1.98411	
.083	2.2763	.4911	.46587	.38037	5.87823	1.68406	.08522	.09440	.10659	.11983	.13451	.17171	
							1.03627	1.18108	1.35906	1.53041	1.69206	1.97511	
.084	2.2686	.4928	.46909	.37846	5.83765	1.67445	.08624	.09550	.10782	.12119	.13604	.17379	
							1.03602	1.17979	1.35635	1.52621	1.68633	1.96626	
.085	2.2610	.4944	.47229	.37657	5.79769	1.66504	.08726	.09661	.10904	.12256	.13757	.17588	
							1.03578	1.17852	1.35369	1.52209	1.68069	1.95756	
.086	2.2536	.4961	.47547	.37469	5.75833	1.65582	.08828	.09772	.11027	.12392	.13910	.17798	
							1.03554	1.17727	1.35107	1.51803	1.67515	1.94901	
.087	2.2462	.4977	.47864	.37282	5.71956	1.64679	.08930	.09882	.11149	.12529	.14064	.18010	
							1.03530	1.17603	1.34849	1.51403	1.66970	1.94060	
.088	2.2389	.4994	.48178	.37096	5.68135	1.63793	.09032	.09993	.11272	.12665	.14217	.18222	
							1.03507	1.17482	1.34596	1.51010	1.66434	1.93233	
.089	2.2317	.5010	.48490	.36911	5.64370	1.62925	.09135	.10103	.11394	.12802	.14371	.18435	
							1.03484	1.17362	1.34346	1.50624	1.65907	1.92419	

TABLE IX. Specific Heat Ratio = 1.40

p/p_t	MACH	T/T_t	GAMMA	K_{crit}	SHOCK P_y/P_x	P_{ty}/P_{tx}	.01	.05	.1	.15	.2	.25	.3
.090	2.2245	.5026	.48800	.36728	5.60659	1.62073	.09237	.10214	.11517	.12938	.14525	.16363	.18649
							1.03461	1.17244	1.34100	1.50243	1.65388	1.79274	1.91617
.091	2.2175	.5042	.49108	.36546	5.57001	1.61238	.09339	.10324	.11639	.13075	.14679	.16540	.18865
							1.03439	1.17128	1.33858	1.49868	1.64877	1.78626	1.90831
.092	2.2105	.5058	.49414	.36364	5.53393	1.60419	.09441	.10435	.11762	.13211	.14833	.16718	.19082
							1.03417	1.17014	1.33619	1.49499	1.64374	1.77987	1.90055
.093	2.2036	.5073	.49718	.36184	5.49836	1.59615	.09543	.10545	.11884	.13348	.14987	.16896	.19300
							1.03395	1.16901	1.33384	1.49135	1.63878	1.77358	1.89291
.094	2.1967	.5089	.50020	.36006	5.46328	1.58826	.09645	.10656	.12007	.13485	.15141	.17075	.19519
							1.03374	1.16790	1.33152	1.48777	1.63391	1.76739	1.88538
.095	2.1900	.5104	.50321	.35828	5.42867	1.58051	.09747	.10766	.12129	.13621	.15296	.17254	.19740
							1.03353	1.16680	1.32924	1.48425	1.62910	1.76129	1.87797
.096	2.1833	.5119	.50619	.35651	5.39453	1.57291	.09849	.10877	.12252	.13758	.15451	.17433	.19962
							1.03332	1.16572	1.32699	1.48077	1.62437	1.75528	1.87067
.097	2.1767	.5135	.50916	.35475	5.36084	1.56544	.09951	.10988	.12375	.13895	.15605	.17613	.20186
							1.03311	1.16466	1.32477	1.47734	1.61971	1.74937	1.86347
.098	2.1701	.5150	.51211	.35301	5.32760	1.55811	.10053	.11098	.12497	.14032	.15760	.17794	.20411
							1.03291	1.16361	1.32258	1.47397	1.61511	1.74353	1.85638
.099	2.1636	.5165	.51504	.35127	5.29479	1.55091	.10155	.11209	.12620	.14169	.15916	.17974	.20638
							1.03271	1.16257	1.32043	1.47064	1.61058	1.73779	1.84939
.100	2.1572	.5179	.51795	.34954	5.26240	1.54383	.10257	.11319	.12742	.14306	.16071	.18156	.20867
							1.03251	1.16155	1.31830	1.46736	1.60611	1.73212	1.84249
.101	2.1508	.5194	.52085	.34783	5.23043	1.53688	.10359	.11430	.12865	.14443	.16227	.18337	.21098
							1.03232	1.16054	1.31620	1.46412	1.60171	1.72653	1.83569
.102	2.1445	.5209	.52373	.34612	5.19886	1.53005	.10461	.11540	.12988	.14580	.16382	.18520	.21331
							1.03212	1.15954	1.31413	1.46093	1.59737	1.72103	1.82899
.103	2.1383	.5223	.52659	.34442	5.16769	1.52334	.10562	.11651	.13110	.14718	.16538	.18703	.21565
							1.03193	1.15856	1.31209	1.45778	1.59309	1.71559	1.82237
.104	2.1321	.5238	.52943	.34273	5.13690	1.51674	.10664	.11761	.13233	.14855	.16694	.18886	.21802
							1.03175	1.15759	1.31008	1.45467	1.58886	1.71023	1.81584
.105	2.1260	.5252	.53226	.34106	5.10650	1.51025	.10767	.11872	.13356	.14992	.16851	.19070	.22042
							1.03159	1.15663	1.30809	1.45161	1.58470	1.70495	1.80940
.106	2.1199	.5266	.53507	.33939	5.07646	1.50387	.10869	.11982	.13479	.15130	.17007	.19255	.22284
							1.03141	1.15568	1.30612	1.44858	1.58059	1.69973	1.80304
.107	2.1139	.5281	.53787	.33772	5.04678	1.49759	.10971	.12093	.13601	.15267	.17164	.19440	.22528
							1.03123	1.15475	1.30419	1.44560	1.57653	1.69459	1.79676
.108	2.1080	.5295	.54065	.33607	5.01746	1.49142	.11073	.12203	.13724	.15405	.17321	.19626	.22775
							1.03105	1.15382	1.30227	1.44265	1.57252	1.68951	1.79056
.109	2.1021	.5309	.54341	.33443	4.98849	1.48535	.11175	.12314	.13847	.15543	.17478	.19812	.23026
							1.03088	1.15291	1.30038	1.43974	1.56857	1.68449	1.78444
.110	2.0962	.5322	.54616	.33279	4.95985	1.47938	.11277	.12424	.13970	.15681	.17636	.19999	.23279
							1.03070	1.15201	1.29852	1.43687	1.56467	1.67954	1.77839
.111	2.0904	.5336	.54889	.33117	4.93155	1.47350	.11379	.12535	.14093	.15818	.17793	.20187	.23536
							1.03053	1.15108	1.29667	1.43403	1.56082	1.67465	1.77242
.112	2.0847	.5350	.55161	.32955	4.90357	1.46772	.11481	.12646	.14215	.15956	.17951	.20376	.23796
							1.03036	1.15020	1.29485	1.43123	1.55701	1.66983	1.76651
.113	2.0790	.5364	.55431	.32794	4.87592	1.46203	.11583	.12756	.14338	.16095	.18109	.20565	.24061
							1.03019	1.14934	1.29305	1.42847	1.55326	1.66503	1.76068
.114	2.0733	.5377	.55699	.32634	4.84857	1.45643	.11685	.12867	.14461	.16233	.18268	.20755	.24330
							1.03003	1.14848	1.29128	1.42573	1.54955	1.66032	1.75491
.115	2.0678	.5390	.55966	.32474	4.82154	1.45091	.11787	.12977	.14584	.16371	.18427	.20946	.24604
							1.02987	1.14763	1.28952	1.42303	1.54588	1.65567	1.74920
.116	2.0622	.5404	.56232	.32316	4.79480	1.44548	.11889	.13088	.14707	.16509	.18586	.21138	.24883
							1.02970	1.14679	1.28779	1.42037	1.54226	1.65108	1.74356
.117	2.0567	.5417	.56496	.32158	4.76836	1.44014	.11991	.13199	.14831	.16648	.18745	.21330	.25167
							1.02954	1.14596	1.28607	1.41773	1.53868	1.64654	1.73798
.118	2.0512	.5430	.56759	.32001	4.74220	1.43487	.12093	.13309	.14954	.16787	.18905	.21524	.25458
							1.02938	1.14514	1.28438	1.41513	1.53515	1.64205	1.73246
.119	2.0458	.5443	.57020	.31845	4.71633	1.42968	.12195	.13420	.15077	.16925	.19065	.21718	.25756
							1.02923	1.14433	1.28270	1.41255	1.53165	1.63761	1.72699

TABLE IX. Specific Heat Ratio = 1.40

p/p_t	MACH	T/T_t	GAMMA	K_crit	P_y/P_x	P_ty/P_ty	.01	.05	.1	.15	.2	.25	.3
					SHOCK				R₁ and TPR versus Loss Coefficient K				
.120	2.0405	.5456	.57280	.31689	4.69074	1.42458	.12297	.13530	.15200	.17064	.19225	.21914	.26062
							1.02907	1.14353	1.28105	1.41001	1.52820	1.63324	1.72158
.121	2.0351	.5469	.57538	.31534	4.66542	1.41954	.12399	.13641	.15323	.17203	.19385	.22111	.26378
							1.02892	1.14274	1.27941	1.40749	1.52478	1.62891	1.71624
.122	2.0299	.5482	.57795	.31380	4.64037	1.41459	.12501	.13752	.15446	.17342	.19546	.22308	.26703
							1.02877	1.14196	1.27779	1.40501	1.52141	1.62462	1.71094
.123	2.0246	.5495	.58050	.31227	4.61559	1.40970	.12603	.13862	.15570	.17482	.19707	.22507	.27041
							1.02862	1.14118	1.27619	1.40255	1.51807	1.62038	1.70568
.124	2.0194	.5508	.58305	.31074	4.59105	1.40489	.12705	.13973	.15693	.17621	.19869	.22706	.27393
							1.02847	1.14041	1.27460	1.40012	1.51477	1.61619	1.70048
.125	2.0143	.5520	.58557	.30922	4.56678	1.40015	.12807	.14084	.15817	.17760	.20031	.22907	.27763
							1.02832	1.13966	1.27304	1.39771	1.51151	1.61205	1.69532
.126	2.0091	.5533	.58809	.30771	4.54275	1.39547	.12909	.14194	.15940	.17900	.20193	.23109	.28155
							1.02818	1.13890	1.27149	1.39533	1.50828	1.60794	1.69020
.127	2.0041	.5546	.59059	.30620	4.51896	1.39086	.13011	.14305	.16064	.18040	.20356	.23312	.28575
							1.02803	1.13816	1.26995	1.39298	1.50509	1.60389	1.68512
.128	1.9990	.5558	.59308	.30470	4.49542	1.38632	.13113	.14416	.16187	.18180	.20519	.23517	.29035
							1.02789	1.13743	1.26844	1.39065	1.50193	1.59987	1.68007
.129	1.9940	.5570	.59555	.30321	4.47211	1.38184	.13215	.14527	.16311	.18320	.20682	.23722	.29557
							1.02775	1.13670	1.26693	1.38835	1.49881	1.59590	1.67507
.130	1.9890	.5583	.59801	.30173	4.44903	1.37743	.13317	.14637	.16434	.18460	.20846	.23930	.30189
							1.02761	1.13598	1.26545	1.38607	1.49572	1.59196	1.67007
.131	1.9841	.5595	.60046	.30025	4.42618	1.37308	.13419	.14748	.16558	.18600	.21010	.24138	.31181
							1.02747	1.13527	1.26398	1.38382	1.49266	1.58807	1.66508
.132	1.9792	.5607	.60290	.29877	4.40355	1.36878	.13521	.14859	.16682	.18741	.21175	.24349	
							1.02733	1.13456	1.26252	1.38159	1.48963	1.58421	
.133	1.9744	.5619	.60532	.29731	4.38114	1.36455	.13623	.14970	.16806	.18881	.21340	.24560	
							1.02720	1.13386	1.26108	1.37938	1.48664	1.58040	
.134	1.9695	.5631	.60773	.29585	4.35894	1.36037	.13725	.15081	.16930	.19022	.21505	.24774	
							1.02706	1.13317	1.25966	1.37719	1.48367	1.57662	
.135	1.9648	.5643	.61013	.29439	4.33696	1.35625	.13827	.15191	.17054	.19163	.21671	.24989	
							1.02693	1.13248	1.25825	1.37503	1.48074	1.57288	
.136	1.9600	.5655	.61252	.29295	4.31519	1.35219	.13929	.15302	.17178	.19304	.21838	.25206	
							1.02680	1.13181	1.25685	1.37289	1.47783	1.56917	
.137	1.9553	.5667	.61489	.29150	4.29362	1.34818	.14031	.15413	.17302	.19446	.22005	.25425	
							1.02667	1.13113	1.25547	1.37077	1.47495	1.56550	
.138	1.9506	.5679	.61725	.29007	4.27225	1.34423	.14133	.15524	.17426	.19587	.22172	.25647	
							1.02654	1.13047	1.25410	1.36867	1.47211	1.56187	
.139	1.9459	.5690	.61960	.28864	4.25108	1.34032	.14235	.15635	.17550	.19729	.22340	.25870	
							1.02641	1.12981	1.25274	1.36659	1.46929	1.55826	
.140	1.9413	.5702	.62194	.28721	4.23011	1.33647	.14337	.15746	.17675	.19871	.22508	.26095	
							1.02628	1.12916	1.25140	1.36453	1.46649	1.55469	
.141	1.9367	.5714	.62426	.28580	4.20933	1.33267	.14439	.15857	.17799	.20013	.22677	.26323	
							1.02616	1.12851	1.25007	1.36250	1.46372	1.55115	
.142	1.9321	.5725	.62658	.28438	4.18873	1.32892	.14541	.15968	.17923	.20155	.22847	.26554	
							1.02603	1.12787	1.24875	1.36048	1.46098	1.54765	
.143	1.9276	.5737	.62888	.28298	4.16832	1.32522	.14643	.16079	.18048	.20298	.23017	.26787	
							1.02591	1.12723	1.24744	1.35848	1.45827	1.54417	
.144	1.9231	.5748	.63117	.28158	4.14810	1.32157	.14745	.16190	.18172	.20440	.23188	.27023	
							1.02579	1.12661	1.24615	1.35650	1.45558	1.54073	
.145	1.9186	.5760	.63345	.28018	4.12805	1.31796	.14847	.16301	.18297	.20583	.23359	.27262	
							1.02567	1.12598	1.24487	1.35454	1.45292	1.53731	
.146	1.9142	.5771	.63571	.27879	4.10818	1.31440	.14949	.16412	.18422	.20726	.23531	.27504	
							1.02555	1.12537	1.24360	1.35259	1.45028	1.53392	
.147	1.9098	.5782	.63797	.27741	4.08849	1.31089	.15052	.16523	.18547	.20870	.23703	.27750	
							1.02543	1.12475	1.24234	1.35067	1.44766	1.53056	
.148	1.9054	.5793	.64021	.27603	4.06897	1.30742	.15154	.16634	.18671	.21013	.23876	.28000	
							1.02531	1.12415	1.24110	1.34876	1.44507	1.52723	
.149	1.9010	.5805	.64245	.27466	4.04961	1.30399	.15256	.16745	.18796	.21157	.24050	.28253	
							1.02519	1.12355	1.23986	1.34687	1.44250	1.52392	

TABLE IX. Specific Heat Ratio = 1.40

p/p_t	MACH	T/T_t	GAMMA	K_{crit}	SHOCK P_y/P_x	P_{tx}/P_{ty}	.01	.05	.07	.1	.15	.2	.25
.150	1.8967	.5816	.64467	.27329	4.03043	1.30061	.15358	.16856	.17653	.18922	.21301	.24225	.28511
							1.02508	1.12295	1.17022	1.23864	1.34499	1.43995	1.52064
.151	1.8924	.5827	.64688	.27193	4.01140	1.29726	.15460	.16967	.17769	.19047	.21445	.24400	.28774
							1.02496	1.12236	1.16938	1.23743	1.34313	1.43743	1.51738
.152	1.8881	.5838	.64908	.27057	3.99254	1.29397	.15562	.17079	.17885	.19172	.21589	.24576	.29042
							1.02485	1.12177	1.16855	1.23622	1.34129	1.43492	1.51415
.153	1.8839	.5849	.65127	.26922	3.97383	1.29071	.15664	.17190	.18002	.19297	.21734	.24752	.29315
							1.02473	1.12119	1.16773	1.23503	1.33947	1.43244	1.51094
.154	1.8797	.5860	.65345	.26787	3.95529	1.28749	.15766	.17301	.18118	.19423	.21879	.24931	.29595
							1.02462	1.12062	1.16691	1.23385	1.33766	1.43000	1.50775
.155	1.8755	.5870	.65562	.26653	3.93689	1.28431	.15868	.17412	.18235	.19548	.22024	.25109	.29882
							1.02451	1.12005	1.16611	1.23268	1.33587	1.42756	1.50458
.156	1.8713	.5881	.65777	.26519	3.91865	1.28117	.15970	.17524	.18351	.19674	.22170	.25288	.30178
							1.02440	1.11948	1.16531	1.23152	1.33409	1.42514	1.50145
.157	1.8671	.5892	.65992	.26386	3.90056	1.27806	.16072	.17635	.18468	.19799	.22315	.25468	.30482
							1.02429	1.11892	1.16451	1.23037	1.33233	1.42274	1.49833
.158	1.8630	.5903	.66205	.26253	3.88262	1.27500	.16174	.17746	.18585	.19925	.22461	.25649	.30796
							1.02418	1.11837	1.16372	1.22923	1.33058	1.42036	1.49523
.159	1.8589	.5913	.66418	.26121	3.86482	1.27197	.16276	.17858	.18701	.20051	.22607	.25831	.31122
							1.02408	1.11781	1.16294	1.22809	1.32885	1.41801	1.49215
.160	1.8548	.5924	.66630	.25989	3.84716	1.26898	.16379	.17969	.18818	.20177	.22754	.26013	.31462
							1.02397	1.11727	1.16217	1.22697	1.32713	1.41567	1.48908
.161	1.8508	.5934	.66840	.25858	3.82965	1.26602	.16481	.18081	.18935	.20303	.22901	.26197	.31818
							1.02386	1.11672	1.16140	1.22586	1.32542	1.41334	1.48603
.162	1.8468	.5945	.67049	.25727	3.81228	1.26310	.16583	.18192	.19052	.20429	.23048	.26381	.32195
							1.02376	1.11619	1.16064	1.22476	1.32373	1.41104	1.48300
.163	1.8428	.5955	.67258	.25597	3.79504	1.26022	.16685	.18304	.19169	.20556	.23195	.26567	.32599
							1.02365	1.11565	1.15988	1.22366	1.32206	1.40875	1.47997
.164	1.8388	.5966	.67465	.25467	3.77794	1.25736	.16787	.18415	.19286	.20682	.23343	.26754	.33038
							1.02355	1.11512	1.15913	1.22257	1.32040	1.40649	1.47696
.165	1.8348	.5976	.67672	.25338	3.76097	1.25454	.16889	.18527	.19403	.20808	.23491	.26942	.33531
							1.02345	1.11460	1.15839	1.22150	1.31875	1.40424	1.47396
.166	1.8309	.5987	.67877	.25209	3.74413	1.25176	.16991	.18638	.19520	.20935	.23639	.27131	.34107
							1.02335	1.11408	1.15765	1.22043	1.31711	1.40200	1.47096
.167	1.8270	.5997	.68082	.25080	3.72742	1.24900	.17093	.18750	.19637	.21062	.23788	.27321	.34877
							1.02324	1.11356	1.15692	1.21937	1.31549	1.39979	1.46796
.168	1.8231	.6007	.68285	.24952	3.71085	1.24628	.17196	.18862	.19754	.21189	.23937	.27513	
							1.02314	1.11305	1.15619	1.21832	1.31387	1.39759	
.169	1.8192	.6017	.68488	.24825	3.69439	1.24359	.17298	.18973	.19871	.21316	.24086	.27705	
							1.02305	1.11254	1.15547	1.21727	1.31228	1.39540	
.170	1.8153	.6027	.68689	.24698	3.67807	1.24093	.17400	.19085	.19989	.21443	.24236	.27899	
							1.02295	1.11203	1.15476	1.21624	1.31069	1.39324	
.171	1.8115	.6037	.68890	.24571	3.66186	1.23830	.17502	.19197	.20106	.21570	.24386	.28095	
							1.02285	1.11153	1.15405	1.21521	1.30912	1.39108	
.172	1.8077	.6048	.69089	.24445	3.64578	1.23570	.17604	.19309	.20223	.21697	.24536	.28292	
							1.02275	1.11103	1.15334	1.21419	1.30756	1.38895	
.173	1.8039	.6058	.69288	.24319	3.62982	1.23313	.17706	.19421	.20341	.21825	.24687	.28490	
							1.02266	1.11054	1.15265	1.21318	1.30601	1.38682	
.174	1.8002	.6068	.69486	.24194	3.61397	1.23059	.17808	.19532	.20458	.21952	.24838	.28690	
							1.02256	1.11005	1.15195	1.21218	1.30447	1.38472	
.175	1.7964	.6078	.69682	.24069	3.59824	1.22808	.17911	.19644	.20576	.22080	.24990	.28892	
							1.02246	1.10956	1.15126	1.21118	1.30294	1.38262	
.176	1.7927	.6087	.69878	.23944	3.58263	1.22560	.18013	.19756	.20694	.22208	.25142	.29096	
							1.02237	1.10908	1.15058	1.21019	1.30142	1.38055	
.177	1.7890	.6097	.70073	.23820	3.56713	1.22314	.18115	.19868	.20811	.22336	.25294	.29301	
							1.02228	1.10860	1.14990	1.20921	1.29992	1.37848	
.178	1.7853	.6107	.70267	.23697	3.55174	1.22071	.18217	.19980	.20929	.22464	.25447	.29508	
							1.02218	1.10813	1.14923	1.20824	1.29843	1.37643	
.179	1.7816	.6117	.70460	.23573	3.53647	1.21831	.18319	.20092	.21047	.22592	.25600	.29717	
							1.02209	1.10766	1.14856	1.20727	1.29694	1.37439	

TABLE IX. Specific Heat Ratio = 1.40

					SHOCK		R₁ and TPR versus Loss Coefficient K						
p/p_t	MACH	T/T_t	GAMMA	K_{crit}	P_y/P_x	P_{ty}/P_{tx}	.01	.05	.07	.09	.1	.15	.2
.180	1.7780	.6127	.70652	.23451	3.52130	1.21594	.18422	.20205	.21165	.22185	.22720	.25754	.29929
							1.02200	1.10719	1.14790	1.18722	1.20631	1.29547	1.37237
.181	1.7743	.6136	.70843	.23328	3.50624	1.21359	.18524	.20317	.21283	.22310	.22849	.25908	.30142
							1.02191	1.10672	1.14724	1.18637	1.20535	1.29401	1.37035
.182	1.7707	.6146	.71034	.23206	3.49129	1.21126	.18626	.20429	.21401	.22435	.22977	.26062	.30358
							1.02182	1.10626	1.14659	1.18552	1.20441	1.29256	1.36835
.183	1.7671	.6156	.71223	.23085	3.47644	1.20897	.18728	.20541	.21519	.22560	.23106	.26217	.30576
							1.02173	1.10580	1.14594	1.18468	1.20347	1.29112	1.36637
.184	1.7635	.6165	.71412	.22963	3.46169	1.20669	.18830	.20653	.21638	.22685	.23235	.26373	.30797
							1.02164	1.10535	1.14530	1.18384	1.20254	1.28968	1.36439
.185	1.7600	.6175	.71599	.22843	3.44705	1.20445	.18933	.20766	.21756	.22810	.23364	.26529	.31021
							1.02155	1.10489	1.14466	1.18301	1.20161	1.28826	1.36243
.186	1.7564	.6184	.71786	.22722	3.43251	1.20222	.19035	.20878	.21874	.22935	.23494	.26685	.31248
							1.02146	1.10445	1.14402	1.18219	1.20069	1.28685	1.36048
.187	1.7529	.6194	.71972	.22602	3.41807	1.20002	.19137	.20991	.21993	.23061	.23623	.26842	.31478
							1.02138	1.10400	1.14339	1.18137	1.19978	1.28545	1.35853
.188	1.7494	.6203	.72157	.22483	3.40373	1.19785	.19239	.21103	.22111	.23186	.23753	.27000	.31711
							1.02129	1.10356	1.14276	1.18056	1.19887	1.28406	1.35660
.189	1.7459	.6213	.72341	.22363	3.38949	1.19570	.19342	.21216	.22230	.23312	.23882	.27158	.31948
							1.02120	1.10312	1.14214	1.17975	1.19797	1.28267	1.35468
.190	1.7424	.6222	.72524	.22245	3.37535	1.19357	.19444	.21328	.22349	.23438	.24012	.27316	.32188
							1.02112	1.10268	1.14152	1.17895	1.19708	1.28130	1.35277
.191	1.7390	.6231	.72707	.22126	3.36130	1.19146	.19546	.21441	.22468	.23564	.24142	.27475	.32433
							1.02103	1.10225	1.14093	1.17816	1.19619	1.27993	1.35087
.192	1.7355	.6241	.72888	.22008	3.34734	1.18938	.19648	.21553	.22587	.23690	.24273	.27635	.32682
							1.02095	1.10182	1.14032	1.17737	1.19531	1.27858	1.34898
.193	1.7321	.6250	.73069	.21890	3.33348	1.18731	.19751	.21666	.22706	.23816	.24403	.27795	.32936
							1.02087	1.10139	1.13972	1.17658	1.19443	1.27723	1.34710
.194	1.7287	.6259	.73249	.21773	3.31970	1.18527	.19853	.21779	.22825	.23943	.24534	.27957	.33196
							1.02078	1.10097	1.13912	1.17580	1.19356	1.27590	1.34523
.195	1.7253	.6268	.73428	.21656	3.30602	1.18326	.19955	.21892	.22944	.24069	.24665	.28119	.33461
							1.02070	1.10054	1.13852	1.17503	1.19270	1.27457	1.34336
.196	1.7219	.6278	.73606	.21540	3.29243	1.18126	.20058	.22005	.23064	.24196	.24796	.28281	.33735
							1.02062	1.10012	1.13793	1.17426	1.19184	1.27325	1.34152
.197	1.7185	.6287	.73783	.21424	3.27893	1.17928	.20160	.22118	.23183	.24323	.24927	.28444	.34015
							1.02054	1.09971	1.13734	1.17349	1.19098	1.27194	1.33967
.198	1.7152	.6296	.73960	.21308	3.26552	1.17733	.20262	.22231	.23302	.24450	.25058	.28608	.34303
							1.02046	1.09929	1.13675	1.17273	1.19014	1.27063	1.33783
.199	1.7119	.6305	.74136	.21192	3.25219	1.17539	.20365	.22344	.23422	.24577	.25190	.28772	.34600
							1.02037	1.09888	1.13617	1.17198	1.18929	1.26933	1.33600
.200	1.7085	.6314	.74311	.21077	3.23895	1.17348	.20467	.22457	.23542	.24705	.25322	.28937	.34907
							1.02029	1.09847	1.13559	1.17123	1.18846	1.26804	1.33417
.201	1.7052	.6323	.74485	.20962	3.22579	1.17158	.20569	.22570	.23661	.24832	.25454	.29103	.35228
							1.02022	1.09807	1.13502	1.17048	1.18762	1.26676	1.33235
.202	1.7019	.6332	.74658	.20848	3.21272	1.16971	.20672	.22683	.23781	.24960	.25586	.29269	.35562
							1.02014	1.09767	1.13445	1.16974	1.18680	1.26549	1.33053
.203	1.6987	.6341	.74830	.20734	3.19973	1.16785	.20774	.22796	.23901	.25088	.25718	.29437	.35915
							1.02006	1.09727	1.13388	1.16901	1.18598	1.26422	1.32872
.204	1.6954	.6350	.75002	.20620	3.18682	1.16602	.20876	.22910	.24021	.25216	.25851	.29605	.36289
							1.01998	1.09687	1.13332	1.16828	1.18516	1.26296	1.32691
.205	1.6922	.6359	.75173	.20507	3.17400	1.16420	.20979	.23023	.24141	.25344	.25984	.29774	.36691
							1.01990	1.09647	1.13276	1.16755	1.18435	1.26171	1.32511
.206	1.6889	.6367	.75343	.20394	3.16125	1.16240	.21081	.23136	.24262	.25472	.26117	.29944	.37133
							1.01983	1.09608	1.13221	1.16683	1.18355	1.26047	1.32331
.207	1.6857	.6376	.75512	.20282	3.14859	1.16062	.21183	.23250	.24382	.25601	.26250	.30115	.37630
							1.01975	1.09569	1.13165	1.16611	1.18274	1.25923	1.32150
.208	1.6825	.6385	.75681	.20169	3.13600	1.15886	.21286	.23363	.24502	.25729	.26383	.30287	.38222
							1.01967	1.09530	1.13110	1.16540	1.18195	1.25800	1.31969
.209	1.6793	.6394	.75848	.20057	3.12349	1.15711	.21388	.23477	.24623	.25858	.26517	.30460	.39042
							1.01960	1.09492	1.13056	1.16469	1.18116	1.25677	1.31787

167

TABLE IX. Specific Heat Ratio = 1.40

p/P₁	MACH	T/T₁	GAMMA	K_crit	SHOCK P_y/P_x	P_ty/P_ty	.01	.03	.05	.07	.09	.1	.15
.210	1.6761	.6402	.76015	.19946	3.11105	1.15539	.21491	.22510	.23591	.24744	.25987	.26651	.30633
							1.01952	1.05766	1.09453	1.13002	1.16399	1.18037	1.25556
.211	1.6730	.6411	.76181	.19835	3.09869	1.15368	.21593	.22618	.23705	.24864	.26117	.26785	.30808
							1.01945	1.05743	1.09415	1.12948	1.16329	1.17959	1.25434
.212	1.6698	.6420	.76347	.19724	3.08641	1.15199	.21695	.22726	.23818	.24985	.26246	.26920	.30984
							1.01937	1.05721	1.09377	1.12894	1.16259	1.17881	1.25314
.213	1.6667	.6428	.76511	.19613	3.07420	1.15031	.21798	.22834	.23932	.25106	.26376	.27055	.31161
							1.01930	1.05698	1.09340	1.12841	1.16190	1.17804	1.25194
.214	1.6636	.6437	.76675	.19503	3.06207	1.14866	.21900	.22941	.24046	.25228	.26506	.27190	.31339
							1.01923	1.05676	1.09302	1.12788	1.16122	1.17727	1.25075
.215	1.6605	.6446	.76838	.19393	3.05000	1.14702	.22003	.23049	.24160	.25349	.26636	.27325	.31518
							1.01915	1.05654	1.09265	1.12736	1.16053	1.17651	1.24957
.216	1.6574	.6454	.77001	.19284	3.03801	1.14539	.22105	.23157	.24274	.25470	.26766	.27461	.31699
							1.01908	1.05632	1.09228	1.12683	1.15985	1.17575	1.24839
.217	1.6543	.6463	.77162	.19174	3.02609	1.14378	.22208	.23265	.24389	.25592	.26897	.27596	.31881
							1.01901	1.05610	1.09191	1.12629	1.15918	1.17500	1.24721
.218	1.6512	.6471	.77323	.19065	3.01424	1.14219	.22310	.23373	.24503	.25713	.27027	.27732	.32064
							1.01894	1.05588	1.09155	1.12578	1.15851	1.17425	1.24605
.219	1.6482	.6480	.77483	.18957	3.00246	1.14062	.22413	.23481	.24617	.25835	.27158	.27869	.32248
							1.01887	1.05567	1.09119	1.12526	1.15784	1.17350	1.24488
.220	1.6451	.6488	.77642	.18849	2.99075	1.13906	.22515	.23589	.24731	.25957	.27289	.28006	.32434
							1.01880	1.05545	1.09082	1.12475	1.15718	1.17276	1.24373
.221	1.6421	.6497	.77801	.18741	2.97911	1.13751	.22618	.23697	.24846	.26079	.27421	.28143	.32622
							1.01873	1.05524	1.09047	1.12424	1.15652	1.17202	1.24258
.222	1.6390	.6505	.77959	.18633	2.96754	1.13598	.22720	.23805	.24960	.26201	.27552	.28280	.32811
							1.01866	1.05503	1.09011	1.12374	1.15586	1.17129	1.24143
.223	1.6360	.6513	.78116	.18526	2.95603	1.13447	.22823	.23913	.25075	.26324	.27684	.28417	.33002
							1.01859	1.05482	1.08975	1.12324	1.15521	1.17056	1.24029
.224	1.6330	.6522	.78272	.18419	2.94459	1.13297	.22925	.24022	.25190	.26446	.27816	.28555	.33194
							1.01852	1.05463	1.08940	1.12274	1.15456	1.16983	1.23916
.225	1.6300	.6530	.78428	.18312	2.93321	1.13149	.23028	.24130	.25305	.26569	.27949	.28694	.33388
							1.01845	1.05442	1.08905	1.12224	1.15391	1.16911	1.23803
.226	1.6271	.6538	.78583	.18206	2.92190	1.13002	.23130	.24238	.25419	.26692	.28081	.28832	.33584
							1.01838	1.05421	1.08870	1.12175	1.15327	1.16839	1.23691
.227	1.6241	.6546	.78737	.18100	2.91065	1.12856	.23233	.24347	.25534	.26815	.28214	.28971	.33782
							1.01831	1.05400	1.08836	1.12126	1.15263	1.16768	1.23579
.228	1.6211	.6555	.78890	.17995	2.89947	1.12712	.23336	.24455	.25649	.26938	.28347	.29110	.33983
							1.01825	1.05380	1.08801	1.12077	1.15200	1.16697	1.23467
.229	1.6182	.6563	.79043	.17889	2.88835	1.12570	.23438	.24563	.25764	.27061	.28481	.29250	.34185
							1.01818	1.05360	1.08767	1.12029	1.15137	1.16626	1.23356
.230	1.6153	.6571	.79195	.17784	2.87729	1.12429	.23541	.24672	.25880	.27184	.28614	.29390	.34390
							1.01811	1.05339	1.08733	1.11980	1.15074	1.16556	1.23246
.231	1.6124	.6579	.79347	.17679	2.86629	1.12289	.23643	.24780	.25995	.27308	.28748	.29530	.34597
							1.01805	1.05319	1.08699	1.11932	1.15011	1.16486	1.23135
.232	1.6094	.6587	.79497	.17575	2.85536	1.12150	.23746	.24889	.26110	.27431	.28883	.29671	.34807
							1.01798	1.05299	1.08665	1.11885	1.14949	1.16417	1.23026
.233	1.6065	.6595	.79647	.17471	2.84448	1.12013	.23849	.24997	.26226	.27555	.29017	.29812	.35019
							1.01791	1.05279	1.08632	1.11837	1.14887	1.16347	1.22916
.234	1.6037	.6604	.79796	.17367	2.83367	1.11877	.23951	.25106	.26341	.27679	.29152	.29953	.35234
							1.01785	1.05260	1.08598	1.11790	1.14826	1.16279	1.22808
.235	1.6008	.6612	.79945	.17264	2.82291	1.11743	.24054	.25214	.26457	.27803	.29287	.30095	.35453
							1.01778	1.05240	1.08565	1.11743	1.14765	1.16210	1.22699
.236	1.5979	.6620	.80093	.17161	2.81221	1.11610	.24157	.25323	.26573	.27928	.29422	.30238	.35675
							1.01772	1.05220	1.08532	1.11696	1.14704	1.16142	1.22591
.237	1.5951	.6628	.80240	.17058	2.80157	1.11478	.24259	.25432	.26689	.28052	.29558	.30380	.35900
							1.01765	1.05201	1.08499	1.11650	1.14643	1.16074	1.22483
.238	1.5922	.6636	.80386	.16955	2.79099	1.11348	.24362	.25541	.26804	.28177	.29694	.30523	.36129
							1.01759	1.05181	1.08467	1.11603	1.14583	1.16007	1.22376
.239	1.5894	.6644	.80532	.16853	2.78046	1.11219	.24465	.25649	.26920	.28302	.29831	.30667	.36363
							1.01753	1.05162	1.08434	1.11557	1.14523	1.15940	1.22268

TABLE IX. Specific Heat Ratio = 1.40

p/p_t	MACH	T/T_t	GAMMA	K_crit	SHOCK P_y/P_x	P_tx/P_ty	.01	.03	R₂ and TPR versus Loss Coefficient K .05	.07	.09	.1	.15
.240	1.5865	.6651	.80677	.16751	2.77000	1.11091	.24568	.25758	.27037	.28427	.29967	.30811	.36600
							1.01746	1.05143	1.08402	1.11512	1.14464	1.15873	1.22162
.241	1.5837	.6659	.80821	.16649	2.75958	1.10964	.24670	.25867	.27153	.28552	.30104	.30956	.36843
							1.01740	1.05124	1.08370	1.11466	1.14404	1.15807	1.22055
.242	1.5809	.6667	.80965	.16548	2.74923	1.10838	.24773	.25976	.27269	.28678	.30242	.31101	.37093
							1.01734	1.05105	1.08338	1.11421	1.14345	1.15740	1.21950
.243	1.5781	.6675	.81108	.16447	2.73892	1.10714	.24876	.26085	.27386	.28804	.30379	.31246	.37347
							1.01728	1.05086	1.08306	1.11376	1.14286	1.15675	1.21844
.244	1.5753	.6683	.81250	.16346	2.72868	1.10591	.24979	.26194	.27502	.28929	.30518	.31392	.37608
							1.01721	1.05067	1.08274	1.11331	1.14228	1.15609	1.21738
.245	1.5726	.6691	.81392	.16246	2.71848	1.10469	.25081	.26303	.27619	.29055	.30656	.31538	.37875
							1.01715	1.05049	1.08243	1.11286	1.14170	1.15544	1.21633
.246	1.5698	.6699	.81533	.16145	2.70834	1.10349	.25184	.26412	.27736	.29182	.30795	.31685	.38151
							1.01709	1.05030	1.08211	1.11242	1.14112	1.15479	1.21527
.247	1.5671	.6706	.81673	.16046	2.69825	1.10229	.25287	.26522	.27852	.29308	.30934	.31833	.38437
							1.01703	1.05012	1.08180	1.11198	1.14054	1.15414	1.21422
.248	1.5643	.6714	.81813	.15946	2.68822	1.10111	.25390	.26631	.27969	.29435	.31074	.31981	.38732
							1.01697	1.04993	1.08149	1.11154	1.13997	1.15350	1.21317
.249	1.5616	.6722	.81952	.15847	2.67823	1.09993	.25493	.26740	.28087	.29562	.31214	.32130	.39040
							1.01691	1.04975	1.08118	1.11110	1.13940	1.15286	1.21212
.250	1.5588	.6729	.82090	.15748	2.66830	1.09877	.25596	.26849	.28204	.29689	.31354	.32279	.39362
							1.01685	1.04957	1.08088	1.11067	1.13883	1.15222	1.21107
.251	1.5561	.6737	.82228	.15649	2.65842	1.09762	.25699	.26959	.28321	.29816	.31495	.32429	.39702
							1.01679	1.04939	1.08057	1.11023	1.13826	1.15159	1.21003
.252	1.5534	.6745	.82365	.15551	2.64859	1.09649	.25801	.27068	.28439	.29944	.31636	.32579	.40063
							1.01673	1.04921	1.08027	1.10980	1.13770	1.15096	1.20897
.253	1.5507	.6752	.82502	.15452	2.63881	1.09536	.25904	.27178	.28556	.30072	.31778	.32730	.40450
							1.01667	1.04903	1.07997	1.10937	1.13714	1.15033	1.20792
.254	1.5480	.6760	.82637	.15355	2.62908	1.09424	.26007	.27287	.28674	.30200	.31921	.32882	.40879
							1.01661	1.04885	1.07967	1.10895	1.13658	1.14970	1.20687
.255	1.5453	.6768	.82772	.15257	2.61940	1.09314	.26110	.27397	.28792	.30328	.32063	.33034	.41358
							1.01655	1.04867	1.07937	1.10852	1.13603	1.14908	1.20582
.256	1.5427	.6775	.82907	.15160	2.60976	1.09204	.26213	.27507	.28910	.30457	.32206	.33188	.41925
							1.01650	1.04849	1.07907	1.10810	1.13547	1.14847	1.20475
.257	1.5400	.6783	.83041	.15063	2.60018	1.09096	.26316	.27617	.29028	.30586	.32350	.33342	.42692
							1.01644	1.04832	1.07877	1.10768	1.13492	1.14785	1.20368
.258	1.5373	.6790	.83174	.14966	2.59064	1.08988	.26419	.27726	.29146	.30715	.32494	.33497	
							1.01638	1.04814	1.07848	1.10726	1.13437	1.14724	
.259	1.5347	.6798	.83306	.14870	2.58115	1.08882	.26522	.27836	.29264	.30844	.32639	.33652	
							1.01632	1.04797	1.07818	1.10684	1.13383	1.14662	
.260	1.5321	.6805	.83438	.14774	2.57171	1.08777	.26625	.27946	.29383	.30974	.32784	.33808	
							1.01627	1.04780	1.07789	1.10643	1.13329	1.14601	
.261	1.5294	.6813	.83569	.14678	2.56231	1.08672	.26728	.28056	.29502	.31104	.32930	.33965	
							1.01621	1.04762	1.07760	1.10601	1.13274	1.14541	
.262	1.5268	.6820	.83700	.14582	2.55296	1.08569	.26831	.28166	.29620	.31234	.33076	.34123	
							1.01615	1.04745	1.07731	1.10560	1.13221	1.14480	
.263	1.5242	.6828	.83830	.14487	2.54366	1.08467	.26934	.28277	.29739	.31364	.33223	.34281	
							1.01610	1.04728	1.07702	1.10519	1.13167	1.14420	
.264	1.5216	.6835	.83959	.14392	2.53440	1.08365	.27037	.28387	.29858	.31495	.33371	.34440	
							1.01604	1.04711	1.07673	1.10478	1.13113	1.14360	
.265	1.5190	.6842	.84088	.14297	2.52519	1.08265	.27141	.28497	.29978	.31626	.33519	.34601	
							1.01599	1.04694	1.07645	1.10438	1.13060	1.14300	
.266	1.5164	.6850	.84216	.14203	2.51602	1.08166	.27244	.28607	.30097	.31757	.33668	.34762	
							1.01593	1.04678	1.07617	1.10397	1.13007	1.14240	
.267	1.5138	.6857	.84344	.14109	2.50689	1.08067	.27347	.28718	.30217	.31889	.33817	.34924	
							1.01588	1.04661	1.07588	1.10357	1.12954	1.14181	
.268	1.5112	.6865	.84471	.14015	2.49781	1.07970	.27450	.28828	.30336	.32021	.33967	.35087	
							1.01582	1.04644	1.07560	1.10317	1.12902	1.14121	
.269	1.5087	.6872	.84597	.13921	2.48877	1.07873	.27553	.28939	.30456	.32153	.34118	.35252	
							1.01577	1.04628	1.07532	1.10277	1.12849	1.14062	

TABLE IX. Specific Heat Ratio = 1.40

p/p_t	MACH	T/T_t	GAMMA	K_{crit}	SHOCK P_y/P_x	P_{ty}/P_{tx}	.01	.03	.05	.07	.08	.09	.1
.270	1.5061	.6879	.84722	.13828	2.47978	1.07778	.27656	.29050	.30576	.32286	.33234	.34269	.35417
							1.01573	1.04611	1.07504	1.10237	1.11540	1.12797	1.14004
.271	1.5036	.6886	.84848	.13735	2.47082	1.07683	.27760	.29160	.30696	.32419	.33376	.34421	.35584
							1.01568	1.04595	1.07476	1.10198	1.11494	1.12745	1.13945
.272	1.5010	.6894	.84972	.13642	2.46192	1.07589	.27863	.29271	.30817	.32552	.33517	.34574	.35751
							1.01562	1.04578	1.07449	1.10159	1.11449	1.12693	1.13887
.273	1.4985	.6901	.85096	.13550	2.45305	1.07496	.27966	.29382	.30937	.32686	.33660	.34728	.35920
							1.01557	1.04562	1.07421	1.10119	1.11402	1.12642	1.13828
.274	1.4960	.6908	.85219	.13458	2.44422	1.07404	.28069	.29493	.31058	.32820	.33803	.34882	.36090
							1.01552	1.04546	1.07394	1.10080	1.11357	1.12590	1.13770
.275	1.4934	.6915	.85341	.13366	2.43544	1.07313	.28173	.29604	.31179	.32954	.33946	.35038	.36262
							1.01546	1.04530	1.07366	1.10041	1.11312	1.12539	1.13713
.276	1.4909	.6922	.85463	.13274	2.42669	1.07223	.28276	.29715	.31300	.33089	.34091	.35194	.36434
							1.01541	1.04514	1.07339	1.10003	1.11268	1.12488	1.13655
.277	1.4884	.6930	.85585	.13183	2.41799	1.07134	.28379	.29826	.31421	.33224	.34236	.35351	.36609
							1.01536	1.04498	1.07312	1.09964	1.11224	1.12437	1.13597
.278	1.4859	.6937	.85706	.13092	2.40933	1.07045	.28483	.29937	.31543	.33360	.34381	.35509	.36784
							1.01530	1.04482	1.07285	1.09925	1.11179	1.12386	1.13540
.279	1.4834	.6944	.85826	.13001	2.40070	1.06958	.28586	.30049	.31664	.33496	.34527	.35667	.36961
							1.01525	1.04466	1.07258	1.09887	1.11135	1.12336	1.13483
.280	1.4810	.6951	.85945	.12910	2.39212	1.06871	.28690	.30160	.31786	.33633	.34674	.35827	.37140
							1.01520	1.04450	1.07232	1.09849	1.11091	1.12286	1.13426
.281	1.4785	.6958	.86064	.12820	2.38358	1.06785	.28793	.30272	.31908	.33769	.34821	.35988	.37321
							1.01515	1.04434	1.07205	1.09811	1.11047	1.12235	1.13369
.282	1.4760	.6965	.86183	.12730	2.37507	1.06700	.28896	.30383	.32030	.33907	.34969	.36150	.37503
							1.01510	1.04419	1.07179	1.09773	1.11003	1.12185	1.13313
.283	1.4736	.6972	.86301	.12640	2.36661	1.06616	.29000	.30495	.32153	.34044	.35117	.36313	.37687
							1.01505	1.04403	1.07152	1.09735	1.10959	1.12135	1.13256
.284	1.4711	.6979	.86418	.12551	2.35818	1.06532	.29103	.30607	.32276	.34183	.35266	.36477	.37874
							1.01499	1.04388	1.07126	1.09698	1.10916	1.12086	1.13200
.285	1.4687	.6986	.86534	.12462	2.34979	1.06450	.29207	.30719	.32399	.34321	.35416	.36642	.38062
							1.01494	1.04372	1.07100	1.09661	1.10872	1.12036	1.13144
.286	1.4662	.6993	.86650	.12373	2.34144	1.06368	.29310	.30831	.32522	.34461	.35567	.36808	.38252
							1.01489	1.04357	1.07074	1.09623	1.10829	1.11987	1.13088
.287	1.4638	.7000	.86766	.12284	2.33312	1.06287	.29414	.30943	.32645	.34600	.35718	.36976	.38445
							1.01484	1.04342	1.07048	1.09586	1.10786	1.11938	1.13032
.288	1.4614	.7007	.86881	.12196	2.32485	1.06207	.29518	.31055	.32769	.34740	.35870	.37145	.38640
							1.01479	1.04326	1.07022	1.09549	1.10743	1.11888	1.12976
.289	1.4589	.7014	.86995	.12108	2.31661	1.06127	.29621	.31167	.32892	.34881	.36023	.37315	.38838
							1.01474	1.04311	1.06996	1.09512	1.10701	1.11839	1.12920
.290	1.4565	.7021	.87109	.12020	2.30840	1.06049	.29725	.31280	.33016	.35022	.36177	.37487	.39039
							1.01469	1.04296	1.06970	1.09475	1.10658	1.11791	1.12864
.291	1.4541	.7028	.87222	.11932	2.30023	1.05971	.29828	.31392	.33141	.35164	.36332	.37660	.39242
							1.01464	1.04281	1.06945	1.09439	1.10615	1.11742	1.12809
.292	1.4517	.7035	.87334	.11845	2.29210	1.05894	.29932	.31505	.33265	.35307	.36488	.37835	.39449
							1.01459	1.04266	1.06919	1.09402	1.10573	1.11693	1.12753
.293	1.4493	.7042	.87446	.11758	2.28401	1.05817	.30036	.31617	.33390	.35450	.36644	.38011	.39659
							1.01454	1.04251	1.06894	1.09366	1.10531	1.11645	1.12698
.294	1.4470	.7049	.87558	.11671	2.27595	1.05742	.30140	.31730	.33515	.35593	.36802	.38190	.39873
							1.01450	1.04236	1.06869	1.09330	1.10489	1.11596	1.12643
.295	1.4446	.7055	.87669	.11585	2.26792	1.05667	.30243	.31843	.33640	.35737	.36960	.38369	.40091
							1.01445	1.04221	1.06844	1.09294	1.10447	1.11548	1.12588
.296	1.4422	.7062	.87779	.11499	2.25993	1.05593	.30347	.31956	.33766	.35882	.37120	.38551	.40313
							1.01440	1.04207	1.06819	1.09258	1.10405	1.11500	1.12532
.297	1.4398	.7069	.87889	.11413	2.25198	1.05519	.30451	.32069	.33892	.36028	.37281	.38735	.40539
							1.01435	1.04192	1.06794	1.09222	1.10363	1.11452	1.12477
.298	1.4375	.7076	.87998	.11327	2.24405	1.05447	.30555	.32182	.34018	.36174	.37442	.38921	.40771
							1.01430	1.04177	1.06769	1.09186	1.10322	1.11404	1.12422
.299	1.4351	.7083	.88106	.11242	2.23617	1.05375	.30659	.32296	.34145	.36321	.37605	.39109	.41008
							1.01425	1.04163	1.06744	1.09150	1.10280	1.11356	1.12367

TABLE IX. Specific Heat Ratio = 1.40

p/p_t	MACH	T/T_t	GAMMA	K_crit	SHOCK P_y/P_x	P_ty/P_tx	.01	.03	.05	.07	.08	.09	.1
.300	1.4328	.7089	.88214	.11157	2.22831	1.05303	.30763	.32409	.34271	.36468	.37770	.39299	.41251
							1.01421	1.04148	1.06719	1.09115	1.10239	1.11308	1.12312
.301	1.4304	.7096	.88322	.11072	2.22049	1.05233	.30867	.32523	.34398	.36617	.37935	.39492	.41501
							1.01416	1.04134	1.06695	1.09079	1.10197	1.11261	1.12257
.302	1.4281	.7103	.88429	.10987	2.21271	1.05163	.30971	.32637	.34526	.36766	.38102	.39688	.41759
							1.01411	1.04119	1.06670	1.09044	1.10156	1.11213	1.12202
.303	1.4258	.7110	.88535	.10903	2.20495	1.05094	.31075	.32750	.34653	.36916	.38270	.39886	.42025
							1.01406	1.04105	1.06646	1.09009	1.10115	1.11166	1.12147
.304	1.4234	.7116	.88641	.10819	2.19723	1.05026	.31179	.32864	.34781	.37067	.38440	.40087	.42301
							1.01402	1.04091	1.06621	1.08974	1.10074	1.11118	1.12091
.305	1.4211	.7123	.88746	.10735	2.18955	1.04958	.31283	.32979	.34910	.37218	.38611	.40292	.42588
							1.01397	1.04077	1.06597	1.08939	1.10033	1.11071	1.12036
.306	1.4188	.7130	.88851	.10652	2.18189	1.04891	.31387	.33093	.35038	.37371	.38784	.40500	.42892
							1.01392	1.04062	1.06573	1.08904	1.09993	1.11024	1.11981
.307	1.4165	.7136	.88955	.10568	2.17427	1.04824	.31491	.33207	.35168	.37524	.38958	.40712	.43209
							1.01388	1.04048	1.06549	1.08869	1.09952	1.10976	1.11926
.308	1.4142	.7143	.89058	.10486	2.16668	1.04759	.31595	.33322	.35297	.37678	.39134	.40927	.43546
							1.01383	1.04034	1.06525	1.08834	1.09911	1.10929	1.11870
.309	1.4119	.7149	.89161	.10403	2.15912	1.04693	.31699	.33436	.35427	.37834	.39312	.41147	.43909
							1.01379	1.04020	1.06501	1.08799	1.09871	1.10882	1.11814
.310	1.4096	.7156	.89264	.10320	2.15159	1.04629	.31803	.33551	.35557	.37990	.39492	.41372	.44304
							1.01374	1.04006	1.06477	1.08765	1.09830	1.10834	1.11758
.311	1.4074	.7163	.89366	.10238	2.14409	1.04565	.31908	.33666	.35687	.38148	.39674	.41602	.44746
							1.01370	1.03992	1.06453	1.08730	1.09790	1.10787	1.11701
.312	1.4051	.7169	.89467	.10156	2.13662	1.04502	.32012	.33781	.35818	.38306	.39858	.41838	.45267
							1.01365	1.03979	1.06430	1.08696	1.09749	1.10740	1.11645
.313	1.4028	.7176	.89568	.10075	2.12919	1.04440	.32116	.33896	.35950	.38466	.40044	.42079	.45932
							1.01360	1.03965	1.06406	1.08662	1.09709	1.10693	1.11587
.314	1.4005	.7182	.89668	.09993	2.12178	1.04378	.32221	.34012	.36081	.38627	.40233	.42328	
							1.01356	1.03951	1.06383	1.08627	1.09669	1.10645	
.315	1.3983	.7189	.89768	.09912	2.11441	1.04317	.32325	.34127	.36213	.38789	.40424	.42585	
							1.01351	1.03937	1.06359	1.08593	1.09629	1.10598	
.316	1.3960	.7195	.89867	.09832	2.10706	1.04256	.32430	.34243	.36346	.38952	.40618	.42851	
							1.01347	1.03924	1.06336	1.08559	1.09589	1.10550	
.317	1.3938	.7202	.89965	.09751	2.09975	1.04196	.32534	.34359	.36479	.39117	.40815	.43127	
							1.01343	1.03910	1.06313	1.08525	1.09549	1.10503	
.318	1.3915	.7208	.90064	.09671	2.09246	1.04137	.32639	.34475	.36613	.39283	.41016	.43418	
							1.01338	1.03897	1.06289	1.08491	1.09508	1.10456	
.319	1.3893	.7215	.90161	.09591	2.08520	1.04078	.32743	.34591	.36747	.39451	.41219	.43721	
							1.01334	1.03883	1.06266	1.08457	1.09468	1.10408	
.320	1.3871	.7221	.90258	.09511	2.07798	1.04020	.32848	.34708	.36881	.39620	.41426	.44041	
							1.01329	1.03870	1.06243	1.08424	1.09428	1.10361	
.321	1.3849	.7228	.90355	.09432	2.07078	1.03962	.32952	.34824	.37016	.39791	.41637	.44383	
							1.01325	1.03856	1.06220	1.08390	1.09388	1.10313	
.322	1.3826	.7234	.90451	.09352	2.06361	1.03905	.33057	.34941	.37151	.39963	.41852	.44753	
							1.01321	1.03843	1.06197	1.08356	1.09348	1.10264	
.323	1.3804	.7241	.90546	.09273	2.05647	1.03849	.33162	.35058	.37287	.40137	.42072	.45160	
							1.01316	1.03830	1.06174	1.08322	1.09308	1.10215	
.324	1.3782	.7247	.90641	.09195	2.04936	1.03793	.33266	.35175	.37424	.40313	.42296	.45622	
							1.01312	1.03816	1.06151	1.08289	1.09268	1.10166	
.325	1.3760	.7253	.90735	.09116	2.04227	1.03738	.33371	.35292	.37561	.40492	.42526	.46186	
							1.01308	1.03803	1.06129	1.08255	1.09228	1.10117	
.326	1.3738	.7260	.90829	.09038	2.03522	1.03683	.33476	.35410	.37699	.40672	.42763	.46969	
							1.01303	1.03790	1.06106	1.08222	1.09188	1.10066	
.327	1.3716	.7266	.90922	.08960	2.02819	1.03629	.33581	.35528	.37837	.40856	.43006		
							1.01299	1.03777	1.06083	1.08189	1.09148		
.328	1.3694	.7272	.91015	.08883	2.02119	1.03576	.33686	.35645	.37976	.41041	.43256		
							1.01295	1.03764	1.06061	1.08156	1.09108		
.329	1.3672	.7279	.91107	.08806	2.01421	1.03523	.33791	.35764	.38116	.41229	.43516		
							1.01290	1.03751	1.06038	1.08122	1.09068		

TABLE IX. Specific Heat Ratio = 1.40

p/p_t	MACH	T/T_t	GAMMA	K_crit	SHOCK p_y/p_x	p_ty/p_tx	.01	.03	R_2 and TPR versus Loss Coefficient K .04	.05	.06	.07	.08
.330	1.3651	.7285	.91199	.08729	2.00727	1.03471	.33896	.35882	.37006	.38256	.39689	.41419	.43785
							1.01286	1.03738	1.04900	1.06016	1.07081	1.08089	1.09027
.331	1.3629	.7291	.91290	.08652	2.00035	1.03419	.34001	.36001	.37134	.38397	.39848	.41612	.44069
							1.01282	1.03725	1.04883	1.05994	1.07054	1.08056	1.08988
.332	1.3607	.7298	.91381	.08575	1.99346	1.03367	.34106	.36119	.37262	.38538	.40010	.41808	.44364
							1.01278	1.03712	1.04865	1.05971	1.07026	1.08022	1.08947
.333	1.3585	.7304	.91471	.08499	1.98659	1.03317	.34211	.36238	.37391	.38681	.40172	.42008	.44676
							1.01274	1.03699	1.04848	1.05949	1.06999	1.07989	1.08906
.334	1.3564	.7310	.91561	.08423	1.97975	1.03267	.34316	.36358	.37520	.38824	.40336	.42211	.45009
							1.01269	1.03686	1.04830	1.05927	1.06971	1.07956	1.08866
.335	1.3542	.7316	.91650	.08348	1.97294	1.03217	.34422	.36477	.37650	.38968	.40502	.42418	.45368
							1.01265	1.03674	1.04813	1.05905	1.06944	1.07922	1.08824
.336	1.3521	.7323	.91739	.08272	1.96615	1.03168	.34527	.36597	.37780	.39112	.40669	.42629	.45762
							1.01261	1.03661	1.04796	1.05883	1.06916	1.07889	1.08783
.337	1.3499	.7329	.91827	.08197	1.95939	1.03119	.34632	.36717	.37911	.39258	.40838	.42844	.46206
							1.01257	1.03648	1.04779	1.05861	1.06889	1.07856	1.08741
.338	1.3478	.7335	.91914	.08122	1.95266	1.03071	.34738	.36837	.38042	.39404	.41009	.43065	.46743
							1.01253	1.03636	1.04761	1.05839	1.06862	1.07822	1.08699
.339	1.3457	.7341	.92001	.08048	1.94595	1.03024	.34843	.36958	.38174	.39551	.41181	.43291	.47462
							1.01249	1.03623	1.04744	1.05817	1.06835	1.07789	1.08656
.340	1.3435	.7347	.92088	.07973	1.93926	1.02977	.34949	.37078	.38306	.39700	.41356	.43523	
							1.01245	1.03610	1.04727	1.05795	1.06808	1.07756	
.341	1.3414	.7354	.92174	.07899	1.93260	1.02930	.35054	.37199	.38438	.39849	.41532	.43762	
							1.01241	1.03598	1.04710	1.05773	1.06780	1.07722	
.342	1.3393	.7360	.92259	.07826	1.92597	1.02884	.35160	.37321	.38571	.39999	.41711	.44009	
							1.01237	1.03585	1.04693	1.05751	1.06753	1.07688	
.343	1.3372	.7366	.92344	.07752	1.91936	1.02839	.35266	.37442	.38705	.40150	.41893	.44264	
							1.01233	1.03573	1.04676	1.05729	1.06726	1.07655	
.344	1.3351	.7372	.92429	.07679	1.91277	1.02794	.35372	.37564	.38839	.40302	.42076	.44532	
							1.01229	1.03560	1.04659	1.05707	1.06699	1.07622	
.345	1.3330	.7378	.92513	.07606	1.90622	1.02749	.35477	.37687	.38974	.40456	.42263	.44810	
							1.01225	1.03548	1.04642	1.05686	1.06672	1.07588	
.346	1.3308	.7384	.92596	.07533	1.89968	1.02706	.35583	.37809	.39109	.40610	.42452	.45102	
							1.01221	1.03536	1.04625	1.05664	1.06645	1.07554	
.347	1.3287	.7390	.92679	.07461	1.89317	1.02662	.35689	.37932	.39245	.40766	.42644	.45410	
							1.01217	1.03523	1.04609	1.05642	1.06618	1.07520	
.348	1.3267	.7396	.92762	.07389	1.88668	1.02619	.35795	.38055	.39382	.40923	.42840	.45740	
							1.01213	1.03511	1.04592	1.05621	1.06590	1.07486	
.349	1.3246	.7403	.92844	.07317	1.88022	1.02577	.35901	.38179	.39519	.41082	.43039	.46096	
							1.01209	1.03499	1.04575	1.05599	1.06563	1.07451	
.350	1.3225	.7409	.92925	.07246	1.87378	1.02535	.36007	.38303	.39657	.41241	.43242	.46490	
							1.01205	1.03487	1.04558	1.05578	1.06536	1.07416	
.351	1.3204	.7415	.93006	.07174	1.86736	1.02493	.36114	.38427	.39795	.41403	.43449	.46936	
							1.01201	1.03474	1.04542	1.05556	1.06509	1.07380	
.352	1.3183	.7421	.93087	.07103	1.86097	1.02452	.36220	.38551	.39935	.41566	.43661	.47486	
							1.01197	1.03462	1.04525	1.05535	1.06482	1.07345	
.353	1.3163	.7427	.93167	.07033	1.85460	1.02411	.36326	.38676	.40075	.41730	.43878	.48266	
							1.01193	1.03450	1.04509	1.05513	1.06455	1.07309	
.354	1.3142	.7433	.93246	.06962	1.84826	1.02371	.36433	.38802	.40216	.41897	.44100		
							1.01189	1.03438	1.04492	1.05492	1.06427		
.355	1.3121	.7439	.93325	.06892	1.84193	1.02331	.36539	.38927	.40357	.42065	.44328		
							1.01185	1.03426	1.04476	1.05470	1.06400		
.356	1.3101	.7445	.93403	.06822	1.83563	1.02292	.36646	.39054	.40500	.42235	.44564		
							1.01181	1.03414	1.04459	1.05449	1.06372		
.357	1.3080	.7451	.93481	.06753	1.82936	1.02253	.36752	.39180	.40643	.42407	.44807		
							1.01177	1.03402	1.04443	1.05427	1.06345		
.358	1.3060	.7457	.93559	.06683	1.82310	1.02215	.36859	.39307	.40788	.42581	.45062		
							1.01174	1.03390	1.04426	1.05406	1.06318		
.359	1.3039	.7462	.93636	.06614	1.81687	1.02177	.36966	.39435	.40933	.42758	.45325		
							1.01170	1.03378	1.04410	1.05385	1.06290		

TABLE IX. Specific Heat Ratio = 1.40

p/p_t	MACH	T/T_t	GAMMA	K_crit	SHOCK P_y/P_x	P_ty/P_ty	.005	.01	.02	.03	.04	.05	.06
.360	1.3019	.7468	.93712	.06546	1.81066	1.02140	.36526	.37074	.38249	.39563	.41079	.42937	.45600
							1.00588	1.01167	1.02289	1.03366	1.04393	1.05363	1.06263
.361	1.2998	.7474	.93788	.06477	1.80447	1.02103	.36629	.37181	.38365	.39691	.41226	.43119	.45890
							1.00586	1.01163	1.02281	1.03355	1.04377	1.05342	1.06235
.362	1.2978	.7480	.93864	.06409	1.79830	1.02066	.36732	.37288	.38482	.39820	.41375	.43303	.46198
							1.00585	1.01159	1.02274	1.03343	1.04361	1.05320	1.06207
.363	1.2958	.7486	.93939	.06341	1.79216	1.02030	.36836	.37395	.38598	.39950	.41524	.43491	.46528
							1.00583	1.01156	1.02266	1.03331	1.04344	1.05299	1.06179
.364	1.2937	.7492	.94013	.06274	1.78604	1.01995	.36939	.37502	.38715	.40080	.41675	.43682	.46888
							1.00581	1.01152	1.02258	1.03319	1.04328	1.05277	1.06150
.365	1.2917	.7498	.94087	.06207	1.77994	1.01959	.37043	.37610	.38832	.40210	.41827	.43876	.47288
							1.00579	1.01148	1.02251	1.03307	1.04312	1.05256	1.06121
.366	1.2897	.7504	.94161	.06140	1.77386	1.01925	.37146	.37717	.38949	.40341	.41980	.44074	.47758
							1.00577	1.01144	1.02243	1.03296	1.04296	1.05235	1.06092
.367	1.2877	.7510	.94234	.06073	1.76780	1.01890	.37250	.37825	.39067	.40473	.42134	.44277	.48344
							1.00575	1.01141	1.02236	1.03284	1.04279	1.05213	1.06062
.368	1.2857	.7515	.94306	.06006	1.76176	1.01856	.37353	.37932	.39184	.40605	.42290	.44483	.49366
							1.00574	1.01137	1.02228	1.03272	1.04263	1.05191	1.06031
.369	1.2837	.7521	.94379	.05940	1.75575	1.01823	.37457	.38040	.39302	.40738	.42447	.44695	
							1.00572	1.01133	1.02220	1.03261	1.04247	1.05170	
.370	1.2817	.7527	.94450	.05874	1.74975	1.01789	.37561	.38148	.39421	.40872	.42606	.44913	
							1.00570	1.01130	1.02213	1.03249	1.04231	1.05148	
.371	1.2797	.7533	.94521	.05809	1.74378	1.01757	.37664	.38256	.39539	.41006	.42767	.45136	
							1.00568	1.01126	1.02205	1.03237	1.04215	1.05126	
.372	1.2777	.7539	.94592	.05744	1.73783	1.01724	.37768	.38364	.39658	.41141	.42929	.45367	
							1.00566	1.01122	1.02198	1.03226	1.04198	1.05104	
.373	1.2757	.7545	.94662	.05679	1.73189	1.01692	.37872	.38472	.39777	.41276	.43093	.45609	
							1.00564	1.01119	1.02190	1.03214	1.04182	1.05083	
.374	1.2737	.7550	.94732	.05614	1.72598	1.01661	.37976	.38580	.39897	.41412	.43259	.45857	
							1.00563	1.01115	1.02183	1.03203	1.04166	1.05062	
.375	1.2717	.7556	.94801	.05550	1.72009	1.01630	.38080	.38689	.40017	.41550	.43427	.46116	
							1.00561	1.01111	1.02176	1.03191	1.04150	1.05040	
.376	1.2697	.7562	.94869	.05486	1.71422	1.01599	.38183	.38797	.40137	.41687	.43598	.46387	
							1.00559	1.01108	1.02168	1.03180	1.04134	1.05017	
.377	1.2677	.7568	.94938	.05422	1.70837	1.01569	.38287	.38906	.40258	.41826	.43770	.46674	
							1.00557	1.01104	1.02161	1.03168	1.04118	1.04995	
.378	1.2658	.7573	.95005	.05358	1.70253	1.01539	.38392	.39014	.40378	.41966	.43945	.46979	
							1.00556	1.01101	1.02153	1.03157	1.04101	1.04973	
.379	1.2638	.7579	.95073	.05295	1.69672	1.01509	.38496	.39123	.40500	.42106	.44123	.47309	
							1.00554	1.01097	1.02146	1.03145	1.04085	1.04950	
.380	1.2618	.7585	.95139	.05232	1.69093	1.01480	.38600	.39232	.40621	.42248	.44304	.47670	
							1.00552	1.01093	1.02139	1.03134	1.04069	1.04927	
.381	1.2599	.7590	.95206	.05170	1.68516	1.01451	.38704	.39341	.40743	.42390	.44488	.48075	
							1.00550	1.01090	1.02131	1.03122	1.04053	1.04904	
.382	1.2579	.7596	.95271	.05108	1.67940	1.01423	.38808	.39451	.40866	.42534	.44675	.48566	
							1.00548	1.01086	1.02124	1.03111	1.04037	1.04881	
.383	1.2560	.7602	.95337	.05046	1.67367	1.01395	.38913	.39560	.40988	.42678	.44866	.49208	
							1.00547	1.01083	1.02117	1.03099	1.04020	1.04856	
.384	1.2540	.7607	.95402	.04984	1.66795	1.01367	.39017	.39670	.41112	.42824	.45061		
							1.00545	1.01079	1.02110	1.03088	1.04004		
.385	1.2521	.7613	.95466	.04922	1.66226	1.01340	.39121	.39779	.41235	.42971	.45260		
							1.00543	1.01076	1.02102	1.03077	1.03988		
.386	1.2501	.7619	.95530	.04861	1.65658	1.01313	.39226	.39889	.41360	.43119	.45464		
							1.00542	1.01072	1.02095	1.03065	1.03971		
.387	1.2482	.7624	.95593	.04801	1.65092	1.01286	.39331	.39999	.41484	.43268	.45673		
							1.00540	1.01069	1.02088	1.03054	1.03955		
.388	1.2462	.7630	.95656	.04740	1.64528	1.01260	.39435	.40109	.41609	.43419	.45888		
							1.00538	1.01065	1.02081	1.03042	1.03938		
.389	1.2443	.7636	.95719	.04680	1.63966	1.01234	.39540	.40219	.41735	.43571	.46114		
							1.00536	1.01062	1.02073	1.03031	1.03922		

TABLE IX. Specific Heat Ratio = 1.40

p/p_t	MACH	T/T_t	GAMMA	K_crit	SHOCK P_y/P_x	P_ly/P_ly	.001	.005	.01	.015	.02	.03	.04
.390	1.2424	.7641	.95781	.04620	1.63406	1.01209	.39126	.39645	.40330	.41064	.41860	.43725	.46344
							1.00108	1.00535	1.01058	1.01568	1.02066	1.03020	1.03906
.391	1.2404	.7647	.95842	.04561	1.62848	1.01184	.39227	.39750	.40441	.41182	.41987	.43881	.46583
							1.00108	1.00533	1.01055	1.01563	1.02059	1.03008	1.03889
.392	1.2385	.7652	.95903	.04501	1.62291	1.01159	.39328	.39855	.40551	.41300	.42114	.44038	.46832
							1.00107	1.00531	1.01051	1.01557	1.02051	1.02997	1.03872
.393	1.2366	.7658	.95964	.04443	1.61737	1.01134	.39429	.39960	.40662	.41418	.42242	.44197	.47094
							1.00107	1.00530	1.01048	1.01552	1.02044	1.02985	1.03855
.394	1.2347	.7664	.96024	.04384	1.61184	1.01110	.39530	.40065	.40773	.41537	.42370	.44358	.47370
							1.00107	1.00528	1.01044	1.01547	1.02037	1.02974	1.03838
.395	1.2328	.7669	.96084	.04326	1.60633	1.01086	.39631	.40170	.40885	.41656	.42499	.44521	.47664
							1.00106	1.00526	1.01041	1.01542	1.02030	1.02963	1.03821
.396	1.2309	.7675	.96143	.04268	1.60083	1.01063	.39732	.40275	.40996	.41775	.42629	.44686	.47982
							1.00106	1.00525	1.01038	1.01536	1.02023	1.02951	1.03804
.397	1.2289	.7680	.96202	.04210	1.59536	1.01040	.39833	.40380	.41108	.41895	.42759	.44853	.48330
							1.00106	1.00523	1.01034	1.01531	1.02016	1.02940	1.03786
.398	1.2270	.7686	.96260	.04153	1.58990	1.01017	.39934	.40486	.41220	.42015	.42890	.45023	.48729
							1.00105	1.00521	1.01031	1.01526	1.02009	1.02928	1.03769
.399	1.2251	.7691	.96318	.04096	1.58446	1.00995	.40035	.40591	.41332	.42136	.43022	.45196	.49200
							1.00105	1.00520	1.01027	1.01521	1.02001	1.02917	1.03751
.400	1.2232	.7697	.96375	.04039	1.57904	1.00973	.40136	.40697	.41445	.42257	.43154	.45372	.49828
							1.00105	1.00518	1.01024	1.01516	1.01994	1.02906	1.03732
.401	1.2213	.7702	.96432	.03983	1.57363	1.00951	.40237	.40802	.41557	.42379	.43287	.45551	
							1.00104	1.00516	1.01021	1.01510	1.01987	1.02894	
.402	1.2195	.7708	.96488	.03926	1.56825	1.00929	.40338	.40908	.41670	.42500	.43422	.45733	
							1.00104	1.00515	1.01017	1.01505	1.01980	1.02882	
.403	1.2176	.7713	.96544	.03871	1.56288	1.00908	.40439	.41014	.41783	.42623	.43557	.45919	
							1.00104	1.00513	1.01014	1.01500	1.01973	1.02871	
.404	1.2157	.7719	.96599	.03815	1.55752	1.00887	.40540	.41120	.41896	.42746	.43692	.46110	
							1.00104	1.00511	1.01010	1.01495	1.01966	1.02859	
.405	1.2138	.7724	.96654	.03760	1.55219	1.00867	.40641	.41226	.42010	.42869	.43831	.46304	
							1.00103	1.00510	1.01007	1.01490	1.01960	1.02848	
.406	1.2119	.7729	.96709	.03705	1.54687	1.00847	.40742	.41332	.42124	.42993	.43969	.46508	
							1.00103	1.00508	1.01004	1.01485	1.01953	1.02837	
.407	1.2100	.7735	.96763	.03651	1.54156	1.00827	.40844	.41438	.42238	.43117	.44108	.46714	
							1.00102	1.00506	1.01000	1.01480	1.01946	1.02825	
.408	1.2082	.7740	.96817	.03597	1.53628	1.00807	.40945	.41545	.42352	.43242	.44249	.46927	
							1.00102	1.00505	1.00997	1.01474	1.01939	1.02813	
.409	1.2063	.7746	.96870	.03543	1.53101	1.00788	.41046	.41651	.42467	.43368	.44390	.47147	
							1.00102	1.00503	1.00994	1.01469	1.01932	1.02802	
.410	1.2044	.7751	.96923	.03490	1.52576	1.00769	.41147	.41758	.42582	.43494	.44533	.47376	
							1.00101	1.00502	1.00990	1.01464	1.01925	1.02790	
.411	1.2026	.7757	.96975	.03436	1.52052	1.00751	.41248	.41865	.42697	.43621	.44677	.47615	
							1.00101	1.00500	1.00987	1.01459	1.01918	1.02778	
.412	1.2007	.7762	.97027	.03384	1.51530	1.00732	.41350	.41971	.42813	.43748	.44822	.47866	
							1.00101	1.00498	1.00984	1.01454	1.01911	1.02766	
.413	1.1988	.7767	.97078	.03331	1.51010	1.00714	.41451	.42078	.42929	.43877	.44969	.48131	
							1.00100	1.00497	1.00981	1.01449	1.01903	1.02754	
.414	1.1970	.7773	.97129	.03279	1.50491	1.00696	.41552	.42186	.43045	.44006	.45117	.48414	
							1.00100	1.00495	1.00977	1.01444	1.01896	1.02741	
.415	1.1951	.7778	.97179	.03227	1.49974	1.00679	.41653	.42293	.43162	.44137	.45267	.48720	
							1.00100	1.00494	1.00974	1.01440	1.01889	1.02729	
.416	1.1933	.7783	.97229	.03176	1.49458	1.00662	.41755	.42400	.43279	.44268	.45419	.49058	
							1.00100	1.00492	1.00971	1.01435	1.01882	1.02716	
.417	1.1914	.7789	.97278	.03125	1.48944	1.00645	.41856	.42508	.43396	.44400	.45572	.49451	
							1.00099	1.00490	1.00967	1.01429	1.01875	1.02704	
.418	1.1896	.7794	.97327	.03074	1.48432	1.00628	.41958	.42615	.43514	.44532	.45728	.49920	
							1.00099	1.00489	1.00964	1.01424	1.01868	1.02691	
.419	1.1878	.7799	.97376	.03023	1.47921	1.00612	.42059	.42723	.43632	.44665	.45886	.50597	
							1.00099	1.00487	1.00961	1.01419	1.01861	1.02677	

TABLE IX. Specific Heat Ratio = 1.40

					SHOCK		R: and TPR versus Loss Coefficient K						
p/p_t	MACH	T/T_t	GAMMA	K_{crit}	p_y/p_x	P_{ty}/P_{ty}	.001	.002	.003	.005	.01	.015	.02
.420	1.1859	.7805	.97424	.02973	1.47412	1.00596	.42160	.42324	.42490	.42631	.43751	.44800	.46046
							1.00098	1.00196	1.00293	1.00486	1.00958	1.01414	1.01854
.421	1.1841	.7810	.97472	.02924	1.46904	1.00580	.42262	.42427	.42594	.42940	.43870	.44935	.46209
							1.00098	1.00195	1.00292	1.00484	1.00954	1.01409	1.01847
.422	1.1822	.7815	.97519	.02874	1.46398	1.00565	.42363	.42530	.42699	.43048	.43989	.45071	.46374
							1.00098	1.00195	1.00291	1.00483	1.00951	1.01404	1.01840
.423	1.1804	.7821	.97566	.02825	1.45894	1.00549	.42465	.42633	.42804	.43156	.44109	.45209	.46543
							1.00097	1.00194	1.00290	1.00481	1.00948	1.01399	1.01833
.424	1.1786	.7826	.97612	.02777	1.45391	1.00535	.42566	.42736	.42909	.43265	.44230	.45348	.46715
							1.00097	1.00194	1.00289	1.00480	1.00944	1.01394	1.01826
.425	1.1768	.7831	.97658	.02728	1.44889	1.00520	.42668	.42839	.43014	.43374	.44353	.45488	.46890
							1.00097	1.00193	1.00289	1.00478	1.00942	1.01389	1.01819
.426	1.1749	.7836	.97703	.02680	1.44389	1.00505	.42770	.42942	.43119	.43483	.44475	.45629	.47069
							1.00096	1.00192	1.00288	1.00477	1.00939	1.01384	1.01811
.427	1.1731	.7842	.97748	.02633	1.43890	1.00491	.42871	.43046	.43224	.43592	.44597	.45772	.47253
							1.00096	1.00192	1.00287	1.00475	1.00936	1.01379	1.01804
.428	1.1713	.7847	.97793	.02585	1.43393	1.00477	.42973	.43149	.43329	.43702	.44720	.45917	.47441
							1.00096	1.00191	1.00286	1.00473	1.00933	1.01374	1.01797
.429	1.1695	.7852	.97837	.02538	1.42898	1.00464	.43075	.43253	.43435	.43811	.44844	.46063	.47635
							1.00096	1.00190	1.00285	1.00472	1.00929	1.01369	1.01790
.430	1.1677	.7857	.97880	.02492	1.42404	1.00450	.43176	.43356	.43540	.43921	.44968	.46211	.47835
							1.00095	1.00190	1.00284	1.00470	1.00926	1.01364	1.01782
.431	1.1659	.7863	.97923	.02445	1.41911	1.00437	.43278	.43460	.43646	.44031	.45093	.46361	.48042
							1.00095	1.00189	1.00283	1.00469	1.00923	1.01359	1.01775
.432	1.1641	.7868	.97966	.02400	1.41420	1.00424	.43380	.43564	.43752	.44142	.45218	.46513	.48257
							1.00095	1.00189	1.00282	1.00467	1.00920	1.01354	1.01767
.433	1.1623	.7873	.98008	.02354	1.40931	1.00412	.43482	.43668	.43858	.44252	.45345	.46667	.48482
							1.00094	1.00188	1.00281	1.00466	1.00917	1.01349	1.01760
.434	1.1605	.7878	.98050	.02309	1.40442	1.00399	.43584	.43771	.43964	.44363	.45472	.46823	.48719
							1.00094	1.00187	1.00280	1.00464	1.00913	1.01344	1.01752
.435	1.1587	.7883	.98091	.02264	1.39956	1.00387	.43686	.43875	.44070	.44474	.45600	.46982	.48969
							1.00094	1.00187	1.00279	1.00463	1.00910	1.01338	1.01744
.436	1.1569	.7889	.98132	.02220	1.39470	1.00375	.43788	.43979	.44176	.44586	.45729	.47144	.49244
							1.00093	1.00186	1.00279	1.00461	1.00907	1.01333	1.01737
.437	1.1551	.7894	.98173	.02176	1.38987	1.00364	.43890	.44084	.44284	.44698	.45859	.47310	.49539
							1.00093	1.00186	1.00278	1.00460	1.00904	1.01328	1.01729
.438	1.1533	.7899	.98213	.02132	1.38504	1.00352	.43992	.44188	.44388	.44810	.45990	.47478	.49867
							1.00093	1.00185	1.00276	1.00458	1.00901	1.01323	1.01721
.439	1.1515	.7904	.98252	.02089	1.38023	1.00341	.44094	.44294	.44496	.44922	.46123	.47651	.50247
							1.00093	1.00185	1.00276	1.00457	1.00897	1.01318	1.01713
.440	1.1497	.7909	.98291	.02046	1.37543	1.00330	.44196	.44396	.44603	.45035	.46256	.47827	.50734
							1.00092	1.00183	1.00275	1.00455	1.00894	1.01313	1.01705
.441	1.1479	.7914	.98330	.02003	1.37065	1.00320	.44300	.44500	.44710	.45148	.46390	.48008	.51600
							1.00093	1.00183	1.00274	1.00454	1.00891	1.01307	1.01696
.442	1.1461	.7919	.98368	.01961	1.36588	1.00309	.44400	.44606	.44818	.45261	.46526	.48195	
							1.00092	1.00183	1.00273	1.00452	1.00888	1.01302	
.443	1.1444	.7925	.98406	.01920	1.36113	1.00299	.44503	.44711	.44925	.45375	.46663	.48388	
							1.00091	1.00182	1.00272	1.00451	1.00885	1.01297	
.444	1.1426	.7930	.98443	.01878	1.35639	1.00289	.44605	.44816	.45033	.45489	.46802	.48588	
							1.00091	1.00182	1.00271	1.00449	1.00881	1.01291	
.445	1.1408	.7935	.98480	.01837	1.35166	1.00279	.44707	.44921	.45141	.45604	.46942	.48795	
							1.00091	1.00181	1.00270	1.00448	1.00878	1.01286	
.446	1.1390	.7940	.98516	.01797	1.34695	1.00269	.44810	.45026	.45249	.45719	.47084	.49013	
							1.00090	1.00180	1.00270	1.00446	1.00875	1.01280	
.447	1.1373	.7945	.98552	.01756	1.34225	1.00260	.44912	.45131	.45357	.45834	.47228	.49242	
							1.00090	1.00180	1.00269	1.00445	1.00872	1.01275	
.448	1.1355	.7950	.98588	.01717	1.33756	1.00251	.45015	.45237	.45466	.45950	.47374	.49493	
							1.00090	1.00179	1.00268	1.00443	1.00869	1.01270	
.449	1.1337	.7955	.98623	.01677	1.33289	1.00242	.45117	.45342	.45574	.46066	.47522	.49758	
							1.00090	1.00179	1.00267	1.00442	1.00865	1.01264	

175

TABLE IX. Specific Heat Ratio = 1.40

p/p_t	MACH	T/T_t	GAMMA	K_crit	SHOCK P_y/P_x	P_ty/P_tx	.001	.002	.003	.004	.005	.01	.015
.450	1.1320	.7960	.98657	.01638	1.32823	1.00233	.45220	.45448	.45683	.45928	.46183	.47673	.50049
							1.00089	1.00178	1.00266	1.00353	1.00440	1.00862	1.01258
.451	1.1302	.7965	.98691	.01599	1.32358	1.00225	.45323	.45554	.45793	.46041	.46300	.47826	.50378
							1.00089	1.00177	1.00265	1.00352	1.00439	1.00859	1.01252
.452	1.1284	.7970	.98725	.01561	1.31895	1.00216	.45426	.45660	.45902	.46155	.46418	.47982	.50768
							1.00089	1.00177	1.00264	1.00351	1.00437	1.00856	1.01246
.453	1.1267	.7975	.98758	.01523	1.31433	1.00208	.45529	.45766	.46012	.46269	.46537	.48141	.51317
							1.00088	1.00176	1.00263	1.00350	1.00436	1.00852	1.01240
.454	1.1249	.7980	.98791	.01486	1.30972	1.00200	.45632	.45872	.46122	.46383	.46656	.48304	
							1.00088	1.00176	1.00263	1.00349	1.00434	1.00849	
.455	1.1232	.7985	.98824	.01449	1.30513	1.00193	.45735	.45979	.46233	.46498	.46777	.48470	
							1.00088	1.00175	1.00262	1.00347	1.00433	1.00846	
.456	1.1214	.7990	.98856	.01412	1.30055	1.00185	.45838	.46086	.46344	.46614	.46897	.48641	
							1.00088	1.00175	1.00261	1.00346	1.00431	1.00842	
.457	1.1197	.7995	.98887	.01376	1.29598	1.00178	.45941	.46193	.46455	.46730	.47019	.48818	
							1.00087	1.00174	1.00260	1.00345	1.00429	1.00839	
.458	1.1179	.8000	.98918	.01340	1.29143	1.00171	.46045	.46300	.46567	.46846	.47141	.48999	
							1.00087	1.00173	1.00259	1.00344	1.00428	1.00835	
.459	1.1162	.8005	.98949	.01305	1.28688	1.00164	.46148	.46407	.46678	.46963	.47264	.49188	
							1.00087	1.00173	1.00258	1.00343	1.00426	1.00832	
.460	1.1145	.8010	.98979	.01270	1.28236	1.00157	.46252	.46515	.46791	.47081	.47389	.49384	
							1.00086	1.00172	1.00257	1.00341	1.00425	1.00828	
.461	1.1127	.8015	.99009	.01235	1.27784	1.00150	.46355	.46623	.46904	.47200	.47514	.49596	
							1.00086	1.00172	1.00256	1.00340	1.00423	1.00826	
.462	1.1110	.8020	.99038	.01201	1.27333	1.00144	.46459	.46731	.47017	.47319	.47640	.49815	
							1.00086	1.00171	1.00255	1.00339	1.00422	1.00822	
.463	1.1093	.8025	.99067	.01167	1.26884	1.00138	.46563	.46839	.47131	.47439	.47768	.50049	
							1.00085	1.00170	1.00255	1.00338	1.00420	1.00818	
.464	1.1075	.8030	.99095	.01134	1.26436	1.00132	.46667	.46948	.47245	.47559	.47900	.50304	
							1.00085	1.00170	1.00254	1.00337	1.00420	1.00815	
.465	1.1058	.8035	.99123	.01101	1.25990	1.00126	.46771	.47057	.47360	.47681	.48030	.50586	
							1.00085	1.00169	1.00253	1.00335	1.00418	1.00811	
.466	1.1041	.8040	.99151	.01068	1.25544	1.00120	.46876	.47167	.47475	.47804	.48162	.50909	
							1.00085	1.00169	1.00252	1.00334	1.00417	1.00807	
.467	1.1023	.8045	.99178	.01036	1.25100	1.00115	.46980	.47276	.47591	.47931	.48296	.51326	
							1.00084	1.00168	1.00251	1.00334	1.00415	1.00803	
.468	1.1006	.8050	.99204	.01005	1.24657	1.00109	.47085	.47386	.47707	.48056	.48432	.52015	
							1.00084	1.00167	1.00250	1.00333	1.00414	1.00799	
.469	1.0989	.8055	.99231	.00973	1.24215	1.00104	.47189	.47497	.47825	.48182	.48569		
							1.00084	1.00167	1.00249	1.00332	1.00412		
.470	1.0972	.8060	.99256	.00943	1.23775	1.00099	.47294	.47608	.47946	.48310	.48709		
							1.00083	1.00166	1.00249	1.00330	1.00411		
.471	1.0955	.8064	.99282	.00912	1.23335	1.00094	.47399	.47719	.48066	.48439	.48851		
							1.00083	1.00166	1.00248	1.00329	1.00409		
.472	1.0937	.8069	.99306	.00883	1.22897	1.00089	.47504	.47831	.48186	.48570	.48996		
							1.00083	1.00165	1.00247	1.00328	1.00408		
.473	1.0920	.8074	.99331	.00853	1.22460	1.00085	.47610	.47943	.48307	.48702	.49144		
							1.00082	1.00164	1.00247	1.00327	1.00406		
.474	1.0903	.8079	.99355	.00824	1.22024	1.00080	.47716	.48059	.48429	.48836	.49295		
							1.00082	1.00165	1.00246	1.00326	1.00404		
.475	1.0886	.8084	.99378	.00796	1.21590	1.00076	.47821	.48173	.48553	.48973	.49450		
							1.00082	1.00164	1.00245	1.00324	1.00403		
.476	1.0869	.8089	.99402	.00767	1.21156	1.00072	.47928	.48287	.48677	.49111	.49610		
							1.00082	1.00164	1.00244	1.00323	1.00401		
.477	1.0852	.8094	.99424	.00740	1.20724	1.00068	.48037	.48402	.48803	.49253	.49775		
							1.00082	1.00163	1.00243	1.00322	1.00400		
.478	1.0835	.8099	.99447	.00713	1.20293	1.00064	.48144	.48518	.48931	.49397	.49946		
							1.00082	1.00162	1.00242	1.00321	1.00398		
.479	1.0818	.8103	.99468	.00686	1.19863	1.00061	.48252	.48635	.49060	.49545	.50124		
							1.00081	1.00162	1.00241	1.00320	1.00396		

TABLE IX. Specific Heat Ratio = 1.40

p/p_t	MACH	T/T_t	GAMMA	K_{crit}	SHOCK p_y/p_x	P_{tx}/P_{ty}	.001	.002	R_t and TPR versus Loss Coefficient K .003	.004	.005	.01	.015
.480	1.0801	.8108	.99490	.00660	1.19434	1.00057	.48359 1.00081	.48752 1.00161	.49191 1.00240	.49696 1.00318	.50311 1.00395		
.481	1.0784	.8113	.99511	.00634	1.19006	1.00054	.48467 1.00081	.48871 1.00161	.49324 1.00240	.49852 1.00317	.50509 1.00393		
.482	1.0767	.8118	.99531	.00609	1.18580	1.00050	.48575 1.00081	.48990 1.00160	.49460 1.00239	.50013 1.00316	.50721 1.00391		
.483	1.0750	.8123	.99551	.00584	1.18154	1.00047	.48684 1.00080	.49111 1.00160	.49598 1.00238	.50180 1.00315	.50964 1.00390		
.484	1.0733	.8127	.99571	.00559	1.17730	1.00044	.48793 1.00080	.49233 1.00159	.49738 1.00237	.50354 1.00313	.51230 1.00388		
.485	1.0716	.8132	.99590	.00535	1.17307	1.00041	.48903 1.00080	.49356 1.00159	.49883 1.00236	.50537 1.00312	.51548 1.00386		
.486	1.0699	.8137	.99609	.00512	1.16885	1.00039	.49013 1.00080	.49481 1.00158	.50031 1.00235	.50732 1.00310	.52001 1.00384		
.487	1.0682	.8142	.99627	.00489	1.16464	1.00036	.49124 1.00079	.49607 1.00157	.50184 1.00234	.50952 1.00310			
.488	1.0665	.8147	.99645	.00467	1.16044	1.00034	.49235 1.00079	.49736 1.00157	.50341 1.00233	.51188 1.00308			
.489	1.0649	.8151	.99662	.00445	1.15626	1.00031	.49348 1.00079	.49866 1.00156	.50506 1.00232	.51458 1.00307			
.490	1.0632	.8156	.99679	.00423	1.15208	1.00029	.49460 1.00078	.50000 1.00156	.50678 1.00231	.51792 1.00305			
.491	1.0615	.8161	.99696	.00402	1.14792	1.00027	.49574 1.00078	.50135 1.00155	.50868 1.00231	.52363 1.00303			
.492	1.0598	.8166	.99712	.00382	1.14376	1.00025	.49689 1.00078	.50275 1.00154	.51067 1.00230				
.493	1.0581	.8170	.99727	.00362	1.13962	1.00023	.49804 1.00078	.50418 1.00154	.51285 1.00229				
.494	1.0565	.8175	.99743	.00342	1.13549	1.00021	.49921 1.00077	.50565 1.00153	.51534 1.00228				
.495	1.0548	.8180	.99757	.00323	1.13136	1.00019	.50039 1.00077	.50718 1.00152	.51833 1.00226				
.496	1.0531	.8185	.99772	.00305	1.12725	1.00018	.50158 1.00077	.50877 1.00152	.52306 1.00225				
.497	1.0515	.8189	.99786	.00287	1.12315	1.00016	.50279 1.00076	.51056 1.00152					
.498	1.0498	.8194	.99799	.00269	1.11906	1.00015	.50402 1.00076	.51240 1.00151					
.499	1.0481	.8199	.99812	.00252	1.11498	1.00013	.50527 1.00076	.51440 1.00150					
.500	1.0465	.8203	.99825	.00236	1.11092	1.00012	.50654 1.00075	.51666 1.00150					
.501	1.0448	.8208	.99837	.00220	1.10686	1.00011	.50784 1.00075	.51934 1.00149					
.502	1.0431	.8213	.99849	.00204	1.10281	1.00010	.50918 1.00075	.52352 1.00148					
.503	1.0415	.8217	.99860	.00190	1.09877	1.00009	.51066 1.00075						
.504	1.0398	.8222	.99871	.00175	1.09475	1.00008	.51213 1.00075						
.505	1.0382	.8227	.99881	.00161	1.09073	1.00007	.51366 1.00074						
.506	1.0365	.8231	.99891	.00148	1.08672	1.00006	.51530 1.00074						
.507	1.0348	.8236	.99901	.00135	1.08273	1.00005	.51709 1.00074						
.508	1.0332	.8241	.99910	.00123	1.07874	1.00004	.51909 1.00073						
.509	1.0315	.8245	.99919	.00112	1.07476	1.00004	.52172 1.00073						

TABLE X. Specific Heat Ratio = 1.67

p/p_t	MACH	T/T_t	GAMMA	K_{crit}	SHOCK P_y/P_x	P_{ty}/P_{ty}	.01	.05	.1	.15	.2	.3	.4
.001	6.6872	.0626	.04754	.46815	55.68865	15.09948	.00127	.00235	.00375	.00522	.00678	.01033	.01507
							1.39045	3.20330	5.75877	8.39562	10.97025	15.61286	19.27892
.002	5.7565	.0826	.07122	.45553	41.20163	10.16432	.00240	.00403	.00616	.00839	.01077	.01623	.02377
							1.28549	2.55531	4.28735	6.04428	7.74203	10.76969	13.12107
.003	5.2644	.0972	.09007	.44625	34.41776	8.08775	.00351	.00559	.00830	.01116	.01421	.02125	.03128
							1.23713	2.26920	3.64919	5.03394	6.36252	8.71354	10.51556
.004	4.9364	.1091	.10630	.43861	30.23163	6.88852	.00461	.00708	.01030	.01370	.01734	.02581	.03820
							1.20761	2.09826	3.27265	4.44144	5.55720	7.51801	9.00391
.005	4.6931	.1194	.12079	.43200	27.30150	6.08884	.00569	.00852	.01220	.01610	.02028	.03006	.04481
							1.18709	1.98125	3.01725	4.04150	5.01495	6.71625	7.99183
.006	4.5013	.1284	.13403	.42610	25.09468	5.50924	.00677	.00992	.01403	.01839	.02308	.03410	.05123
							1.17171	1.89459	2.82945	3.74881	4.61883	6.13235	7.25539
.007	4.3436	.1366	.14630	.42074	23.35073	5.06554	.00785	.01130	.01581	.02060	.02577	.03797	.05759
							1.15961	1.82701	2.68387	3.52255	4.31345	5.68334	6.68950
.008	4.2104	.1441	.15779	.41579	21.92482	4.71249	.00891	.01266	.01755	.02275	.02837	.04171	.06399
							1.14967	1.77236	2.56690	3.34101	4.06890	5.32442	6.23706
.009	4.0953	.1511	.16863	.41117	20.72894	4.42334	.00998	.01400	.01925	.02485	.03090	.04535	.07057
							1.14148	1.72695	2.47001	3.19122	3.86744	5.02932	5.86582
.010	3.9942	.1576	.17892	.40683	19.70604	4.18118	.01105	.01532	.02093	.02690	.03338	.04890	.07754
							1.13447	1.68841	2.38811	3.06488	3.69779	4.78125	5.55326
.011	3.9042	.1638	.18873	.40273	18.81724	3.97472	.01211	.01664	.02258	.02892	.03580	.05238	.08549
							1.12838	1.65516	2.31759	2.95648	3.55241	4.56900	5.28541
.012	3.8233	.1696	.19814	.39883	18.03497	3.79613	.01317	.01794	.02420	.03090	.03817	.05580	
							1.12302	1.62607	2.25620	2.86213	3.42604	4.38477	
.013	3.7499	.1751	.20717	.39511	17.33908	3.63976	.01422	.01923	.02581	.03285	.04051	.05917	
							1.11826	1.60034	2.20205	2.77904	3.31487	4.22292	
.014	3.6827	.1804	.21587	.39154	16.71440	3.50143	.01528	.02052	.02740	.03477	.04282	.06250	
							1.11399	1.57735	2.15380	2.70514	3.21610	4.07930	
.015	3.6208	.1855	.22428	.38811	16.14930	3.37800	.01633	.02179	.02898	.03668	.04509	.06579	
							1.11012	1.55665	2.11045	2.63884	3.12759	3.95073	
.016	3.5636	.1903	.23242	.38480	15.63466	3.26701	.01739	.02306	.03054	.03856	.04734	.06904	
							1.10661	1.53788	2.07122	2.57894	3.04768	3.83477	
.017	3.5103	.1950	.24032	.38160	15.16322	3.16656	.01844	.02433	.03209	.04042	.04956	.07227	
							1.10339	1.52075	2.03550	2.52445	2.97506	3.72950	
.018	3.4605	.1995	.24799	.37850	14.72911	3.07510	.01949	.02559	.03363	.04227	.05176	.07548	
							1.10042	1.50503	2.00279	2.47461	2.90870	3.63338	
.019	3.4138	.2039	.25545	.37550	14.32754	2.99141	.02054	.02684	.03515	.04410	.05394	.07866	
							1.09769	1.49054	1.97268	2.42880	2.84774	3.54515	
.020	3.3698	.2082	.26271	.37258	13.95454	2.91445	.02158	.02809	.03667	.04592	.05610	.08183	
							1.09514	1.47712	1.94485	2.38649	2.79148	3.46380	
.021	3.3283	.2123	.26980	.36974	13.60679	2.84341	.02263	.02933	.03818	.04772	.05825	.08499	
							1.09277	1.46464	1.91901	2.34727	2.73936	3.38847	
.022	3.2891	.2163	.27671	.36697	13.28151	2.77756	.02368	.03057	.03968	.04951	.06038	.08814	
							1.09056	1.45300	1.89495	2.31077	2.69089	3.31847	
.023	3.2517	.2202	.28347	.36427	12.97631	2.71633	.02472	.03180	.04117	.05129	.06249	.09128	
							1.08848	1.44211	1.87246	2.27669	2.64566	3.25318	
.024	3.2163	.2239	.29008	.36163	12.68915	2.65922	.02577	.03304	.04266	.05306	.06459	.09441	
							1.08653	1.43188	1.85138	2.24471	2.60333	3.19211	
.025	3.1824	.2276	.29655	.35905	12.41829	2.60578	.02681	.03426	.04413	.05482	.06668	.09754	
							1.08468	1.42226	1.83157	2.21475	2.56360	3.13482	
.026	3.1501	.2313	.30288	.35652	12.16220	2.55566	.02786	.03549	.04561	.05657	.06876	.10067	
							1.08294	1.41326	1.81290	2.18653	2.52622	3.08093	
.027	3.1192	.2348	.30910	.35405	11.91955	2.50853	.02890	.03671	.04707	.05831	.07083	.10380	
							1.08129	1.40468	1.79527	2.15990	2.49095	3.03012	
.028	3.0895	.2382	.31519	.35162	11.68918	2.46412	.02994	.03793	.04853	.06005	.07289	.10694	
							1.07973	1.39654	1.77858	2.13472	2.45761	2.98210	
.029	3.0610	.2416	.32117	.34924	11.47005	2.42217	.03098	.03915	.04999	.06178	.07494	.11009	
							1.07824	1.38882	1.76275	2.11085	2.42603	2.93662	

TABLE X. Specific Heat Ratio = 1.67

p/p_t	MACH	T/T_t	GAMMA	K_crit	SHOCK P_y/P_x	P_tx/P_ty	.01	.05	.1	.15	.2	.25	.3
.030	3.0336	.2449	.32704	.34690	11.26127	2.38249	.03202	.04036	.05144	.06350	.07698	.09278	.11325
							1.07683	1.38148	1.74771	2.08819	2.39606	2.66615	2.89349
.031	3.0073	.2482	.33280	.34461	11.06201	2.34487	.03307	.04157	.05289	.06521	.07901	.09524	.11641
							1.07547	1.37448	1.73340	2.06663	2.36757	2.63116	2.85247
.032	2.9818	.2513	.33847	.34235	10.87157	2.30915	.03411	.04278	.05433	.06692	.08104	.09769	.11960
							1.07418	1.36780	1.71975	2.04609	2.34043	2.59785	2.81342
.033	2.9573	.2545	.34405	.34013	10.68928	2.27517	.03515	.04399	.05576	.06862	.08306	.10014	.12280
							1.07295	1.36142	1.70672	2.02650	2.31454	2.56608	2.77617
.034	2.9336	.2575	.34953	.33795	10.51457	2.24282	.03619	.04520	.05720	.07031	.08508	.10259	.12602
							1.07176	1.35531	1.69426	2.00777	2.28982	2.53574	2.74059
.035	2.9107	.2605	.35493	.33580	10.34693	2.21196	.03722	.04640	.05863	.07200	.08709	.10503	.12926
							1.07063	1.34946	1.68234	1.98985	2.26617	2.50672	2.70655
.036	2.8885	.2635	.36025	.33368	10.18586	2.18248	.03826	.04760	.06005	.07369	.08909	.10747	.13253
							1.06954	1.34384	1.67090	1.97268	2.24351	2.47893	2.67395
.037	2.8669	.2664	.36548	.33160	10.03094	2.15430	.03930	.04880	.06148	.07537	.09109	.10991	.13584
							1.06849	1.33845	1.65993	1.95621	2.22179	2.45229	2.64269
.038	2.8461	.2693	.37064	.32954	9.88177	2.12731	.04034	.05000	.06290	.07705	.09309	.11235	.13917
							1.06748	1.33326	1.64938	1.94039	2.20094	2.42671	2.61266
.039	2.8258	.2721	.37572	.32751	9.73801	2.10145	.04138	.05120	.06431	.07872	.09508	.11479	.14255
							1.06650	1.32827	1.63924	1.92518	2.18089	2.40213	2.58380
.040	2.8061	.2749	.38073	.32551	9.59933	2.07664	.04241	.05239	.06573	.08039	.09707	.11724	.14598
							1.06557	1.32346	1.62947	1.91055	2.16157	2.37848	2.55602
.041	2.7870	.2776	.38567	.32354	9.46542	2.05281	.04345	.05358	.06714	.08206	.09905	.11968	.14945
							1.06466	1.31882	1.62006	1.89645	2.14299	2.35571	2.52925
.042	2.7684	.2803	.39055	.32159	9.33601	2.02990	.04449	.05478	.06855	.08372	.10103	.12213	.15299
							1.06378	1.31434	1.61098	1.88285	2.12509	2.33376	2.50344
.043	2.7502	.2830	.39536	.31966	9.21084	2.00785	.04552	.05597	.06995	.08538	.10301	.12458	.15660
							1.06294	1.31001	1.60221	1.86973	2.10782	2.31258	2.47851
.044	2.7326	.2856	.40010	.31776	9.08969	1.98662	.04656	.05716	.07135	.08703	.10499	.12703	.16028
							1.06212	1.30583	1.59374	1.85706	2.09114	2.29213	2.45443
.045	2.7153	.2882	.40479	.31588	8.97234	1.96616	.04760	.05834	.07276	.08869	.10696	.12949	.16406
							1.06133	1.30178	1.58555	1.84481	2.07502	2.27237	2.43113
.046	2.6985	.2907	.40941	.31402	8.85858	1.94642	.04863	.05953	.07415	.09034	.10894	.13195	.16796
							1.06056	1.29786	1.57762	1.83295	2.05942	2.25325	2.40860
.047	2.6822	.2933	.41398	.31219	8.74824	1.92736	.04967	.06072	.07555	.09199	.11091	.13443	.17199
							1.05981	1.29406	1.56994	1.82148	2.04433	2.23478	2.38675
.048	2.6661	.2957	.41850	.31037	8.64114	1.90895	.05070	.06190	.07695	.09363	.11288	.13690	.17618
							1.05915	1.29037	1.56250	1.81036	2.02971	2.21686	2.36555
.049	2.6505	.2982	.42296	.30858	8.53712	1.89116	.05174	.06308	.07834	.09528	.11485	.13939	.18059
							1.05845	1.28679	1.55528	1.79958	2.01554	2.19949	2.34497
.050	2.6352	.3006	.42736	.30680	8.43604	1.87395	.05277	.06427	.07973	.09692	.11682	.14188	.18527
							1.05777	1.28332	1.54827	1.78912	2.00179	2.18263	2.32498
.051	2.6203	.3030	.43172	.30504	8.33775	1.85728	.05381	.06545	.08112	.09856	.11879	.14438	.19035
							1.05711	1.27995	1.54147	1.77897	1.98845	2.16628	2.30552
.052	2.6056	.3054	.43603	.30330	8.24212	1.84114	.05484	.06663	.08251	.10020	.12076	.14690	.19607
							1.05647	1.27667	1.53486	1.76911	1.97549	2.15039	2.28659
.053	2.5913	.3077	.44028	.30158	8.14903	1.82550	.05588	.06781	.08390	.10183	.12273	.14942	.20302
							1.05584	1.27348	1.52844	1.75953	1.96290	2.13495	2.26810
.054	2.5773	.3101	.44449	.29987	8.05838	1.81034	.05691	.06899	.08528	.10347	.12470	.15195	
							1.05523	1.27038	1.52219	1.75021	1.95066	2.11994	
.055	2.5636	.3123	.44866	.29818	7.97004	1.79562	.05795	.07017	.08667	.10510	.12667	.15450	
							1.05464	1.26735	1.51610	1.74115	1.93875	2.10534	
.056	2.5501	.3146	.45278	.29650	7.88393	1.78134	.05898	.07134	.08805	.10674	.12865	.15706	
							1.05406	1.26441	1.51018	1.73232	1.92716	2.09112	
.057	2.5369	.3169	.45685	.29485	7.79995	1.76747	.06001	.07252	.08943	.10837	.13062	.15964	
							1.05349	1.26154	1.50441	1.72373	1.91588	2.07727	
.058	2.5240	.3191	.46088	.29320	7.71800	1.75399	.06105	.07370	.09081	.11000	.13259	.16223	
							1.05294	1.25874	1.49879	1.71535	1.90488	2.06377	
.059	2.5113	.3213	.46487	.29157	7.63802	1.74089	.06208	.07487	.09219	.11163	.13457	.16484	
							1.05241	1.25601	1.49331	1.70719	1.89416	2.05059	

TABLE X. Specific Heat Ratio = 1.67

p/p_t	MACH	T/T_t	GAMMA	K_{crit}	SHOCK p_y/p_x	p_{ty}/p_{tx}	.01	.05	.07	.1	.15	.2	.25
.060	2.4988	.3234	.46882	.28996	7.55991	1.72814	.06311	.07605	.08285	.09357	.11326	.13655	.16746
							1.05188	1.25335	1.34972	1.48796	1.69923	1.88371	2.03776
.061	2.4866	.3256	.47273	.28836	7.48360	1.71575	.06415	.07722	.08410	.09494	.11489	.13853	.17011
							1.05137	1.25075	1.34607	1.48274	1.69146	1.87351	2.02523
.062	2.4745	.3277	.47660	.28677	7.40902	1.70368	.06518	.07839	.08535	.09632	.11652	.14051	.17278
							1.05087	1.24821	1.34251	1.47764	1.68388	1.86356	2.01301
.063	2.4627	.3298	.48044	.28520	7.33612	1.69192	.06621	.07957	.08660	.09770	.11815	.14250	.17547
							1.05038	1.24573	1.33902	1.47267	1.67648	1.85384	2.00107
.064	2.4511	.3319	.48423	.28364	7.26481	1.68047	.06724	.08074	.08785	.09907	.11978	.14449	.17819
							1.04990	1.24330	1.33562	1.46780	1.66925	1.84435	1.98940
.065	2.4397	.3340	.48799	.28209	7.19505	1.66932	.06828	.08191	.08909	.10044	.12140	.14648	.18093
							1.04943	1.24093	1.33229	1.46305	1.66218	1.83508	1.97799
.066	2.4285	.3360	.49171	.28055	7.12678	1.65844	.06931	.08308	.09034	.10182	.12303	.14847	.18370
							1.04898	1.23861	1.32903	1.45840	1.65527	1.82601	1.96683
.067	2.4175	.3381	.49540	.27903	7.05994	1.64783	.07034	.08425	.09159	.10319	.12466	.15047	.18651
							1.04853	1.23634	1.32585	1.45386	1.64852	1.81714	1.95592
.068	2.4067	.3401	.49905	.27752	6.99449	1.63748	.07137	.08542	.09283	.10456	.12629	.15248	.18934
							1.04809	1.23412	1.32273	1.44941	1.64191	1.80846	1.94523
.069	2.3960	.3421	.50267	.27601	6.93037	1.62738	.07241	.08659	.09408	.10593	.12791	.15448	.19222
							1.04766	1.23195	1.31968	1.44506	1.63544	1.79998	1.93477
.070	2.3855	.3441	.50626	.27453	6.86754	1.61751	.07344	.08776	.09532	.10730	.12954	.15649	.19513
							1.04724	1.22982	1.31670	1.44079	1.62911	1.79166	1.92452
.071	2.3751	.3460	.50981	.27305	6.80595	1.60788	.07447	.08893	.09657	.10867	.13117	.15851	.19809
							1.04682	1.22773	1.31377	1.43662	1.62291	1.78353	1.91448
.072	2.3650	.3480	.51333	.27158	6.74557	1.59848	.07550	.09010	.09781	.11004	.13280	.16053	.20110
							1.04642	1.22568	1.31090	1.43253	1.61684	1.77555	1.90463
.073	2.3549	.3499	.51682	.27012	6.68635	1.58928	.07653	.09127	.09906	.11141	.13443	.16256	.20416
							1.04602	1.22368	1.30809	1.42852	1.61088	1.76774	1.89498
.074	2.3450	.3518	.52028	.26868	6.62826	1.58030	.07757	.09244	.10030	.11278	.13606	.16459	.20728
							1.04563	1.22171	1.30534	1.42459	1.60505	1.76009	1.88550
.075	2.3353	.3537	.52371	.26724	6.57126	1.57151	.07860	.09360	.10154	.11415	.13769	.16662	.21047
							1.04525	1.21979	1.30263	1.42074	1.59934	1.75258	1.87621
.076	2.3257	.3556	.52711	.26581	6.51531	1.56292	.07963	.09477	.10279	.11552	.13933	.16867	.21373
							1.04487	1.21790	1.29998	1.41696	1.59373	1.74522	1.86708
.077	2.3163	.3575	.53048	.26439	6.46039	1.55452	.08066	.09594	.10403	.11689	.14096	.17072	.21710
							1.04451	1.21604	1.29738	1.41326	1.58823	1.73799	1.85814
.078	2.3069	.3593	.53383	.26299	6.40645	1.54629	.08169	.09710	.10527	.11826	.14259	.17278	.22055
							1.04415	1.21422	1.29483	1.40962	1.58283	1.73093	1.84933
.079	2.2977	.3612	.53714	.26159	6.35348	1.53825	.08272	.09827	.10651	.11962	.14423	.17484	.22411
							1.04379	1.21240	1.29233	1.40605	1.57753	1.72398	1.84068
.080	2.2887	.3630	.54043	.26020	6.30144	1.53037	.08375	.09944	.10775	.12099	.14586	.17692	.22781
							1.04344	1.21064	1.28987	1.40255	1.57234	1.71715	1.83216
.081	2.2797	.3648	.54369	.25882	6.25030	1.52265	.08478	.10060	.10899	.12236	.14750	.17900	.23168
							1.04310	1.20892	1.28745	1.39911	1.56723	1.71044	1.82379
.082	2.2709	.3666	.54692	.25745	6.20003	1.51509	.08581	.10177	.11023	.12373	.14914	.18108	.23575
							1.04276	1.20722	1.28508	1.39573	1.56222	1.70385	1.81555
.083	2.2622	.3684	.55012	.25608	6.15062	1.50769	.08685	.10293	.11148	.12510	.15078	.18318	.24007
							1.04243	1.20556	1.28274	1.39240	1.55729	1.69738	1.80743
.084	2.2536	.3702	.55331	.25473	6.10204	1.50043	.08788	.10410	.11272	.12646	.15242	.18529	.24475
							1.04211	1.20392	1.28045	1.38914	1.55245	1.69102	1.79943
.085	2.2451	.3720	.55646	.25338	6.05426	1.49332	.08891	.10526	.11396	.12783	.15407	.18740	.24997
							1.04179	1.20231	1.27820	1.38593	1.54769	1.68477	1.79155
.086	2.2367	.3737	.55959	.25205	6.00726	1.48635	.08994	.10643	.11520	.12920	.15571	.18953	.25603
							1.04147	1.20073	1.27598	1.38278	1.54302	1.67862	1.78375
.087	2.2284	.3754	.56270	.25072	5.96102	1.47952	.09097	.10759	.11644	.13057	.15736	.19166	.26415
							1.04116	1.19917	1.27380	1.37968	1.53842	1.67257	1.77604
.088	2.2202	.3772	.56578	.24940	5.91552	1.47281	.09200	.10876	.11768	.13194	.15901	.19381	
							1.04086	1.19764	1.27166	1.37663	1.53390	1.66662	
.089	2.2122	.3789	.56883	.24808	5.87074	1.46624	.09303	.10992	.11892	.13331	.16066	.19597	
							1.04056	1.19614	1.26955	1.37363	1.52945	1.66077	

TABLE X. Specific Heat Ratio = 1.67

p/p_t	MACH	T/T_t	GAMMA	K_{crit}	SHOCK P_y/P_x	P_{ty}/P_{tx}	.01	.05	.07	.09	.1	.15	.2
.090	2.2042	.3806	.57187	.24678	5.82666	1.45979	.09406	.11109	.12016	.12970	.13468	.16232	.19814
							1.04026	1.19466	1.26748	1.33710	1.37068	1.52507	1.65501
.091	2.1963	.3823	.57488	.24548	5.78326	1.45346	.09509	.11225	.12140	.13102	.13604	.16397	.20032
							1.03997	1.19320	1.26544	1.33448	1.36777	1.52077	1.64934
.092	2.1885	.3839	.57786	.24419	5.74052	1.44725	.09612	.11342	.12264	.13234	.13741	.16563	.20251
							1.03969	1.19176	1.26343	1.33190	1.36491	1.51653	1.64376
.093	2.1808	.3856	.58083	.24290	5.69843	1.44115	.09715	.11458	.12388	.13367	.13879	.16729	.20472
							1.03941	1.19035	1.26145	1.32936	1.36210	1.51236	1.63827
.094	2.1732	.3873	.58377	.24162	5.65697	1.43516	.09818	.11574	.12512	.13499	.14016	.16896	.20695
							1.03913	1.18896	1.25950	1.32686	1.35933	1.50825	1.63285
.095	2.1657	.3889	.58669	.24035	5.61612	1.42928	.09921	.11691	.12636	.13632	.14153	.17062	.20919
							1.03885	1.18759	1.25759	1.32440	1.35660	1.50420	1.62752
.096	2.1582	.3906	.58959	.23909	5.57587	1.42351	.10024	.11807	.12760	.13764	.14290	.17229	.21144
							1.03858	1.18624	1.25570	1.32198	1.35391	1.50022	1.62226
.097	2.1509	.3922	.59247	.23784	5.53620	1.41784	.10127	.11924	.12884	.13897	.14427	.17396	.21372
							1.03832	1.18491	1.25383	1.31959	1.35126	1.49629	1.61709
.098	2.1436	.3938	.59532	.23659	5.49710	1.41226	.10230	.12040	.13008	.14029	.14564	.17564	.21601
							1.03806	1.18360	1.25200	1.31724	1.34866	1.49243	1.61198
.099	2.1364	.3954	.59816	.23534	5.45856	1.40679	.10333	.12156	.13132	.14162	.14702	.17732	.21831
							1.03780	1.18230	1.25019	1.31492	1.34609	1.48862	1.60695
.100	2.1293	.3970	.60097	.23411	5.42056	1.40141	.10437	.12273	.13256	.14295	.14839	.17900	.22064
							1.03755	1.18103	1.24841	1.31264	1.34355	1.48486	1.60199
.101	2.1222	.3986	.60377	.23288	5.38309	1.39612	.10540	.12389	.13380	.14427	.14977	.18068	.22299
							1.03730	1.17977	1.24666	1.31039	1.34106	1.48116	1.59709
.102	2.1153	.4002	.60654	.23165	5.34613	1.39092	.10643	.12506	.13504	.14561	.15114	.18237	.22537
							1.03705	1.17854	1.24492	1.30820	1.33860	1.47751	1.59226
.103	2.1084	.4017	.60930	.23044	5.30968	1.38581	.10746	.12622	.13628	.14694	.15252	.18406	.22776
							1.03680	1.17732	1.24322	1.30601	1.33617	1.47391	1.58750
.104	2.1015	.4033	.61203	.22923	5.27372	1.38078	.10849	.12739	.13753	.14826	.15389	.18576	.23018
							1.03656	1.17611	1.24153	1.30385	1.33377	1.47036	1.58280
.105	2.0948	.4049	.61475	.22802	5.23824	1.37584	.10952	.12855	.13877	.14959	.15527	.18745	.23263
							1.03633	1.17492	1.23987	1.30172	1.33141	1.46685	1.57816
.106	2.0881	.4064	.61745	.22682	5.20323	1.37098	.11055	.12971	.14001	.15092	.15665	.18916	.23511
							1.03609	1.17375	1.23823	1.29963	1.32909	1.46340	1.57358
.107	2.0815	.4079	.62013	.22563	5.16868	1.36620	.11158	.13088	.14125	.15225	.15803	.19086	.23762
							1.03586	1.17260	1.23662	1.29755	1.32679	1.45999	1.56906
.108	2.0749	.4095	.62279	.22444	5.13458	1.36149	.11261	.13204	.14249	.15359	.15941	.19258	.24016
							1.03563	1.17146	1.23502	1.29551	1.32452	1.45663	1.56459
.109	2.0684	.4110	.62543	.22326	5.10092	1.35686	.11364	.13321	.14374	.15492	.16079	.19429	.24273
							1.03541	1.17033	1.23345	1.29349	1.32228	1.45331	1.56018
.110	2.0620	.4125	.62805	.22209	5.06769	1.35231	.11467	.13437	.14498	.15625	.16217	.19601	.24534
							1.03519	1.16922	1.23190	1.29150	1.32008	1.45003	1.55582
.111	2.0556	.4140	.63066	.22092	5.03488	1.34783	.11570	.13554	.14622	.15758	.16356	.19774	.24800
							1.03497	1.16812	1.23036	1.28954	1.31790	1.44680	1.55151
.112	2.0493	.4155	.63325	.21975	5.00249	1.34341	.11673	.13670	.14747	.15892	.16494	.19947	.25069
							1.03475	1.16704	1.22885	1.28760	1.31575	1.44361	1.54725
.113	2.0430	.4170	.63582	.21859	4.97049	1.33907	.11776	.13787	.14871	.16025	.16633	.20120	.25344
							1.03454	1.16597	1.22736	1.28568	1.31362	1.44045	1.54304
.114	2.0368	.4184	.63837	.21744	4.93889	1.33479	.11879	.13903	.14996	.16159	.16771	.20294	.25623
							1.03433	1.16491	1.22588	1.28379	1.31153	1.43734	1.53888
.115	2.0307	.4199	.64091	.21629	4.90768	1.33058	.11982	.14020	.15120	.16292	.16910	.20468	.25909
							1.03412	1.16387	1.22442	1.28193	1.30946	1.43426	1.53476
.116	2.0246	.4214	.64343	.21515	4.87684	1.32643	.12085	.14137	.15245	.16426	.17049	.20644	.26202
							1.03391	1.16284	1.22298	1.28008	1.30741	1.43124	1.53071
.117	2.0186	.4228	.64593	.21401	4.84638	1.32235	.12188	.14253	.15370	.16560	.17188	.20820	.26502
							1.03371	1.16183	1.22159	1.27826	1.30539	1.42824	1.52668
.118	2.0126	.4243	.64842	.21288	4.81627	1.31833	.12291	.14370	.15495	.16694	.17327	.20996	.26809
							1.03351	1.16082	1.22019	1.27646	1.30339	1.42528	1.52269
.119	2.0067	.4257	.65089	.21175	4.78652	1.31436	.12394	.14487	.15620	.16828	.17466	.21173	.27125
							1.03331	1.15983	1.21880	1.27468	1.30142	1.42235	1.51875

TABLE X. Specific Heat Ratio = 1.67

p/p_t	MACH	T/T_t	GAMMA	K_crit	SHOCK P_y/P_x	P_ty/P_ty	.01	.05	.07	.09	.1	.15	.2
.120	2.0009	.4271	.65334	.21063	4.75713	1.31046	.12497	.14603	.15744	.16962	.17606	.21350	.27452
							1.03311	1.15885	1.21743	1.27292	1.29947	1.41945	1.51484
.121	1.9950	.4286	.65578	.20951	4.72807	1.30661	.12600	.14720	.15869	.17096	.17745	.21529	.27791
							1.03292	1.15788	1.21607	1.27119	1.29755	1.41659	1.51098
.122	1.9893	.4300	.65820	.20840	4.69934	1.30282	.12703	.14837	.15994	.17231	.17885	.21707	.28144
							1.03273	1.15692	1.21473	1.26947	1.29564	1.41376	1.50714
.123	1.9836	.4314	.66061	.20729	4.67095	1.29909	.12806	.14954	.16119	.17365	.18025	.21887	.28514
							1.03254	1.15597	1.21341	1.26778	1.29376	1.41096	1.50334
.124	1.9779	.4328	.66300	.20619	4.64287	1.29541	.12909	.15070	.16244	.17500	.18165	.22067	.28905
							1.03235	1.15503	1.21210	1.26610	1.29190	1.40819	1.49957
.125	1.9723	.4342	.66537	.20509	4.61511	1.29178	.13012	.15187	.16369	.17634	.18305	.22248	.29322
							1.03216	1.15411	1.21081	1.26444	1.29007	1.40546	1.49583
.126	1.9667	.4356	.66773	.20400	4.58767	1.28821	.13115	.15304	.16494	.17769	.18445	.22430	.29775
							1.03198	1.15319	1.20953	1.26280	1.28825	1.40275	1.49212
.127	1.9612	.4370	.67008	.20291	4.56052	1.28468	.13218	.15421	.16620	.17904	.18586	.22612	.30284
							1.03180	1.15229	1.20827	1.26118	1.28645	1.40007	1.48844
.128	1.9557	.4383	.67241	.20182	4.53367	1.28121	.13321	.15538	.16745	.18039	.18726	.22795	.30874
							1.03162	1.15139	1.20702	1.25958	1.28467	1.39743	1.48478
.129	1.9503	.4397	.67473	.20075	4.50712	1.27778	.13424	.15655	.16870	.18174	.18867	.22979	.31651
							1.03144	1.15051	1.20578	1.25800	1.28291	1.39481	1.48112
.130	1.9449	.4411	.67703	.19967	4.48085	1.27440	.13527	.15772	.16996	.18310	.19008	.23164	
							1.03127	1.14963	1.20456	1.25643	1.28118	1.39222	
.131	1.9395	.4424	.67931	.19860	4.45486	1.27107	.13630	.15889	.17121	.18445	.19149	.23350	
							1.03109	1.14877	1.20335	1.25488	1.27946	1.38965	
.132	1.9342	.4438	.68159	.19753	4.42915	1.26779	.13733	.16006	.17247	.18581	.19291	.23537	
							1.03092	1.14791	1.20215	1.25335	1.27775	1.38711	
.133	1.9290	.4451	.68385	.19647	4.40371	1.26455	.13836	.16123	.17372	.18716	.19432	.23725	
							1.03075	1.14706	1.20097	1.25183	1.27607	1.38460	
.134	1.9237	.4465	.68609	.19541	4.37854	1.26135	.13940	.16240	.17498	.18852	.19574	.23914	
							1.03058	1.14622	1.19979	1.25033	1.27441	1.38211	
.135	1.9186	.4478	.68832	.19436	4.35363	1.25820	.14043	.16357	.17624	.18988	.19716	.24104	
							1.03042	1.14539	1.19864	1.24884	1.27276	1.37965	
.136	1.9134	.4491	.69054	.19331	4.32897	1.25509	.14146	.16475	.17750	.19124	.19858	.24294	
							1.03025	1.14457	1.19749	1.24737	1.27113	1.37721	
.137	1.9083	.4505	.69274	.19227	4.30457	1.25202	.14249	.16592	.17876	.19261	.20001	.24487	
							1.03009	1.14376	1.19635	1.24591	1.26951	1.37480	
.138	1.9033	.4518	.69493	.19123	4.28042	1.24899	.14352	.16709	.18002	.19397	.20143	.24680	
							1.02993	1.14295	1.19523	1.24447	1.26791	1.37241	
.139	1.8982	.4531	.69711	.19019	4.25651	1.24601	.14455	.16827	.18128	.19534	.20286	.24874	
							1.02977	1.14216	1.19412	1.24305	1.26633	1.37004	
.140	1.8932	.4544	.69927	.18916	4.23284	1.24306	.14558	.16944	.18254	.19671	.20429	.25070	
							1.02961	1.14137	1.19302	1.24164	1.26476	1.36770	
.141	1.8883	.4557	.70142	.18813	4.20941	1.24015	.14661	.17061	.18381	.19808	.20572	.25267	
							1.02945	1.14059	1.19193	1.24024	1.26321	1.36537	
.142	1.8834	.4570	.70356	.18711	4.18621	1.23728	.14764	.17179	.18507	.19945	.20716	.25466	
							1.02930	1.13982	1.19085	1.23886	1.26168	1.36307	
.143	1.8785	.4583	.70568	.18608	4.16324	1.23445	.14867	.17297	.18633	.20082	.20860	.25665	
							1.02914	1.13905	1.18978	1.23749	1.26016	1.36079	
.144	1.8736	.4596	.70779	.18507	4.14049	1.23165	.14970	.17414	.18760	.20220	.21004	.25867	
							1.02899	1.13830	1.18872	1.23613	1.25865	1.35853	
.145	1.8688	.4608	.70989	.18406	4.11796	1.22889	.15074	.17532	.18887	.20358	.21148	.26070	
							1.02884	1.13755	1.18767	1.23479	1.25716	1.35629	
.146	1.8640	.4621	.71198	.18305	4.09565	1.22617	.15177	.17650	.19014	.20496	.21293	.26274	
							1.02869	1.13680	1.18664	1.23346	1.25569	1.35408	
.147	1.8593	.4634	.71405	.18204	4.07355	1.22348	.15280	.17767	.19141	.20634	.21437	.26481	
							1.02855	1.13607	1.18561	1.23214	1.25422	1.35188	
.148	1.8546	.4646	.71611	.18104	4.05166	1.22082	.15383	.17885	.19268	.20772	.21582	.26689	
							1.02840	1.13534	1.18459	1.23083	1.25277	1.34969	
.149	1.8499	.4659	.71816	.18004	4.02997	1.21820	.15486	.18003	.19395	.20911	.21728	.26899	
							1.02826	1.13462	1.18356	1.22954	1.25134	1.34753	

TABLE X. Specific Heat Ratio = 1.67

p/pₜ	MACH	T/Tₜ	GAMMA	K_crit	SHOCK P_y/P_x	P_tx/P_ty	.01	.03	.05	.07	.09	.1	.15
.150	1.8453	.4671	.72020	.17905	4.00849	1.21561	.15589	.16816	.18121	.19522	.21049	.21873	.27111
							1.02811	1.08235	1.13390	1.18256	1.22826	1.24991	1.34539
.151	1.8406	.4684	.72222	.17806	3.98721	1.21305	.15692	.16927	.18239	.19650	.21188	.22019	.27325
							1.02797	1.08195	1.13319	1.18157	1.22699	1.24851	1.34326
.152	1.8361	.4696	.72423	.17707	3.96613	1.21053	.15796	.17037	.18357	.19777	.21328	.22166	.27541
							1.02783	1.08153	1.13249	1.18059	1.22573	1.24711	1.34115
.153	1.8315	.4709	.72623	.17609	3.94524	1.20804	.15899	.17147	.18475	.19905	.21467	.22312	.27760
							1.02769	1.08111	1.13180	1.17962	1.22449	1.24573	1.33906
.154	1.8270	.4721	.72822	.17511	3.92454	1.20557	.16002	.17257	.18593	.20033	.21607	.22459	.27981
							1.02755	1.08070	1.13111	1.17866	1.22325	1.24435	1.33699
.155	1.8225	.4733	.73020	.17414	3.90403	1.20314	.16105	.17367	.18712	.20160	.21747	.22606	.28204
							1.02741	1.08029	1.13043	1.17771	1.22203	1.24299	1.33493
.156	1.8180	.4746	.73217	.17317	3.88370	1.20074	.16208	.17477	.18830	.20289	.21887	.22754	.28431
							1.02728	1.07988	1.12975	1.17676	1.22081	1.24165	1.33288
.157	1.8136	.4758	.73412	.17220	3.86355	1.19837	.16311	.17588	.18948	.20417	.22027	.22902	.28660
							1.02714	1.07948	1.12908	1.17583	1.21961	1.24031	1.33086
.158	1.8092	.4770	.73606	.17123	3.84358	1.19603	.16415	.17698	.19067	.20545	.22168	.23050	.28892
							1.02701	1.07908	1.12841	1.17490	1.21842	1.23899	1.32885
.159	1.8048	.4782	.73800	.17027	3.82379	1.19371	.16518	.17808	.19185	.20673	.22309	.23198	.29128
							1.02688	1.07868	1.12776	1.17398	1.21724	1.23767	1.32685
.160	1.8005	.4794	.73992	.16931	3.80418	1.19142	.16621	.17919	.19304	.20802	.22450	.23347	.29368
							1.02675	1.07829	1.12710	1.17307	1.21607	1.23637	1.32486
.161	1.7961	.4806	.74183	.16836	3.78473	1.18917	.16724	.18029	.19423	.20931	.22591	.23497	.29611
							1.02662	1.07790	1.12646	1.17216	1.21490	1.23508	1.32289
.162	1.7918	.4818	.74372	.16741	3.76545	1.18693	.16828	.18139	.19542	.21060	.22733	.23646	.29858
							1.02649	1.07751	1.12581	1.17126	1.21375	1.23380	1.32094
.163	1.7876	.4830	.74561	.16646	3.74634	1.18473	.16931	.18250	.19660	.21189	.22875	.23797	.30110
							1.02636	1.07713	1.12518	1.17038	1.21261	1.23253	1.31900
.164	1.7833	.4842	.74749	.16552	3.72739	1.18255	.17034	.18360	.19779	.21318	.23017	.23947	.30366
							1.02624	1.07675	1.12455	1.16949	1.21148	1.23127	1.31707
.165	1.7791	.4854	.74935	.16457	3.70860	1.18040	.17137	.18471	.19898	.21447	.23160	.24098	.30630
							1.02611	1.07638	1.12392	1.16862	1.21035	1.23003	1.31516
.166	1.7749	.4865	.75121	.16364	3.68998	1.17827	.17241	.18581	.20017	.21577	.23303	.24249	.30898
							1.02599	1.07600	1.12330	1.16775	1.20924	1.22879	1.31326
.167	1.7708	.4877	.75305	.16270	3.67151	1.17617	.17344	.18692	.20137	.21706	.23446	.24401	.31173
							1.02586	1.07563	1.12269	1.16689	1.20814	1.22756	1.31137
.168	1.7666	.4889	.75489	.16177	3.65319	1.17409	.17447	.18802	.20256	.21836	.23589	.24553	.31455
							1.02574	1.07527	1.12208	1.16604	1.20704	1.22634	1.30949
.169	1.7625	.4900	.75671	.16084	3.63502	1.17204	.17550	.18913	.20375	.21966	.23733	.24706	.31745
							1.02562	1.07490	1.12147	1.16520	1.20595	1.22513	1.30762
.170	1.7584	.4912	.75852	.15992	3.61701	1.17001	.17654	.19024	.20495	.22096	.23878	.24859	.32043
							1.02550	1.07454	1.12087	1.16436	1.20488	1.22393	1.30576
.171	1.7544	.4924	.76032	.15900	3.59914	1.16800	.17757	.19134	.20614	.22227	.24022	.25013	.32353
							1.02538	1.07419	1.12028	1.16353	1.20381	1.22274	1.30391
.172	1.7503	.4935	.76212	.15808	3.58142	1.16602	.17860	.19245	.20734	.22357	.24167	.25167	.32674
							1.02526	1.07383	1.11969	1.16270	1.20274	1.22156	1.30207
.173	1.7463	.4947	.76390	.15716	3.56384	1.16406	.17964	.19356	.20853	.22488	.24312	.25321	.33009
							1.02514	1.07348	1.11910	1.16188	1.20169	1.22039	1.30023
.174	1.7423	.4958	.76567	.15625	3.54641	1.16212	.18067	.19467	.20973	.22619	.24458	.25477	.33361
							1.02503	1.07313	1.11852	1.16107	1.20065	1.21923	1.29841
.175	1.7383	.4969	.76743	.15534	3.52911	1.16021	.18170	.19578	.21093	.22750	.24604	.25632	.33733
							1.02491	1.07279	1.11795	1.16026	1.19961	1.21808	1.29659
.176	1.7344	.4981	.76919	.15444	3.51195	1.15831	.18274	.19689	.21213	.22881	.24750	.25789	.34131
							1.02480	1.07244	1.11738	1.15947	1.19858	1.21693	1.29477
.177	1.7304	.4992	.77093	.15354	3.49493	1.15644	.18377	.19800	.21333	.23013	.24897	.25945	.34567
							1.02468	1.07210	1.11681	1.15867	1.19756	1.21579	1.29297
.178	1.7265	.5003	.77266	.15264	3.47804	1.15459	.18481	.19911	.21453	.23144	.25044	.26103	.35049
							1.02457	1.07176	1.11625	1.15788	1.19655	1.21467	1.29117
.179	1.7227	.5015	.77438	.15174	3.46129	1.15277	.18584	.20022	.21573	.23276	.25192	.26262	.35607
							1.02446	1.07143	1.11569	1.15710	1.19554	1.21356	1.28936

183

TABLE X. Specific Heat Ratio = 1.67

p/p_t	MACH	T/T_t	GAMMA	K_{crit}	SHOCK P_y/P_x	P_{tx}/P_{ty}	.01	.03	.05	.07	.09	.1	.15
.180	1.7188	.5026	.77609	.15085	3.44466	1.15096	.18687	.20133	.21694	.23408	.25339	.26420	.36318
							1.02435	1.07110	1.11513	1.15633	1.19454	1.21245	1.28755
.181	1.7150	.5037	.77780	.14995	3.42817	1.14917	.18791	.20244	.21814	.23541	.25488	.26580	
							1.02424	1.07077	1.11458	1.15556	1.19355	1.21135	
.182	1.7111	.5048	.77949	.14907	3.41180	1.14740	.18894	.20355	.21935	.23673	.25637	.26740	
							1.02413	1.07044	1.11404	1.15480	1.19257	1.21025	
.183	1.7073	.5059	.78117	.14818	3.39555	1.14566	.18998	.20466	.22056	.23806	.25786	.26900	
							1.02402	1.07011	1.11350	1.15404	1.19159	1.20916	
.184	1.7036	.5070	.78285	.14730	3.37944	1.14393	.19101	.20578	.22177	.23939	.25936	.27061	
							1.02391	1.06979	1.11296	1.15329	1.19062	1.20808	
.185	1.6998	.5081	.78451	.14642	3.36344	1.14222	.19205	.20689	.22297	.24072	.26086	.27223	
							1.02381	1.06947	1.11243	1.15254	1.18966	1.20701	
.186	1.6961	.5092	.78617	.14555	3.34756	1.14053	.19308	.20801	.22419	.24206	.26237	.27386	
							1.02370	1.06915	1.11190	1.15180	1.18870	1.20595	
.187	1.6923	.5103	.78782	.14467	3.33181	1.13886	.19412	.20912	.22540	.24339	.26388	.27550	
							1.02359	1.06884	1.11137	1.15106	1.18775	1.20489	
.188	1.6887	.5114	.78945	.14380	3.31617	1.13721	.19515	.21023	.22661	.24473	.26540	.27714	
							1.02349	1.06853	1.11085	1.15033	1.18681	1.20384	
.189	1.6850	.5125	.79108	.14294	3.30064	1.13557	.19619	.21135	.22782	.24607	.26692	.27879	
							1.02339	1.06821	1.11033	1.14960	1.18587	1.20279	
.190	1.6813	.5136	.79270	.14207	3.28523	1.13396	.19722	.21247	.22904	.24742	.26845	.28045	
							1.02328	1.06791	1.10982	1.14888	1.18494	1.20176	
.191	1.6777	.5147	.79431	.14121	3.26994	1.13236	.19826	.21358	.23026	.24877	.26998	.28211	
							1.02318	1.06760	1.10931	1.14817	1.18402	1.20073	
.192	1.6741	.5158	.79591	.14035	3.25476	1.13078	.19929	.21470	.23147	.25012	.27152	.28379	
							1.02308	1.06730	1.10880	1.14746	1.18310	1.19970	
.193	1.6705	.5169	.79750	.13950	3.23968	1.12922	.20033	.21582	.23269	.25147	.27307	.28547	
							1.02298	1.06699	1.10830	1.14675	1.18219	1.19868	
.194	1.6669	.5179	.79909	.13864	3.22472	1.12767	.20136	.21694	.23391	.25282	.27462	.28717	
							1.02288	1.06670	1.10780	1.14605	1.18129	1.19767	
.195	1.6633	.5190	.80066	.13779	3.20986	1.12614	.20240	.21806	.23514	.25418	.27618	.28887	
							1.02278	1.06640	1.10730	1.14536	1.18039	1.19667	
.196	1.6598	.5201	.80223	.13695	3.19511	1.12463	.20344	.21918	.23636	.25554	.27774	.29059	
							1.02268	1.06610	1.10681	1.14467	1.17949	1.19567	
.197	1.6562	.5211	.80378	.13610	3.18047	1.12314	.20447	.22030	.23758	.25691	.27931	.29231	
							1.02258	1.06581	1.10632	1.14398	1.17860	1.19468	
.198	1.6527	.5222	.80533	.13526	3.16593	1.12166	.20551	.22142	.23881	.25828	.28089	.29405	
							1.02249	1.06552	1.10584	1.14330	1.17772	1.19369	
.199	1.6492	.5232	.80687	.13442	3.15149	1.12020	.20655	.22254	.24004	.25965	.28248	.29580	
							1.02239	1.06523	1.10535	1.14262	1.17685	1.19271	
.200	1.6457	.5243	.80840	.13358	3.13715	1.11875	.20758	.22366	.24127	.26102	.28407	.29755	
							1.02229	1.06494	1.10487	1.14195	1.17597	1.19174	
.201	1.6423	.5253	.80993	.13275	3.12291	1.11732	.20862	.22478	.24250	.26240	.28567	.29932	
							1.02220	1.06466	1.10440	1.14128	1.17511	1.19077	
.202	1.6388	.5264	.81144	.13192	3.10877	1.11590	.20966	.22591	.24373	.26378	.28727	.30111	
							1.02210	1.06437	1.10393	1.14062	1.17425	1.18981	
.203	1.6354	.5274	.81295	.13109	3.09473	1.11450	.21069	.22703	.24497	.26516	.28889	.30290	
							1.02201	1.06409	1.10346	1.13996	1.17339	1.18885	
.204	1.6320	.5285	.81444	.13026	3.08079	1.11311	.21173	.22816	.24620	.26655	.29051	.30471	
							1.02192	1.06381	1.10299	1.13930	1.17254	1.18790	
.205	1.6286	.5295	.81593	.12944	3.06694	1.11174	.21277	.22928	.24744	.26794	.29214	.30654	
							1.02183	1.06353	1.10253	1.13865	1.17170	1.18695	
.206	1.6252	.5305	.81741	.12862	3.05319	1.11039	.21381	.23041	.24868	.26933	.29378	.30838	
							1.02173	1.06326	1.10207	1.13800	1.17086	1.18601	
.207	1.6219	.5316	.81888	.12780	3.03953	1.10905	.21485	.23153	.24992	.27073	.29543	.31023	
							1.02164	1.06298	1.10161	1.13736	1.17002	1.18508	
.208	1.6185	.5326	.82035	.12699	3.02596	1.10772	.21588	.23266	.25116	.27213	.29709	.31210	
							1.02155	1.06271	1.10115	1.13672	1.16919	1.18414	
.209	1.6152	.5336	.82180	.12618	3.01248	1.10641	.21692	.23379	.25241	.27354	.29876	.31399	
							1.02146	1.06244	1.10070	1.13608	1.16837	1.18322	

TABLE X. Specific Heat Ratio = 1.67

p/p,	MACH	T/T,	GAMMA	K_crit	SHOCK P_y/P_x	P_tx/P_ty	.01	.03	.05	.07	.08	.09	.1
.210	1.6119	.5347	.82325	.12537	2.99909	1.10511	.21796	.23492	.25365	.27495	.28701	.30044	.31589
							1.02137	1.06217	1.10025	1.13545	1.15190	1.16755	1.18230
.211	1.6086	.5357	.82469	.12456	2.98579	1.10382	.21900	.23605	.25490	.27636	.28854	.30213	.31782
							1.02128	1.06191	1.09981	1.13483	1.15118	1.16673	1.18138
.212	1.6053	.5367	.82612	.12375	2.97258	1.10255	.22004	.23718	.25615	.27778	.29007	.30382	.31976
							1.02120	1.06164	1.09937	1.13420	1.15047	1.16592	1.18047
.213	1.6020	.5377	.82755	.12295	2.95946	1.10129	.22108	.23831	.25740	.27921	.29162	.30553	.32173
							1.02111	1.06138	1.09893	1.13358	1.14975	1.16511	1.17956
.214	1.5987	.5387	.82896	.12215	2.94642	1.10005	.22212	.23944	.25865	.28063	.29317	.30726	.32371
							1.02102	1.06112	1.09849	1.13297	1.14905	1.16431	1.17866
.215	1.5955	.5397	.83037	.12136	2.93347	1.09882	.22316	.24058	.25991	.28207	.29473	.30899	.32572
							1.02093	1.06086	1.09805	1.13235	1.14834	1.16351	1.17776
.216	1.5923	.5407	.83177	.12056	2.92060	1.09760	.22420	.24171	.26117	.28350	.29629	.31074	.32775
							1.02085	1.06060	1.09762	1.13174	1.14764	1.16271	1.17686
.217	1.5891	.5417	.83316	.11977	2.90781	1.09639	.22524	.24285	.26243	.28494	.29786	.31249	.32981
							1.02076	1.06034	1.09719	1.13114	1.14694	1.16192	1.17597
.218	1.5859	.5427	.83455	.11898	2.89511	1.09520	.22628	.24398	.26369	.28639	.29945	.31427	.33190
							1.02068	1.06009	1.09677	1.13054	1.14625	1.16114	1.17508
.219	1.5827	.5437	.83593	.11819	2.88249	1.09402	.22732	.24512	.26495	.28784	.30103	.31605	.33401
							1.02061	1.05983	1.09634	1.12994	1.14556	1.16036	1.17420
.220	1.5795	.5447	.83730	.11741	2.86995	1.09285	.22836	.24625	.26622	.28930	.30263	.31785	.33615
							1.02053	1.05958	1.09592	1.12934	1.14488	1.15958	1.17332
.221	1.5763	.5457	.83866	.11663	2.85749	1.09170	.22940	.24739	.26749	.29076	.30424	.31967	.33833
							1.02044	1.05933	1.09550	1.12875	1.14420	1.15880	1.17245
.222	1.5732	.5467	.84001	.11585	2.84511	1.09055	.23044	.24853	.26876	.29222	.30585	.32150	.34054
							1.02036	1.05908	1.09508	1.12816	1.14352	1.15803	1.17157
.223	1.5701	.5477	.84136	.11507	2.83280	1.08942	.23148	.24967	.27003	.29370	.30747	.32335	.34279
							1.02028	1.05884	1.09467	1.12758	1.14285	1.15726	1.17070
.224	1.5670	.5487	.84270	.11430	2.82057	1.08830	.23252	.25081	.27131	.29518	.30911	.32521	.34508
							1.02020	1.05859	1.09426	1.12700	1.14218	1.15650	1.16984
.225	1.5639	.5497	.84403	.11353	2.80842	1.08720	.23356	.25196	.27258	.29666	.31075	.32710	.34742
							1.02012	1.05835	1.09385	1.12642	1.14151	1.15574	1.16897
.226	1.5608	.5506	.84535	.11276	2.79635	1.08610	.23461	.25310	.27386	.29815	.31240	.32900	.34980
							1.02003	1.05810	1.09344	1.12584	1.14084	1.15498	1.16811
.227	1.5577	.5516	.84667	.11199	2.78435	1.08502	.23565	.25424	.27515	.29965	.31407	.33092	.35223
							1.01995	1.05786	1.09304	1.12527	1.14018	1.15423	1.16726
.228	1.5546	.5526	.84798	.11123	2.77242	1.08395	.23669	.25539	.27643	.30115	.31574	.33287	.35473
							1.01987	1.05762	1.09264	1.12470	1.13953	1.15348	1.16640
.229	1.5516	.5536	.84928	.11047	2.76056	1.08288	.23773	.25653	.27772	.30266	.31743	.33484	.35728
							1.01979	1.05738	1.09224	1.12413	1.13887	1.15273	1.16555
.230	1.5485	.5545	.85058	.10971	2.74878	1.08183	.23878	.25768	.27901	.30418	.31913	.33683	.35990
							1.01972	1.05715	1.09184	1.12357	1.13822	1.15198	1.16470
.231	1.5455	.5555	.85186	.10895	2.73707	1.08080	.23982	.25883	.28031	.30570	.32084	.33884	.36260
							1.01964	1.05691	1.09144	1.12301	1.13757	1.15124	1.16385
.232	1.5425	.5565	.85314	.10820	2.72543	1.07977	.24086	.25998	.28160	.30723	.32256	.34089	.36539
							1.01956	1.05668	1.09105	1.12245	1.13693	1.15050	1.16300
.233	1.5395	.5574	.85442	.10744	2.71386	1.07875	.24191	.26113	.28290	.30877	.32430	.34296	.36828
							1.01948	1.05644	1.09066	1.12189	1.13629	1.14977	1.16215
.234	1.5365	.5584	.85568	.10669	2.70236	1.07774	.24295	.26228	.28420	.31031	.32605	.34506	.37132
							1.01940	1.05621	1.09027	1.12134	1.13565	1.14904	1.16132
.235	1.5335	.5593	.85694	.10595	2.69093	1.07675	.24399	.26343	.28551	.31187	.32781	.34719	.37447
							1.01933	1.05598	1.08988	1.12079	1.13501	1.14830	1.16047
.236	1.5306	.5603	.85819	.10520	2.67957	1.07576	.24504	.26459	.28682	.31343	.32959	.34936	.37778
							1.01925	1.05575	1.08950	1.12024	1.13438	1.14758	1.15963
.237	1.5276	.5612	.85944	.10446	2.66827	1.07479	.24608	.26574	.28813	.31500	.33139	.35156	.38129
							1.01918	1.05552	1.08911	1.11970	1.13375	1.14685	1.15879
.238	1.5247	.5622	.86067	.10372	2.65704	1.07382	.24713	.26690	.28944	.31658	.33320	.35381	.38505
							1.01910	1.05530	1.08873	1.11916	1.13312	1.14613	1.15794
.239	1.5217	.5631	.86190	.10298	2.64588	1.07286	.24817	.26805	.29076	.31817	.33503	.35610	.38915
							1.01903	1.05507	1.08835	1.11862	1.13249	1.14539	1.15710

TABLE X. Specific Heat Ratio = 1.67

p/p_1	MACH	T/T_1	GAMMA	K_{crit}	SHOCK p_y/p_x	p_{tx}/p_{ty}	.01	.03	.05	.07	.08	.09	.1
.240	1.5188	.5641	.86313	.10225	2.63478	1.07192	.24922	.26921	.29208	.31977	.33687	.35843	.39369
							1.01895	1.05485	1.08798	1.11808	1.13187	1.14467	1.15625
.241	1.5159	.5650	.86434	.10152	2.62374	1.07098	.25026	.27037	.29341	.32137	.33874	.36082	.39900
							1.01888	1.05463	1.08760	1.11754	1.13125	1.14395	1.15540
.242	1.5130	.5660	.86555	.10078	2.61277	1.07006	.25131	.27153	.29473	.32299	.34063	.36326	.40562
							1.01880	1.05440	1.08723	1.11701	1.13063	1.14324	1.15454
.243	1.5101	.5669	.86676	.10006	2.60187	1.06914	.25236	.27270	.29607	.32462	.34253	.36576	.41728
							1.01873	1.05418	1.08686	1.11648	1.13001	1.14252	1.15366
.244	1.5073	.5678	.86795	.09933	2.59102	1.06824	.25340	.27386	.29740	.32626	.34446	.36833	
							1.01866	1.05397	1.08649	1.11595	1.12940	1.14181	
.245	1.5044	.5688	.86914	.09861	2.58024	1.06734	.25445	.27503	.29874	.32791	.34642	.37098	
							1.01859	1.05375	1.08612	1.11543	1.12879	1.14110	
.246	1.5016	.5697	.87032	.09789	2.56952	1.06645	.25550	.27619	.30008	.32957	.34840	.37371	
							1.01851	1.05353	1.08576	1.11491	1.12818	1.14038	
.247	1.4987	.5706	.87150	.09717	2.55886	1.06558	.25655	.27736	.30143	.33124	.35040	.37654	
							1.01844	1.05332	1.08539	1.11438	1.12757	1.13967	
.248	1.4959	.5716	.87267	.09645	2.54826	1.06471	.25759	.27853	.30278	.33293	.35244	.37953	
							1.01837	1.05310	1.08503	1.11387	1.12696	1.13897	
.249	1.4931	.5725	.87383	.09574	2.53772	1.06385	.25864	.27970	.30414	.33463	.35450	.38261	
							1.01830	1.05289	1.08467	1.11335	1.12636	1.13826	
.250	1.4903	.5734	.87499	.09503	2.52724	1.06300	.25969	.28087	.30550	.33634	.35660	.38586	
							1.01823	1.05268	1.08431	1.11283	1.12576	1.13755	
.251	1.4875	.5743	.87613	.09432	2.51682	1.06216	.26074	.28205	.30686	.33807	.35873	.38930	
							1.01816	1.05246	1.08396	1.11232	1.12516	1.13684	
.252	1.4847	.5752	.87728	.09361	2.50646	1.06132	.26179	.28322	.30823	.33981	.36089	.39298	
							1.01809	1.05225	1.08360	1.11181	1.12456	1.13613	
.253	1.4819	.5761	.87841	.09291	2.49615	1.06050	.26284	.28440	.30960	.34157	.36310	.39698	
							1.01802	1.05205	1.08325	1.11130	1.12396	1.13542	
.254	1.4791	.5771	.87954	.09221	2.48590	1.05969	.26389	.28558	.31098	.34334	.36535	.40142	
							1.01795	1.05184	1.08290	1.11079	1.12336	1.13470	
.255	1.4764	.5780	.88066	.09151	2.47571	1.05888	.26494	.28676	.31236	.34514	.36765	.40658	
							1.01788	1.05163	1.08255	1.11029	1.12277	1.13400	
.256	1.4736	.5789	.88178	.09081	2.46557	1.05808	.26599	.28794	.31375	.34695	.37000	.41296	
							1.01782	1.05143	1.08220	1.10978	1.12217	1.13327	
.257	1.4709	.5798	.88289	.09012	2.45549	1.05729	.26704	.28912	.31514	.34877	.37240	.42344	
							1.01775	1.05122	1.08185	1.10928	1.12158	1.13253	
.258	1.4682	.5807	.88399	.08942	2.44546	1.05651	.26809	.29031	.31654	.35062	.37487		
							1.01768	1.05102	1.08151	1.10878	1.12099		
.259	1.4654	.5816	.88509	.08873	2.43549	1.05574	.26914	.29150	.31794	.35251	.37740		
							1.01761	1.05082	1.08116	1.10829	1.12040		
.260	1.4627	.5825	.88618	.08805	2.42557	1.05497	.27020	.29269	.31935	.35441	.38001		
							1.01755	1.05061	1.08082	1.10780	1.11980		
.261	1.4600	.5834	.88726	.08736	2.41571	1.05422	.27125	.29388	.32076	.35632	.38271		
							1.01748	1.05041	1.08048	1.10730	1.11921		
.262	1.4574	.5843	.88834	.08668	2.40589	1.05347	.27230	.29507	.32218	.35827	.38555		
							1.01741	1.05021	1.08014	1.10681	1.11863		
.263	1.4547	.5852	.88941	.08600	2.39613	1.05273	.27336	.29627	.32361	.36024	.38847		
							1.01735	1.05002	1.07981	1.10632	1.11804		
.264	1.4520	.5861	.89048	.08532	2.38643	1.05200	.27441	.29746	.32504	.36223	.39153		
							1.01728	1.04982	1.07947	1.10582	1.11745		
.265	1.4493	.5870	.89154	.08464	2.37677	1.05127	.27546	.29866	.32648	.36426	.39476		
							1.01722	1.04962	1.07913	1.10533	1.11686		
.266	1.4467	.5878	.89259	.08397	2.36716	1.05055	.27652	.29986	.32793	.36632	.39818		
							1.01715	1.04943	1.07880	1.10485	1.11627		
.267	1.4440	.5887	.89363	.08330	2.35761	1.04984	.27757	.30106	.32938	.36842	.40186		
							1.01709	1.04923	1.07847	1.10436	1.11567		
.268	1.4414	.5896	.89467	.08263	2.34810	1.04914	.27863	.30227	.33084	.37055	.40587		
							1.01703	1.04904	1.07814	1.10387	1.11508		
.269	1.4388	.5905	.89571	.08196	2.33864	1.04845	.27969	.30348	.33230	.37272	.41035		
							1.01696	1.04884	1.07781	1.10339	1.11448		

TABLE X. Specific Heat Ratio = 1.67

p/p_t	MACH	T/T_t	GAMMA	K_{crit}	SHOCK p_y/p_x	p_{tx}/p_{ty}	.01	.03	.04	.05	.06	.07	.08
.270	1.4362	.5914	.89674	.08130	2.32924	1.04776	.28074	.30469	.31837	.33378	.35187	.37493	.41564
							1.01690	1.04865	1.06346	1.07748	1.09067	1.10290	1.11388
.271	1.4336	.5923	.89776	.08063	2.31988	1.04708	.28180	.30590	.31969	.33527	.35361	.37720	.42234
							1.01684	1.04846	1.06320	1.07716	1.09027	1.10241	1.11327
.272	1.4310	.5931	.89877	.07997	2.31057	1.04641	.28286	.30711	.32102	.33676	.35536	.37951	
							1.01677	1.04827	1.06294	1.07683	1.08987	1.10193	
.273	1.4284	.5940	.89978	.07932	2.30131	1.04574	.28391	.30833	.32235	.33825	.35713	.38188	
							1.01671	1.04808	1.06269	1.07651	1.08947	1.10144	
.274	1.4258	.5949	.90078	.07866	2.29209	1.04509	.28497	.30955	.32369	.33976	.35892	.38432	
							1.01665	1.04789	1.06243	1.07618	1.08907	1.10096	
.275	1.4232	.5957	.90178	.07801	2.28292	1.04444	.28603	.31077	.32503	.34128	.36073	.38683	
							1.01659	1.04771	1.06218	1.07586	1.08868	1.10047	
.276	1.4207	.5966	.90277	.07736	2.27380	1.04379	.28709	.31199	.32638	.34280	.36256	.38941	
							1.01653	1.04752	1.06193	1.07554	1.08828	1.09999	
.277	1.4181	.5975	.90376	.07671	2.26473	1.04316	.28815	.31322	.32773	.34434	.36441	.39213	
							1.01646	1.04733	1.06168	1.07522	1.08789	1.09952	
.278	1.4156	.5983	.90474	.07606	2.25570	1.04253	.28921	.31445	.32909	.34588	.36629	.39492	
							1.01640	1.04715	1.06143	1.07490	1.08750	1.09903	
.279	1.4130	.5992	.90571	.07542	2.24672	1.04190	.29027	.31568	.33045	.34744	.36819	.39783	
							1.01634	1.04696	1.06118	1.07459	1.08710	1.09855	
.280	1.4105	.6001	.90668	.07478	2.23778	1.04129	.29133	.31691	.33181	.34900	.37012	.40088	
							1.01628	1.04678	1.06093	1.07427	1.08671	1.09807	
.281	1.4080	.6009	.90764	.07414	2.22888	1.04068	.29240	.31815	.33319	.35058	.37207	.40411	
							1.01622	1.04660	1.06068	1.07396	1.08632	1.09758	
.282	1.4055	.6018	.90859	.07350	2.22003	1.04008	.29346	.31939	.33456	.35217	.37406	.40756	
							1.01616	1.04642	1.06044	1.07364	1.08593	1.09709	
.283	1.4029	.6026	.90954	.07287	2.21123	1.03948	.29452	.32063	.33595	.35377	.37608	.41128	
							1.01610	1.04623	1.06019	1.07333	1.08554	1.09660	
.284	1.4004	.6035	.91049	.07224	2.20246	1.03889	.29559	.32188	.33734	.35538	.37813	.41537	
							1.01604	1.04605	1.05995	1.07302	1.08515	1.09611	
.285	1.3980	.6043	.91143	.07161	2.19375	1.03831	.29665	.32313	.33873	.35701	.38022	.42006	
							1.01598	1.04587	1.05970	1.07270	1.08477	1.09562	
.286	1.3955	.6052	.91236	.07098	2.18507	1.03773	.29771	.32438	.34014	.35865	.38236	.42565	
							1.01593	1.04569	1.05946	1.07239	1.08438	1.09512	
.287	1.3930	.6060	.91328	.07035	2.17643	1.03716	.29878	.32564	.34154	.36031	.38453	.43329	
							1.01587	1.04552	1.05922	1.07208	1.08399	1.09460	
.288	1.3905	.6069	.91421	.06973	2.16784	1.03660	.29985	.32690	.34296	.36198	.38676		
							1.01581	1.04534	1.05898	1.07177	1.08360		
.289	1.3881	.6077	.91512	.06911	2.15929	1.03604	.30091	.32816	.34438	.36366	.38903		
							1.01575	1.04516	1.05874	1.07147	1.08321		
.290	1.3856	.6086	.91603	.06849	2.15078	1.03549	.30198	.32943	.34581	.36536	.39137		
							1.01569	1.04499	1.05850	1.07116	1.08283		
.291	1.3832	.6094	.91693	.06787	2.14231	1.03494	.30305	.33070	.34725	.36709	.39377		
							1.01564	1.04481	1.05826	1.07085	1.08244		
.292	1.3807	.6103	.91783	.06726	2.13389	1.03440	.30412	.33197	.34869	.36882	.39624		
							1.01558	1.04464	1.05802	1.07055	1.08205		
.293	1.3783	.6111	.91872	.06665	2.12550	1.03387	.30519	.33325	.35015	.37058	.39884		
							1.01552	1.04446	1.05779	1.07024	1.08167		
.294	1.3759	.6119	.91961	.06604	2.11715	1.03334	.30626	.33453	.35161	.37236	.40149		
							1.01547	1.04429	1.05755	1.06994	1.08129		
.295	1.3735	.6128	.92049	.06543	2.10884	1.03282	.30733	.33581	.35308	.37416	.40426		
							1.01541	1.04412	1.05732	1.06963	1.08090		
.296	1.3711	.6136	.92137	.06483	2.10058	1.03231	.30840	.33710	.35456	.37598	.40715		
							1.01535	1.04394	1.05708	1.06933	1.08051		
.297	1.3687	.6144	.92223	.06423	2.09235	1.03180	.30948	.33840	.35605	.37783	.41020		
							1.01531	1.04377	1.05685	1.06903	1.08012		
.298	1.3663	.6153	.92310	.06363	2.08416	1.03130	.31056	.33969	.35754	.37971	.41344		
							1.01526	1.04360	1.05661	1.06872	1.07973		
.299	1.3639	.6161	.92396	.06303	2.07601	1.03080	.31163	.34100	.35905	.38161	.41690		
							1.01520	1.04343	1.05638	1.06842	1.07933		

TABLE X. Specific Heat Ratio = 1.67

					SHOCK		R₂ and TPR versus Loss Coefficient K						
p/p_t	MACH	T/T_t	GAMMA	K_{crit}	P_y/P_x	P_{ty}/P_{tx}	.005	.01	.02	.03	.04	.05	.06
.300	1.3615	.6169	.92481	.06244	2.06789	1.03031	.30622	.31270	.32665	.34231	.36057	.38354	.42068
							1.00766	1.01515	1.02957	1.04327	1.05615	1.06812	1.07894
.301	1.3591	.6177	.92566	.06184	2.05982	1.02982	.30725	.31378	.32782	.34362	.36210	.38550	.42486
							1.00763	1.01509	1.02946	1.04310	1.05592	1.06782	1.07854
.302	1.3568	.6186	.92650	.06125	2.05178	1.02934	.30829	.31486	.32900	.34494	.36364	.38750	.42980
							1.00760	1.01504	1.02935	1.04293	1.05569	1.06752	1.07814
.303	1.3544	.6194	.92734	.06066	2.04378	1.02886	.30933	.31593	.33018	.34626	.36519	.38954	.43590
							1.00758	1.01498	1.02924	1.04276	1.05546	1.06722	1.07773
.304	1.3520	.6202	.92817	.06008	2.03581	1.02839	.31036	.31701	.33136	.34759	.36676	.39161	.44619
							1.00755	1.01493	1.02913	1.04259	1.05523	1.06692	1.07731
.305	1.3497	.6210	.92899	.05950	2.02788	1.02793	.31140	.31809	.33254	.34892	.36834	.39373	
							1.00752	1.01488	1.02903	1.04243	1.05500	1.06662	
.306	1.3474	.6218	.92981	.05891	2.01999	1.02747	.31244	.31917	.33373	.35026	.36993	.39589	
							1.00750	1.01482	1.02892	1.04226	1.05477	1.06631	
.307	1.3450	.6226	.93063	.05834	2.01214	1.02702	.31347	.32025	.33492	.35160	.37153	.39811	
							1.00747	1.01477	1.02881	1.04210	1.05455	1.06601	
.308	1.3427	.6235	.93144	.05776	2.00432	1.02657	.31451	.32133	.33611	.35295	.37315	.40039	
							1.00744	1.01472	1.02870	1.04193	1.05432	1.06571	
.309	1.3404	.6243	.93224	.05719	1.99653	1.02613	.31555	.32242	.33730	.35431	.37479	.40276	
							1.00742	1.01466	1.02860	1.04177	1.05409	1.06542	
.310	1.3381	.6251	.93304	.05661	1.98879	1.02569	.31659	.32350	.33850	.35567	.37644	.40518	
							1.00739	1.01461	1.02849	1.04160	1.05387	1.06512	
.311	1.3358	.6259	.93383	.05604	1.98107	1.02526	.31763	.32458	.33970	.35704	.37811	.40769	
							1.00736	1.01456	1.02838	1.04144	1.05364	1.06482	
.312	1.3335	.6267	.93462	.05548	1.97339	1.02483	.31867	.32567	.34090	.35842	.37980	.41029	
							1.00734	1.01451	1.02828	1.04128	1.05341	1.06452	
.313	1.3312	.6275	.93540	.05491	1.96575	1.02441	.31971	.32676	.34210	.35980	.38150	.41299	
							1.00731	1.01446	1.02817	1.04112	1.05319	1.06421	
.314	1.3289	.6283	.93618	.05435	1.95814	1.02399	.32075	.32784	.34331	.36119	.38323	.41583	
							1.00729	1.01440	1.02807	1.04095	1.05296	1.06391	
.315	1.3266	.6291	.93695	.05379	1.95056	1.02358	.32179	.32893	.34452	.36259	.38498	.41883	
							1.00726	1.01435	1.02796	1.04079	1.05274	1.06360	
.316	1.3243	.6299	.93772	.05323	1.94302	1.02317	.32283	.33002	.34574	.36399	.38675	.42201	
							1.00724	1.01430	1.02786	1.04063	1.05252	1.06330	
.317	1.3221	.6307	.93848	.05268	1.93551	1.02277	.32388	.33111	.34695	.36540	.38855	.42544	
							1.00721	1.01425	1.02775	1.04047	1.05229	1.06299	
.318	1.3198	.6315	.93923	.05213	1.92803	1.02237	.32492	.33220	.34817	.36683	.39037	.42918	
							1.00719	1.01420	1.02765	1.04031	1.05207	1.06267	
.319	1.3175	.6323	.93998	.05158	1.92059	1.02198	.32596	.33330	.34940	.36826	.39222	.43344	
							1.00716	1.01415	1.02755	1.04015	1.05184	1.06237	
.320	1.3153	.6331	.94073	.05103	1.91318	1.02159	.32700	.33439	.35062	.36970	.39410	.43841	
							1.00714	1.01410	1.02744	1.03999	1.05162	1.06205	
.321	1.3130	.6339	.94147	.05048	1.90580	1.02121	.32805	.33549	.35185	.37115	.39601	.44480	
							1.00711	1.01405	1.02734	1.03983	1.05140	1.06172	
.322	1.3108	.6347	.94220	.04994	1.89845	1.02083	.32909	.33658	.35309	.37260	.39795		
							1.00709	1.01400	1.02724	1.03968	1.05117		
.323	1.3086	.6355	.94293	.04940	1.89114	1.02046	.33014	.33768	.35433	.37407	.39994		
							1.00706	1.01395	1.02714	1.03952	1.05095		
.324	1.3064	.6363	.94366	.04886	1.88385	1.02009	.33118	.33878	.35557	.37555	.40196		
							1.00704	1.01390	1.02704	1.03936	1.05073		
.325	1.3041	.6370	.94438	.04833	1.87660	1.01973	.33223	.33988	.35682	.37705	.40403		
							1.00701	1.01385	1.02694	1.03920	1.05050		
.326	1.3019	.6378	.94509	.04779	1.86938	1.01937	.33328	.34098	.35807	.37855	.40615		
							1.00699	1.01380	1.02683	1.03905	1.05028		
.327	1.2997	.6386	.94580	.04726	1.86219	1.01901	.33432	.34208	.35932	.38006	.40832		
							1.00696	1.01375	1.02673	1.03889	1.05005		
.328	1.2975	.6394	.94651	.04673	1.85503	1.01866	.33537	.34319	.36058	.38159	.41059		
							1.00694	1.01370	1.02663	1.03873	1.04984		
.329	1.2953	.6402	.94720	.04621	1.84790	1.01832	.33642	.34429	.36184	.38313	.41289		
							1.00692	1.01365	1.02653	1.03858	1.04961		

TABLE X. Specific Heat Ratio = 1.67

p/p_t	MACH	T/T_t	GAMMA	K_crit	SHOCK P_y/P_x	P_ty/P_ty	.001	.005	.01	.015	.02	.03	.04
.330	1.2931	.6410	.94790	.04568	1.84081	1.01798	.33146	.33747	.34540	.35390	.36310	.38469	.41527
							1.00139	1.00689	1.01360	1.02011	1.02643	1.03842	1.04939
.331	1.2909	.6417	.94859	.04516	1.83374	1.01764	.33247	.33852	.34651	.35508	.36438	.38626	.41774
							1.00139	1.00687	1.01356	1.02004	1.02633	1.03826	1.04916
.332	1.2887	.6425	.94927	.04464	1.82670	1.01731	.33348	.33957	.34762	.35626	.36565	.38784	.42032
							1.00138	1.00684	1.01351	1.01997	1.02623	1.03811	1.04894
.333	1.2866	.6433	.94995	.04413	1.81969	1.01698	.33449	.34062	.34873	.35745	.36694	.38945	.42301
							1.00138	1.00682	1.01346	1.01989	1.02613	1.03795	1.04871
.334	1.2844	.6441	.95062	.04361	1.81271	1.01665	.33550	.34167	.34985	.35864	.36822	.39107	.42584
							1.00137	1.00680	1.01341	1.01982	1.02603	1.03780	1.04848
.335	1.2822	.6448	.95129	.04310	1.80576	1.01634	.33651	.34273	.35096	.35983	.36952	.39271	.42884
							1.00137	1.00677	1.01336	1.01975	1.02593	1.03764	1.04825
.336	1.2801	.6456	.95196	.04259	1.79884	1.01602	.33752	.34378	.35208	.36103	.37082	.39436	.43206
							1.00137	1.00675	1.01332	1.01968	1.02584	1.03749	1.04802
.337	1.2779	.6464	.95261	.04209	1.79195	1.01571	.33853	.34483	.35320	.36223	.37212	.39604	.43556
							1.00136	1.00673	1.01327	1.01960	1.02574	1.03733	1.04779
.338	1.2758	.6471	.95327	.04158	1.78509	1.01540	.33954	.34589	.35432	.36343	.37343	.39774	.43952
							1.00136	1.00670	1.01322	1.01953	1.02564	1.03718	1.04756
.339	1.2736	.6479	.95392	.04108	1.77825	1.01510	.34055	.34694	.35544	.36464	.37475	.39947	.44406
							1.00135	1.00668	1.01318	1.01946	1.02554	1.03702	1.04732
.340	1.2715	.6487	.95456	.04058	1.77145	1.01480	.34156	.34800	.35656	.36585	.37608	.40122	.44969
							1.00135	1.00666	1.01313	1.01939	1.02545	1.03687	1.04707
.341	1.2694	.6494	.95520	.04009	1.76467	1.01450	.34257	.34905	.35769	.36706	.37741	.40299	.45897
							1.00134	1.00664	1.01308	1.01932	1.02535	1.03672	1.04682
.342	1.2672	.6502	.95583	.03959	1.75792	1.01421	.34358	.35011	.35882	.36828	.37875	.40480	
							1.00134	1.00661	1.01304	1.01925	1.02525	1.03656	
.343	1.2651	.6510	.95646	.03910	1.75119	1.01393	.34459	.35117	.35995	.36950	.38009	.40664	
							1.00133	1.00659	1.01299	1.01918	1.02516	1.03641	
.344	1.2630	.6517	.95709	.03861	1.74450	1.01364	.34560	.35223	.36108	.37073	.38144	.40851	
							1.00133	1.00657	1.01294	1.01911	1.02506	1.03625	
.345	1.2609	.6525	.95771	.03813	1.73783	1.01336	.34661	.35329	.36222	.37196	.38281	.41041	
							1.00132	1.00654	1.01290	1.01904	1.02496	1.03609	
.346	1.2588	.6532	.95832	.03764	1.73119	1.01309	.34763	.35435	.36335	.37320	.38420	.41236	
							1.00132	1.00652	1.01285	1.01897	1.02488	1.03594	
.347	1.2567	.6540	.95893	.03716	1.72457	1.01282	.34864	.35541	.36449	.37443	.38558	.41439	
							1.00132	1.00650	1.01281	1.01890	1.02478	1.03579	
.348	1.2546	.6548	.95953	.03668	1.71798	1.01255	.34965	.35648	.36563	.37568	.38697	.41643	
							1.00131	1.00648	1.01276	1.01883	1.02469	1.03564	
.349	1.2525	.6555	.96013	.03621	1.71142	1.01229	.35066	.35754	.36678	.37693	.38837	.41853	
							1.00131	1.00645	1.01271	1.01876	1.02459	1.03548	
.350	1.2504	.6563	.96073	.03573	1.70489	1.01203	.35167	.35860	.36792	.37818	.38978	.42069	
							1.00130	1.00643	1.01267	1.01869	1.02450	1.03533	
.351	1.2483	.6570	.96132	.03526	1.69838	1.01177	.35268	.35967	.36907	.37944	.39120	.42293	
							1.00130	1.00641	1.01262	1.01862	1.02440	1.03517	
.352	1.2462	.6578	.96190	.03479	1.69190	1.01152	.35370	.36074	.37022	.38071	.39263	.42523	
							1.00129	1.00639	1.01258	1.01855	1.02431	1.03501	
.353	1.2442	.6585	.96248	.03433	1.68544	1.01127	.35471	.36180	.37138	.38198	.39407	.42763	
							1.00129	1.00637	1.01253	1.01848	1.02422	1.03486	
.354	1.2421	.6593	.96306	.03386	1.67901	1.01102	.35572	.36287	.37253	.38325	.39552	.43014	
							1.00129	1.00634	1.01249	1.01841	1.02412	1.03470	
.355	1.2400	.6600	.96363	.03340	1.67260	1.01078	.35673	.36394	.37369	.38453	.39699	.43277	
							1.00128	1.00632	1.01244	1.01834	1.02403	1.03454	
.356	1.2380	.6608	.96420	.03294	1.66622	1.01055	.35775	.36501	.37485	.38584	.39847	.43555	
							1.00128	1.00630	1.01240	1.01828	1.02393	1.03438	
.357	1.2359	.6615	.96476	.03249	1.65986	1.01031	.35876	.36609	.37602	.38714	.39996	.43851	
							1.00127	1.00628	1.01236	1.01822	1.02384	1.03421	
.358	1.2339	.6622	.96531	.03204	1.65353	1.01008	.35977	.36716	.37719	.38844	.40147	.44171	
							1.00127	1.00626	1.01231	1.01815	1.02375	1.03405	
.359	1.2318	.6630	.96587	.03159	1.64722	1.00985	.36079	.36823	.37836	.38975	.40300	.44530	
							1.00126	1.00624	1.01227	1.01808	1.02365	1.03389	

TABLE X. Specific Heat Ratio = 1.67

p/pt	MACH	T/Tt	GAMMA	K_crit	Py/Px	Ptx/Pty	.001	.002	.003	.005	.01	.015	.02
					SHOCK		\-\-			R₂ and TPR versus Loss Coefficient K			
.360	1.2298	.6637	.96641	.03114	1.64094	1.00963	.36180	.36363	.36549	.36931	.37954	.39107	.40454
							1.00126	1.00251	1.00375	1.00622	1.01222	1.01801	1.02356
.361	1.2277	.6645	.96695	.03069	1.63468	1.00941	.36281	.36466	.36654	.37039	.38072	.39240	.40610
							1.00126	1.00250	1.00374	1.00620	1.01218	1.01795	1.02346
.362	1.2257	.6652	.96749	.03025	1.62845	1.00919	.36383	.36569	.36758	.37147	.38190	.39373	.40768
							1.00125	1.00249	1.00373	1.00617	1.01213	1.01788	1.02337
.363	1.2237	.6659	.96802	.02981	1.62224	1.00898	.36484	.36672	.36863	.37255	.38308	.39507	.40928
							1.00125	1.00249	1.00372	1.00615	1.01209	1.01781	1.02328
.364	1.2217	.6667	.96855	.02937	1.61606	1.00877	.36586	.36775	.36967	.37363	.38427	.39643	.41090
							1.00124	1.00248	1.00370	1.00613	1.01205	1.01774	1.02318
.365	1.2197	.6674	.96908	.02894	1.60989	1.00856	.36687	.36878	.37072	.37471	.38547	.39779	.41254
							1.00124	1.00247	1.00369	1.00611	1.01200	1.01768	1.02309
.366	1.2176	.6681	.96959	.02851	1.60376	1.00836	.36789	.36981	.37177	.37580	.38667	.39916	.41421
							1.00123	1.00246	1.00368	1.00609	1.01196	1.01761	1.02300
.367	1.2156	.6689	.97011	.02808	1.59764	1.00816	.36890	.37084	.37282	.37688	.38787	.40054	.41591
							1.00123	1.00245	1.00367	1.00607	1.01192	1.01754	1.02290
.368	1.2136	.6696	.97062	.02765	1.59155	1.00796	.36992	.37187	.37387	.37797	.38910	.40193	.41764
							1.00123	1.00244	1.00366	1.00605	1.01189	1.01748	1.02281
.369	1.2116	.6703	.97112	.02723	1.58548	1.00776	.37093	.37291	.37491	.37906	.39031	.40333	.41939
							1.00122	1.00244	1.00364	1.00603	1.01184	1.01741	1.02271
.370	1.2096	.6711	.97162	.02681	1.57944	1.00757	.37195	.37395	.37595	.38015	.39153	.40475	.42119
							1.00122	1.00243	1.00362	1.00601	1.01180	1.01734	1.02262
.371	1.2076	.6718	.97212	.02639	1.57341	1.00739	.37297	.37497	.37702	.38124	.39275	.40618	.42302
							1.00121	1.00242	1.00362	1.00599	1.01176	1.01728	1.02253
.372	1.2057	.6725	.97261	.02598	1.56741	1.00720	.37400	.37600	.37807	.38234	.39398	.40762	.42489
							1.00122	1.00241	1.00360	1.00597	1.01171	1.01721	1.02243
.373	1.2037	.6732	.97310	.02556	1.56144	1.00702	.37500	.37704	.37913	.38343	.39522	.40907	.42681
							1.00121	1.00240	1.00359	1.00595	1.01167	1.01714	1.02234
.374	1.2017	.6740	.97358	.02515	1.55548	1.00684	.37600	.37808	.38018	.38453	.39646	.41054	.42877
							1.00119	1.00239	1.00358	1.00592	1.01163	1.01708	1.02224
.375	1.1997	.6747	.97405	.02475	1.54955	1.00667	.37703	.37911	.38124	.38563	.39770	.41203	.43080
							1.00120	1.00239	1.00357	1.00590	1.01159	1.01701	1.02214
.376	1.1977	.6754	.97453	.02434	1.54364	1.00649	.37805	.38015	.38229	.38674	.39895	.41354	.43289
							1.00119	1.00238	1.00356	1.00588	1.01155	1.01694	1.02205
.377	1.1958	.6761	.97499	.02394	1.53775	1.00632	.37907	.38119	.38335	.38784	.40021	.41506	.43506
							1.00119	1.00237	1.00354	1.00586	1.01150	1.01688	1.02195
.378	1.1938	.6768	.97546	.02354	1.53189	1.00616	.38009	.38223	.38441	.38895	.40148	.41660	.43731
							1.00119	1.00236	1.00353	1.00584	1.01146	1.01681	1.02185
.379	1.1919	.6776	.97592	.02315	1.52604	1.00599	.38111	.38326	.38547	.39005	.40275	.41816	.43966
							1.00118	1.00235	1.00352	1.00582	1.01142	1.01674	1.02175
.380	1.1899	.6783	.97637	.02275	1.52022	1.00583	.38213	.38430	.38653	.39117	.40403	.41975	.44213
							1.00118	1.00235	1.00351	1.00580	1.01138	1.01668	1.02165
.381	1.1880	.6790	.97682	.02236	1.51442	1.00567	.38315	.38534	.38760	.39228	.40532	.42136	.44481
							1.00117	1.00234	1.00350	1.00578	1.01134	1.01661	1.02156
.382	1.1860	.6797	.97727	.02198	1.50864	1.00552	.38417	.38639	.38866	.39340	.40661	.42299	.44763
							1.00117	1.00233	1.00348	1.00576	1.01129	1.01654	1.02146
.383	1.1841	.6804	.97771	.02159	1.50288	1.00537	.38519	.38743	.38973	.39451	.40792	.42466	.45070
							1.00117	1.00232	1.00347	1.00574	1.01125	1.01648	1.02136
.384	1.1821	.6811	.97814	.02121	1.49714	1.00522	.38621	.38847	.39079	.39564	.40923	.42635	.45411
							1.00116	1.00232	1.00346	1.00572	1.01121	1.01641	1.02126
.385	1.1802	.6818	.97857	.02083	1.49142	1.00507	.38723	.38951	.39186	.39676	.41055	.42808	.45802
							1.00116	1.00231	1.00345	1.00570	1.01117	1.01634	1.02115
.386	1.1783	.6826	.97900	.02046	1.48572	1.00493	.38825	.39056	.39293	.39789	.41189	.42985	.46297
							1.00115	1.00230	1.00344	1.00568	1.01113	1.01628	1.02105
.387	1.1763	.6833	.97942	.02008	1.48005	1.00479	.38927	.39160	.39400	.39902	.41323	.43166	.47082
							1.00115	1.00229	1.00342	1.00566	1.01109	1.01621	1.02093
.388	1.1744	.6840	.97984	.01971	1.47439	1.00465	.39029	.39265	.39507	.40015	.41459	.43351	
							1.00115	1.00228	1.00341	1.00564	1.01104	1.01614	
.389	1.1725	.6847	.98025	.01935	1.46876	1.00451	.39132	.39370	.39615	.40129	.41595	.43542	
							1.00114	1.00228	1.00340	1.00562	1.01100	1.01607	

TABLE X. Specific Heat Ratio = 1.67

p/p_t	MACH	T/T_t	GAMMA	K_crit	SHOCK P_y/P_x	P_tx/P_ty	.001	.002	.003	.004	.005	.01	.015
.390	1.1706	.6854	.98066	.01898	1.46314	1.00438	.39234	.39475	.39722	.39978	.40242	.41733	.43738
							1.00114	1.00227	1.00339	1.00450	1.00560	1.01096	1.01600
.391	1.1687	.6861	.98107	.01862	1.45755	1.00425	.39336	.39580	.39830	.40089	.40356	.41873	.43941
							1.00113	1.00226	1.00338	1.00449	1.00558	1.01092	1.01593
.392	1.1668	.6868	.98147	.01826	1.45197	1.00412	.39439	.39685	.39938	.40200	.40471	.42013	.44151
							1.00113	1.00225	1.00337	1.00447	1.00556	1.01088	1.01586
.393	1.1648	.6875	.98186	.01791	1.44642	1.00399	.39541	.39790	.40046	.40311	.40586	.42156	.44370
							1.00113	1.00225	1.00335	1.00445	1.00554	1.01084	1.01579
.394	1.1629	.6882	.98226	.01756	1.44089	1.00387	.39644	.39895	.40155	.40423	.40701	.42300	.44600
							1.00112	1.00224	1.00334	1.00444	1.00552	1.01080	1.01572
.395	1.1610	.6889	.98264	.01721	1.43537	1.00375	.39746	.40000	.40263	.40535	.40817	.42445	.44850
							1.00112	1.00223	1.00333	1.00442	1.00550	1.01075	1.01566
.396	1.1592	.6896	.98302	.01686	1.42987	1.00363	.39849	.40106	.40372	.40647	.40934	.42593	.45111
							1.00112	1.00222	1.00332	1.00441	1.00549	1.01071	1.01559
.397	1.1573	.6903	.98340	.01652	1.42440	1.00352	.39952	.40212	.40481	.40760	.41050	.42743	.45393
							1.00111	1.00222	1.00331	1.00439	1.00547	1.01067	1.01551
.398	1.1554	.6910	.98378	.01618	1.41894	1.00340	.40054	.40317	.40590	.40873	.41168	.42895	.45703
							1.00111	1.00221	1.00330	1.00438	1.00545	1.01063	1.01544
.399	1.1535	.6917	.98415	.01584	1.41350	1.00329	.40157	.40423	.40699	.40986	.41285	.43049	.46054
							1.00110	1.00220	1.00329	1.00436	1.00543	1.01059	1.01536
.400	1.1516	.6924	.98451	.01551	1.40808	1.00319	.40260	.40530	.40809	.41100	.41403	.43206	.46487
							1.00110	1.00219	1.00327	1.00435	1.00541	1.01055	1.01529
.401	1.1497	.6931	.98487	.01518	1.40268	1.00308	.40363	.40636	.40919	.41214	.41522	.43365	.47079
							1.00110	1.00218	1.00326	1.00433	1.00539	1.01050	1.01521
.402	1.1479	.6938	.98523	.01485	1.39730	1.00298	.40466	.40742	.41029	.41328	.41642	.43528	
							1.00109	1.00218	1.00325	1.00432	1.00537	1.01046	
.403	1.1460	.6945	.98558	.01452	1.39194	1.00287	.40569	.40849	.41140	.41443	.41762	.43694	
							1.00109	1.00217	1.00324	1.00430	1.00535	1.01042	
.404	1.1441	.6952	.98593	.01420	1.38660	1.00278	.40672	.40955	.41250	.41559	.41882	.43865	
							1.00108	1.00216	1.00323	1.00428	1.00533	1.01038	
.405	1.1423	.6958	.98627	.01388	1.38127	1.00268	.40775	.41062	.41361	.41674	.42003	.44039	
							1.00108	1.00215	1.00322	1.00427	1.00531	1.01033	
.406	1.1404	.6965	.98661	.01357	1.37597	1.00258	.40879	.41169	.41473	.41791	.42125	.44218	
							1.00108	1.00215	1.00321	1.00425	1.00529	1.01029	
.407	1.1386	.6972	.98694	.01326	1.37068	1.00249	.40982	.41277	.41584	.41907	.42248	.44402	
							1.00107	1.00214	1.00319	1.00424	1.00527	1.01025	
.408	1.1367	.6979	.98727	.01295	1.36541	1.00240	.41086	.41384	.41696	.42025	.42372	.44593	
							1.00107	1.00213	1.00318	1.00422	1.00525	1.01020	
.409	1.1349	.6986	.98760	.01264	1.36015	1.00231	.41189	.41492	.41809	.42143	.42496	.44791	
							1.00107	1.00212	1.00317	1.00421	1.00523	1.01016	
.410	1.1330	.6993	.98792	.01234	1.35492	1.00223	.41293	.41600	.41922	.42261	.42625	.45004	
							1.00106	1.00212	1.00316	1.00419	1.00522	1.01012	
.411	1.1312	.7000	.98823	.01204	1.34970	1.00214	.41397	.41708	.42035	.42380	.42751	.45223	
							1.00106	1.00211	1.00315	1.00418	1.00520	1.01008	
.412	1.1293	.7006	.98855	.01174	1.34450	1.00206	.41500	.41816	.42148	.42500	.42879	.45454	
							1.00105	1.00210	1.00314	1.00416	1.00518	1.01003	
.413	1.1275	.7013	.98885	.01145	1.33932	1.00198	.41604	.41925	.42262	.42620	.43008	.45702	
							1.00105	1.00209	1.00313	1.00415	1.00517	1.00998	
.414	1.1257	.7020	.98916	.01116	1.33416	1.00191	.41708	.42034	.42377	.42745	.43137	.45971	
							1.00105	1.00209	1.00312	1.00414	1.00515	1.00994	
.415	1.1238	.7027	.98946	.01087	1.32901	1.00183	.41813	.42143	.42492	.42868	.43269	.46270	
							1.00104	1.00208	1.00310	1.00413	1.00513	1.00989	
.416	1.1220	.7034	.98975	.01059	1.32389	1.00176	.41917	.42252	.42607	.42991	.43401	.46612	
							1.00104	1.00207	1.00309	1.00411	1.00511	1.00984	
.417	1.1202	.7040	.99004	.01031	1.31877	1.00169	.42021	.42362	.42727	.43115	.43535	.47060	
							1.00104	1.00206	1.00309	1.00410	1.00509	1.00979	
.418	1.1184	.7047	.99033	.01003	1.31368	1.00162	.42126	.42472	.42844	.43240	.43671	.47812	
							1.00103	1.00206	1.00308	1.00408	1.00507	1.00974	
.419	1.1166	.7054	.99061	.00976	1.30860	1.00155	.42231	.42582	.42962	.43367	.43808		
							1.00103	1.00205	1.00307	1.00407	1.00505		

TABLE X. Specific Heat Ratio = 1.67

p/p_t	MACH	T/T_t	GAMMA	K_crit	SHOCK P_y/P_x	P_tx/P_ty	.001	.002	.003	.004	.005	.01	.015
.420	1.1147	.7061	.99089	.00949	1.30354	1.00148	.42335	.42693	.43080	.43494	.43947		
							1.00102	1.00204	1.00306	1.00405	1.00503		
.421	1.1129	.7067	.99116	.00922	1.29850	1.00142	.42440	.42808	.43199	.43622	.44089		
							1.00102	1.00204	1.00305	1.00404	1.00501		
.422	1.1111	.7074	.99143	.00896	1.29347	1.00136	.42546	.42920	.43319	.43752	.44232		
							1.00102	1.00204	1.00304	1.00402	1.00499		
.423	1.1093	.7081	.99170	.00870	1.28846	1.00130	.42651	.43032	.43439	.43884	.44378		
							1.00101	1.00203	1.00303	1.00401	1.00497		
.424	1.1075	.7088	.99196	.00844	1.28347	1.00124	.42756	.43145	.43561	.44017	.44527		
							1.00101	1.00202	1.00302	1.00399	1.00496		
.425	1.1057	.7094	.99222	.00819	1.27849	1.00118	.42866	.43258	.43683	.44151	.44679		
							1.00101	1.00202	1.00300	1.00398	1.00494		
.426	1.1039	.7101	.99247	.00794	1.27353	1.00113	.42972	.43372	.43807	.44288	.44835		
							1.00101	1.00201	1.00299	1.00396	1.00492		
.427	1.1021	.7108	.99272	.00769	1.26859	1.00107	.43078	.43486	.43931	.44427	.44994		
							1.00101	1.00200	1.00298	1.00395	1.00490		
.428	1.1003	.7114	.99296	.00745	1.26366	1.00102	.43185	.43601	.44057	.44567	.45158		
							1.00100	1.00200	1.00297	1.00393	1.00488		
.429	1.0986	.7121	.99320	.00721	1.25875	1.00097	.43292	.43717	.44184	.44711	.45327		
							1.00100	1.00199	1.00296	1.00392	1.00486		
.430	1.0968	.7128	.99344	.00698	1.25385	1.00092	.43399	.43833	.44313	.44857	.45502		
							1.00100	1.00198	1.00295	1.00390	1.00484		
.431	1.0950	.7134	.99367	.00674	1.24897	1.00087	.43506	.43950	.44443	.45006	.45684		
							1.00099	1.00197	1.00294	1.00389	1.00481		
.432	1.0932	.7141	.99389	.00652	1.24411	1.00083	.43614	.44068	.44574	.45159	.45875		
							1.00099	1.00197	1.00293	1.00387	1.00479		
.433	1.0914	.7148	.99412	.00629	1.23926	1.00079	.43722	.44186	.44708	.45316	.46075		
							1.00099	1.00196	1.00292	1.00386	1.00477		
.434	1.0897	.7154	.99434	.00607	1.23442	1.00074	.43830	.44306	.44844	.45477	.46299		
							1.00098	1.00195	1.00291	1.00384	1.00476		
.435	1.0879	.7161	.99455	.00585	1.22961	1.00070	.43939	.44426	.44981	.45644	.46534		
							1.00098	1.00195	1.00290	1.00382	1.00474		
.436	1.0861	.7167	.99476	.00564	1.22481	1.00066	.44048	.44548	.45122	.45817	.46795		
							1.00098	1.00194	1.00288	1.00381	1.00471		
.437	1.0844	.7174	.99497	.00543	1.22002	1.00063	.44158	.44671	.45265	.45998	.47096		
							1.00097	1.00193	1.00287	1.00379	1.00469		
.438	1.0826	.7181	.99517	.00522	1.21525	1.00059	.44268	.44795	.45412	.46187	.47468		
							1.00097	1.00193	1.00286	1.00377	1.00466		
.439	1.0809	.7187	.99537	.00502	1.21049	1.00056	.44378	.44921	.45562	.46399	.48117		
							1.00097	1.00192	1.00285	1.00376	1.00464		
.440	1.0791	.7194	.99556	.00482	1.20575	1.00052	.44489	.45048	.45716	.46619			
							1.00096	1.00191	1.00284	1.00375			
.441	1.0774	.7200	.99575	.00462	1.20103	1.00049	.44601	.45177	.45875	.46862			
							1.00096	1.00190	1.00283	1.00373			
.442	1.0756	.7207	.99593	.00443	1.19632	1.00046	.44713	.45308	.46040	.47138			
							1.00096	1.00190	1.00281	1.00371			
.443	1.0739	.7213	.99611	.00424	1.19162	1.00043	.44826	.45441	.46212	.47472			
							1.00095	1.00189	1.00280	1.00369			
.444	1.0721	.7220	.99629	.00406	1.18694	1.00040	.44939	.45576	.46403	.47989			
							1.00095	1.00188	1.00280	1.00367			
.445	1.0704	.7226	.99646	.00388	1.18227	1.00037	.45054	.45714	.46597				
							1.00095	1.00187	1.00278				
.446	1.0686	.7233	.99663	.00370	1.17762	1.00035	.45169	.45856	.46808				
							1.00094	1.00187	1.00277				
.447	1.0669	.7239	.99680	.00353	1.17299	1.00032	.45285	.46001	.47039				
							1.00094	1.00186	1.00276				
.448	1.0652	.7246	.99696	.00336	1.16836	1.00030	.45402	.46150	.47304				
							1.00094	1.00185	1.00274				
.449	1.0634	.7252	.99711	.00319	1.16376	1.00028	.45520	.46304	.47623				
							1.00093	1.00184	1.00273				

TABLE X. Specific Heat Ratio = 1.67

p/p_t	MACH	T/T_t	GAMMA	K_crit	SHOCK P_y/P_x	P_tx/P_ty	.001	.002	R_t and TPR versus Loss Coefficient K .003	.004	.005	.01	.015
.450	1.0617	.7259	.99726	.00303	1.15916	1.00026	.45640	.46475	.48149				
							1.00093	1.00184	1.00271				
.451	1.0600	.7265	.99741	.00287	1.15458	1.00024	.45761	.46646					
							1.00092	1.00183					
.452	1.0583	.7272	.99756	.00272	1.15002	1.00022	.45883	.46828					
							1.00092	1.00183					
.453	1.0565	.7278	.99770	.00257	1.14547	1.00020	.46008	.47023					
							1.00092	1.00182					
.454	1.0548	.7285	.99783	.00242	1.14093	1.00018	.46134	.47239					
							1.00091	1.00181					
.455	1.0531	.7291	.99796	.00228	1.13641	1.00017	.46263	.47484					
							1.00091	1.00180					
.456	1.0514	.7298	.99809	.00214	1.13190	1.00015	.46394	.47804					
							1.00090	1.00179					
.457	1.0497	.7304	.99821	.00201	1.12740	1.00014	.46539	.48356					
							1.00091	1.00178					
.458	1.0480	.7310	.99833	.00188	1.12292	1.00012	.46680						
							1.00090						
.459	1.0463	.7317	.99845	.00175	1.11845	1.00011	.46827						
							1.00090						
.460	1.0446	.7323	.99856	.00163	1.11400	1.00010	.46980						
							1.00090						
.461	1.0429	.7330	.99866	.00152	1.10956	1.00009	.47141						
							1.00089						
.462	1.0412	.7336	.99876	.00140	1.10513	1.00008	.47314						
							1.00089						
.463	1.0395	.7342	.99886	.00129	1.10072	1.00007	.47503						
							1.00088						
.464	1.0378	.7349	.99896	.00119	1.09632	1.00006	.47715						
							1.00088						
.465	1.0361	.7355	.99905	.00109	1.09193	1.00005	.48009						
							1.00088						